Organic Structural Analysis

A Series of Books in Organic Chemistry
Andrew Streitwieser, Jr., Editor

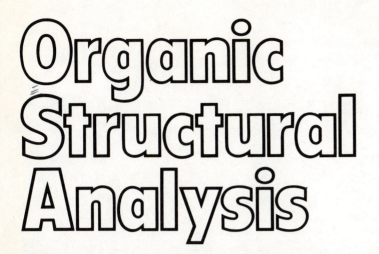

Organic Structural Analysis

Joseph B. Lambert
Northwestern University

Herbert F. Shurvell
Queen's University

Lawrence Verbit
State University of New York at Binghamton

R. Graham Cooks
Purdue University

George H. Stout
University of Washington

Macmillan Publishing Co., Inc.
New York
Collier Macmillan Publishers
London

Permission for the publication herein of Sadtler Standard Spectra®
has been granted, and all rights are reserved, by Sadtler Research
Laboratories, Inc.

Macmillan Publishing Co., Inc.
866 Third Avenue, New York, New York 10022

Collier Macmillan Canada, Ltd.

Library of Congress Cataloging in Publication Data

Main entry under title:

Organic structural analysis.

   (A Series of books in organic chemistry)
   Includes bibliographies and index.
   1. Chemistry, Organic. 2. Spectrum analysis.
I. Lambert, Joseph B.
QD272.S6073   547   74-27595
ISBN 0-02-367290-0

Printing: 3 4 5 6 7 8   Year: 8 9 0 1 2

Preface

The organic chemist today must be master of many fields of spectroscopy. The traditional areas of nuclear magnetic resonance, infrared, ultraviolet–visible, and mass spectroscopies continue to grow richer, with the result that we now must become familiar with such techniques as Fourier transform nmr and ir, ion cyclotron resonance, chemical ionization mass spectrometry, and magnetic circular dichroism. Other fields, once peripheral to the organic chemist, have become necessary components in a structural analysis: Raman spectroscopy, spurred by the development of laser technology, and X-ray crystallography, streamlined by the advent of automated instrumentation and more sophisticated computer software. In this text it has been our aim to describe the methods generally available to most chemists for the solution of structural problems. Although our examples are drawn primarily from organic systems, the principles apply equally well to inorganic and biochemical problems.

The five parts of the book are intended to have the flavor of both the elementary and the state-of-the-art, so that the practicing chemist can obtain authoritative treatments of all subjects under one cover, but the beginner can still expect a clear introduction. Accordingly, we hope that the material may be used by advanced undergraduates in their first commitment to research, by graduate students, and by chemists in full-time practice in industrial, government, and educational laboratories. Certain spectroscopic areas have been excluded from this text, we hope not arbitrarily: methods that are not generally available (microwave, electron and neutron diffraction); methods that are not strictly applicable to the majority of organic compounds (electron paramagnetic resonance); and methods that are still too embryonic for a general discussion (electron spectroscopy). In the near future some of these exclusions may be rendered untenable. Photoelectron spectroscopy (X-ray and uv) has already found numerous applications in organic structural problems, and future texts of this type may be able to include it as a mature field. Because there are numerous sources today of structural or spectral problems, we have chosen to incorporate only a minimum of such material into this text. The reader is encouraged to select a companion textbook of problems if additional experience is desired.

Joseph B. Lambert
Herbert F. Shurvell
Lawrence Verbit
R. Graham Cooks
George H. Stout

v

To MWPL

Contents

Part One

NUCLEAR MAGNETIC
RESONANCE SPECTROSCOPY

1

INTRODUCTION

1-1 Nuclear Magnetic Resonance

After the discovery of the nuclear magnetic resonance phenomenon in the physics laboratories of Purcell at Harvard and Bloch at Stanford in 1945, the field was developed into a major structural tool by chemists during the 1950's. Today nmr spectroscopy stands as a mature discipline studied in itself by dozens of research groups throughout the world and utilized to solve structural and kinetic problems by investigators in almost all chemical disciplines.

The nmr phenomenon is based on the fact that all nuclei are charged particles and some in addition have spin angular momentum. Those without spin properties (^{12}C, ^{16}O) are assigned a spin quantum number I of zero. A spinning, charged particle generates a magnetic field, so that the nuclei of interest are those for which $I \neq 0$. Those spinning nuclei that have a spherically shaped charge distribution are assigned a spin of $\frac{1}{2}$ (^1H, ^3H, ^{13}C, ^{15}N, ^{19}F, ^{31}P). The remaining *quadrupolar* nuclei comprise the larger set and may have a spin of 1 or larger, in divisions of $\frac{1}{2}$ (^2H, ^{14}N, ^{11}B, ^{17}O).

In the absence of an external magnetic field, all orientations of a nuclear spin are equivalent. The small effect of the earth's magnetic field can be ignored. If a macroscopic collection of spinning nuclei is immersed in a strong laboratory magnetic field B_0, the nuclei will tend to align themselves with the B_0 field by reason of their own spin-generated magnetic fields. For nuclei of spin $\frac{1}{2}$, there is in addition a less stable orientation against the field. Because only these two orientations are permitted, the spin is said to be quantized. For the stable orientation, the z component of I has a value of $-\frac{1}{2}$ (see below), and for the unstable orientation of I_z of $+\frac{1}{2}$ (Figure 1-1). For any arbitrary magnitude of B_0, there is a well-defined energy difference, $\Delta E = E_2 - E_1$, between the two populated spin states. This energy difference is zero at zero field and increases linearly with the field. For nuclei with other values of I, a different number of spin states is available. Thus for $I = 1$ (^2H, ^{14}N), the allowed arrangements of the

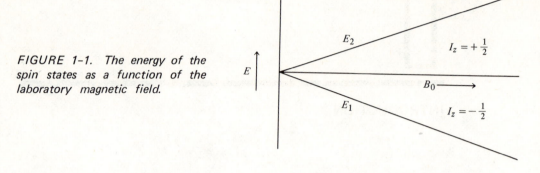

FIGURE 1-1. The energy of the spin states as a function of the laboratory magnetic field.

nuclear magnet are with, perpendicular to, and against the laboratory magnetic field ($I_z = -1$, $0, +1$). In general, the magnitude of ΔE is quite small, so that at equilibrium the Boltzmann distribution of nuclei produces an excess of only about six per million in the stabler spin state for $I = \frac{1}{2}$.

If a second rf field B_1 is applied at the frequency corresponding to the energy difference between spin states, $\Delta E = h\nu_0$, a nucleus in the lower spin state may absorb energy and be promoted to the higher state ($-\frac{1}{2} \longrightarrow +\frac{1}{2}$). This absorption of energy, or spin flip, is termed a nuclear magnetic resonance. The value of the resonance frequency ν_0 is dependent on the nature of the nucleus and on the magnitude of the laboratory magnetic field B_0. For the widely used magnetic field of 1.41 tesla[1] (14,100 gauss), the proton resonates at 60.00 MHz, ^{19}F at 56.4 MHz, ^{13}C at 15.1 MHz, and ^{15}N at 6.1 MHz. This wide separation of resonance frequencies means that one particular nucleus may be examined without interference from resonances of other nuclei. If the magnitude of B_0 is raised (Figure 1-1), the value of ΔE and hence ν_0 is raised in linear proportion. For a field of 2.11 tesla, the proton will therefore resonate at 90 MHz.

The nmr experiment thus consists of (1) selecting nuclei with spin properties, (2) immersing them in a strong magnetic field B_0, and (3) applying a second field B_1 that corresponds to the natural resonance frequency ν_0. In the following paragraphs, the description of the nmr phenomenon will be placed on a more quantitative basis. The reader without a background in quantum mechanics may proceed directly to the final result (eq. 1-9) without undue loss.

Spinning nuclei produce a magnetic field with a magnetic moment μ that is proportional to the magnitude of the spin angular momentum $\mathbf{J}$ (eq. 1-1). The proportionality constant

$$\mu = \gamma \mathbf{J} \qquad\qquad 1\text{-}1$$

between these two vector quantities is γ, the gyromagnetic ratio, which is a characteristic of the nucleus under consideration. The spin angular momentum $\mathbf{J}$ may be converted to a dimensionless spin $\mathbf{I}$ by the substitution given in eq. 1-2. The spin $\mathbf{I}$ is quantized, and according to the

$$\mathbf{J} = \hbar \mathbf{I} \qquad\qquad 1\text{-}2$$

$$\mu = \gamma \hbar \mathbf{I} \qquad\qquad 1\text{-}3$$

[1]The units tesla or gauss refer to B, the magnetic flux density. The commonly used symbol H refers to the magnetic field intensity, whose units are amperes meter^{-1} or oersteds. Although the quantities are proportional, the physical observables should be expressed in terms of B in SI units (tesla).

quantum mechanical properties of angular momentum only the magnitude (I^2) and one component (I_z) may be specified. The eigenvalue of I^2 is $I(I + 1)$, where the quantity I, simply called "spin" earlier in this section, may have the values $0, \frac{1}{2}, 1, \frac{3}{2}, \ldots$. The component of I along the direction of B_0 (arbitrarily the z coordinate) is selected to be the other good quantum number, with eigenvalue I_z. The quantity I_z assumes the values $+I, +I - 1, +I - 2, \ldots, -I + 1,$ $-I$, or a total of $(2I + 1)$ different values. It is thus clear why a nucleus with a spin of 1, for example, can assume three orientations in the magnetic field.

The value of the resonance frequency may be obtained in a straightforward manner from the above equations. The Hamiltonian for the interaction between the magnetic moment of the nucleus μ and the magnetic field $\mathbf{B}$ must be a dot product because the energy, which is the eigenvalue of the Hamiltonian, is a scalar (eq. 1-4). Substitution from eq. 1-3 and reduction of

$$\mathcal{H} = \mu \cdot \mathbf{B} \qquad\qquad 1\text{-}4$$

the dot product to one term, because $\mathbf{B}$ has magnitude only in the z direction, give eq. 1-6.

$$\mathcal{H} = \gamma \hbar \mathbf{I} \cdot \mathbf{B} \qquad\qquad 1\text{-}5$$

$$\mathcal{H} = \gamma \hbar I_z B_0 \qquad\qquad 1\text{-}6$$

The eigenvalue of $\mathcal{H}$ is E and of I_z is I_z (also called m_I in some books), so the energy for a given spin energy level is given by eq. 1-7. For E_1, I_z is $-\frac{1}{2}$, and for E_2, I_z is $+\frac{1}{2}$. Substitution of

$$E = \gamma \hbar I_z B_0 \qquad\qquad 1\text{-}7$$

these values into eq. 1-7 and subtraction gives the energy difference ΔE (eq. 1-8). Since $\Delta E =$

$$\Delta E = \frac{1}{2}\gamma \hbar B_0 - (-\frac{1}{2}\gamma \hbar B_0) = \gamma \hbar B_0 \qquad\qquad 1\text{-}8$$

$h\nu_0$, the resonance frequency may be calculated (eq. 1-9, in which $\gamma = \gamma/2\pi$). The angular frequency ω_0 is related to ν_0 by the relationship $\omega_0 = 2\pi\nu_0$.

$$\nu_0 = \gamma B_0 \qquad \omega_0 = \gamma B_0 \qquad\qquad 1\text{-}9$$

According to eq. 1-9, the resonance frequency depends only upon the gyromagnetic ratio and the choice of magnetic field B_0. For a constant value of B_0, the resonance frequencies of various nuclei are directly proportional to their gyromagnetic ratios, which are constant for a given isotope. For a given isotope, the resonance frequency is directly proportional to the value of B_0.

1-2   Magnetic Shielding and the Chemical Shift

An examination of the ^{14}N magnetic resonance spectrum of ammonium nitrate by Proctor and Yu in 1950 revealed not one but two resonances. Since the gyromagnetic ratio of ^{14}N is a constant, only a single peak should have been observed according to eq. 1-9. It was concluded that the two peaks must be separate resonances for the ammonium nitrogen and the nitrate nitrogen.

It was argued that the actual field experienced by the nuclei is not B_0, but a quantity some-what attenuated by the diamagnetic electron cloud surrounding the nuclei (eq. 1-10). Because

$$B_{local} = B_0(1 - \sigma) \qquad\qquad 1\text{-}10$$

the electronic structure about ^{14}N in the ammonium ion is different from that in the nitrate ion, the attenuation of the applied field is different, and distinct resonances occur. The resonance frequency of a given nucleus (eq. 1-11) therefore depends not only on the gyromagnetic ratio

$$\nu_0 = \gamma B_0(1 - \sigma) \qquad\qquad 1\text{-}11$$

and the applied field, but also on the electronic structure about the nucleus. The nuclei are said to be *shielded* or *screened* from the applied field by the electrons, and the variation of resonance frequency with electronic structure has been termed the *chemical shift*. This latter phrase has come to be synonymous with the resonance frequency.

The discovery of the chemical shift brought nmr spectroscopy from the physicist's laboratory to that of the chemist. If in theory each different type of ^{14}N, ^{13}C, or ^{1}H has a different resonance position, the job of the structural chemist will be vastly simplified. Furthermore, high resolution nmr spectra are taken on a solution rather than on the solid, so measurements may be made on the phase in which most chemical reactions occur. The method is also nondestructive, so that the sample may be retrieved after completion of the experiment.

The ability to observe separate signals for chemically distinct nuclei of a given isotope depends on the bandwidth of the resonance and the resolution of the spectrometer. The nmr experiment has been very successful particularly because the natural linewidths of proton resonances are usually well below 1 Hz, and spectrometers have been developed that can readily resolve peaks separated by as little as 0.1 Hz. The range of proton resonances is over

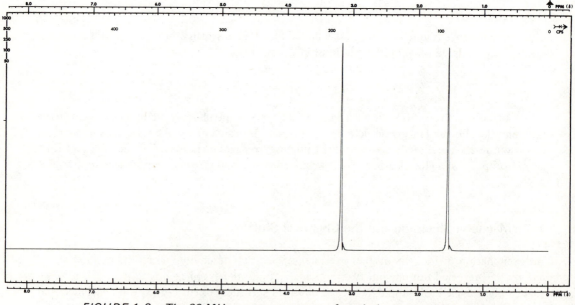

FIGURE 1-2. The 60 MHz proton spectrum of methyl acetate without solvent.

1000 Hz at 1.41 tesla, and the ranges for other nuclei (^{19}F, ^{13}C, ^{14}N) are many thousands of hertz.

It might be expected that the proton magnetic resonance (pmr) spectrum of a compound contains a separate resonance for each chemical type of hydrogen. Figure 1-2 illustrates such an experiment for the pmr spectrum of methyl acetate ($CH_3CO_2CH_3$). To observe the different resonances, the frequency of the B_1 field is held constant, and the B_0 field is swept over the region in which resonances are normally found. The B_0 field always increases from left to right in the chart representation. Alternatively, the B_0 field can be held constant and the frequency of the B_1 field varied. These methods will be compared in Chapter 2. Two resonances are observed in the pmr spectrum of methyl acetate, one at lower field (on the left) from the methoxyl methyl and one at higher field from the acetyl methyl.

Chemically distinct protons need not necessarily have different chemical shifts. The spectrum of toluene ($C_6H_5CH_3$) at 60 MHz (Figure 1-3) is such an example. Although the methyl group is well shifted to higher field, there is only one peak on the left from the ortho, meta, and para aromatic protons. The electronic environment of these three sets of protons are not sufficiently different to cause differential shielding. This problem occurs frequently in pmr spectra. In a steroid, for example, little differentiation is made among the resonance positions of the many similar methylene protons.

Greater clarification is frequently found in natural abundance (1.1%) carbon-13 magnetic resonance (cmr) spectra. For reasons to be discussed in Chapters 2 and 3, the range of carbon chemical shifts is much larger than that in proton spectra, but the experiment is more difficult to carry out. A cmr spectrum can produce a very nice map of the backbone of a compound, just as the pmr spectrum can tell much about proton-containing functional groups. Figure 1-4 gives the cmr spectrum of sucrose, in which separate resonances are observed for each of the twelve carbon atoms.

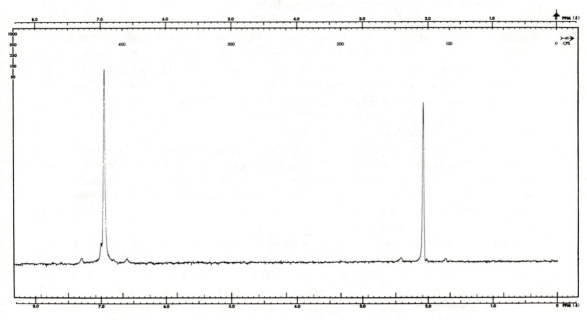

FIGURE 1-3. *The 60 MHz proton spectrum of toluene in CDCl$_3$.*

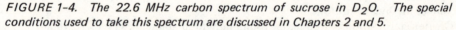

FIGURE 1-4.   The 22.6 MHz carbon spectrum of sucrose in D_2O.   The special conditions used to take this spectrum are discussed in Chapters 2 and 5.

The pmr spectrum of a compound can ideally give the investigator information about each chemically distinct set of protons.   The cmr spectrum can give parallel information on the carbon atoms, and supplementary data may be obtained from the ^{19}F, ^{31}P, ^{14}N, ^{11}B, etc., spectra.   The discussion in later chapters will tell how this knowledge is correlated with specific functionalities.

1-3   Electron-Coupled Nuclear Spin-Spin Interactions

Gutowsky, McCall, and Slichter in 1951 reported another case of peak multiplicity that differed from that described by Proctor and Yu.   The ^{19}F spectrum of $POCl_2F$ was found to contain two peaks.   Since the molecule possesses only one type of fluorine, the chemical shift explanation could not apply.   Because the molecule was tumbling freely in solution, the two peaks could not be due to two environments caused by crystal variations.   Furthermore, the spacing between the two peaks was independent of the applied field B_0.   The phenomenon was clearly distinct from magnetic shielding.

The molecule under consideration has a second magnetic nucleus, ^{31}P.   This nucleus has a spin of $\frac{1}{2}$ and therefore exists in two spin states in the magnetic field ($I_z = \pm\frac{1}{2}$).   The ^{19}F nucleus experiences one overall magnetic environment when the ^{31}P nucleus has $I_z = +\frac{1}{2}$, and another when $I_z = -\frac{1}{2}$.   The ^{19}F resonance therefore is a doublet, even though only one chemical type of ^{19}F is present.   When a nucleus experiences different magnetic environments because a second nucleus in the molecule exists in more than one spin state, the nuclei are said to be coupled. The spacing between the two ^{19}F peaks is a measure of the strength of the interaction and is called J, the coupling constant.   The ^{31}P spectrum of $POCl_2F$ should also be split into two peaks separated by the same amount, J, because the ^{19}F nucleus likewise exists in two spin states.   The chlorine nuclei, although magnetic, do not perturb the spectrum because of their quadrupolar properties (Chapter 7).   Figure 1-5 shows the spectrum of the two-spin system of dichlorofluoromethane, in which each nucleus splits the resonance of the other into two peaks.

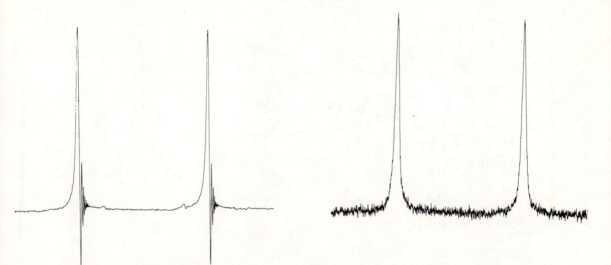

FIGURE 1-5. The 90 MHz proton and the 84.6 MHz fluorine-19 magnetic resonance spectra of CHFCl$_2$ (neat).

Such a pair of nuclei is said to produce an AX spectrum. The coupling constant between hydrogen and fluorine in CHCl$_2$F is 50.7 Hz.

Spin information is not transferred from one nucleus to the other by a direct dipole-dipole interaction between the two magnetic dipoles. Such an interaction has an angular dependence that is averaged to zero because the molecule is tumbling rapidly in solution. In order that one nucleus be sensitive to the spin states of the other, the nuclear dipole must spin-polarize the immediate electrons, which pass the spin information through the bonds and finally to the other nucleus. This indirect mechanism is the source of the weighty name given to the phenomenon and found in the title of this section. A more detailed discussion of the mechanism may be found in Chapter 4.

If a nucleus is coupled to more than one other nucleus, the resonance of the former will be more complicated than the doublet of the AX spectrum. The molecule cyclopropene (1)

1

contains two sets of two nuclei each (A$_2$X$_2$). Each X proton is equally coupled to two A protons. The spins of the two A nuclei can exist in four possible I_z combinations: $(+\frac{1}{2}, +\frac{1}{2})$, $(+\frac{1}{2}, -\frac{1}{2})$, $(-\frac{1}{2}, +\frac{1}{2})$, and $(-\frac{1}{2}, -\frac{1}{2})$. The $(+ -)$ and $(- +)$ orientations are equivalent, so the X protons will be split by the A protons into a 1/2/1 triplet (Figure 1-6). The A proton is similarly split, and the distance between any two adjacent peaks gives J (1.8 Hz).

The ethyl groups in ethyl ether present an A$_2$X$_3$ spin system (Figure 1-7). The methyl (X) protons are split by two methylene (A) protons, so the methyl resonance is a triplet. Perturbations from the ideal 1/2/1 intensity ratio are caused by second-order effects discussed

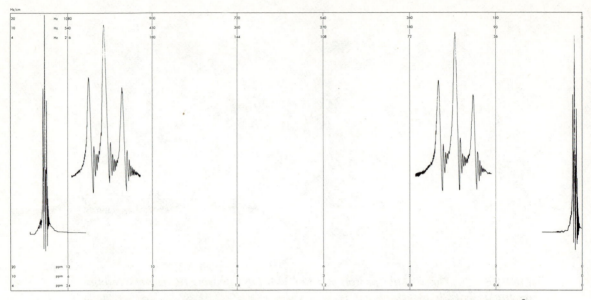

FIGURE 1-6. The 90 MHz proton spectrum of cyclopropene (neat) taken at −70°. The spectrum is 600 Hz wide, with the blowup expanded by a factor of 10. [Reproduced with the permission of the American Chemical Society from J. B. Lambert, A. P. Jovanovich, and W. L. Oliver, Jr., J. Phys. Chem., 74, 2221 (1970).]

in Chapter 4. The methylene protons are split by three methyl protons. The spin arrangements available to a three-nucleus system are $(+ + +)$; $(+ + -)$, $(+ - +)$, $(- + +)$; $(- - +)$, $(- + -)$, $(+ - -)$; $(- - -)$. Because the three arrangements with one $-\frac{1}{2}$ spin and two $+\frac{1}{2}$ spins are equivalent (and the three with one $+\frac{1}{2}$ spin and two $-\frac{1}{2}$ spins as well), the methylene

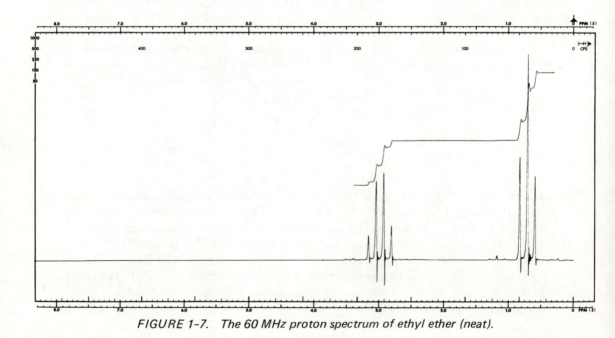

FIGURE 1-7. The 60 MHz proton spectrum of ethyl ether (neat).

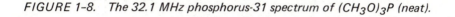

FIGURE 1-8.   The 32.1 MHz phosphorus-31 spectrum of (CH₃O)₃P (neat).

protons coupled to the three methyl protons form a 1/3/3/1 quartet. The familiar triplet-quartet pattern seen in Figure 1-7 is a reasonably reliable indication that an ethyl group is present in a molecule.

Similar arguments apply to larger systems. A nucleus coupled to four equivalent nuclei gives a 1/4/6/4/1 quintet. It is seen that the intensity ratios follow the coefficients of the binomial expansion, and may be readily obtained by use of Pascal's triangle. Thus the methine proton in an isopropyl group, $-CH(CH_3)_2$, is split by six protons to form a 1/6/15/20/15/6/1 septet. The outside lines are frequently too small to observe. Similarly, the ^{31}P spectrum of trimethyl phosphite, $(CH_3O)_3P$, is a decet of appropriate intensities, but only six or eight peaks are usually observed (Figure 1-8).

A proton coupled to a nucleus with spin 1 will be a 1/1/1 triplet, since the latter nucleus has three allowed orientations of I_z $(+1, 0, -1)$. The proton spectrum of ammonium chloride gives such a pattern, since each proton is coupled to a single ^{14}N nucleus (Figure 1-9). Acetone-d_6, a commonly used nmr solvent, contains as impurity a small amount of acetone-d_5 $(CD_3-CO-CD_2H)$. The proton is coupled to the two deuterium atoms on the same carbon, so a 1/2/3/2/1 quintet is formed (Figure 1-10): $(+ +)$; $(+0)$, $(0+)$; $(+ -)$, (00), $(- +)$; (-0), $(0-)$; $(- -)$. In general, a nucleus or set of nuclei coupled to n equivalent nuclei of spin I will be split into a multiplet containing $(2nI + 1)$ components.

The spin-spin splitting pattern of a nucleus gives substantial information about its detailed structural environment in terms of the surrounding magnetic nuclei. It should be emphasized that equivalent nuclei do not split each other. The methyl portion of an ethyl group is a triplet from coupling with the methylene protons, not a quartet. A set of nuclei of a single type therefore gives no splitting. The proton spectra of methane, ethane, benzene, cyclopropane, and ethylene are all singlets.

1-4   Quantitative Aspects

A methyl group, with three protons, absorbs three times the energy of a single proton. The integrated intensities of a series of resonances in general give the relative number of protons in each set. The spectrum of ethyl ether (Figure 1-7) shows an experimentally determined integral.

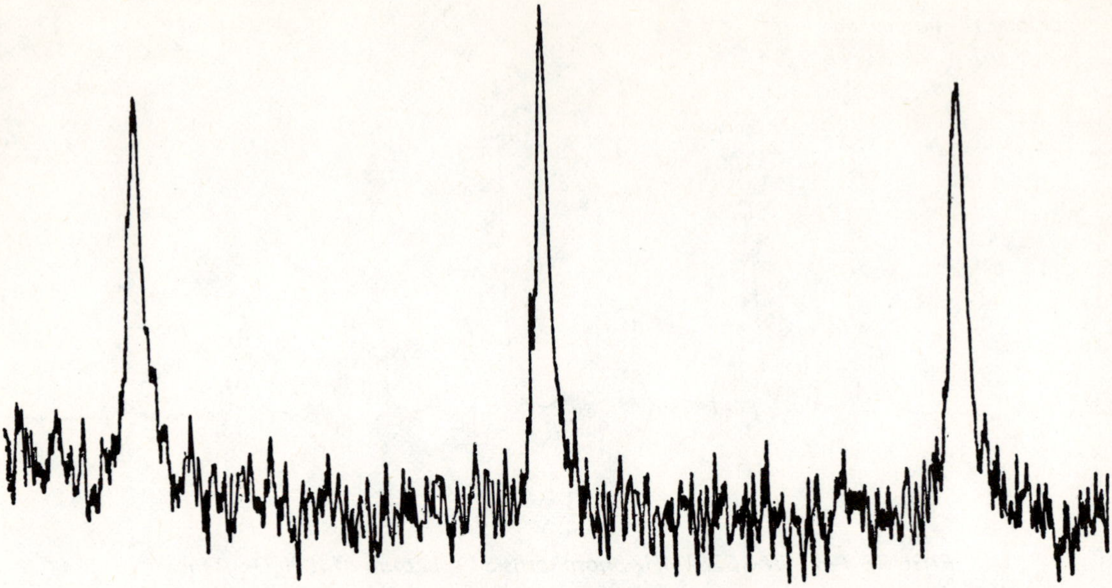

FIGURE 1–9. The 60 MHz proton spectrum of ammonium chloride in H_2O.

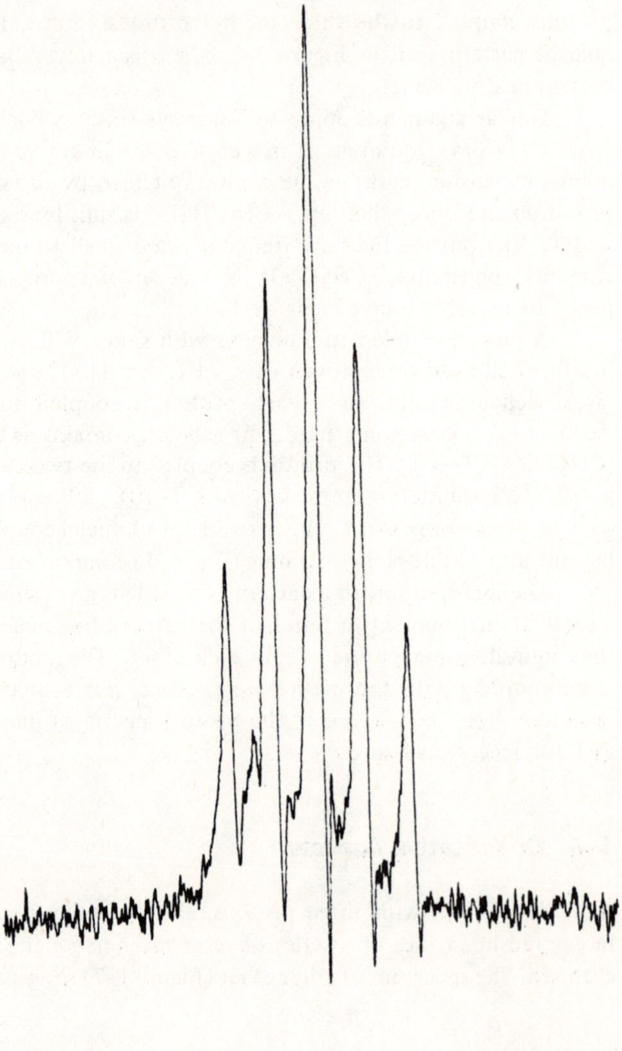

FIGURE 1-10. The 90 MHz proton spectrum of CD_3COCD_3 containing a small amount of CD_3COCD_2H.

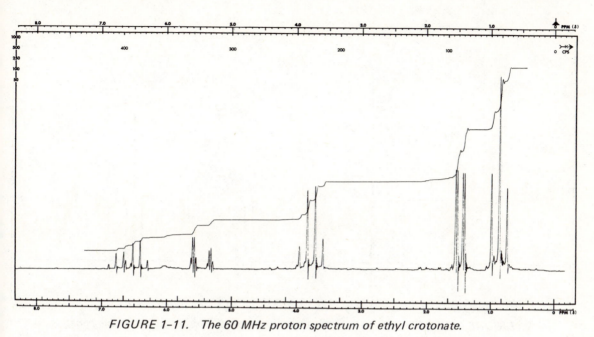

FIGURE 1-11.   The 60 MHz proton spectrum of ethyl crotonate.

The continuous horizontal line gives a measure of the relative area of a peak by its vertical displacement as the field is swept through the resonance. The displacements are in the ratio 3/2, as expected for the ethyl group.

For more complicated molecules, the spectral integral is invaluable in determining molecular structure, or just in assigning resonances for a known structure. Figure 1-11 gives the 60 MHz proton spectrum of ethyl crotonate ($CH_3CH=CHCO_2C_2H_5$). The lowest field set of peaks (a doublet of quartets) has a relative area of unity and may be assigned to one of the alkenic protons (coupled to the methyl group and to the other alkenic proton). The next set of peaks, also a doublet of quartets, comes from the second alkenic proton, which has a smaller coupling to the methyl group and is therefore probably farther away. The quartet at next higher field has an integral of two and may be attributed to the ester methylene group. The doublet of doublets (split by two nonequivalent protons) has an integral of three and splittings corresponding to those of the alkenic protons, so the resonance must arise from the methyl group on the double bond. The triplet (split by two protons) at highest field also has an integral of three, but the splitting corresponds to that in the quartet. This resonance is therefore due to the ester methyl group. The exact values of the chemical shifts and coupling constants give further information about molecular structure, but these will be considered in Chapters 3 and 4.

The quantitative analysis of mixtures of compounds can also be carried out by nmr integration. Figure 1-12 is the spectrum of a mixture of bromobenzene (C_6H_5Br), methylene chloride (CH_2Cl_2), and ethyl iodide (CH_3CH_2I). The peaks may be assigned by their splitting patterns and by comparison with spectra of the pure compounds. The multiplet at lowest field is due to C_6H_5Br, the singlet to CH_2Cl_2, and the quartet and triplet at highest field to CH_3CH_2I. From the integrals, with allowance for the number of protons causing a given resonance, the $C_6H_5Br/CH_2Cl_2/CH_3CH_2I$ ratio may be calculated to be 18/34/48. The error on an integration permits an accuracy of no better than about ±1.5%.

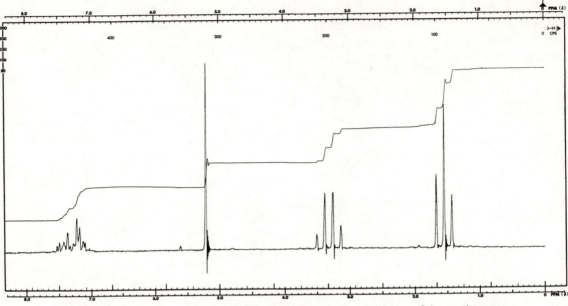

FIGURE 1-12. The 60 MHz proton spectrum of a mixture of bromobenzene, methylene chloride, and ethyl iodide.

1-5 Relaxation and Saturation

When a collection of nuclei is at equilibrium in the laboratory magnetic field B_0, there is a slight excess of nuclei in the more stable $-\frac{1}{2}$ state over those in the $+\frac{1}{2}$ state. The exact proportion is given by the Boltzmann distribution (eq. 1-12), in which the energy difference corresponds

$$\frac{n_-}{n_+} = e^{-(\Delta E/kT)} \qquad\qquad 1\text{-}12$$

to the quantities in eqs. 1-8 and 1-10 (eq. 1-13). When the oscillating field B_1 is applied at the

$$\frac{n_-}{n_+} = e^{-[\gamma\hbar B_0(1-\sigma)/kT]} \qquad\qquad 1\text{-}13$$

resonance frequency, nuclear spin flips occur in both directions (absorption, $-\frac{1}{2} \rightarrow +\frac{1}{2}$; emission, $+\frac{1}{2} \rightarrow -\frac{1}{2}$). Because there is a greater population of nuclei in the $-\frac{1}{2}$ state, there are more spin flips from $-\frac{1}{2}$ to $+\frac{1}{2}$, and the net effect is therefore absorption. When the excess $-\frac{1}{2}$ nuclei have been exhausted, however, the rates of absorption and emission become equal. The sample under these conditions is said to be "saturated," and no further resonance can be observed. Mechanisms must therefore exist to return the nuclei continuously to the equilibrium spin states with the excess of spin $-\frac{1}{2}$ nuclei.

Relaxation processes of this type derive mostly from interaction of the nuclei with local fields in the sample. Random oscillatory fields caused by other magnetic nuclei in motion or from unpaired electrons, which have a spin of $\frac{1}{2}$, can have the same frequency and phase as the excited spin and therefore accept its excess energy. By this means spin energy is converted to translational and rotational energy of the entire lattice. If this process is reasonably efficient, the nuclei may undergo spin flips, transfer the excess spin energy to the lattice, return to the ground state, and be available for another spin flip to continue the resonance experiment.

If the relaxation time is too long (several seconds), the spin equilibrium cannot be maintained, and there is degradation of the nmr signal. On the other hand, if the relaxation time is too short (a small fraction of a second), the signal may also be broadened by an interesting application of the uncertainty principle. According to this principle, the product of ΔE and Δt must remain constant. In this context Δt is the lifetime of the spin state and ΔE is the spread of energies that is associated with the resonance condition. If the lifetime of the state is extremely short (rapid relaxation), the spread of resonance energies, reflected in the width of the peak, must correspondingly increase. Thus, excessively short or long relaxation times can broaden linewidths. The investigator is generally desirous of peaks that are as narrow as possible in order to resolve distinct resonances. Ordinarily, *uncertainty broadening* is not a serious problem. The conditions under which it becomes important are discussed in Chapters 2 and 7. Chapter 7 contains a more complete description of relaxation processes and demonstrates how they may be exploited to solve structural problems.

Relaxation processes serve not only to maintain the spin population at equilibrium following application of the B_1 field, but also to generate that equilibrium condition when the sample is originally placed in the B_0 field. Without the laboratory field, the $+\frac{1}{2}$ and $-\frac{1}{2}$ spin states necessarily have the same energy and population. When the sample is placed in the B_0 field, the same relaxation processes described above will drive the sample to the equilibrium condition with an excess of $-\frac{1}{2}$ spin.

1-6 Rate-Dependent Phenomena

According to the principles outlined in the previous sections, the spectrum of methanol should contain a doublet of integral three (CH_3 coupled to OH) and a quartet of integral one (OH coupled to CH_3). Under conditions for which the hydroxyl proton does not exchange with the medium, this type of spectrum is observed (Figure 1-13). The presence of a small amount of acidic or basic impurity, however, can catalyze the intermolecular exchange of the hydroxyl proton. If this proton becomes detached from the methanol molecule by any mechanism, information about its spin states is no longer available to the rest of the molecule, and the methyl group will no longer be coupled to it. Even if the proton is continuously returned to the methanol molecule, its spin information is randomized. Under such conditions the methyl resonance will be a singlet.

The spectral changes associated with an exchange phenomenon of this sort are rate dependent. If the rate of exchange is very fast, no coupling is observed between the exchanging proton and other nuclei in the molecule. If the rate is slowed by lowering the temperature or decreasing the amount of acidic or basic catalyst, the coupling continues to be washed out until it reaches a critical value at which the proton resides sufficiently long on oxygen to permit the methyl group to detect the unaveraged spin states. For the coupling to be observed, the rate of exchange (sec^{-1}) must be approximately slower than the magnitude of the coupling in Hz. Thus a proton could exchange a few times per second and still maintain coupling. The

FIGURE 1-13.   The 60 MHz proton spectrum of CH₃OH at +50° (upper) and at −30° (lower).

spectra of methanol under fast and slow exchange conditions are given in Figure 1-13. In this case the transition through the critical rate was accomplished by changing the temperature 80°.

Under most spectral conditions hydroxyl protons do not exhibit couplings to other nuclei because of chemical exchange. A very small amount of catalyst can effect this exchange, so normally the unsplit hydroxyl resonance still has an integral of one. Related to the averaging of coupling constants is the averaging of chemical shifts. A mixture of acetic acid and benzoic acid contains only three resonances: methyl, phenyl, and one carboxyl peak. In this case the carboxyl protons of the two compounds interchange so rapidly that the spectrum exhibits only the average of the two. Similarly, if water is used as the solvent, the spectrum of a molecule containing an exchangeable hydrogen can appear to be devoid of the resonance from that proton. The spectrum of acetic acid in water contains two peaks, not three. The water and carboxyl protons appear in a single, weighted-average resonance. The position of the peak can serve as a quantitative guide to the proportions of the mixture.

Theoretically, if the rate of exchange of these protons were slowed down sufficiently, separate resonances could be observed. Dimethyl sulfoxide as a solvent can serve in this way to slow exchange. DMSO forms hydrogen bonds with an hydroxyl proton, but does not bring about its exchange. An alcohol dissolved in DMSO therefore is expected to exhibit couplings with the hydroxyl proton, and two alcohols should give separate hydroxyl resonances.

Intermolecular proton exchange reactions were involved in all the above examples. Unimolecular structural reorganizations, through their kinetic properties, can also influence the appearance of an nmr spectrum. It is known, for example, that cyclohexane contains distinct axial and equatorial protons, yet the room temperature spectrum contains only a sharp singlet. The process of ring reversal must occur so rapidly that the resonances of the axial and equatorial protons are observed at the average position. If the rate of this process could be slowed, distinct resonances might be observed. Such an experiment was carried out about 1960. Later investigators used the commercially available cyclohexane-d_{11} (eq. 1-14).

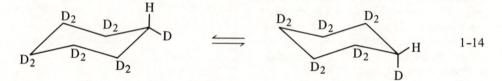

$$1\text{-}14$$

The couplings to deuterium may be removed (Chapter 5) to give a clarified spectrum. As the temperature is lowered, the rate of ring reversal slows, and the singlet broadens considerably and re-forms as a doublet (Figure 1-14). At $-89°$ the rate of ring reversal is so slow that separate resonances are observed for the axial and equatorial protons.

The nmr spectrum is sensitive to unimolecular processes of this sort when the rate of exchange is on the order of the distance between the slow exchange resonances ($1-100 \text{ sec}^{-1}$). The nmr experiment thus has a very sensitive time scale. A wide variety of unimolecular processes (particularly hindered rotation about bonds) may be followed in this fashion, and their kinetics determined. A more complete discussion of this subject may be found in Chapter 6.

BIBLIOGRAPHY

1. J. D. Roberts, *Nuclear Magnetic Resonance*, McGraw-Hill, New York, 1959.
2. J. A. Pople, W. G. Schneider, and H. J. Bernstein, *High-resolution Nuclear Magnetic Resonance*, McGraw-Hill, New York, 1959.
3. J. W. Emsley, J. Feeney, and L. H. Sutcliffe, *High Resolution Nuclear Magnetic Resonance Spectroscopy*, Pergamon, Oxford, 1965.
4. E. D. Becker, *High Resolution NMR*, Academic, New York, 1969.
5. F. A. Bovey, *Nuclear Magnetic Resonance Spectroscopy*, Academic, New York, 1969.
6. L. M. Jackman and S. Sternhell, *Applications of Nuclear Magnetic Resonance Spectroscopy in Organic Chemistry*, 2nd ed., Pergamon, Oxford, 1969.
7. B. I. Ionin and B. A. Ershov, *NMR Spectroscopy in Organic Chemistry*, Plenum, New York, 1970.
8. W. W. Paudler, *Nuclear Magnetic Resonance*, 2nd ed., Allyn and Bacon, Boston, 1974.

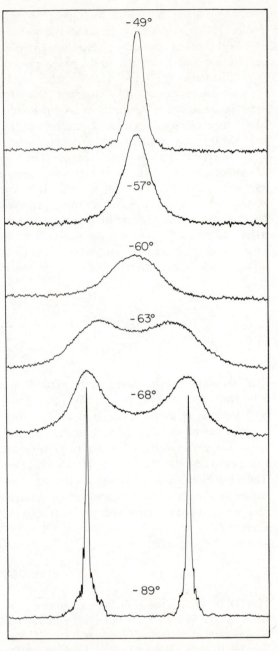

*FIGURE 1–14.   The 60 MHz proton spectrum of cyclohexane-d_{11} as a function of temperature.   [Reproduced with permission from F. A. Bovey, F. P. Hood III, E. W. Anderson, and R. L. Kornegay, J. Chem. Phys., **41**, 2041 (1964).]*

9. Spectroscopic Problems: R. M. Silverstein and G. C. Bassler, *Spectrometric Identification of Organic Compounds*, 3rd ed., Wiley, New York, 1974; D. H. Williams and I. Fleming, *Spectroscopic Problems in Organic Chemistry*, McGraw-Hill, London, 1969; R. H. Shapiro, *Spectral Exercises in Structural Determination of Organic Compounds*, Holt, New York, 1969; J. R. Dyer, *Organic Spectral Problems*, Prentice-Hall, Englewood Cliffs, N.J., 1972.

2

EXPERIMENTAL METHODS

2-1 Nuclear Magnetic Resonance Equipment

During the early 1970's, new nmr instruments have been introduced with B_0 fields that range from 0.7 to 7.0 tesla and prices from \$5000 to \$350,000. The most important aspect that is common to these instruments is their solid state circuitry. The breakdown rate of solid state components is much lower than that of corresponding vacuum tube circuits, and much solid state repair work is done by simply replacing whole circuitboard modules. The complexity of the solid state electronics is also considerably greater. As a result, the necessity and the possibility of home repairs have lessened, and the spectrometer owner need attend much less to the detailed electronics of his spectrometer than he did prior to 1970. For this reason, we will describe only the essential elements of the modern spectrometer and omit consideration of the electronics involved in measuring a nuclear magnetic resonance signal. The major components of an nmr machine are

(a) the laboratory magnet to supply the B_0 field;
(b) the radiofrequency unit to transmit the B_1 field and to receive the nmr signal;
(c) sweep coils to vary the B_0 field or the B_1 frequency;
(d) stabilization methods to produce as stable and homogeneous a field as possible; and
(e) accessories to improve the performance of the machine and to increase its capabilities.

These components will be discussed in turn. Figures 2-1 and 2-2 show two types of recent nmr spectrometers.

Three types of magnets are available: permanent, superconducting, and electromagnet. The superconducting magnet is used to achieve very high B_0 fields, above about 2.5 tesla. Despite the expense of purchasing and operating such a magnet, superconducting systems are becoming increasingly common. They are particularly well suited for obtaining very sensitive,

19

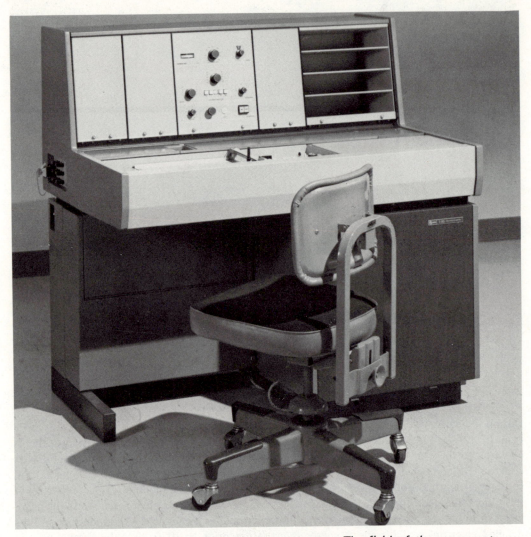

FIGURE 2-1. *The Varian T-60 nmr spectrometer. The field of the permanent magnet is 1.41 tesla. A small instrument of this type is used principally for proton spectra at room temperature.*

routine proton spectra. The permanent magnet can be the least expensive to own and operate, but it is also the least flexible. Most of today's machines are equipped with an electromagnet. The B_0 field may be changed by altering the current that passes through the magnet coils, and machine operation is not too sensitive to temperature. The permanent magnet has neither of these characteristics.

Stability and homogeneity are the most important requirements of the magnet, so that small differences between resonance frequencies can be differentiated. The sample is placed in the most homogeneous region between the magnet pole faces on an adjustable probe. This probe contains, in addition to a holder for the sample and mechanical means for adjusting its

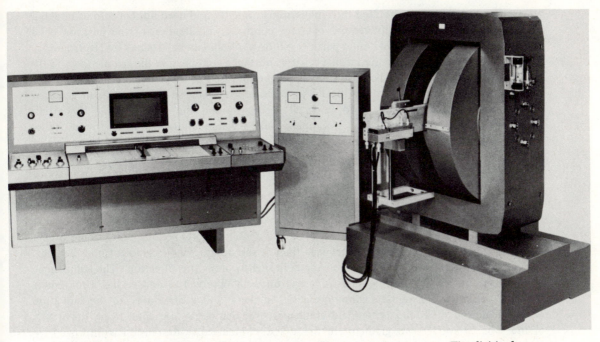

FIGURE 2-2.   The Bruker HFX-10 multinuclear nmr spectrometer.   The field of
the electromagnet is 2.12 tesla.   With suitable accessories, an instrument of this
type can carry out all the experiments described in this chapter.

position in the field, devices for improving magnetic homogeneity, for supplying the B_1
frequency, and for receiving the nmr signal.

Usually separate from the magnet (Figure 2-2) is a console that contains, inter alia, the
radiofrequency (rf) unit.   The B_1 field oscillating at the resonance frequency is supplied either
from a crystal-controlled oscillator in this unit or from a frequency synthesizer and is transmitted to a small coil that encircles the sample in the probe.   The nmr signal is detected by
the rf unit (see Chapter 7 for a more detailed description), amplified, and displayed either on an
oscilloscope for immediate observation or on an X–Y recorder for permanent record.

Because of electronic shielding, structurally different examples of a given isotope resonate
over a range of frequencies.   Many instruments hold the frequency constant and observe the
different resonances by slowly altering the field.   This *field sweep* method has been more
widely used in the past, but it suffers from three drawbacks.   (1) The sweep coil can give a
nonlinear signal that produces an incorrect read-out on the X–Y recorder.   (2) Homonuclear
decoupling experiments (Chapter 5) are extremely difficult.   (3)   Direct spectral calibration
can only be done with an attached audio oscillator (next section).   Although field sweep
spectra are adequate for some purposes, highly accurate work and certain decoupling experiments require the type of instrument that holds the field constant and sweeps the frequency of
the B_1 field.   Such *frequency sweep* capabilities are standard on most current instrumentation.

The entire range of proton resonances is contained in about 15 parts per million (ppm)
of the field at constant frequency.   Strenuous efforts have therefore been made to produce a
field that is sufficiently stable and homogeneous to permit differentiation of resonances with
small field separations.   Numerous methods have been devised to create a stable field.   These

are for the most part incorporated into modern instruments, and need not concern us in detail because they do not generally require operator control. Stabilization of the position of a resonance on the recorder to within 1 part per billion per hour can be attained by electronically locking the magnetic field to the resonance of a substance contained in the sample specifically for this purpose. Tetramethylsilane (TMS) is commonly used as the source of a resonance for this *internal lock*. With the field locked to TMS, a frequency sweep process can generate the resonance spectrum, except for a small region around the lock signal. In certain instruments, the field is locked to a sample, usually water, contained in a separate tube. A field sweep spectrum of the sample to be observed is conveniently obtained by means of sweep coils that only affect the field around this sample and not around the *external lock* reference sample located elsewhere in the probe. Field stabilization by this method is about one part in 10^8.

For best spectral results, the magnetic field should be completely homogeneous within the sample under examination. Gradients are present in the B_0 field that militate against optimal results. For example, the field along the z direction (from one pole to the other) might be slightly higher at one magnet face than at the other. Such a gradient may be compensated for by an electric coil that alters the field slightly along this direction. The adjustment is made empirically by altering this *shim* coil until an observed resonance has a minimum linewidth. Similar coils correct inhomogeneities in the x and y directions, in the curvature of the field, and in second order field gradients. The sample is also spun along the y axis of the tube to average out gradients in the x and z directions. Resolution is improved by spinning because a molecule at a particular point in the tube experiences a field that is averaged over its circular path. No averaging is effected in this manner along the vertical axis of the tube (the y axis), so considerable use of the y gradient shim coil is necessary.

The above elements are essential to the execution of an nmr experiment. In addition, there are numerous accessories that expand the capability of an instrument. (1) Variable temperature operation permits reaction kinetics to be carried out on the machine (Chapter 6). Higher temperatures can sometimes allow the examination of poorly soluble materials, and lower temperatures, unstable compounds. (2) Signal enhancement accessories (section 2-5) can appreciably decrease the amount of sample necessary to obtain a spectrum. (3) More than one type of nucleus can be observed on a machine if the B_1 frequency can be altered and the transmitter-receiver coils replaced. It is better to observe all nuclei at the highest possible field because chemical shifts are largest and sensitivity best under these conditions. A change from one nucleus to another is therefore best done by altering B_1 rather than B_0. In some machines it is inconvenient or impossible to interchange the coils, whereas in others the components are modular to facilitate a change. For machines with a multinuclear capability, separate rf units and coils must still be available for each nucleus. (4) The capability to decouple one nucleus from another, i.e., to remove the coupling (Chapter 5), is very useful, but separate modular components are necessary for each nucleus for which decoupling is desired. (5) Instead of applying a continuous B_1 field, the nmr experiment can be carried out by applying short bursts (pulses) of rf energy. The pulse method finds particular utility in certain signal enhancement experiments involving the Fourier transform method (section 2-5). (6) High resolution spectra may only be taken on a nonviscous liquid. Because of dipole-dipole interactions, molecules in the solid give very broad resonances. Certain pieces of useful information, e.g., second moments, can sometimes be extracted from the spectra of solid materials. To obtain these spectra, however, accessories are required for sweeping over huge ranges of the magnetic field. This technique is called *wideline* or *broadline* nmr spectroscopy.

2-2 Sample Preparation

The ideal shape for an nmr sample container is a sphere. Homogeneity problems are minimized because the container is isotropic, and material is used most effectively. Spherical sample cells, however, have found little popularity because they are expensive, difficult to fill, and awkward to place in the probe, since the transmitter-receiver coils are not visible to the operator. Most spectrometers have been designed to accept a 5 mm cylindrical sample cell. Precision-ground cells of variable length (15–20 cm) are commercially available. If about 25 mm of sample is placed in the tube, end effects from the cylindrical shape are minimized, and there is considerable leeway in locating the tube in the coils. More sample than necessary is needed to fill the tube, so the common cylindrical tube is very inefficient with sample material. Nonetheless, it is by far the most convenient to use. Special spherical containers imbedded in a 5 mm cylindrical tube are also available. When properly positioned in the coil, these *microtubes* require only about 35 μl of sample and achieve almost the same resolution as the cylindrical tube, which requires 350 μl. A good spectrum of 1 mg of a material (MW <400) can be obtained with a microtube. Many spectrometers are now designed to accept 8, 10, 12, and 15 mm sample tubes. Sensitivity is increased by the larger cross section, but resolution is impaired by the larger volume over which the homogeneity must be optimized. The natural sensitivity of the ^{13}C nucleus is only 1.6% that of ^{1}H, and the nucleus is only 1.1% abundant. Therefore 10 mm tubes are used routinely for cmr at natural abundance because of the increased sensitivity requirements. Peaks are generally well separated in carbon spectra, so the impaired homogeneity is not a serious drawback.

A compound may be examined as a neat liquid, provided the material is not too viscous. Highly viscous materials give broad, poorly resolved spectra because of unaveraged dipolar interactions and short relaxation times. These effects are extreme in the nmr spectra of solids. Normally, the sample is dissolved in a suitable solvent for examination. The amount of material required is conserved by this means, and even very dilute solutions (<0.1 *M*) give respectable spectra. The solvent is chosen not only for its solubility characteristics but also for its magnetic properties. It should not possess resonances in the regions of interest. For this reason CCl_4 and $CDCl_3$ have long been the most popular nmr solvents. Deuterated solvents ($CDCl_3$, D_2O, C_6D_6, CD_3COCD_3, CD_3SOCD_3, CD_3OD, CD_3CN) have now become so inexpensive that competing solvent resonances are rarely a problem. Solvents do exert an effect on the chemical shift (Chapter 3) so that choice of solvent is sometimes dictated by the desire to control or exploit these effects.

If spectra are to be taken at nonambient temperatures, the melting and boiling points of the solvent must be considered. Among the good low temperature solvents are CD_3OD, CS_2, CD_3COCD_3, cyclopentane, cyclopropane, $C_6D_5CD_3$, $(CH_3)_2O$, $CHFCl_2$, CF_2Cl_2, CH_2=CHCl, CH_2FCl, SO_2, CH_2Cl_2, SO_2FCl, and CH_2=CHCH$_3$. If the solvent is gaseous, it is condensed over the solute into the tube, which is then sealed. For high temperature experiments, frequently no solvent is used if boiling point permits. Otherwise biphenyl, naphthalene, dimethylformamide, DMSO, or sulfolane might serve as solvent.

If an internal lock is used to stabilize the field, a material with a sharp singlet resonance must be present in the sample. Frequently, a resonance of the compound under investigation is used for this purpose if it is not of interest. Otherwise, a material such as TMS or benzene is added. Spectra are frequently taken of nuclei other than proton with a lock on a proton signal. Some spectrometers also use ^{19}F or ^{2}H for a lock signal for other nuclei. Hexafluorobenzene is a common lock compound for ^{13}C spectra. Because of its poor solubility, C_6F_6 is normally contained in a 2 mm capillary that rests in the 10 mm tube.

2-3 Calibration and Referencing

The common unit for expressing chemical shifts is parts per million of the magnetic field. Thus at 1.41 tesla, 1 ppm corresponds to 1.41×10^{-6} tesla. Since the proton resonates at 60.0 MHz in this field, 1 ppm is 60 Hz in frequency (energy) units. At 100 MHz, 1 ppm is 100 Hz. Since the difference between the chemical shifts of two nuclei is proportional to the magnitude of the field, the separation expressed in ppm is independent of the value of the field. Two nuclei may resonate 60 Hz apart at 60 MHz, or 100 Hz at 100 MHz, but in either case the separation is 1 ppm.

The coupling constant is the measure of an interaction between two nuclei and is therefore independent of B_0. A J of 6.0 Hz (0.1 ppm) at 60 MHz is still 6.0 Hz (0.06 ppm) at 100 MHz. If it is not known whether a particular separation is a chemical shift difference or a coupling constant, the problem can be readily resolved by taking the spectrum at two different fields. A chemical shift separation in Hz will change with field, but a coupling constant will remain the same.

The proton resonance of TMS occurs at a very high field, almost at the extreme of observed resonances. In addition, TMS is volatile, soluble in most organics, almost chemically inert, and gives a strong signal. For these reasons, TMS has been chosen to serve as the reference for all proton resonances. The distance in ppm of a resonance from that of TMS is the accepted method to express a chemical shift. If the distance is measured in Hz, it may be converted to ppm by eq. 2-1. The value of δ by convention is positive in the downfield

$$\delta \text{ (ppm)} = \frac{\text{distance from TMS in Hz}}{\text{value of } B_1 \text{ in MHz}} \qquad 2\text{-}1$$

direction (right to left) from TMS. Resonances at higher field than TMS therefore have a negative δ. Coupling constants are always expressed in Hz. Standards have also been chosen for nuclei other than hydrogen. Carbon shifts are expressed in ppm downfield from the ^{13}C resonance of TMS; ^{31}P from 85% aqueous H_3PO_4; ^{19}F from CCl_3F; ^{14}N and ^{15}N from $(CH_3)_4N^+$; and ^{11}B from BF_3-ethyl etherate. These values are also defined to be positive in the downfield direction.

To report differences between peak positions, the spectrum must be properly calibrated. The most widespread and convenient method is to record the spectrum on precalibrated chart paper. This method is only possible for spectrometers that have internal or external field-frequency lock. The chart paper may be calibrated by taking the spectrum of a sample containing two peaks separated by a known distance. The TMS-$CHCl_3$ separation of 436 Hz (at 60 MHz) is widely used. A potentiometer available for the purpose on the recorder is adjusted until the distance between the peaks corresponds to 436 Hz on the chart paper.

Alternatively, the spectrum of a single component can be modulated with an audio oscillator. This device creates a peak on either side of the resonance, separated from the centerband by the frequency of the oscillation. Thus the spectrum of TMS might be modulated with a 50 Hz oscillation. Three peaks are observed—the TMS resonance and 50 Hz sidebands on either side. The recorder potentiometer is then adjusted until the 50 Hz separation corresponds to 50 Hz on the chart paper. The audio oscillator can also be used for the direct measurement of peak separations from TMS. For this purpose the oscillator frequency is adjusted until the TMS sideband falls directly on top of the peak in question. For example, a 436 Hz oscillation should put a TMS sideband directly on top of the $CHCl_3$ resonance (and vice versa). This method was initially used to establish the TMS–CHCl$_3$ distance that now serves to calibrate the recorder directly.

An internal lock, frequency sweep spectrometer has a built-in reference in the lock signal. As the spectrum is recorded on chart paper, a frequency counter can read out the distance in Hz from the lock of the point being recorded. It is therefore not necessary to rely on the accuracy of the precalibrated chart paper, which does drift with time. The exact frequency of each peak can be obtained from the counter as the spectrum is recorded. If the lock is TMS, then this number may be converted directly to units of δ by eq. 2-1.

2–4   Optimization of Spectrometer Performance

To produce best results, the spectrometer requires optimization of resolution and sensitivity. In this section we will discuss the methods of optimization that are common to most spectrometers. The quartet of acetaldehyde serves as the standard for resolution measurements, although any sharp peak in a spectrum may be used for tuning purposes. A linewidth at half-height less than 0.1 Hz is quite good, but broader lines are acceptable for many purposes. The sample is set spinning at a rate of 30–60 revolutions per second. A slower rate may not be sufficient to average the x and z gradients, and a higher rate can set up a vortex in the tube that distorts the signal. If the magnet homogeneity is not optimal, if the tube is unsymmetrical, or if the tube is not rotating on its axis, the spinning process can modulate the signal at the frequency of spinning in the same way that an audio oscillation does. Spinning modulation peaks occur on either side of the centerband, separated from it by the frequency of spinning in Hz (sec^{-1}). Examples may be seen in the spectrum of toluene (Figure 1-3). The sidebands may be identified and moved simply by changing the spinning rate. Spinning sidebands are a particular nuisance if an undeuterated solvent is used, since its sidebands can be quite large. Sidebands may be reduced by faster spinning, by improving the resolution, or by using a more symmetrical sample tube.

Spinning does not improve resolution in the y direction, so the y gradient shim coil is most frequently tuned. The x and z gradient shims are generally adjusted first without spinning, followed by the y gradient and the curvature with spinning. Some machines also have second order and higher shim controls (xz, xy, yz) for additional tuning. The controls are adjusted while a peak is monitored for maximum height, minimum linewidth, or maximum ringout (the beat pattern that follows a sharp peak; see Chapter 7). On some instruments the shim controls may be prevented from drifting by an automatic shimming device.

In older magnets, the center of the field has a tendency to decay slowly. At some time the field would "collapse" at the center, and resolution would be unacceptable. A similar situation would exist when the magnet was turned on from zero field. To compensate for this low point in the center of the magnetic field, the current to the magnet would be raised to a higher than normal value for a short period of time. When returned to the normal current value, the magnetic field would have regained its normal, flat contour across the pole caps. This process of "cycling" is either automatic or unnecessary in the newer magnet systems.

Once the magnet has been cycled, the sample set spinning, and the shim controls tuned, there is little more the operator can do with the instrument to improve resolution. Attention to the sample can yield additional gains. Solid particles in a sample reduce homogeneity just as a vortex does. Filtering the solution can remove this problem. Any paramagnetic materials can have an extremely deleterious effect on resolution. The strong magnetic moment from the unpaired electron spin serves to reduce the nuclear relaxation time very effectively. Short relaxation times seriously impair the resolution through uncertainty broadening. Even dissolved atmospheric oxygen has a small broadening effect, so samples for which highest resolution is required must be degassed. Paramagnetic materials can produce other effects that are discussed in Chapter 7.

Even the spectrum of a nucleus with a favorable relaxation time can be saturated if too strong a B_1 field (rf power) is used. When the power level is too high, the equilibrium situation cannot be maintained, and both homogeneity and sensitivity are impaired. To optimize sensitivity, the strongest B_1 field is used that does not permit saturation. To make this adjustment, the B_1 power is slowly increased as the field is repeatedly swept through a resonance. The intensity of the resonance increases until the power reaches the saturation level. The B_1 power is then set slightly below this value. When sensitivity is not critical, the lowest possible power settings are used. For spectrometers that utilize an internal lock, the power used for this purpose is also critical. Too low a level can make for a weak lock that if lost will naturally remove the field-frequency stabilization. Too high a power level can saturate the signal and give rise to oscillations that can remove the lock. Optimum results therefore require careful setting of the lock power at an intermediate value.

All spectrometers are equipped with gain and filtering devices. The gain control multiplies signal and noise together. It is generally kept at the lowest value that gives a good display of the spectrum. The filter serves to remove high frequency noise. If the spectrum is too highly filtered, however, the signal as well as the noise is removed. For this reason the time constant of the filter is set at a high value (little filtering), unless the investigator is confronted with an extremely weak spectrum. The sweep rate is also important in determining sensitivity. A very fast rate decreases saturation but may lose a significant portion of the signal. A slow sweep rate is therefore to be preferred.

When sensitivity is a problem, four controls must be balanced against one another to maximize response:

(a) B_1 is placed at the highest level (lowest attenuation) that does not cause saturation.
(b) Signal content per unit time is maximized by slowing the sweep rate.
(c) The gain is set to the highest possible level permitted without overloading the system.
(d) The filter time constant is decreased (more filtering) to the point of diminishing returns at which too much signal is being removed.

The methylene quartet of a 1% solution of ethylbenzene in CCl_4 serves as the standard for measuring the sensitivity of a spectrometer. The height of the tallest peak in the quartet multiplied by $2\sqrt{2}$ (from noise theory) is compared to the height of the noise. Figure 2-3 shows a sensitivity measurement with a S/N of about 50/1.

2-5 Sensitivity Enhancement

Sensitivity becomes critical if a compound is poorly soluble in all solvents, if it is in very short supply, or if the molecule is so large that the signal from any single nucleus is very weak. Many nuclei do not have a very high natural abundance, and synthesis of materials labeled with the nucleus is usually inconvenient and expensive (^{10}B, 18.83%; ^{13}C, 1.11%; ^{15}N, 0.365%; ^{29}Si, 4.70%). Sensitivity is therefore extremely critical in examining the spectra of these nuclei at natural abundance. In addition, almost all other nuclei are inherently less sensitive than the proton to the magnetic resonance experiment, even at the same field (^2H, 0.96%; ^7Li, 29.4%;

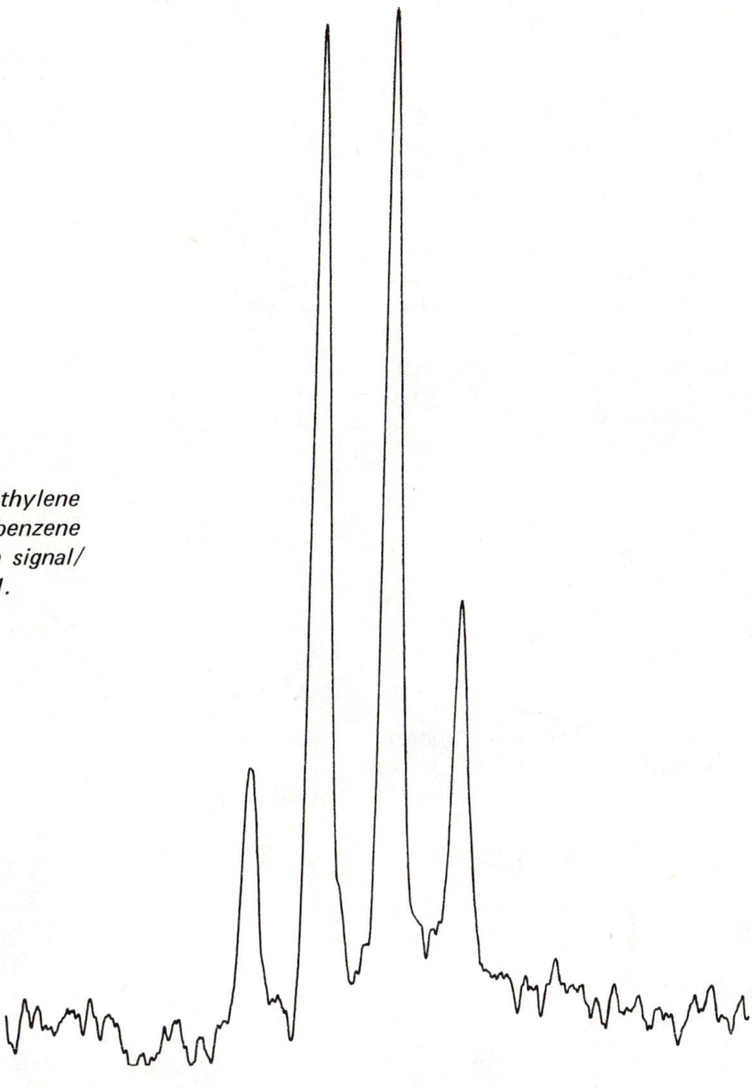

FIGURE 2-3. The methylene quartet of a 1% ethylbenzene solution in CCl_4, with a signal/noise ratio of about 50/1.

^{11}B, 16.5%; ^{13}C, 1.59%; ^{14}N and ^{15}N, 0.10%; ^{17}O, 2.91%; ^{19}F, 83.4%; ^{29}Si, 7.85%; ^{31}P, 6.64%). When both natural abundance and natural sensitivity are taken into account, nuclei like ^{13}C and ^{15}N pose a very difficult experimental problem.

For optimum sensitivity, the spectrometer is first tuned according to the procedures of the previous section for best performance. The sample size and concentration are increased as much as is practical. The 10 mm tube is used routinely with spinning for ^{13}C spectra. If these methods are insufficient, devices developed for artificially expanding the sensitivity capabilities of the spectrometer must be used. Experience has taught us that certain experiments, such as natural abundance ^{13}C spectra, are usually not even attempted without "artificial" signal enhancement.

The nmr spectrum that is read out onto the recorder may also be stored in the memory of a small computer, after the analog signal is converted to digital information. If the computer has 4000 storage locations, the spectrum is divided up into that many points, and the magnitude of the signal at each point is stored in the corresponding channel. Reproduction of a

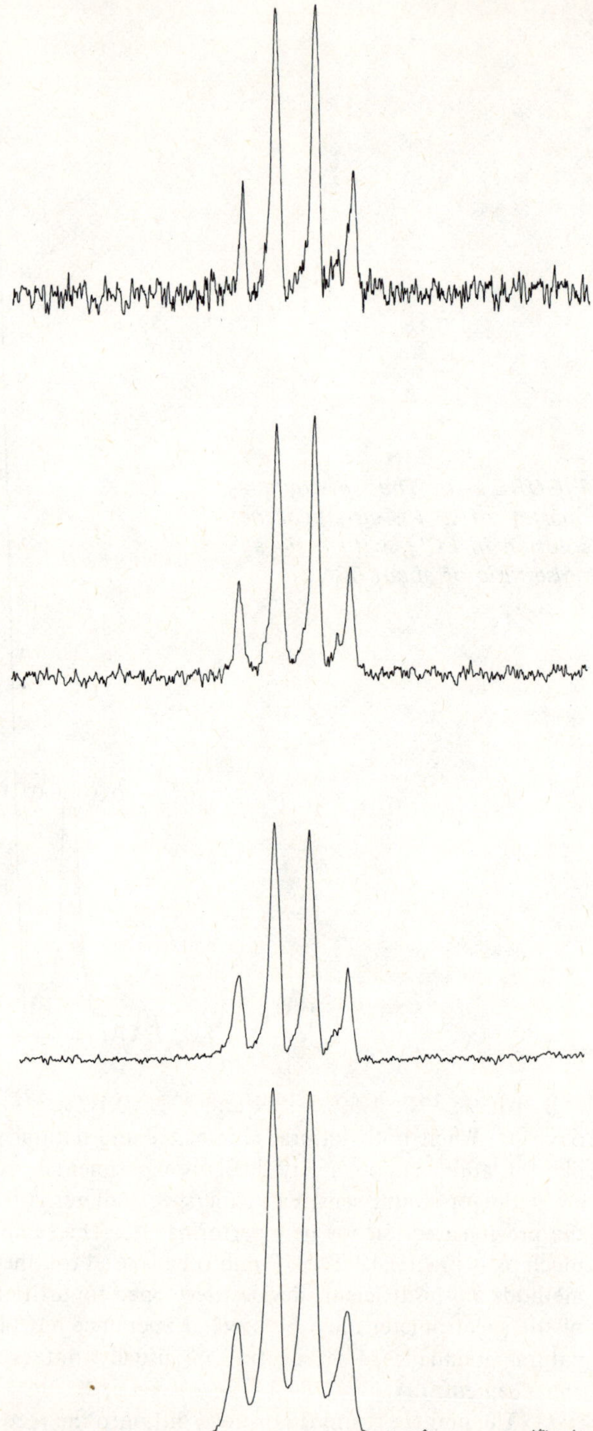

FIGURE 2-4. The same sample of ethylbenzene as in Figure 2-3, with accumulations of 1, 4, 16, and 128 scans.

digital readout of 4000 points is effectively indistinguishable from the usual analog chart display. Older computers have 400 or 1024 channels, but these numbers are insufficient for wide sweeps such as in ^{13}C spectra. Computers with more than 4000 locations are also available. If the same spectrum is swept a second time and stored in the same locations, any signal present will be reinforced, but noise will tend to cancel out. If n such sweeps are carried out and stored, the theory of random processes tells us that the signal amplitude is proportional to n, but the noise is proportional to $\sqrt{n}$. The signal/noise ratio therefore increases as $n/\sqrt{n}$, or $\sqrt{n}$. Thus 100 sweeps added together will enhance the S/N by a factor of 10. For averages of longer duration, the stability of a field-frequency lock is necessary. Figure 2-4 shows progressive averaging of up to 128 scans of the methylene quartet of ethylbenzene.

The technology of computer enhancement of nmr sensitivity has improved to the point that the process may be routinely carried out. The maximum factor of $\sqrt{n}$ is not reached in practice, but averaging 10–100 spectra of a weak but visible signal is usually sufficient. Averages of 100–1000 spectra are not uncommon. The most serious drawback to this method is the amount of time required for the experiment. If a single spectrum requires 250 sec, then 100 spectra require 25,000 sec (about 7 hr). Spectrometer time can thus be expended rather extravagantly to obtain the spectrum of a single substance. Nonetheless, the method of computer averaging is very useful for weak ^{1}H and ^{19}F spectra and a necessity for natural abundance ^{13}C spectra. The spectrum of sucrose in Figure 1-4, for example, utilized 128 scans.

An entirely different method may be used to take the spectrum many times for computer averaging without utilizing excessive amounts of time. In an ordinary frequency sweep experiment, one point after another in the spectrum is examined as the B_1 frequency is swept over the spectral region of interest. A far more efficient way to take the spectrum is to apply a sharp rf pulse simultaneously over the entire frequency range. The pulse might last 100 μsec, and all the spectral information is present in its decay pattern (Figure 2-5). The Fourier transform of this interference pattern gives the normal continuous wave (cw) spectral representation. The advantage of the method is that a new spectrum may be taken every second or so. A thousand spectra are typically recorded point by point in a small computer every ten minutes. When the desired number of spectra have been averaged in the computer in the form of the decay pattern, the Fourier transform operation is carried out by computer to give the normal representation. Even a computer with only 4000 memory locations can perform the transform in less than 5 minutes, given the appropriate software, and larger computers accomplish the task in a few seconds.

The Fourier transform procedure has become the method of choice for obtaining ^{13}C spectra. This method was used to obtain the sucrose spectrum in Figure 1-4. The required accessories are the pulse equipment and a small computer or computers for signal averaging and for performing the Fourier transform. Proton spectra of very high sensitivity may also be obtained by this method.

Finally, it should be pointed out that an effective gain in signal/noise ratio may be obtained if all the couplings between nuclei are removed. The fewer remaining peaks are more readily discernible. *Decoupling* of this type, which was used in the sucrose spectrum, is described in Chapter 5. A related technique can provide an actual increase in the integral of a resonance through the nuclear Overhauser effect. Signal enhancement by this technique will be discussed in relation to other relaxation phenomena in Chapter 7. Both of these methods have been particularly useful in obtaining ^{13}C resonances at natural abundance.

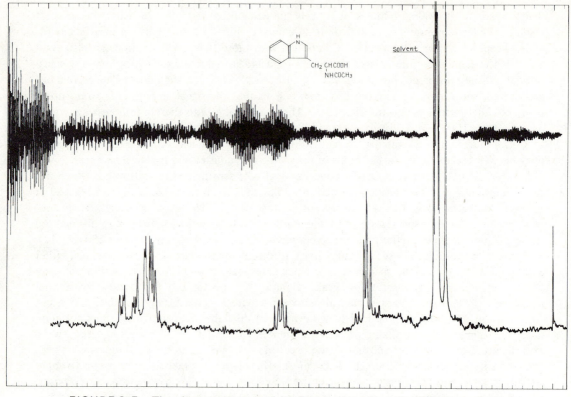

FIGURE 2-5. *The decay pattern and its Fourier transform for 35 μsec pulses of the 100 MHz proton spectrum of N-acetyl-dl-tryptophan in acetone-d$_6$, accumulated for 256 scans. [Reproduced with the permission of JEOL.]*

BIBLIOGRAPHY

1. Practical Aspects: D. Chapman and P. D. Magnus, *Introduction to Practical High Resolution Nuclear Magnetic Resonance Spectroscopy*, Academic, New York, 1966; D. G. Gillies, *Nucl. Magn. Resonance* (*Spec. Period. Report*), **3**, 156 (1974).
2. Applications at High Magnetic Fields: W. Naegele, *Determ. Org. Struct. Phys. Meth.*, **4**, 1 (1971); A. A. Grey, *Canad. J. Spectrosc.*, **17**, 82 (1972).
3. Fourier Transform Methods: D. A. Netzel, *Appl. Spectrosc.*, **26**, 430 (1972); T. C. Farrar and E. D. Becker, *Pulse and Fourier Transform NMR*, Academic, New York, 1971; N. Boden, *Determ. Org. Struct. Phys. Meth.*, **4**, 51 (1971); D. G. Gillies and D. Shaw, *Ann. Reports NMR Spectrosc.*, **5A**, 560 (1972); E. D. Becker and T. C. Farrar, *Science*, **178**, 361 (1972); A. G. Redfield and R. K. Gupta, *Advan. Magn. Resonance*, **5**, 82 (1971).
4. Quantitative Methods: F. Kasler, *Quantitative Analysis by NMR Spectroscopy*, Academic, London, 1973.

3

THE CHEMICAL SHIFT

3-1 Atomic and Molecular Shielding

Up to this point, a spin $\frac{1}{2}$ nucleus has been described as lining up either with or against the B_0 field. The field, however, is not successful in effecting a perfect alignment, so the nucleus precesses about the direction of the magnetic field. For $I_z = -\frac{1}{2}$, the nuclei precess about the $-z$ direction,[1] and for $I_z = +\frac{1}{2}$ the $+z$ direction. The angular frequency of precession corresponds to the natural resonance frequency ω_0 (eq. 1-9). The process is termed Larmor precession, and ω_0 is often called the Larmor frequency. An isolated atom experiences a local magnetic field somewhat different from B_0 because of shielding by the surrounding electrons. In the magnetic field, the entire electron cloud also undergoes a Larmor precession. By Lenz's law the magnetic field created by this motion of the diamagnetic electrons opposes the B_0 field. It is this opposing field, experienced at the nucleus, that gives rise to the shielding phenomenon. For the effect to be maximal, i.e., unhindered electron precession, the electron cloud must be approximately spherical. This condition is strictly adhered to in free atoms, and in some molecules with tetrahedral symmetry, e.g., $^+NH_4$ and $^-BH_4$. This diamagnetic contribution to the shielding (σ_D) is sometimes called the Lamb term after its discoverer. To compensate for the effect of diamagnetic shielding, the magnetic field B_0 must be raised (upfield shift) to bring a nucleus into resonance.

The diamagnetic shielding term is strongly dependent on the electron density about the nucleus. Any substituent within a molecule that reduces the electron density will reduce σ_D and produce a shift to lower field. An electronegative group attached to a molecule therefore gives rise to such an effect. The series CH_4, CH_3Cl, CH_2Cl_2, $CHCl_3$, with chemical shifts in δ of 0.2, 2.7, 5.3, and 7.3, shows a progressive downfield shift as hydrogen is replaced by chlorine.

---

[1] Thus the B_0 field is in the negative z direction, since the stabler spin state precesses with the field.

Because diamagnetic shielding is so closely related to electronegativity, studies of proton chemical shifts have produced several empirical electronegativity scales. The downfield shifts caused by replacement of hydrogen by a more electronegative group (halogen, nitro, cyano, carbonyl, alkoxyl, hydroxyl, amino) are due in part to a reduction in the diamagnetic term. Conversely, a more electropositive group (organosilicon, organomagnesium) reinforces σ_D and causes an upfield shift, as exemplified by the high field resonance of TMS.

Resonance (mesomerism) as well as inductive effects can influence electron density. Compounds **1-3** illustrate electron donation (**1**, upfield shift) and electron withdrawal (**3**,

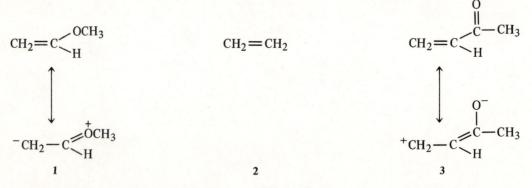

downfield shift) by resonance. Aromatic proton resonances are very sensitive to substitution for the same reasons. The five aromatic protons of toluene have similar chemical shifts (Figure 1-3), whereas the para proton in nitrobenzene is shifted downfield or in anisole upfield. The chemical shifts of meta and para fluorine atoms have been used extensively by Taft as probes to measure σ_I and σ_R for a wide variety of substituents. The effect of charge density shows up very clearly in a comparison of the chemical shift of benzene (δ 7.3) with those of the cyclopentadienyl anion (δ 5.6) and the tropylium cation (δ 9.2). A charge of $-\frac{1}{5}$ on the adjacent carbon atom causes an upfield shift of 1.7 ppm, and a charge of $+\frac{1}{7}$, causes a downfield shift of 1.9 ppm. After corrections for ring anisotropy (section 3-3), the effect of one electronic charge unit is a shift of about 10 ppm. In carbon-13 nmr, the variation in charge occurs directly on the atom of interest, and one unit of charge can provide a shift of 160 ppm.

In addition to substituent electronegativity and formal charge density on carbon, the chemical shift of a proton is influenced by the hybridization of the attached carbon. As s character increases from 25% (sp^3) to 50% (sp), the bonding electrons move closer to carbon, thereby deshielding the proton. In the absence of other effects, this 25% change in s character can cause a downfield shift of over 5 ppm.

Few molecules possess the perfect spherical symmetry required for the Lamb term. The precessional motion of the electrons that gives rise to the diamagnetic shielding is impeded in unsymmetrical molecules. A second term in the shielding expression allows for asymmetries in the electron cloud. These asymmetries may be expressed by mixing the ground state wave function with excited states. This effect is in the opposite direction to σ_D and is called the paramagnetic shielding term, σ_P.

Because protons possess only s electrons, the surrounding electron cloud is reasonably symmetrical and the contribution of σ_P is rather small. For other nuclei (^{13}C, ^{14}N, ^{17}O, ^{19}F, ^{31}P), the presence of p and d electrons can cause large deviations from spherical symmetry and large paramagnetic shifts. In general, σ_D rarely covers a range larger than a hundred ppm, but σ_P can extend over a range of many hundreds of ppm. Because of the dominance of σ_P in the shielding of nuclei other than protons, their ranges of chemical shifts are extremely large. Furthermore, simple electronegativity considerations are generally washed out by the paramagnetic effects. The excited states of lowest energy are best able to mix with the ground state. Nitrogen chemical shifts have been found to be directly proportional to the λ_{max} of the lowest excited state (usually $n-\pi^*$). The ^{15}N chemical shifts downfield from $^+$NH$_4$ ($\sigma_P = 0$ by symmetry) for $C_6H_5NH_2$, $C_6H_5NO_2$, and C_6H_5NO are 34, 348, and 889 ppm. The extremely low field shift of nitrosobenzene is due to the very low-lying excited state that is the cause of its green color. The ^{13}C chemical shifts downfield from TMS of ethane, benzene, and acetone (carbonyl group) are 8, 128, and 205 ppm, again reflecting dependence on excited-state configuration. The dominance of the paramagnetic term for nuclei other than hydrogen complicates the interpretation of chemical shifts.

3–2 Van der Waals and Electric Dipole Shielding

As discussed above, a substituent may deshield a neighboring proton (decrease σ_D) by inductive electron withdrawal. If the substituent is distant by a sufficient number of bonds, the effect becomes negligible. When a substituent atom is held rigidly at a distance from the resonating nucleus that is less than the sum of the van der Waals radii, the substituent atom will repel electrons from the vicinity of the resonating atom. The net effect is therefore a decrease in σ_D (or possibly an increase in σ_P), so the nucleus is deshielded and its resonance is shifted downfield. The phenomenon arises from the mutual repulsion of induced dipoles (dispersion forces). The magnitude of the effect falls off very rapidly with increasing internuclear distance (r), and it depends critically on the size and polarizability of the nuclei. A proton is not deshielded by another proton until it is only about 2.5 Å away. A bromine atom can deshield a proton from a much greater distance, and a fluorine atom is intermediate. The van der Waals shielding for an H $\cdots$ X interaction has been assessed by eq. 3–1, where the minus sign

$$\Delta\sigma_W = -\frac{60Q}{r^6} \qquad\qquad 3\text{–}1$$

implies deshielding and Q is a measure of the effectiveness of the atom X (1 for H, 3.1 for F, 14 for Br) in deshielding the proton.

The 0.2 ppm downfield shift of the *t*-butyl protons in *ortho*-di-*t*-butylbenzene with respect to the *t*-butyl resonances in the meta and para isomers may be attributed to a van der Waals effect. Possibly the most dramatic example has been observed in the cage compound **4**,

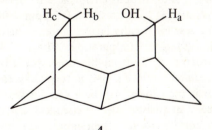

4

in which the chemical shifts of H$_b$ and H$_c$ are δ 3.55 and 0.88, respectively. By comparison, the methylene protons of cyclohexane resonate at about δ 1.4. The oxygen atom therefore deshields H_b by over 2 ppm. Interestingly, the electron density displaced from H_b is in part shifted to H_c, which is shielded by about 0.5 ppm.

A polar bond in a molecule generates an electric field that can have an appreciable value at the position of a nearby resonating nucleus. This electric field distorts the electronic structure around the nucleus and causes a deshielding by the decreased σ_D (or increased σ_P). Unlike the inductive effect, the electric field effect can be derived from a polar group that is many bonds removed from the resonating nucleus. For a significant value of $\Delta\sigma_E$, the polar bond must be reasonably close to the nucleus but need not be in van der Waals contact. Buckingham and co-workers expressed the electric field shift (eq. 3-2) in terms of the magnitude of the field

$$\Delta\sigma_E = -(a \times 10^{-12})E_z - (b \times 10^{-18})E^2 \qquad \text{3-2}$$

(E^2) and the component along the bond between the resonating nucleus and the atom (X) to which it is attached. For protons, $a \cong 3$ and $b \cong 1$; for fluorine $a \cong -10$ and $b \cong 40$. Since the first term is negative (deshielding) for a proton, the effect of the E_z component is to withdraw electrons from the X–H bond. The reverse is true for an X–F bond, since the coefficient a is negative. The sign of b is always positive, so the E^2 term invariably provides a downfield effect. The E_z term is angular dependent, and both E^2 and E_z are radially dependent, so the orientation and distance of the polar bond from the X–H bond influence the value of $\Delta\sigma_E$.

Many examples of electric field shielding have come from ^{19}F spectra, since the larger shifts magnify the effect. The ^{19}F resonance of *ortho*-chlorofluorobenzene (**5**) is more than

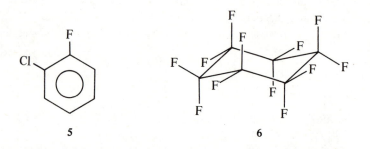

5   6

20 ppm further downfield than a simple inductive effect could explain. The interpretation that this large additional shift is due to the electric field of the C–Cl bond has been substantiated by calculations using eq. 3-2. The 18 ppm chemical shift between the axial and equatorial fluorines in perfluorocyclohexane (**6**) compares to a 0.5 ppm difference for the corresponding protons in cyclohexane. The reasons for the latter figure, to be discussed in the next section, are not sufficient to explain the large ^{19}F separation. The axial and equatorial fluorine atoms in **6** experience different electric fields from the other polar bonds in the molecule. The magnitude of the effect calculated from eq. 3-2 is very close to the observed chemical shift difference. In both these examples, there may be a van der Waals contribution as well, since the nuclei are relatively close. An analysis of ^{19}F chemical shifts is not complete unless both these effects are taken into consideration. Their importance has not been fully explored in many systems or for many nuclei, so much work remains to be done.

3-3 Diamagnetic Anisotropy

The circulation of electrons about nuclei that gives rise to diamagnetic shielding can also shield neighboring nuclei, provided the source of the circulation is anisotropic. To see why an isotropic source gives no effect, consider a proton attached to a spherical distribution of electrons. The sphere represents a functional group elsewhere in the molecule. For configuration 7, the proton

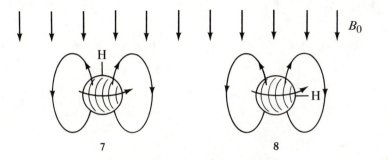

is in that part of the induced field that opposes the B_0 field, so a higher field is required for resonance. On the other hand, in configuration **8**, the proton is in the part of the induced field that reinforces B_0, so resonance is at a lower field. Any other orientation can be considered to be a weighted average of these two configurations. Averaging over all orientations, since the molecule is tumbling freely, results in no net shielding of the proton by the spherical electron cloud.

Anisotropic electron clouds with an ellipsoidal shape may be classified as oblate (dish-like) or prolate (rod-like). The former is a good model for aromatic rings and the latter for single or triple bonds. For a proton situated at the edge of an oblate ellipsoid, two extreme configurations must again be considered. When the flat portion of the ellipsoid is perpendicular to the applied field (configuration **9**), the proton at the edge will resonate at lower field

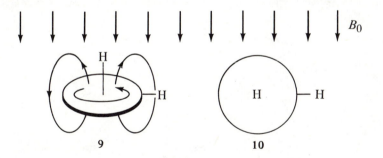

because it is in the reinforced region of the field. A proton situated over the ellipsoid must resonate at higher field because it is in the opposed region of the field. If the plane of the ellipsoid is rotated so that it is parallel to the field (**10**), little or no field is induced in comparison to that in **9**. Configuration **9**, which presents the largest portion of the cloud perpendicular to the B_0 field, has a greater susceptibility to induction of a current. In such a case, the electron cloud is said to have *diamagnetic anisotropy*, in the sense that different orientations in the field set up different induced currents. For the oblate ellipsoid, a proton at the edge is deshielded and one at the center is shielded. Any other orientation may be expressed as a weighted average of **9** and **10**. The effect is not averaged to zero by tumbling because one configuration does not completely cancel out the effect of another. A benzene ring is well described as an oblate ellipsoid, and we shall later give several examples of the shielding properties of aromatic systems.

The two configurations of the prolate ellipsoid (**11** and **12**) may be considered in a

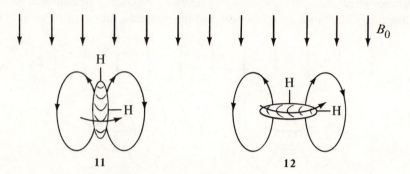

similar fashion. If the diamagnetic susceptibility is greatest when the long axis is parallel to the field (**11**), then a proton at the end is shielded and one at the side is deshielded. If the susceptibility is greatest when the long axis is perpendicular to the field (**12**), the reverse is true. The prolate ellipsoid is a good model for a chemical bond, and as we shall see, either case (**11** or **12**) can dominate, depending on the nature of the bond.

The π electrons above and below the benzene ring are particularly susceptible to the influence of an applied magnetic field. The circulation of electrons is strongest in configuration **9**. Such a *ring current* therefore deshields a proton at the edge of the ring. The anisotropy of the diamagnetic susceptibility, therefore, helps explain the low field resonance position of benzene (δ 7.3) compared to that of ethylene (5.3).

A proton situated over the benzene ring would be more shielded (configuration **9**) than otherwise expected. One of the early examples of an upfield aromatic shift was in [2,2]meta-cyclophane (**13**), in which the indicated protons are held over the ring. A 10 π electron system

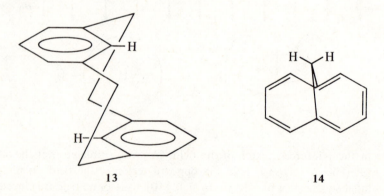

such as the methano[10]annulene (**14**) can also support a ring current. The methylene protons over the π electron cloud resonate at δ −0.5 (the minus sign indicates a resonance at higher field than TMS), compared to an ordinary methylene resonance at about 1.5. Porphyrins and other annulenes with $(4n + 2)\pi$ electrons show similar diamagnetic properties. Thus in [18]annulene (**15**), the inside protons are shielded (δ −3.0) and the outside ones are deshielded

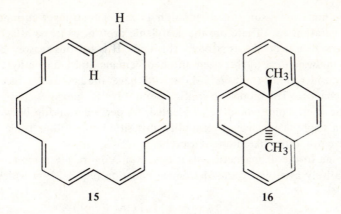

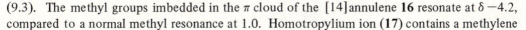

15   16

(9.3). The methyl groups imbedded in the π cloud of the [14]annulene **16** resonate at δ −4.2, compared to a normal methyl resonance at 1.0. Homotropylium ion (**17**) contains a methylene

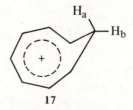

17

group in which one proton is shielded (H_a, δ −0.6) and the other deshielded (H_b, 5.2). The differential effect of the anisotropic aromatic ring produces a chemical shift difference of about 5.8 ppm between the two protons bonded to the same carbon atom.

The presence of $(4n + 2)\pi$ electrons is a requirement for the existence of a diamagnetic circulation of electrons. Pople showed that an external magnetic field can induce a paramagnetic circulation in a $4n$ π electron system. Under such circumstances, the conclusions drawn from configuration **9** must be reversed, i.e., outer protons are shielded and inner protons deshielded. The spectrum of [16]annulene is consistent with this interpretation (inner protons at δ 10.3, outer at 5.2), but the most dramatic example is the [12]annulene (**18**). The bromine

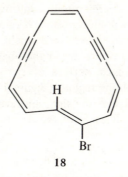

18

atom was included to prevent a conformational interconversion of the inner and outer protons. The indicated inner proton of **18** resonates at δ 16.4, compared to δ −3.0 for the inner protons of [18]annulene (**15**).

The π electrons in acetylene are cylindrically arranged about the carbon–carbon triple bond. When the molecule assumes configuration **11**, a much stronger current can be induced than from configuration **12**. Therefore the acetylenic proton, which is attached to the end of the cylindrical array of electrons, is strongly shielded. It is for this reason that the acetylene resonance falls midway (δ 2.7) between those of ethane (0.8) and ethylene (5.3). Strict hybridization considerations (section 3-1) would have predicted that acetylene have the lowest field resonance. The model of configuration **11** also indicates that a proton situated at the side of the triple bond would be deshielded. A peri relationship between a triple bond and a proton on naphthalene would exemplify this situation. Other triple bonds, as in the nitrile group, have properties like those of acetylene.

Shielding due to the diamagnetic anisotropy of a cylindrically symmetrical bond (X–Y) may be quantitatively assessed by the McConnell equation (eq. 3-3), in which χ_L and χ_T are

$$\sigma(r, \theta) = \frac{(\chi_L - \chi_T)\,(3 \cos^2 \theta - 1)}{3r^3} \qquad\qquad 3\text{-}3$$

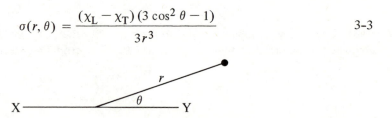

the diamagnetic susceptibilities along the longitude (**11**) and the transverse (**12**) of the bond, r is the distance from the center of the bond to the point in space of interest, θ is the angle subtended by this line and the bond (see diagram), and σ is the value of the effect at the coordinates (r, θ). The requirement of axial symmetry comes from the use of only two susceptibilities in the equation. Three different susceptibilities would have to be included for a double bond. Only the difference ($\Delta\chi$) between χ_L and χ_T enters into the equation, and $\Delta\chi$ can have either sign. For a triple bond $\chi_L > \chi_T$, i.e., configuration **11** dominates, so σ at $0°$ is positive (shielding) and at $90°$ negative (deshielding), in agreement with our description of the triple bond. For a carbon-carbon single bond $\chi_T > \chi_L$, i.e., configuration **12** dominates, so a proton at the end of the bond is deshielded, and one along the side shielded. The equation can thus handle both situations.

At this point a comment about the sign of σ is appropriate. By convention, a positive (increased) shielding shifts a proton to higher field. Also by convention, a positive value of δ is downfield from TMS. We will adhere to both conventions, but it must be kept in mind that a positive contribution to σ makes for a less positive δ value.

Since the triple bond can either shield or deshield a proton, at some point in space the effect must be null. This situation occurs when the angular term ($3 \cos^2 \theta - 1$) goes to zero at $55°44'$. An approximately conical surface of two nappes may be drawn in space that separates the shielding and deshielding regions (Figure 3-1). The same model is used for the carbon-carbon single bond, except that the signs are reversed, since configuration **12** produces the strongest electron circulation. Because of the simplicity of the McConnell equation, there

FIGURE 3-1. The diamagnetic anisotropy of the triple bond, showing the shielded (+) and deshielded (−) regions. For the single bond, the same model is used with the signs reversed.

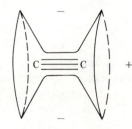

has been a great temptation to overuse it. The equation is based on a point-dipole approximation and only holds rigorously at distances that are too large to be useful. Qualitative conclusions drawn from the model are generally reliable, but it is advisable to avoid quantitative
applications.

One early application of the anisotropy of the carbon–carbon single bond was to the
axial–equatorial chemical shift difference in cyclohexane (**19**). The two protons are equiva-

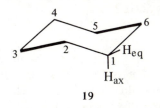

19

lently positioned with respect to the 1,2 and 6,1 bonds, which therefore produce no differential
effect. The axial proton, however, is in the shielding region of the 2,3 and 5,6 bonds (darkened),
whereas the equatorial proton is in the deshielding region. Therefore H_{ax} resonates at a higher
field than H_{eq} ($\delta_{ae} = 0.5$ ppm). The 3,4 and 4,5 C–C bonds and the C–H bonds also have
small contributions. The downfield shift in going from methyl to methylene to methine in a
hydrocarbon series has been attributed to the anisotropy of the added C–C bonds.

The cyclopropane ring has unusual shielding properties that may be explained in terms of
C–C bond anisotropies. The methylene group in a cyclopropane ring is situated directly
opposite a C–C bond (**20**) and is therefore shielded with respect to an ordinary methylene

group (**21**) (δ 0.3 vs. 1.5). The effect is much larger than the indicated 1.2 ppm, because the
sp^2 cyclopropane carbon orbital to hydrogen (compared to the sp^3 orbital in **21**) deshields the
proton. A cyclopropane ring can also shield adjacent nuclei. In spiro[2.5]octane (**22**), the
indicated equatorial proton resonates 1.2 ppm higher field than the axial proton. Since H_{ax}
is normally 0.5 ppm higher field, the differential effect is 1.7 ppm. In **22**, H_{eq} is perched
over the shielding region of the cyclopropane bonds, so it experiences a very strong upfield
shift. The anisotropy of the cyclopropane ring has also been discussed in terms of a ring
current.

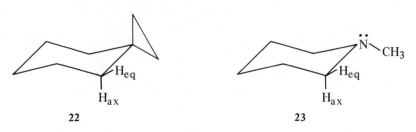

Most single bonds (O–C, N–C) have shielding properties that parallel those of the C–C
bond. In some cases a lone electron pair can have a special effect. In *N*-methylpiperidine (**23**),
the axial lone pair shields the vicinal H_{ax} by an *n–σ** interaction without effect on H_{eq}. As a
result, δ_{ae} increases to about 1.0 ppm or more.

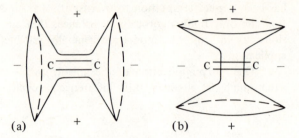

FIGURE 3-2. Models for the anisot-ropy of the carbon–carbon double bond. The plane of the ethylene molecule is perpendicular to the plane of the paper.

The anisotropy of double bonds is much more difficult to assess. Two models have been suggested to describe the shielding effects of a double bond (Figure 3-2). Both models agree that the regions above and below the double bond are shielded and that the regions at either end of the bond are deshielded. Thus the bridge proton in **24** resonates at higher field (δ 3.53) than that in **25** (3.75). An aldehydic proton is in the severely deshielded region at the end of

the carbon–oxygen double bond, so it resonates at a very low field (δ 9.7). The two models disagree, however, on the shielding effect in the plane of and opposite the double bond. Recent experiments have favored the model shown in Figure 3-2(a), but a conclusive answer is not yet available.

Empirical use of double bond anisotropies suffers from two other problems. Lone pairs on a carbonyl group have anisotropic properties over and above those of the double bond itself. The shielding map of a carbonyl group is therefore quite unsymmetrical. With a C—O single bond, this effect is reduced by free rotation. In both cases, electric field effects (previous section) can introduce independent considerations. As a result, conclusions from carbonyl anisotropies should be drawn with considerable caution.

A proton involved in a hydrogen bond experiences a strong downfield shift. Thus the hydroxyl proton in free ethanol resonates at δ 0.7, but in neat ethanol at 5.3. Two explanations of this phenomenon have been given.

(a) The lone pair distorts the electron cloud about the proton, as in the van der Waals effect. The resulting electron withdrawal deshields the proton.
(b) The lone pair may be considered to be a prolate ellipsoid of configuration **11**. The proton is located at the end of this electron cloud and hence is deshielded. The deshielding effect of a hydrogen bond is responsible for the very low field resonance position of acetyl-acetone enol (**26**) and of carboxylic acids that exist as dimers (**27**) or polymers.

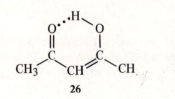

26

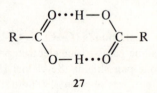

27

Diamagnetic anisotropy is certainly one of the most prominent effects in pmr spectroscopy. It is responsible for special shielding effects in aromatics, acetylenes, aldehydes, cyclopropanes, and hydrogen-bonded species, in addition to numerous other situations. The resonances of other nuclei are dominated by the paramagnetic shielding, so the effects of anisotropic groups, which rarely exceed a few ppm, are not so important.

3-4 Empirical Correlations for Proton Chemical Shifts

Table 3-1 summarizes the chemical shifts of common functional groups. The figures are intended only to serve as a general guide, since substituents can significantly alter the resonance positions. The R groups are saturated and aliphatic. The $-OH$ and $-NH$ chemical shifts are not included, since they vary between δ 1.0 and 6.0, depending on conditions of concentration and acidity.

Table 3-1. *Representative Proton Chemical Shifts*

	δ (ppm)		δ (ppm)
$(CH_3)_4Si$	0.0	CH_3-NR_2	2.8
Cyclopropyl CH_2	0.3	CH_2RBr	3.5
CH_3-R	0.9	CH_2RCl	3.7
CH_2R_2	1.4	CH_3-OR	3.7
CHR_3	1.8	$CH_2=CR_2$	5.0
$CH_3-CR=CR_2$	1.8	$CHR=CR_2$	5.5
$CH_3-C_6H_5$	2.2	Aromatic	7.4
CH_3-CO-R	2.2	$H-CO-R$	9.7
$H-C{\equiv}C-R$	2.5	HO_2C-R	12.0
CH_3-SR	2.6	Enolic	15.5

Various empirical correlations have been developed that relate chemical shifts to structure. Shoolery's rule gives the chemical shift of protons in a $Y-CH_2-X$ group (eq. 3-4) from

$$\delta = 0.23 + \Delta_Y + \Delta_X \qquad\qquad 3\text{-}4$$

substituent parameters Δ_i (Table 3-2) added to the chemical shift of methane. The calculation

Table 3-2. *Substituent Parameters for Shoolery's Rule*

Substituent	Δ_i	*Substituent*	Δ_i
$-CH_3$	0.47	$-CO-R$	1.70
$-CR=CR_2$	1.32	$-I$	1.82
$-C{\equiv}CR$	1.44	$-C_6H_5$	1.85
$-NR_2$	1.57	$-Br$	2.33
$-SR$	1.64	$-OR$	2.36
$-CN$	1.70	$-Cl$	2.53

is reasonably successful for CH_2XY, but less so for CHXYZ. For example, the calculated shift for CH_2Cl_2 is $0.23 + 2 \times 2.53 = 5.29$, and for $CHCl_3$ is $0.23 + 3 \times 2.53 = 7.82$, compared to observed shifts of δ 5.30 and 7.27.

Tobey and Pascual, Meier, and Simon have developed a similar relationship (eq. 3-5) for

$$\delta = 5.28 + Z_{gem} + Z_{cis} + Z_{trans} \qquad\qquad 3\text{-}5$$

Table 3-3. *Substituent Parameters for the Tobey-Simon Rule*

Substituent	Z_{gem}	Z_{cis}	Z_{trans}
–H	0.0	0.0	0.0
–Alkyl	0.44	–0.26	–0.29
–CH_2O, –CH_2I	0.67	–0.02	–0.07
–CH_2S	0.53	–0.15	–0.15
–CH_2Cl, –CH_2Br	0.72	0.12	0.07
–CH_2N	0.66	–0.05	–0.23
–C=C	0.50	0.35	0.10
–C≡N	0.23	0.78	0.58
–C=C (isolated)	0.98	–0.04	–0.21
–C=C (conjugated)	1.26	0.08	–0.01
–C=O (isolated)	1.10	1.13	0.81
–C=O (conjugated)	1.06	1.01	0.95
–CO_2H (isolated)	1.00	1.35	0.74
–CO_2R (isolated)	0.84	1.15	0.56
–CHO	1.03	0.97	1.21
–OR (R aliphatic)	1.18	–1.06	–1.28
–OCOR	2.09	–0.40	–0.67
–Aromatic	1.35	0.37	–0.10
–Cl	1.00	0.19	0.03
–Br	1.04	0.40	0.55
–NR_2 (R aliphatic)	0.69	–1.19	–1.31
–SR	1.00	–0.24	–0.04

the chemical shifts of a proton on a double bond. Substituent constants Z_i (Table 3-3) for the other groups on the double bond are added to the chemical shift of ethylene. For example by this means the stereochemistry for the crotonaldehyde $CH_3-CH_a=CH_b-CHO$ may be assigned. The observed resonances are at 6.87 and 6.03, but an assignment to H_a and H_b cannot be made. Equation 3-5 computes shifts for the cis compound of 6.93 (H_a) and 6.02 (H_b), and for the trans compound of 6.69 (H_a) and 6.05 (H_b). The cis stereochemistry is therefore favored, and the lower field resonance is unambiguously assigned to H_a. The two stilbenes ($C_6H_5-CH=CH-C_6H_5$) give alkenic resonances at δ 7.10 and 6.55, respectively. The calculated *cis*-stilbene resonance is at 6.53, and *trans*-stilbene at 7.00, so the isomers may be readily differentiated.

Aromatic proton resonances may be treated in a similar way, provided no two substituents are ortho to each other. The shift of a particular aromatic proton is obtained (eq. 3-6) by

$$\delta = 7.27 + \Sigma\, S_i \qquad\qquad 3\text{-}6$$

adding substituent parameters to the shift of benzene (Table 3-4, gathered from several sources by Jackman and Sternhell). For example, *p*-chlorobenzaldehyde (**28**) gives an aromatic AB spectrum (with some further complications) with resonances centered at δ 7.50 and 7.75. The

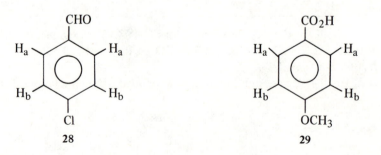

28   29

Table 3–4.   *Substituent Parameters for Aromatic Proton Shifts*

Substituent	S_{ortho}	S_{meta}	S_{para}
CH_3	−0.17	−0.09	−0.18
CH_2CH_3	−0.15	−0.06	−0.18
NO_2	0.95	0.17	0.33
Cl	0.02	−0.06	−0.04
Br	0.22	−0.13	−0.03
I	0.40	−0.26	−0.03
CHO	0.58	0.21	0.27
OH	−0.50	−0.14	−0.4
NH_2	−0.75	−0.24	−0.63
CN	0.27	0.11	0.3
CO_2H	0.8	0.14	0.2
CO_2CH_3	0.74	0.07	0.20
$COCH_3$	0.64	0.09	0.3
OCH_3	−0.43	−0.09	−0.37
$OCOCH_3$	−0.21	−0.02	−
$N(CH_3)_2$	−0.60	−0.10	−0.62
SCH_3	−0.03	0.0	−

calculated position for H_a is 7.79 and for H_b is 7.50. The observed resonances for *p*-methoxy-benzoic acid (**29**) are at δ 8.08 and 6.98. The calculated position for H_a is 7.98 and H_b 6.98. The spectral assignments for these protons can therefore be made with confidence. In some cases eq. 3–6 may also be used for structural assignment.

To the extent that the aromatic substituents cause shifts in proton resonance positions by changes in electron densities rather than by effects of diamagnetic anisotropy or electric fields, these shifts will correlate with Hammett substituent constants. Reasonably good correlations have been found between proton or fluorine chemical shifts and σ_p. Thus a methoxyl group donates electrons to the para position, so the proton is shifted upfield (negative S_{para}), and the nitro group withdraws electrons from the para position, so the proton is shifted downfield (positive S_{para}). The meta effects are predominately inductive, but little can be said for the ortho effects, since direct anisotropic and electric field contributions are likely.

Frequently in spectral analysis, the investigator is faced with differentiating between a pair of closely related structures. There is a strong temptation to try to obtain a definitive answer on the basis of small differences in chemical shifts. Such simple approaches should be resisted. There are too many factors that influence proton chemical shifts, and without the use of numerous model compounds, incorrect conclusions are easily possible if based on chemical shift data alone.

3–5   Medium and Isotope Effects

The observed shielding of a particular nucleus consists of intramolecular components (discussed in sections 3–1 to 3–3) and intermolecular components (eq. 3–7). Buckingham, Schaefer, and

$$\sigma = \sigma_{intra} + \sigma_{inter} \qquad\qquad 3\text{–}7$$

Schneider pointed out five sources of intermolecular shielding (eq. 3–8). We shall consider each contribution in turn, and then give several illustrations of medium effects.

$$\sigma_{inter} = \sigma_B + \sigma_W + \sigma_E + \sigma_A + \sigma_S \qquad\qquad 3\text{–}8$$

The solvent has a bulk diamagnetic susceptibility that is dependent on the shape of the sample container. Thus the solvent in a spherical container shields the solute to a slightly different extent from the solvent in a cylindrical container. This shielding is given by eq. 3-9,

$$\sigma_B = (\tfrac{4}{3} \pi - \alpha) \chi_V \qquad\qquad 3\text{-}9$$

in which α is a geometric parameter and χ_V is the volume susceptibility of the solvent. For a sphere, $\alpha = 4\pi/3$, so there is no effect, but for a cylinder, $\alpha = 2\pi/3$, and σ_B is $4\pi\chi_V/3$. Normally the solute and the standard (TMS) are present in the same solution. Under these circumstances, they experience parallel bulk effects and no correction for σ_B on the relative shift (σ) is necessary. Since internal standards are common now, the effect of bulk susceptibility is largely discounted. A correction would be necessary only if chemical shifts had to be compared between data obtained without an internal standard from containers of different shape, e.g., the normal cylinder and a spherical microtube, or from solvents with different volume susceptibilities.

Close approach of the solute and the solvent can distort the shape of the electron cloud around a proton and deshield it, even when both components are nonpolar. Such a phenomenon (σ_W) is analogous to the van der Waals effect on the chemical shift. The magnitude is rarely more than 0.1 ppm. If chemical shifts are measured from the resonance of an internal standard, the contribution from σ_W should affect solute and standard similarly. Chemical shifts so measured should be largely independent of σ_W.

A polar solute, or even a nonpolar molecule with polar groups, induces an electric field in the surrounding dielectric medium. This reaction field, proportional to $(\epsilon - 1)/(\epsilon + 1)$, can influence the shielding of protons elsewhere in the molecule. Generally, the effect is largest for protons close to the polar group. The sign can be either positive or negative because of an angular dependence, but more often it is negative (deshielding). This effect is not compensated for by use of an internal standard. Even within the solute molecule, the effect can be quite variable for different protons. For polar molecules in solvents of high dielectric constant, σ_E can range up to 1 ppm. It may be minimized by the use of solvents with small dielectric constants.

An anisotropic solvent will not orient itself completely randomly with respect to the solute. Thus even a nonpolar molecule such as methane will be exposed preferentially to the shielding face of benzene or the deshielding side of acetonitrile. In general, aromatic (dish-like) solvents induce upfield shifts and rod-like solvents (acetylene, nitriles, CS_2) induce downfield shifts (σ_A). In the absence of any special solute–solvent interaction (charge transfer, dipole–dipole, hydrogen bond), the internal standard and the solute will experience similar anisotropic shifts, so that σ_A is compensated for in the δ value. As often as not, however, there is a special interaction between the solvent and a polar group in the solute molecule. As a result, the anisotropic solvent will have a different effect on different protons, and solute resonances can experience real shifts up to 0.5 ppm with respect to the internal standard. In some cases only certain protons near a functional group are affected. The solvent effect can then be used to bring about differential shifts within the spectrum of a molecule. This technique is useful in spectral analysis to move the chemical shifts of closely coupled groups apart and to make the spectrum more first order (next chapter). Because dish-like and rod-like solvents cause shifts in opposite directions, the investigator has some control over the movement of the resonances. The chemical shift alterations caused by aromatics have been termed *aromatic solvent-induced shifts*, with the acronym asis. The chemical shift in the aromatic solvent is compared to the resonance position in $CDCl_3$ (eq. 3-10). Because the shift is

$$\Delta_{C_6H_6}^{CDCl_3} = \delta_{CDCl_3} - \delta_{C_6H_6} \qquad\qquad 3\text{-}10$$

usually upfield, Δ is normally positive. Although anisotropic shifts are frequently strongest close to a polar group, they may be differentiated from electric field effects by the dependence of the latter on solvent dielectric constant.

Specific interactions between solvent and solute, such as hydrogen bonding, can cause quite large effects (σ_S). It is not known whether the asis is caused by a time-averaged cluster of solvent molecules about a polar functional group or by a 1/1 solute/solvent charge transfer complex. In the latter case, the asis is more legitimately classified under σ_S than under σ_A.

In an early study of solvent shifts, Buckingham, Schaefer, and Schneider examined the solute methane. For this molecule, σ_E and σ_S are zero and σ_B may be calculated. Thus only σ_W and σ_A should affect the solute chemical shift. These authors plotted the difference between the methane shift in a given solvent from that in the gas phase vs. the heat of vaporization of the solvent at the boiling point. The latter quantity was taken as a measure of the van der Waals interaction. More than a dozen solvents, including neopentane, cyclopentane, n-hexane, cyclohexane, the 2-butenes, ethyl ether, acetone, $SiCl_4$, and $SnCl_4$, fell on a straight line with a small negative slope. The downfield shifts from gaseous methane range from 0.13 (neopentane) to 0.32 ppm ($SnCl_4$). The linear relationship with ΔH_v indicates that these shifts are due solely to σ_W. Well above this line are the dish-shaped aromatic molecules, benzene, toluene, chlorobenzene, but also nitromethane and nitroethane. Below the line are the rod-like molecules acetonitrile, methylacetylene, dimethylacetylene, butadiyne, and carbon disulfide. Deviations from the $\Delta\nu$ vs. ΔH_v line are due to true anisotropic shifts (σ_A), since the nonpolar, isotropic methane molecule has no direct (σ_S) interactions with the solvent. The largest positive displacement from the line was nitrobenzene (0.72 ppm) and the largest negative displacement was $NC-C\equiv C-CN$ (0.53 ppm).

Molecule **30** provides an interesting example of the electric field effect. Table 3–5 shows

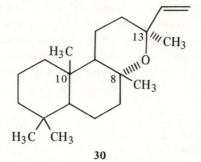

30

the resonance positions of the methyl groups on C-8, C-10, and C-13 in cyclohexane ($\epsilon = 2.02$) and in methylene chloride (9.1). These solvents were chosen for a low σ_A effect; σ_W and σ_B are discounted by use of an internal standard. The methyl groups close to the ether linkage (C-8, C-13) are shifted downfield about 4 Hz by an electric field effect. There is little or no

Table 3–5. *Methyl Chemical Shifts in Hz from TMS at 60 MHz as a Function of Solvent (Laszlo)*

Solvent	ν_{10}	ν_8	ν_{13}
Cyclohexane	51.3	60.4	67.1
CH_2Cl_2	51.1	64.2	71.7

shift of the C-10 methyl resonance, since the reaction field diminishes rapidly with distance. In other systems, effects up to 0.3 ppm have been observed.

There is a large asis on the relative chemical shifts of the nonequivalent methyl groups in *N,N*-dimethylformamide (**31**). The distance between the two methyl peaks increases by up

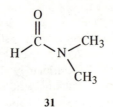

31

to 1.7 ppm on replacement of $CHCl_3$ by benzene. Interestingly, the high field peak is responsible for almost all the shift. This observation was explained in terms of a short-lived 1/1 solvent/solute complex, in which one of the DMF methyl groups is situated over the benzene ring, and the other is directed away from the ring. The model of a 1/1 complex between cyclohexanone and benzene has been used to justify the upfield shift of the 2,6-axial protons (located above the benzene ring) and the negligible shift of the 2,6-equatorial protons (near the region of no effect at $\theta = 55°44'$). These shifts have also been explained in terms of a time-averaged solvent cluster, which produces the same effect without recourse to a short-lived 1/1 complex.

Certain paramagnetic materials are extremely effective in changing chemical shifts. This phenomenon will be discussed in Chapter 7.

Isotopic changes within a molecule can alter the chemical shifts of the neighboring nuclei. The methyl group in toluene is 0.015 ± 0.002 ppm downfield from the corresponding protons in toluene-α-d_1. The protons in cyclohexane are 0.057 ppm downfield from the lone proton in cyclohexane-d_{11}. The effect is larger on other nuclei such as ^{19}F, and it falls off very rapidly with distance. The isotope shift has been explained in terms of zero point vibrational energy differences. An alternative explanation as a purely inductive phenomenon is probably not valid because the effect is not strictly additive. Differential isotope effects have been exploited in a study of the ring reversal of 1,4-dioxane (eq. 3–11). In undeuterated dioxane,

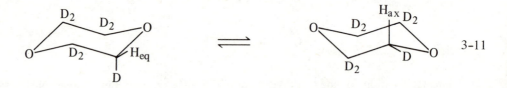

$3-11$

H_{eq} and H_{ax} have exactly the same chemical shift, so they cannot be differentiated at low temperature (see section 1-6 and Chapter 6). In 1,4-dioxane-d_7 (an impurity in commercial 1,4-dioxane-d_8), both H_{ax} and H_{eq} experience upfield isotope shifts, but H_{ax} is shifted somewhat further. As a result, the axial and equatorial protons give separate resonances at low temperatures, in contrast to the undeuterated material. Because of a chlorine isotope effect, chloroform is a poor substance for an internal lock or a resolution standard. At high resolution the chloroform proton resonance shows up as several peaks, due to $CH(^{35}Cl)(^{37}Cl)_2$, $CH(^{35}Cl)_2(^{37}Cl)$, $CH(^{35}Cl)_3$, and $CH(^{37}Cl)_3$.

3-6   Chemical Shifts of Carbon-13 and Other Nuclei

After ^{1}H, the most commonly studied nuclei are ^{11}B, ^{13}C, ^{14}N/^{15}N, ^{19}F, and ^{31}P. Of these, ^{11}B, ^{14}N, ^{19}F, and ^{31}P are sufficiently abundant to pose no serious sensitivity problem. Signal averaging and Fourier transform methods have made natural abundance ^{13}C experiments relatively routine. Recently, these methods have also found some success with ^{15}N, but this nucleus is still difficult to observe without synthesis of labeled materials.

To the organic chemist, the ^{13}C nucleus has become second in importance to the proton. It has a spin of $\frac{1}{2}$ and resonates at 15.1 MHz in a 1.41 tesla field. In the natural abundance spectrum, there is a very small probability that two ^{13}C nuclei will be found in the same molecule. As a result, no ^{13}C–^{13}C couplings complicate the spectra, and the ^{13}C–^{1}H couplings may be removed by double irradiation (see Chapter 5). For these conditions, a ^{13}C spectrum consists of one peak for each carbon in the molecule, as in Figure 1–4. Such a carbon map can give extremely valuable structural information. Because of relaxation effects, intensity information is not always reliable. The dominance of σ_P produces a wide range of ^{13}C chemical shifts (Table 3–6). In the tables, a positive sign always indicates a downfield shift in

Table 3–6. *Representative Carbon-13 Chemical Shifts*

	δ *(ppm)*		δ *(ppm)*
CHI_3	–139	$CHCl_3$	79
CH_3I	– 21	$R_2C{=}CH_2$	110
$(CH_3)_4Si$	0.0	$R{-}C{\equiv}N$	113
CH_3Cl	20	C_6H_6	128
Cyclohexane	27	$CH_3CO_2CH_3$	170
CH_3COCH_3	29	CH_3CO_2H	178
CH_3NR_2	53	CS_2	193
RCH_2OH	60	CH_3CHO	201
$R{-}C{\equiv}C{-}R$	75	CH_3COCH_3	205

ppm, and R is a saturated alkyl group. Even in a molecule such as cholesterol with many structurally similar carbons, 25 peaks have been observed for the 27 carbons. Correlations with electronegativity and anisotropies are not so clearcut as for protons. Thus CHI_3 is at higher field than CH_3I, but $CHCl_3$ is at lower field than CH_3Cl. The extremely low field carbonyl resonance positions are related to the presence of low-lying excited states. For the same reason, alkenes and aromatics have intermediate resonance positions, and saturated carbons resonate at high field. In aromatics, there is a strong dependence of shielding on electron density. Considerable work has now been completed on all carbon functionalities, and chemical shifts are beginning to be reasonably well understood.

The ^{19}F nucleus is 100% abundant. With a spin of $\frac{1}{2}$, a high natural sensitivity, and resonance at a high frequency (56.4 MHz at 1.41 tesla), it has long been an attractive nucleus to both organic and inorganic chemists. Table 3–7 lists some ^{19}F chemical shifts measured from CCl_3F. Taft has used the chemical shifts of ^{19}F atoms meta and para to various groups as a measure of the electronic effect of the group. The ^{19}F chemical shifts may be separated into inductive (σ_I) and resonance (σ_R) parameters that are of use in studying chemical reactivity. Roberts has used ^{19}F nuclei with great success to serve as conformational probes in studies of intramolecular processes. Considerable work is still necessary to come to an understanding of ^{19}F chemical shifts. The nucleus has nonetheless been of great use simply in characterizing fluorine-containing compounds.

Table 3-7. *Representative Fluorine-19 Chemical Shifts*

	δ *(ppm)*		δ *(ppm)*
RCH_2F	-220	$C_2H_5CF_3$	-64
$(CF_3)_3CF$	-190	CF_4	-62
SiF_4	-165	$(CF_3)_3N$	-56
Perfluorocyclohexane	-133	CCl_3F	0.0
$CF_2=CF_2$	-133	$R-CO-F$	20
C_6H_5F	-113	SF_6	57
CF_3CO_2H	-77	$C_6H_5SO_2F$	76
$(CF_3)_3CF$	-75	NF_3	147

Of the nitrogen isotopes, ^{14}N is 99.6% abundant but has a spin of 1; ^{15}N is 0.37% abundant and has a spin of $\frac{1}{2}$. Both nuclei have low sensitivity and low resonance frequencies (^{14}N, 4.1 MHz; ^{15}N, 6.1 MHz at 1.41 tesla). The quadrupolar moment of ^{14}N has precluded its use in measuring coupling constants or accurate chemical shifts. The low abundance and sensitivity of ^{15}N have made it a rarely studied nuclide. Table 3-8 lists representative ^{15}N shifts,

Table 3-8. *Representative Nitrogen-15 Chemical Shifts*

	δ *(ppm)*		δ *(ppm)*
NH_3 (anhydrous)	0.0	CH_3CN	245
CH_3NH_2	2	KCN	279
NH_4Cl	24	$(C_6H_5)_2C=NH$	308
$C_6H_5NH_3Cl$	55	$C_6H_5N=N(O)C_6H_5(trans)$	$324, 328$
$C_6H_5NH_2$	59	HNO_3 (8.57 M)	367
NH_2CONH_2	82	$C_6H_5NO_2$	372
CH_3NCS	93	$C_6H_5N=NC_6H_5$ (trans)	510
$C_6H_5NHNHC_6H_5$	96	n-C_4H_9ONO	572
$C_6H_5CONH_2$	100	$NaNO_2$	608
$C_6H_5NHCOCH_3$	135	C_6H_5NO	913

which are identical to ^{14}N shifts. The paramagnetic term dominates the chemical shift and renders it reasonably well understood. Thus amines resonate at highest field because only high energy n-σ^* transitions are available for mixing. The remaining compounds crudely follow the order of the n-π^* transitions: amide, nitrile, imine, nitrate, nitro, azo, nitrite, and finally nitroso compounds.

With a spin of $\frac{1}{2}$ and an abundance of 100%, ^{31}P has been a favorable nucleus for study. At 1.41 tesla, its resonance frequency is 24.3 MHz. Table 3-9 lists the chemical shifts of some

Table 3-9. *Representative Phosphorus-31 Chemical Shifts*

	δ *(ppm)*		δ *(ppm)*
P_4	-488	$(C_6H_5)_3P(O)$	23
PH_3	-241	$[(CH_3)_2N]_3P(O)$	23
PCl_5	-80	n-$C_4H_9P(O)(OH)_2$	33
$(CH_3)_3P$	-61	$(n$-$C_4H_9)_3P(S)$	51
$(C_6H_5O)_3P(O)$	-18	$(CH_3O)_3P(S)$	73
$(C_6H_5)_3P$	-6	$[(CH_3)_2N]_3P$	122
H_3PO_4 (85%)	0.0	$(CH_3O)_3P$	141
H_3PO_3	5	$CH_3(n$-$C_4H_9O)_2P$	183
$(CH_3O)_2P(O)H$	11	PBr_3	227

phosphorus compounds. The range of shifts for trivalent phosphorus is larger than that for pentavalent phosphorus. The resonance positions have generally been discussed in terms of the ionic and double bond character of the bonds to phosphorus.

The ^{11}B nucleus is 81.2% abundant, but its spin is $\frac{3}{2}$. Its resonance frequency is 19.1 MHz at 1.41 tesla. The resonances are generally broad because of quadrupolar effects and extensive coupling. High field, superconducting magnets have been particularly useful in clarifying ^{11}B chemical shifts. Table 3–10 gives a few ^{11}B resonance positions. The borohydride ion resonates

Table 3–10. *Representative Boron-11 Chemical Shifts*

	δ (ppm)		δ (ppm)
Pentaborane (B-1)	−52	$HCH_3BH_2BCH_3H$	21
$NaBH_4$	−43	$(HBNH)_3$	30
Pentaborane (B-2, 3, 4, 5)	−12	$(CH_3O)_2B-B(OCH_3)_2$	30
$(C_2H_5)_2OBF_3$	0.0	$(CH_3)_2N-B(CH_3)_2$	45
$B(OCH_3)_3$	18	BCl_3	47
$B(OH)_3$	19	$(CH_3)_3B$	86

at high field, since σ_P is zero for tetrahedral molecules. Trialkylboranes are at the lowest field, since they lie at the other extreme of symmetry, planar and sp^2.

In addition to these nuclei, ^2H, ^7Li, ^{29}Si, ^{17}O and ^{59}Co, among others, have received some attention. With the increased number of multinuclear spectrometers coming into use, the nmr spectroscopist can now tailor the instrument to specific chemical needs.

BIBLIOGRAPHY

1. Diamagnetic Anisotropy: R.C. Haddon, *Fortsch. Chem. Forsch.*, **16**, 105 (1971).
2. Empirical Correlations: S. W. Tobey, *J. Org. Chem.*, **34**, 1281 (1969); C. Pascual, J. Meier, and W. Simon, *Helv. Chim. Acta*, **49**, 164 (1966).
3. Medium Effects: J. Ronayne and D. H. Williams, *Ann. Rev. NMR Spectrosc.*, **2**, 83 (1969); P. Laszlo, *Progr. NMR Spectrosc.*, **3**, 231 (1967); R. J. Abraham and E. Bretschneider in *Internal Rotation in Molecules,* W. J. Orville-Thomas, ed., Wiley-Interscience, London, 1974, pp. 481 ff.
4. Isotope Effects: H. Batiz-Hernandez and R. A. Bernheim, *Progr. NMR Spectrosc.*, **3**, 63 (1967).
5. Carbon-13 NMR: J. B. Stothers, *Carbon-13 NMR Spectroscopy,* Academic, New York, 1972; G. C. Levy and G. L. Nelson, *Carbon-13 Magnetic Resonance for Organic Chemists,* Wiley-Interscience, New York, 1972; G. C. Levy, ed., *Top. Carbon-13 NMR Spectrosc.,* **1** (1974).
6. Fluorine-19 NMR: E. F. Mooney, *An Introduction to ^{19}F NMR Spectroscopy,* Heyden and Son, London, 1970.
7. Nitrogen-15 NMR: R. L. Lichter, *Determ. Org. Struct. Phys. Meth.*, **4**, 195 (1971); M. Witanowski and G. A. Webb, *Ann. Reports NMR Spectrosc.*, **5A**, 395 (1972).
8. Boron-11 NMR: W. G. Henderson and E. F. Mooney, *Ann. Rev. NMR Spectrosc.*, **2**, 219 (1969).
9. Phosphorus-31 NMR: *Top. Phosphorus Chem.*, **5** (1967); J. R. Van Wazer, *Determ. Org. Struct. Phys. Meth.*, **4**, 323 (1971); G. Mavel, *Ann. Reports NMR Spectrosc.*, **5B**, 1 (1973).
10. Other Nuclei: P. R. Wells, *Determ. Org. Struct. Phys. Meth.*, **4**, 233 (1971).

4

THE COUPLING CONSTANT

4-1 Chemical and Magnetic Equivalence

For dealing with detailed questions of structure and stereochemistry, the coupling constant is undoubtedly the most useful nmr spectral parameter. In order to obtain and interpret coupling data properly, a relatively sophisticated level of understanding of nmr spectra is required. In this chapter we will first consider the interpretation of coupling constants already in hand, and then we will turn to the theoretical and mechanical aspects of deriving coupling constants from spectra.

It would be an easy matter to obtain couplings if all spectra were of the first-order type that we described in section 1-3. For a spectrum to be first order, the chemical shift difference (Δv) between two nonequivalent nuclei must be much larger than their coupling constant ($\Delta v / J > 10$). Coupling constants may then be read directly from peak spacings. Unfortunately, this is not frequently the case, and in second-order spectra intensities deviate from the binomial pattern and spacings no longer necessarily correspond to couplings. For example, as the AX (first-order) pattern goes to AB (second order), the inner peaks in each doublet grow larger at the expense of the outer peaks (Figure 4-1). The chemical shift difference no longer corresponds to the distance between the midpoints of the doublets. The doublet splitting, however, still corresponds to J.

There is a second requirement that must be satisfied for a spectrum to be first order, namely, that chemically equivalent nuclei must also be magnetically equivalent. In order to understand this criterion, it is necessary to examine in detail the influence of molecular symmetry on spectral characteristics. In the course of this discussion, we will define chemical and magnetic equivalence and then be in a position to give a complete definition of first-order spectra. After these points are made, we shall return to the interpretation of coupling constants.

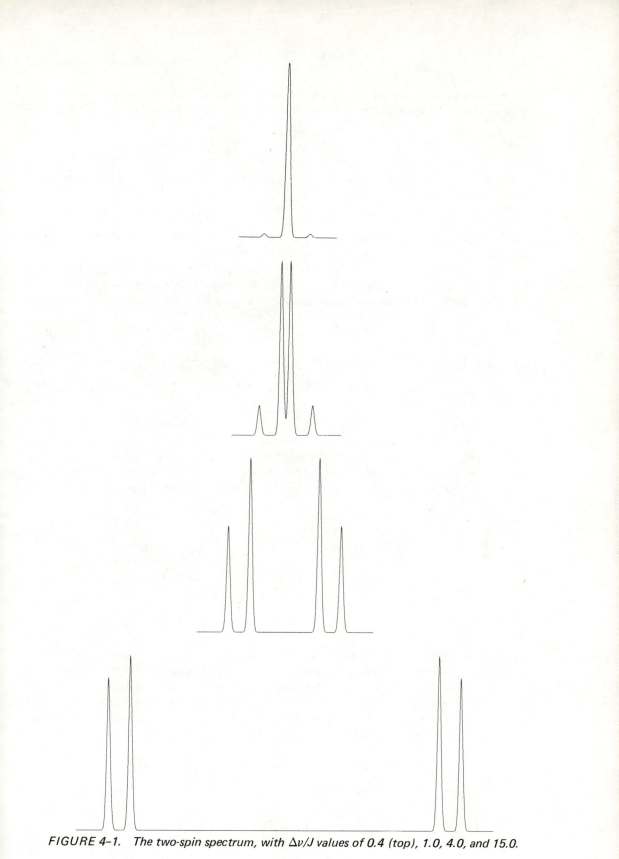

FIGURE 4–1.   The two-spin spectrum, with $\Delta v/J$ values of 0.4 (top), 1.0, 4.0, and 15.0.

Nuclei are chemically equivalent if they can be interchanged by a symmetry operation of the molecule. For example, the two protons in 1,1-difluoroethylene (**1**) or in methylene

1

fluoride (**2**) may be interchanged by a C_2 operation. Nuclei related by rotational symmetry are said to be *homotopic*, and have identical chemical properties.

Nuclei that are related by other symmetry elements (improper rotation, plane of symmetry in the absence of a C_n axis) are called *enantiotopic*. For example, the protons in bromochloromethane (**3**) are chemically equivalent and enantiotopic because they are related

by the plane of symmetry containing C, Br, and Cl. The term enantiotopic arises because replacement of first one proton, then the other, by another group, e.g., deuterium, produces molecules (**3a**, **3b**) that are nonsuperimposable mirror images or enantiomers. Replacement of homotopic nuclei in this fashion produces superimposable mirror images. Thus the alkenic protons in cyclopropene (**4**) are homotopic, but those in 3-methylcyclopropene (**5**) are enantiotopic.

In achiral solvents, enantiotopic nuclei have identical chemical properties. In a chiral environment produced by an optically active solvent or by an enzyme, enantiotopic nuclei are no longer chemically equivalent. They can exhibit coupling in the nmr spectrum or have different chemical properties such as acidity. It is this distinction that gives merit to the terms *homotopic* and *enantiotopic*. Although the terms do not apply to nuclei in different molecules, it is interesting to note that protons in enantiomers (**6**) are equivalent in achiral solvents, but

can give separate resonances in optically active solvents. Chemically equivalent nuclei are represented by the same letter in the spectral shorthand (AX_2, A_2B_3).

In certain molecules, chemically nonequivalent nuclei may be interchanged by an intramolecular rate process. The protons on nitrogen in formamide (**7**), for example, are nonequiva-

lent because amide resonance restricts C–N rotation ($E_a \sim$20-25 kcal mole^{-1}). If the procedure for sampling these protons, however, is slower than the rate of C–N rotation, they would appear to be chemically equivalent by this technique. The axial and equatorial protons in cyclohexane, which can be interchanged by ring reversal, provide another example of nuclei whose equivalence depends on the time scale of observation. If the interconversion is fast on the nmr time scale ($\sim >$1000 sec^{-1}), the nuclei will appear to be chemically equivalent and have the same chemical shift. If the process is slow on the nmr time scale ($\sim <$1 sec^{-1}), they can have different resonance positions. One great advantage of the nmr method is that both extremes may frequently be observed by changing the temperature (see Figure 1-14 and Chapter 6).

Nuclei that resonate at the same frequency are said to be *isochronous* (frequency equivalent). Nuclei that are chemically equivalent, either by symmetry (homotopic or enantiotopic) or by averaging, are necessarily isochronous. In addition, chemically distinct nuclei may be isochronous by accident. The ortho, meta, and para protons in toluene (see Figure 1-3) are nearly isochronous. At one temperature, the hydroxyl and methyl protons in methanol are isochronous. Below room temperature, the hydroxyl proton is downfield from the methyl protons. As the temperature is raised, the resonance is shifted upfield because the hydroxyl group becomes less hydrogen bonded. Above about 100°, the resonance is at higher field than the methyl protons. At the temperature of crossover, the hydroxyl and methyl protons are isochronous and hence do not split each other.

Two criteria must be fulfilled for nuclei to be *magnetically equivalent*. Symmetry-equivalent nuclei are magnetically equivalent only if they are equally coupled to each member of any other symmetry-equivalent set of nuclei. Thus the protons in methylene fluoride (**2**) are magnetically equivalent, because they are equally coupled to the two fluorine nuclei. Such a set of nuclei is represented by the symbol A_2X_2. The same may be said for the alkenic protons in cyclopropene (**4**). Nuclei that are not equally coupled to each member of other sets of nuclei are said to be *magnetically nonequivalent by the coupling constant criterion*. 1,1-Difluoroethylene (**1**) provides a classic example of magnetic nonequivalence. A given fluorine nucleus has different couplings (J_{cis}, J_{trans}) to the two protons, thereby rendering them magnetically nonequivalent. The symbol for this set of nuclei is AA'XX', indicating that an A nucleus has two different couplings with the X nuclei, J_{AX} and $J_{AX'}$. Any spin system that contains magnetically nonequivalent nuclei is second order by definition. The proton part of the AA'XX' spectrum contains ten peaks (Figure 4-2), compared to three in the first-order A_2X_2 spectrum of **2**. One of the interesting properties of the AA'XX' spectrum is the fact that the coupling between the chemically equivalent nuclei ($J_{AA'}$ and $J_{XX'}$) may be obtained directly from the spectrum. These second-order spectra will be discussed further in sections 4-7 and 4-9.

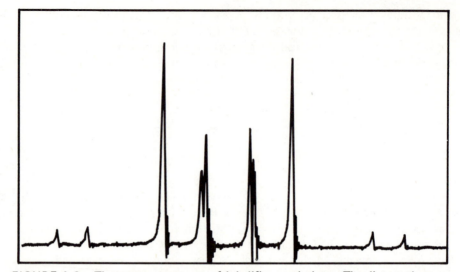

FIGURE 4-2. The proton spectrum of 1,1-difluoroethylene. The distance between the two largest peaks is 34.6 Hz. [Reproduced with the permission of W. A. Benjamin, Inc., from J. D. Roberts, An Introduction to the Analysis of Spin-Spin Splitting in High-Resolution Nuclear Magnetic Resonance Spectra, 1961.]

For a spectrum to be first order, it is required that $\Delta v/J$ be greater than about ten, and that all chemically equivalent nuclei be magnetically equivalent. The ring protons of *p*-nitrotoluene (**8**) fail both tests. A proton next to the methyl group has different couplings with

the two homotopic protons next to the nitro group (J_{ortho}, J_{para}). Since the chemical shifts are also close together, the spectrum is called AA′BB′ rather than AA′XX′. The protons next to the ether oxygen in butyrolactone (**9**) are magnetically nonequivalent, since the adjacent protons are unequally coupled to them (J_{cis}, J_{trans}). Because the two first-order criteria ($\Delta v/J > 10$; magnetic equivalence) are so seldom both met, there are few examples of truly first-order spectra, particularly in spin systems containing only protons.

There is a second, seemingly obvious, requirement to magnetic equivalence, namely that such nuclei have the same chemical shift. Anisochronous nuclei are said to be *magnetically nonequivalent by the chemical shift criterion*. Although most cases of such nonequivalence are self-evident, others are extremely subtle. Consider the methylene protons in ethylbenzene (**10**) and in its β-bromo-β-chloro derivative (**11**). The side chain of **10** produces

$$C_6H_5-CH_2-CH_3$$

10

$$C_6H_5-CH_2-CHBrCl$$

11

an almost first-order spectrum, between A_2B_3 and A_2X_3. The Newman projection **10a** shows

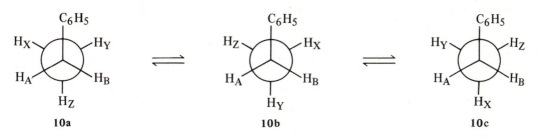

10a   10b   10c

that H_A and H_B are chemically equivalent and enantiotopic by reason of a plane of symmetry. Although H_A and H_B have different couplings to H_X in **10a**, they are magnetically equivalent on ,the average. Rapid C–C rotation (**10a** $\rightleftharpoons$ **10b** $\rightleftharpoons$ **10c**) interchanges the positions of H_A and H_B and results in only one methylene-methyl coupling constant. If the chemical shift difference were large enough, the spectrum would be entirely first order.

The Newman projections for **11** give a contrasting situation. First, it is noted that **11a**,

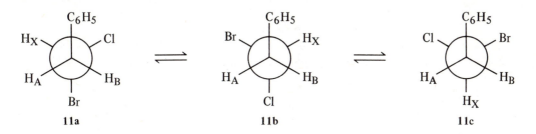

11a   11b   11c

11b, and **11c** are distinct rotational isomers, whereas **10a–c** are identical. In none of the three are H_A and H_B related by a symmetry operation. Even if there were rapid C–C rotation, H_A and H_B would be chemically nonequivalent. Such nuclei are magnetically nonequivalent by the chemical shift criterion unless they are fortuitously isochronous. For the present example (**11**), the spin system would be termed ABX.

The AB protons in **11** exemplify a particular set of chemically and magnetically non-equivalent nuclei that are termed *diastereotopic*. If H_A is replaced by deuterium in **11**, the resulting compound (**11d**) is a diastereomer, not an enantiomer, of the compound (**11e**) formed

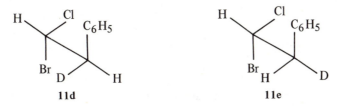

11d   11e

when H_B is replaced by deuterium. Diastereoisomerism results because of the presence of the adjacent asymmetric carbon atom. Introduction of the deuterium atom introduces a second asymmetric center. In general, the protons of a methylene group are diastereotopic when there is a chiral center elsewhere in the molecule. Accidental degeneracy is frequently the case ($\Delta\nu_{AB} = 0$), so diastereotopic protons can appear to be equivalent in the spectrum. Diastereotopic protons in some cases can be interconverted by intramolecular processes. Methyl groups in an isopropyl group can be diastereotopic when a chiral center is present, as in α-thujene (**12**). It is not necessary that a molecule contain a chiral center for methylene protons to be diastereotopic. The diethyl acetal of acetaldehyde (**13**) contains diastereotopic

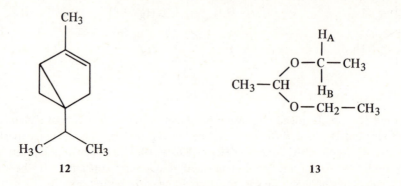

12   13

protons (H$_A$ and H$_B$), even though there is no chiral center. Replacement of H$_A$ by deuterium creates two chiral centers, and the resulting molecule is a diastereomer of the molecule in which H$_B$ is replaced by deuterium.

The term diastereotopic may be applied to a broad range of cases, if diastereoisomers are defined as *any stereoisomers (including geometric) that are not enantiomers.* Thus the protons in *cis*-1,2-dichlorocyclopropane (**14**) are diastereotopic because **14a** and **14b** are diastereomers.

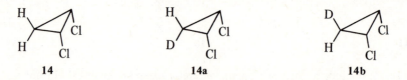

14   14a   14b

By the same token, the axial and equatorial protons on a single carbon atom in ring-frozen cyclohexane are diastereotopic. Molecules **10–13** contain examples of configurationally diastereotopic protons, since, for example, **11d** and **11e** differ in configuration. The protons in **14** are geometrically diastereotopic, since **14a** and **14b** are geometrical isomers. The axial and equatorial protons in cyclohexane are conformationally diastereotopic, since axial cyclohexane-d_1 and equatorial cyclohexane-d_1 are conformational isomers. If the rate of conformational interconversion is fast on the nmr time scale, conformationally diastereotopic protons become spectrally enantiotopic.

4–2 Mechanisms and Signs

Two mechanisms have been suggested to explain how coupling information is passed indirectly by bonding electrons rather than by the direct dipole-dipole interaction. The contact mechanism was derived from a concept originally developed by Fermi to explain hyperfine structure in atomic spectra. An electron in a bond between two nuclei, X and Y, spends a finite amount of time at the same point in space as, say, nucleus X. If nucleus X ($I = \frac{1}{2}$) has a spin of $I_z = +\frac{1}{2}$, then the spin of the electron ($I = \frac{1}{2}$) must be $-\frac{1}{2}$ because the two spins occupy the same space at the same time. In this way the nuclear spin polarizes the electron spin. This electron shares an orbital with a second electron, which must have a spin of $+\frac{1}{2}$ when the spin of the first electron is $-\frac{1}{2}$. This second electron, with a spin of $+\frac{1}{2}$, occupies the same point in space as nucleus Y only when Y has a spin of $-\frac{1}{2}$. Thus the fact the nucleus X has a spin of $+\frac{1}{2}$ favors a spin of $-\frac{1}{2}$ for nucleus Y. Since the bonding electrons were used to pass the spin information, there is no dependence on the orientation of the nuclear dipoles. Such a mechanism of coupling, unlike the direct dipole-dipole term, is not averaged to zero by molecular tumbling. Furthermore, the interaction is not dependent on the presence of an external magnetic field (B_0). Therefore, the same magnitude of coupling is observed for all values of B_0.

According to the second suggested mechanism, the nuclear magnet induces electron currents, which give rise to a small magnetic field by reason of orbital motion. A second nucleus detects this field and thereby gains knowledge of the spin of the first nucleus. Although couplings have from time to time been explained in terms of this *orbital motion* mechanism, the Fermi contact term appears to be responsible for most coupling phenomena.

When the preferred arrangement of two coupled spins is antiparallel, as in the above X–Y example, the coupling constant J by convention is given a positive sign. A negative coupling occurs when the preferred arrangement of the coupled spins is parallel. An example can be found in the coupling between X and Z in an X–Y–Z fragment. Nucleus X with a spin (I_z) of $+\frac{1}{2}$ brings about a preferred spin of $-\frac{1}{2}$ for an electron from the X–Y bond at the same point in space by the contact mechanism. This electron in turn is spin-paired with the second electron in the bond ($I_z = +\frac{1}{2}$). This electron resides in an orbital about nucleus Y. By Hund's rule, an electron in another orbital about Y must preferentially have a parallel spin. Therefore the electron in the X–Y bond conveys a spin of $+\frac{1}{2}$ to the electron in the Y–Z bond. The other electron in the Y–Z bond necessarily has the opposite spin ($-\frac{1}{2}$). This electron spends a finite amount of time at the Z nucleus, which consequently must be opposite in spin ($+\frac{1}{2}$). A $+\frac{1}{2}$ X nucleus, therefore, favors a $+\frac{1}{2}$ Z nucleus. These spin polarizations can be summarized in the following fashion: X(+)e(−)e(+)Ye(+)e(−)Z(+). Thus X and Z are stabler when their spins are parallel, and the coupling constant by convention has a negative sign. The spin of nucleus Y is irrelevant to this process. Although this qualitative argument has little theoretical basis, it is useful as a mnemonic device. Most two-bond proton-proton couplings do, in fact, have a negative sign. If the argument is pursued to the next stage, as in the three-bond W–Z coupling in a W–X–Y–Z fragment, a positive sign is predicted. In the same manner, couplings over an odd number of bonds are expected, in general, to be positive and couplings over an even number negative.

High resolution nmr spectra are not dependent on the absolute signs of the coupling constants. If the sign of every coupling constant in a spin system were reversed, an identical spectrum would result. Many spectra, however, are dependent on the relative signs of the component couplings. There are three coupling constants in an ABX spectrum: J_{AB}, J_{AX}, and J_{BX}. Quite different spectra can be obtained when J_{AX} and J_{BX} have the same sign (either both positive or both negative) from when they have opposite signs. In this way, and by certain double resonance experiments (Chapter 5), the relative signs of all major coupling types have been determined.

Ordinarily, the spectrum is independent of the direct dipole-dipole couplings (D). If a molecule is partially oriented, either by an external electric field or by a liquid crystal solvent (see section 4–11), the D couplings begin to have major effects on the spectrum. The absolute signs of D couplings can be obtained from basic principles if the direction of the molecular orientation is known. The relative signs of J and D can be obtained from the spectrum of a partially oriented molecule, and in this manner the absolute sign of J can be derived. Buckingham and McLauchlan showed by electric field experiments that J_{ortho} in *p*-nitrotoluene is positive.   Snyder and Anderson showed that all three couplings in benzene (J_{ortho}, J_{meta}, J_{para}) are positive by liquid crystal orientation.   The one-bond $^{13}C-H$ coupling is always positive, and many absolute signs have now been obtained by relation to J_{13C-H}.

The usual convention for referring to a coupling constant is to denote the number of bonds between the coupling nuclei by a left-hand superscript and any other descriptive information by a right-hand subscript. The one-bond $^{13}C-H$ coupling is therefore represented by the symbol $^{1}J_{13C-H}$.   A two-bond (geminal) coupling is $^{2}J_{H-C-H}$, and a three-bond (vicinal) coupling $^{3}J_{H-C-C-H}$.   A three-bond coupling between protons that are trans to each other on a double bond would be $^{3}J_{trans}$.   Beyond three bonds, couplings are said to be *long range*.

4-3   Directly Bonded Couplings

Because a p orbital has a node at the nucleus, only electrons in s orbitals can contribute to the contact coupling mechanism.   For protons, all electrons reside in s orbitals, but for other nuclei only that proportion of an orbital that has s character contributes to the coupling.   This fact was demonstrated empirically in 1958 in studies of the one-bond $^{13}C-H$ coupling.   When a proton is attached to an sp^{3} carbon atom (25% s character), $^{1}J_{13C-H}$ should be half as large as that for a proton attached to an sp-hybridized carbon (50%), and the ethylenic coupling should be intermediate.   The values of $^{1}J_{13C-H}$ for methane (sp^{3}), ethylene (sp^{2}), benzene (sp^{2}), and acetylene (sp) were found to be 125, 157, 159, and 249 Hz, respectively. These points define a linear relationship between the s character of the carbon orbital and $^{1}J_{13C-H}$ (eq. 4–1).   The zero intercept of this equation indicates that there would be no coupling if there were no s character, in agreement with the Fermi contact model.

$$s(^{13}C-H) = 0.2 \, ^{1}J_{13C-H} \qquad\qquad 4-1$$

The J–s relationship of eq. 4–1 can be used as an empirical source of hybridization information.   The coupling constant in cyclopropane (160 Hz) demonstrates that the carbon orbital to hydrogen is approximately sp^{2} hybridized.   For the orbitals to be orthonormal, the carbon orbital to carbon in the ring must be sp^{5} hybridized.   The measurement of $^{13}C-H$ coupling constants in the full range from 100 to 260 Hz clearly demonstrated to the organic chemist that fractional hybridization of any amount is permitted.   The coupling constants and s characters for the C–H bonds indicated in the figures are 144 Hz (29%, sp$^{2.4}$) in **15**, 160 (32%, sp^{2}) in cubane (**16**), 179 (36%, sp$^{1.8}$) in norbornane (**17**), and 226 (45%, sp$^{1.2}$) in cyclopropene (**4**).

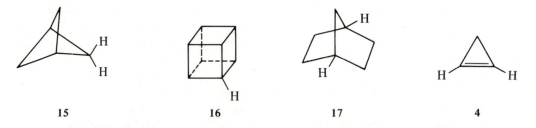

The J–s relationship may also be used as a method for structure proof. The synthesis of the 9,10-methano[10]annulene (**18**) involved the dehydrobromination of **19**. The valence tautomerization **20** $\rightleftharpoons$ **18** was expected to favor **18**, but a definitive differentiation of the two

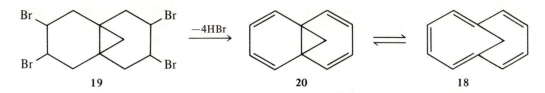

hydrocarbons was desired nonetheless. The ^{13}C–H coupling of the methano protons should be about 160 Hz for the cyclopropane **20** but about 145 for the annulene **18**, cf.$^{1}J_{13C-H} = 144$ Hz for the structurally similar protons in **15**. The observed value of 142 Hz confirmed the structure **18**. Although the J–s relationship has worked well for hydrocarbons, there has been some question as to its applicability to polar molecules. Variations in the effective nuclear charge, in addition to hybridization effects, may alter the coupling constant.

A similar J–s relationship (eq. 4–2) between the ^{15}N–H coupling constant and the hybridization of the nitrogen orbital to hydrogen has been based on observations from $^{+}NH_4$ (sp^3,

$$s(^{15}N-H) = 0.43\ ^{1}J_{15N-H} - 6 \qquad 4-2$$

$^{1}J_{15N-H} = 73$ Hz), $(C_6H_5)_2C=\overset{+}{N}H_2$ (sp^2, 93), and $CH_3-C\equiv\overset{+}{N}-H$ (sp, 130). The significance of the small, but nonzero intercept is not fully understood. Equation 4–2 may be used much as eq. 4–1 to answer questions of hybridization. For many years, the hybridization of ammonia was under discussion. Pauling's valence bond model indicated that the molecule was nearly unhybridized (s = 4%), whereas structural considerations (the bond angle is 107°, not 90°) suggested an s character of 20–25%. The measured $^{1}J_{15N-H}$ of 61 Hz corresponds to a hybridization of sp^5 (s = 20%), according to eq. 4–2. Pauling's model would have predicted a very small coupling.

Coupling constants between a proton and many other types of nuclei have also been measured, but no other reliable J–s relationships have been established. One-bond coupling constants between ^{31}P and ^{1}H range from 186 Hz (PH_2-PH_2) to 707 (H_3PO_3). Boron-11 couplings to proton range from 29 Hz (bridging proton in $H_2BHBH_2\cdot N(CH_3)_3$) to 211 in HBF_2. Some extremely large constants have been observed for ^{119}Sn–H (1931 Hz in SnH_4), ^{195}Pt–H (1307 in $[P(C_2H_5)_3]_2PtHCl$), and ^{207}Pb–H (2379 in $(CH_3)_3PbH$). All one-bond X–H couplings have been found to be positive when the sign of the gyromagnetic ratio of H is taken into account. A coupling constant between two nuclei is directly proportional to the product of their gyromagnetic ratios (γ). If γ for the X nucleus in the X–H bond happens to be negative, then $^{1}J_{X-H}$ is negative. For example, $^{1}J_{15N-H}$ has generally been found to be negative by double resonance experiments (Chapter 5), but γ_{15N} is also negative. When the coupling constant is divided by the component gyromagnetic ratios, the result is the reduced *coupling constant* (K), whose signs may be discussed without reference to the signs of the gyromagnetic ratios. Thus $^{1}K_{15N-H}$, like all other $^{1}K_{X-H}$, is positive.

When neither nucleus is proton, the coupling constant depends on the product of the s character of the orbitals from both nuclei that form the bond. Early studies of $^{1}J_{13C-19F}$ established no simple relationship with hybridization, but for $^{1}J_{13C-13C}$, eq. 4–3 was found

$$s(^{13}C-C)s(C-^{13}C) = 17.4\ J_{13C-13C} + 60 \qquad 4-3$$

to have limited applicability. The range of ^{13}C–^{13}C couplings is 34 Hz ($C_6H_5CH_2CH_3$) to 176 ($C_6H_5C\equiv CH$). A similar relationship was established for $^{1}J_{13C-15N}$ (eq. 4–4), but several

$$s(^{13}C-^{15}N)s(^{13}C-^{15}N) \sim 80\ ^{1}J_{13C-15N} \qquad 4-4$$

notable exceptions were found. It appears that *J*–s relationships will only be of value for one-bond couplings involving hydrogen, and even in this limited set only the couplings to ^{15}N and ^{13}C have worked out well.

Couplings between two nuclei other than proton can be extremely large. Thus $^{1}J_{125Te-19F}$ is 3688 Hz in TeF_6; $^{1}J_{205Tl-19F}$ in Tl–F is about 12,000; $^{1}J_{195Pt-31P}$ is 5700 in $\{[(C_2H_5)_2O]_3P\}_2PtCl_2$; $^{1}J_{199Hg-31P}$ is 4777 in $[(n\text{-}C_4H_9)_3P]_2HgBr_2$. The signs of the reduced X–Y coupling constants (*K*) have been found to be either positive or negative. For all couplings to ^{19}F, *K* is negative, but for couplings to ^{31}P, *K* can have either sign.

4–4 Geminal Couplings

Trends in the two-bond (geminal) H–C–H coupling constant were generally enigmatic until 1965, when Pople and Bothner-By published a molecular orbital approach to the problem. By concentrating on substituent effects, they were able to make several important generalizations. The $^{2}J_{H-C-H}$ in methane is −12.4 Hz, in ethylene +2.3. The 15 Hz difference between these two quantities is due to significant structural/hybridization changes. Rather than attempt to explain these differences, Pople and Bothner-By used these two molecules as class standards and related deviations from these coupling constants caused by substitution to changes in structure. A substituent that withdraws electrons from the CH_2 group by induction (σ withdrawal) was found to make ^{2}J more positive. Similarly, a σ-donating substituent makes ^{2}J less positive. In contrast, a substituent that withdraws electrons hyperconjugatively (π withdrawal) makes ^{2}J less positive, but a π donor makes it more positive. These effects are best understood by examples.

According to the Pople–Bothner-By approach, the coupling constant in any H–C–H fragment with an sp^2 carbon is compared to the coupling in ethylene (+2.3 Hz). For most hydrocarbon alkenes, ^{2}J is close to this value or smaller. As a result, geminal couplings are frequently not observed in exo-methylene or terminal vinyl groups. In formaldehyde the substituent on the methylene group has been changed from $=CH_2$ to $=O$. The more electronegative oxygen atom (a σ withdrawer) makes the coupling constant more positive. In addition, a nonbonding pair of electrons on oxygen has the correct symmetry for hyperconjugative donation (**21**). (There is a second lone pair orthogonal to the first, but it does

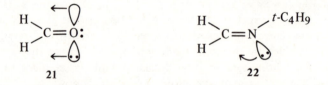

not interact with the CH_2 group.) This π donation also makes ^{2}J more positive. The reinforcement of the σ and π effects therefore brings about the very large and positive coupling constant observed in formaldehyde, $^{2}J = +42$ Hz. Similar considerations apply to imines such as **22**, but the lower electronegativity of nitrogen and less effective π overlap decrease the effects ($^{2}J = +17$ Hz for **22**).

In an allene, the $=CH_2$ group of ethylene has been replaced by $=C=CR_2$. There is only a small inductive effect. The more distant double bond can be strengthened by withdrawing electrons from orbitals of π symmetry in the methylene group, as illustrated in **23**. The result

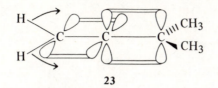

23

of this hyperconjugative electron withdrawal is a more negative geminal coupling, -9 Hz in the case of 1,1-dimethylallene. Most geminal couplings through an sp^2 carbon atom fall between the large positive value for formaldehyde and imines and the moderately negative value for allenes and ketene. For simple vinyl compounds ($CH_2=CXY$), ^2J is directly proportional to the sum of the electronegativities of X and Y (eq. 4-5), with a β electronegative

$$^2J_{H-C-H} = a(E_X + E_Y) + b \qquad\qquad 4\text{-}5$$

substituent making ^2J less positive. When substituents of extreme electronegativities are included, eq. 4-6 produces a more satisfactory fit.

$$^2J_{H-C-H} = \frac{a}{E_X + E_Y} + b \qquad\qquad 4\text{-}6$$

Geminal couplings through an sp^3 carbon atom are compared to the -12.4 Hz value in methane. Simple σ withdrawal of electrons makes ^2J more positive (less negative), as in CH_3OH (-10.8 Hz), CH_3I (-9.2), CH_2Br_2 (-5.5), and **24** (-6); σ donation makes ^2J less

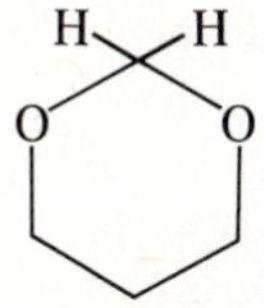

24 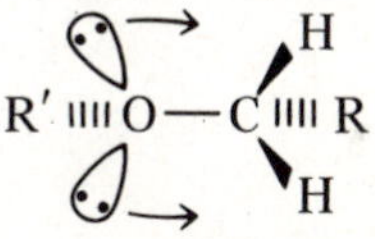   **25**

positive, as in CH_3SiMe_3 (TMS, -14.1 Hz). A lone pair of electrons on oxygen can serve as a π donor to the CH_2 group (**25**) and thereby make ^2J more positive. In open chain systems (**25**), there is free rotation about the O—C bond, so strong π overlap is inhibited. In a planar ring system such as **26**, the filled lone pair orbitals are frozen into the correct orientation for

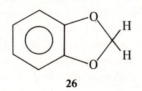

26

overlap, and in addition there are two contributing oxygen atoms. As a result, ^2J is ± 1.5 Hz in **26** (sign not known), an increase of at least 10 Hz from the value in methane.

Consequently, the influence of the filled lone pair orbital is geometry dependent. Best overlap occurs when the dihedral angle between a C—H bond and a single lone pair is close to $0°$ or $180°$ (**27c** and **27e**, shown in a Newman-like perspective). Least favorable overlap

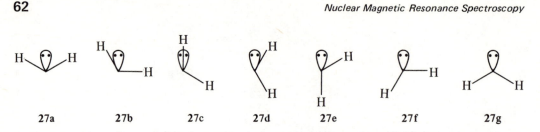

occurs in the staggered forms, **27a**, **27f**, and **27g**, with **27b** and **27d** intermediate. Since overlap makes ^2J more positive, the coupling constant may be used as a structural diagnostic. The two conformations of thiane 1-oxide (**28-ax**, **28-eq**) correspond to **27a** (staggered) and **27e**

(antiperiplanar), respectively. The measured values of ^2J are -11.7 and -13.7 Hz; the more positive value (-11.7) identifies the equatorial conformer (**28-eq**). Overlap in a planar situation such as **26** corresponds to **27c**. The divalent oxygen adds a second lone pair that is also eclipsed.

An adjacent double bond can withdraw π electrons from a CH_2 group and thereby make ^2J less positive. In open chain systems, a free rotation dilutes any overlap, so effects are minimal, e.g., $^2J = -14.9$ in acetone. Overlap is optimal in planar systems (**29**, **30**). The

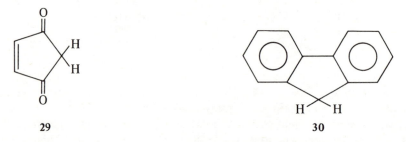

negative contributions of two carbonyl groups produce a coupling constant of -21.5 Hz in **29**, and of two benzene rings a coupling of -22.3 in **30** (fluorene). In order that withdrawal of electrons from the CH_2 group strengthen a double bond optimally, the dihedral arrangement given by **31a** is required (shown in perspective). Least favorable overlap occurs in the almost orthogonal arrangements **31c** and **31d**, with **31b** intermediate. Overlap with triple bonds does not have an angular dependence. As a result, an acetylenic or nitrilic substituent is a very

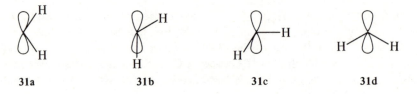

effective π withdrawer, e.g., $CH_3-C\equiv N$ ($^2J = -16.9$ Hz), $CH_2(CN)_2$ ($^2J = -20.4$). Both σ and π effects appear to be qualitatively additive.

The structural factors that bring about the basic difference between ^2J in methane and ethylene also come into play in three-membered rings. The change in hybridization of carbon and/or the increase in the CH_2 valence angle produces a more positive coupling in the parent cyclopropane ($^2J = -4.3$ Hz) than in methane. Couplings in structurally related compounds may be compared to this value. The series **32** exhibits the normal Pople–Bothner-By substituent

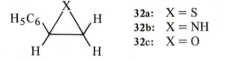

32a:	X = S
32b:	X = NH
32c:	X = O

effects; ^2J becomes more positive in the order X = S (**32a**, -1.4 Hz), NH (**32b**, $+1.0$), O (**32c**, $+5.7$). The oxide has a positive geminal coupling because of strong σ withdrawal and π donation.

In compounds such as **32**, ^2J may be obtained as the AB coupling in the ABX spectrum. In many of the compounds discussed above, couplings were reported between chemically and magnetically equivalent nuclei. Such nuclei give no spectral splitting, so recourse must be made to alternative methods. Normally, one of the protons in the methylene group is replaced by deuterium (H–C–D). The H-D coupling is readily obtained from the proton spectrum. Since a coupling constant is proportional to the product of the gyromagnetic ratios of the coupling nuclei, J_{H-H} may be derived from J_{H-D} by eq. 4-7. In this manner, ^1J for the hydrogen molecule ($+285$ Hz) was determined by analysis of the spectrum of H–D.

$$J_{H-H} = \frac{\gamma_H}{\gamma_D} J_{H-D} = 6.51 J_{H-D} \qquad 4\text{-}7$$

Geminal coupling constants between protons and other nuclei ($^2J_{H-C-X}$) have also been studied. The H–C–^{13}C coupling is thought to respond to substituents in much the same manner as does the H–C–H coupling. Most such couplings are relatively small, e.g., $^{13}CH_3-CH_3$ (-4.5 Hz), $^{13}CHCl_2-CHCl_2$ ($+1.2$), $CH_3-CH_2-^{13}CO_2CH_3$ (6.5). One notable difference between ^{13}C–C–H and H–C–H couplings is that the former can have π bond character between the two carbon atoms along the coupling path. Factors that have not previously been discussed consequently become important. Thus $^2J_{H-C-^{13}C}$ in *cis*-dichloro-ethylene (**33**) is 16.0 Hz, but in the trans isomer (**34**) it is 0.8 Hz. In ethylene ($^{13}CH_2=CH_2$),

^2J is -2.4 Hz. A most extraordinary value of 49.3 Hz was found for ^2J in acetylene (H–^{13}C≡C–H). With the increase in the availability of natural abundance ^{13}C spectra today, couplings between ^{1}H and ^{13}C should begin to take on appreciably more structural significance.

Such is already the case with the H–C–^{31}P coupling. The magnitude of $^2J_{H-C-^{31}P}$ is strongly dependent on the dihedral angle α between the lone pair on trivalent phosphorus and

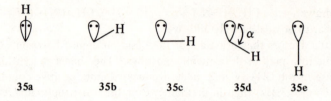

35a   35b   35c   35d   35e

the C–H bond (**35**, shown in projection). The maximum coupling is observed when the two are eclipsed (**35a**), as in **36** (H$_a$, $^2J = +25$ Hz). The coupling constant decreases as the dihedral

36   37   38

angle increases (**37**, $^2J = +18$; **38**, H$_e$, $^2J = 10$). At a dihedral angle close to 90° (**35c**) it goes negative. At about 120°, it attains its most negative value (**35d**; **36**, H$_b$, $^2J = -6$) and then

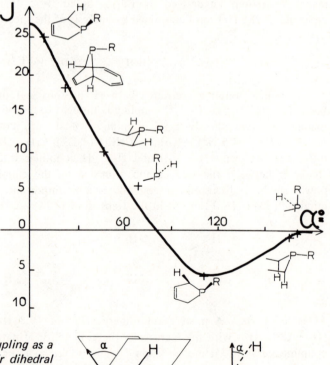

FIGURE 4-3. The H–C–^{31}P coupling as a function of the H–C–P lone pair dihedral angle α. [Reproduced with permission from J. P. Albrand, D. Gagnaire, and J. B. Robert, Chem. Commun., 1469 (1968).]

turns back up at an angle of $180°$ (**35e**; **38**, H_a, $^2J = -1$). The smooth curve (Figure 4–3) permits structural assignments by coupling constant measurements.

Some stereochemical work has also been carried out on the $^{15}N–C–H$ coupling constant. In oximes of the type **39**, $^2J_{15N-C-H}$ was found to be about 3 Hz for the anti form (**39a**) and

39a   **39b**

16 Hz for the syn form (**39b**). The coupling constant may be used therefore as a reliable structural diagnostic for syn-anti isomerism in oximes of this type.

Geminal H–C–F couplings are usually close to $+50$ Hz for an sp^3 carbon and near $+80$ for an sp^2 carbon, e.g., $CH_3–CH_2F$ (47.5 Hz), $CH_3–CHF_2$ (57.2), *cis*-CHF=CHF (72.7), *trans*-CHF=CHF (74.3), CH_2=CHF (84.7). Geminal F–C–F couplings are quite large $(+150-250$ Hz) for saturated carbon, but $<+100$ Hz for unsaturated carbon, e.g., CH_2=CF_2 (35.6 Hz), CHF=CF_2 (87), 1,1-difluorocyclopropanes ($\sim$150), 1,1-difluorocyclobutanes ($\sim$200), 1,1-difluorocyclohexane (240). Considerable stereochemical work must still be done with $^2J_{H-C-F}$ and $^2J_{F-C-F}$ before they can be accepted as reliable structural tools.

4-5   Vicinal Couplings

The widespread application of vicinal proton-proton coupling constants to stereochemistry has been one of the prime reasons for the success of nmr spectroscopy as a structural tool. In 1959, Karplus utilized valence bond calculations to define the mathematical relationship between $^3J_{H-C-C-H}$ and the H–C–C–H dihedral angle, ϕ. The simple form of the relationship (eq. 4–8, Figure 4–4) has provided stereochemists with a general and easily applied tool. The

$$^3J = \begin{cases} A \cos^2 \phi + C & (\phi = 0\text{–}90°) \\ A' \cos^2 \phi + C & (\phi = 90\text{–}180°) \end{cases} \qquad 4\text{–}8$$

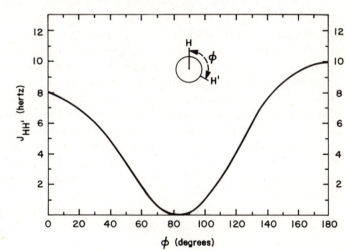

FIGURE 4-4. The vicinal H–C–C–H coupling constant as a function of the dihedral angle ϕ.

cosine-squared form of the dependence indicates that ^3J is largest when the protons are anti-periplanar ($\phi = 150$–$180°$), quite large when eclipsed (0–$30°$), and small when staggered (60–$120°$). If the constants A and C can be evaluated, quantitative work is suggested. The additive constant C is usually neglected, since it is thought to be less than 0.3 Hz. The multiplicative constants A and A' vary from system to system in the range 8–14 Hz. Because of this variation, quantitative applications cannot be transferred from one system to another, unless there is good reason to believe that the Karplus A is constant.

Before giving specific examples, let us examine the general areas of applicability of the Karplus equation. In chair cyclohexanes, J_{aa} is large (8–12 Hz) because ϕ_{aa} is approaching $180°$ (Figure 4-4), whereas J_{ee} (0–4 Hz) and J_{ae} (1–5 Hz) are small because ϕ_{ee} and ϕ_{ae} are close to $60°$. An axial proton that has an axial proton neighbor can easily be identified by its large J_{aa}. In three-membered rings (**40**), J_{cis} ($\phi \cong 0°$) is larger than J_{trans} ($\phi \cong 120°$). For the parent

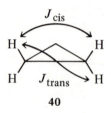

40

cyclopropane, $J_{trans} = 5.58$ and $J_{cis} = 8.97$ Hz. In cyclohexane, ring reversal averages the trans couplings (J_{aa} and J_{ee}) and the cis couplings (J_{ae} and J_{ea}). Because of the large value of J_{aa}, $J_{trans} = \frac{1}{2}(J_{aa} + J_{ee})$ is still larger (8.07 Hz) than $J_{cis} = \frac{1}{2}(J_{ae} + J_{ea})$ (3.73). In five-membered rings, the situation is less clear, and either J_{trans} or J_{cis} can be larger. In alkenes J_{trans} ($\phi = 180°$) is always larger than J_{cis} ($\phi = 0°$). In the parent ethylene, the values are 19.0 and 11.7 Hz, but the effect of substituents, to be discussed later, is extremely large. Thus, for the *cis*- and *trans*-dichloroethylenes, $J_{trans} = 12.1$ and $J_{cis} = 5.3$ Hz. Vicinal couplings have no stereochemical utility in acetylene ($^2J_{H-C-C-H} = 9.8$ Hz). Because ortho protons in aromatic systems have a favorable dihedral angle ($0°$), $^3J_{ortho}$ is generally quite large (6–10 Hz in benzenes), and can be distinguished from J_{meta} (1–4 Hz) and J_{para} (0–1.5 Hz).

Qualitative Karplus considerations lend themselves readily to stereochemical assignments in six-membered rings. For example, the proton spectrum of a 1,4-dinitro-2,3,5,6-tetraacetoxy-cyclohexane is δ 2.02 (6H, s), 2.12 (6H, s), 5.03 (2H, d of d, $J = 11$, 3.5), 5.89 (2H, d of d, $J = 11$, 3.5), 6.22 (2H, t, 3.5). [This shorthand notation is commonly used in experimental sections to describe nmr spectra; the δ value for each set of protons is followed parenthetically by the integrated number of protons and the multiplicity, i.e., singlet (s), doublet (d), triplet (t), multiplet (m).] The two acetoxy peaks suggest but do not prove that there are only two types of acetoxy groups. A doublet of doublets with coupling constants of 3.5 and 11 Hz can only arise from an axial proton flanked by one axial and one equatorial proton. There must be two distinct protons of this type adjacent to each other (δ 5.03, 5.89). The triplet of triplets with $J = 3.5$ Hz indicates an equatorial proton. Each of these three types of protons must occur twice in the molecule. Of the various possibilities, only **41** is consistent with these observations.

If there is a mixture of conformations, the observed coupling constants are weighted averages. For example, the 2 proton in 2-bromocyclohexanone (**42**) may be either axial

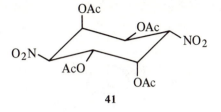

41

(**42**-eq) or equatorial (**42**-ax).   The couplings with the vicinal protons are therefore given by

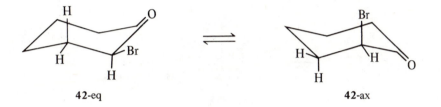

42-eq   **42**-ax

eq. 4–9, in which a is the fraction of **42**-ax and e of **42**-eq.   The spectrum permits the

$$J_{cis} = aJ_{ea} + eJ_{ae}$$

$$J_{trans} = aJ_{ee} + eJ_{aa}$$

4–9

determination of only $(J_{cis} + J_{trans})$.   By using *cis*-2-bromo-4-*t*-butylcyclohexanone as a model for **42**-eq, $(J_{cis} + J_{trans})$ for this extreme was found to be about 18 Hz; the trans bromo-*t*-butyl compound gave a model value of about 5.7 Hz for $(J_{cis} + J_{trans})$ in **42**-ax.   The observed value for **42** varies from 11.5 Hz (no solvent) to 7.0 (in CCl_4 extrapolated to infinite dilution), corresponding to a/e of 0.52/0.48 and 0.91/0.09, respectively.   The observed preponderance of the axial bromine is due to favorable polar interactions in **42**-ax.   Solvent–solute interactions make the equilibrium constant strongly concentration dependent.   It should be cautioned that the *t*-butyl anchoring group distorts the ring and makes a precise comparison impossible.   The semiquantitative result, however, is reliable.

Acyclic stereochemistry may also be obtained from vicinal coupling constant analysis. The differentiation of threo and erythro isomers provides one such example.   Catalytic deuteration of 2′-acetoxystyrene is thought to occur stereospecifically cis (eq. 4–10).   The aliphatic

$$\textit{trans-}C_6H_5CH{=}CHOAc \xrightarrow[\text{PdCl}_2]{D_2} \textit{threo-}C_6H_5CHDCHDOAc$$

4–10a

43-thr

$$\textit{cis-}C_6H_5{-}CH{=}CHOAc \xrightarrow[\text{PdCl}_2]{D_2} \textit{erythro-}C_6H_5CHDCHDOAc$$

4–10b

43-ery

protons in the product 2-phenylethyl-1,2-d_2 acetates (**43**) give AB spectra with different coupling constants, 6.5 Hz for the trans product, 8.5 Hz for the cis product.   The rotamers of the products (eq. 4-11) show that the coupling in the erythro compound should be larger than

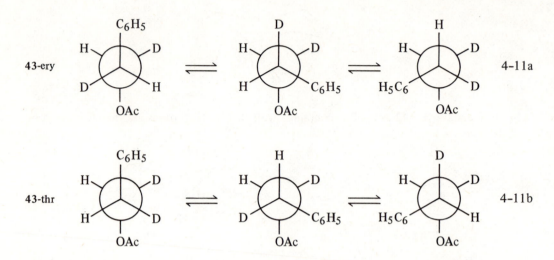

that in the threo compound, since in the former case the couplings are antiperiplanar in the favored anti rotamer at the left. The 8.5 Hz coupling (erythro) is found in the product of the cis alkene, in agreement with a stereospecifically cis deuteration. The erythro and threo couplings differ by only 2 Hz because of significant contributions from the gauche rotamers.

Despite the potentially general application of the Karplus equation to dihedral angle problems, there are limitations that must be considered. The vicinal H–C–C–H coupling constant is dependent on the C–C bond length, the H–C–C valence angle, the electronegativity of substituents, and the orientation of substituents, in addition to the H–C–C–H dihedral angle. We will consider each of these factors in turn. The ^3J is found to decrease as the C–C bond length becomes smaller. Such changes are reflected in the different Karplus A factors found in different systems. If a stereochemical differentiation is being made by examination of coupling constants for molecules thought to have similar C–C bond lengths, then this factor may be neglected. In aromatic compounds, J_{ortho} has been found to depend on the bond order. Thus in naphthalenes, J_{12} (8–9 Hz) is consistently larger than J_{23} (6–7 Hz). The ortho coupling is therefore of interest for its theoretical potential as well as for its utility in structural assignments.

The vicinal coupling decreases as the H–C–C angles increase. This factor is seen most dramatically in the series of cyclic alkenes, for which J_{12} decreases monotonically even though the dihedral angle is constant at $0°$: cycloheptene (10.8 Hz), cyclohexene (8.8), cyclopentene (5.1), cyclobutene (3), cyclopropene (1.3). This factor may frequently be disregarded when comparing members of a structurally similar set.

By far the most difficult factor to deal with is the effect of substituents. If a hydrogen on ethane is replaced by a more electronegative substituent, the coupling constant decreases, e.g., CH_3CH_3 (8.0 Hz), CH_3CH_2Cl (7.2), CH_3CH_2Li (8.9). For molecules of the type CH_3CHXY, the dependence takes the form of eq. 4-12. The general form of the dependence for a fragment

$$^3J = 18.0 - 0.80(E_X + E_Y) \qquad\qquad 4\text{-}12$$

XYHC–CHWZ is given by eq. 4-13, in which $J°$ is a coupling parameter with no dependence

$$^3J = {}^3J^\circ\,(1 - M\Sigma\,\Delta E)\tag{4-13}$$

on electronegativity, M is constant for geometrically related couplings, and ΔE is the sum of the differences between the electronegativity of hydrogen and those of the substituents W, X, Y, and Z. If the values of $^3J^\circ$ and M are known for a given series, this effect can be taken into account. Otherwise, quantitative Karplus considerations are not possible. The orientation of a substituent can also influence ^3J, with the maximum decrease observed when an electronegative substituent is antiperiplanar with any of the bonds on the coupling pathway.

One alternative to assessing the electronegativity effect is to eliminate it by using two coupling constants that depend on the substituents in the same way. In the ring system of eq. 4-14 only two coupling constants are observed because of rapid ring reversal (eq. 4-15).

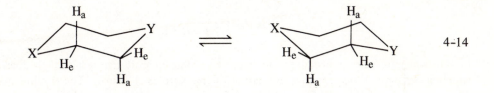

$$J_{\text{trans}} = \tfrac{1}{2}(J_{\text{aa}} + J_{\text{ee}})\tag{4-15a}$$

$$J_{\text{cis}} = \tfrac{1}{2}(J_{\text{ae}} + J_{\text{ea}}) = J_{\text{ae}}\tag{4-15b}$$

The ratio R of these coupling constants is independent of the electronegativities, since the dependence enters as a constant multiplicative factor (eq. 4-16). The orientation effect is

$$R = \frac{J_{\text{trans}}}{J_{\text{cis}}} = \frac{J^\circ_{\text{trans}}\,(1 - M\sum\Delta E)}{J^\circ_{\text{cis}}\,(1 - M\sum\Delta E)} = \frac{J^\circ_{\text{trans}}}{J^\circ_{\text{cis}}}\tag{4-16}$$

also absent in R because the coupling constants (J_{trans} and J_{cis}) (eq. 4-15) are orientationally averaged by reason of having equivalent axial and equatorial contributions. The value of R is thus a direct measure of the conformation of the molecule in eq. 4-14, or of any X–CH$_2$–CH$_2$–Y ring fragment. In an undistorted dihedral arrangement (**44b**), R has been

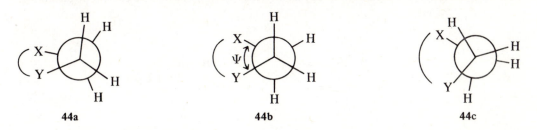

44a	**44b**	**44c**

found to lie in the range 1.9–2.2. Flattening (**44a**) lowers R (<1.8), and puckering (**44c**) increases R (>2.3). Thus R is close to 2.2 for cyclohexane, piperazine (**45**), and 1,4-dioxane

45   46   47

(46). For the 2,3 fragment of the flattened cyclohexanone, it is 1.7; of the puckered thiane **(47)**, 2.6. A valence angle larger than 109°, such as at the carbonyl group in cyclohexanone, flattens the ring, whereas a smaller angle, such as at sulfur in **47**, puckers it. The R value is related to the internal X–C–C–Y torsional angle Ψ **(44)** by eq. 4–17. Because R is independent

$$\cos \Psi = \left(\frac{3}{2 + 4R}\right)^{1/2} \qquad\qquad 4\text{–}17$$

of substituent effects, it produces much more reliable quantitative results than does a single coupling constant, which requires evaluation of the Karplus A factor. The R value analysis of **48** gives $\Psi_{23} = 52°$ $(R = 1.5)$ and $\Psi_{34} = 65°$ $(R = 3.6)$. The trigonal bipyramidal structure

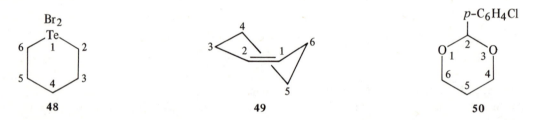

48   49   50

around tellurium produces extreme flattening at the 6,1,2 positions but extreme puckering by reflex at the opposite end. For the 4,5 bond of half-chair cyclohexene **(49)**, $\Psi = 63°$ $(R = 3.83)$, compared to an identical value obtained by microwave. In the 1,3-dioxane **50**, $\Psi_{45} = 55°$ $(R = 1.81)$, compared to an identical X-ray value. Thus cyclohexene is puckered and 1,3-dioxane flattened in the regions under discussion.

The cross-conjugated H–C(sp^2)–C(sp^3)–H or H–C(sp^2)–C(sp^2)–H pathways produce couplings that follow the Karplus curve, although other factors can dominate. In cyclic alkenes, the $-CH_2-CH=CR-$ coupling is strongly dependent on ring size (1.0 Hz for the four-membered ring, 3.1 for six, 7.8 for eight) because of changing dihedral and valence angles. The coupling between H$_2$ and H$_3$ in cycloheptatrienes **(51)** can be used to determine relative degrees of puckering.

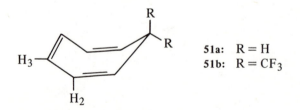

51a: R = H
51b: R = CF$_3$

For R = H, J_{23} is 5.3 Hz, and for R = CF$_3$, 6.9. The larger value in the latter case indicates that the protons are more nearly eclipsed and consequently the ring is flatter. The

–RC=CH–CH=CR′– coupling may be used in general for investigating butadiene conformations.

Considerable data now confirm that vicinal couplings over any H–C–X–H pathway, where X is an atom other than carbon, follow a Karplus-like relationship (see Figure 4-2) in their dependence on dihedral angle. Stereochemical studies have been made for X = O, N, S, Se, Te, and Si. The proton on nitrogen in tetrahydro-1,3-oxazine (**52**) was shown to be axial by

observation of couplings of 13.1 and 2.9 Hz to the 2 protons. The proton on sulfur in protonated thiane (**53**) is also axial because its vicinal couplings are 14.1 and 2.3 Hz. The large coupling in each case must be due to an axial-axial relationship. The H–C–N–H coupling constant has recently found applications in the determination of peptide stereochemistry (R–NH–CO–R′).

The H–C–C–F and H–C–N–F couplings also follow the Karplus relationship, although the magnitudes are much larger than in H–C–X–H couplings. The maximum at $0°$ dihedral angle for the H–C–C–F coupling is about 30 Hz, at $180°$ 45 Hz. Higher substituent electronegativity and smaller ring size decrease $^3J_{H-C-C-F}$. Thus these couplings respond to perturbations in the same way as the proton-proton couplings. The F–C–C–F coupling has also been studied, but it is not dependent on dihedral angles in a simple way. In contrast to H–C–C–H couplings, which are all positive, the F–C–C–F coupling may have either sign.

4-6 Long-Range Couplings

Couplings over more than three bonds are termed *long range*. Ordinarily, spin-spin coupling through the σ framework falls off very rapidly with the number of bonds. Except in certain special cases, σ couplings greater than 1 Hz are rarely observed over more than three bonds. Contributions from the π electrons can become important over these longer pathways. We shall discuss long-range couplings in terms of four broad classifications:

(a) Double-bond pathways (allylic, H–C=C–C–H; homoallylic, H–C–C=C–C–H; conjugated, H–C=C–C=C–H; cross-conjugated; aromatic).
(b) Triple-bond and cumulative pathways (H$+$C≡C$)_n$H, H–C$+$C≡C$)_n$H, H–C$+$C=C$)_n$H, H–C–C$+$C=C$)_n$H, etc.).
(c) Zigzag pathways ($H_{\diagdown}C^{\diagup}C^{\diagdown}C^{\diagup}H$, $H_{\diagdown}C^{\diagup}C^{\diagdown}C^{\diagup}C^{\diagdown}H$).
(d) Through-space pathways mediated by lone pairs.

From valence bond calculations Karplus developed certain generalizations about long-range couplings involving π electron pathways. (1) The magnitude depends on the dihedral angle between the C–H bond and the π orbitals. (2) The sign of the coupling constant follows the model described in section 4-2: couplings over an odd number of bonds are positive and those over an even number of bonds are negative. (3) Replacement of a proton attached to a double bond by a methyl group, e.g., allylic to homoallylic, alters the sign of the coupling but has little effect on the magnitude.

Allylic couplings lie in the range $+1$ to -3 Hz, with typical values close to -1 Hz. The larger values are observed when the saturated $C-H_a$ bond is parallel to the π orbitals, as in **54b**.

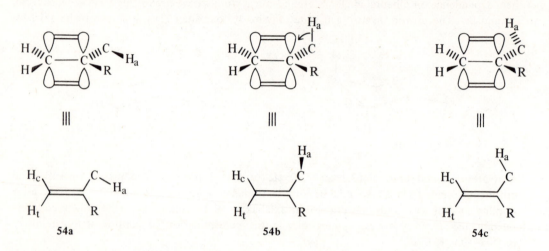

54a   54b   54c

The dihedral angle ϕ between the $C-H_t$ and $C-H_a$ bonds is $90°$ in this arrangement. This $\sigma-\pi$ overlap enables the coupling to be transmitted more effectively. When the $C-H_a$ bond is orthogonal to the π orbitals, as in **54a** and **54c** ($\phi = 0, 180°$), there is no $\sigma-\pi$ contribution and couplings are typically rather small (<1 Hz). The differences between $J_{H_a-H_c}$ (cisoid) and $J_{H_a-H_t}$ (transoid) are usually too small or unsystematic to use in stereochemical work.

In acyclic systems, the dihedral angle is averaged over both favorable and unfavorable arrangements, so an average value of $^4J_{\text{allylic}}$ is found, e.g., 2-methylacrolein (**55**, $|^4J_c| = 1.45$, $|^4J_t| = 1.05$ Hz), 2-bromopropene (**56**, $^4J_c = -1.4$, $^4J_t = -0.8$). Ring constraints can freeze

55   56

the bonds into the favorable ($\phi = 90°$) arrangement, as in the alkaloid himandridine (**57**, $^4J = -3.5$ Hz), or to a lesser extent in indene (**58**, $^4J = -2.0$). When the coupled protons

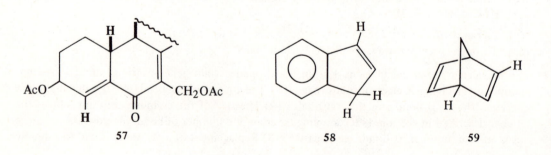

57   58   59

are connected by a planar zigzag arrangement (H_a–H_t in **54a**), a special σ contribution produces a coupling of about +1 Hz, as in norbornadiene (**59**),

The observed coupling may be written as the sum of σ and π contributions (eq. 4–18).

$$^4J_{\text{allylic}} = {}^4J_\sigma + {}^4J_\pi \qquad\qquad 4\text{-}18$$

The π contribution, with a significant magnitude only near $\phi = 90°$, follows a sine-squared dependence on ϕ. The zigzag σ coupling only has magnitude near $\phi = 0°$, so it follows a cosine-squared dependence. Typical values describe the eq. 4–19. Most of the data have come

$$^4J_{\text{allylic}} = \begin{cases} 1.3\,\cos^2\phi - 3.0\,\sin^2\phi & (\phi = 0\text{-}90°) \\[2mm] -3.0\,\sin^2\phi & (\phi = 90\text{-}180°) \end{cases} \qquad 4\text{-}19$$

from transoid allylic couplings (**54a**). The cisoid couplings are frequently not so large at $\phi = 0$, 90, or 180°. Because there is no zigzag pathway available to the cisoid coupling, the special σ contribution is absent.

The homoallylic coupling depends on the orientation of two C–H bonds with the π framework (**60**). For acyclic systems such as *cis*- or *trans*-2-butene, ^5J is typically +2 Hz.

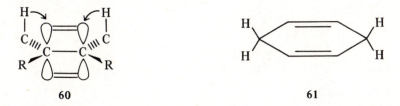

60   **61**

When both coupling protons are properly aligned, as in the nearly planar cyclohexa-1,4-diene (**61**), ^4J can be quite large. For **61**, the coupling between the cis homoallylic protons is 9.63 Hz, and between the trans protons 8.04 Hz. The increased couplings in **61** are also due to the presence of two equivalent coupling pathways. It is not unusual for the homoallylic coupling to be larger than the allylic coupling. Thus $^4J_{\text{CH}_3\text{-H}_a}$ is 1.1 Hz and $^5J_{\text{CH}_3\text{-H}_b}$ is 1.8 Hz in **62**.

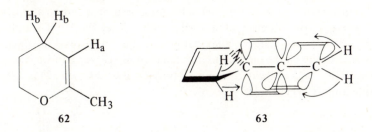

62   **63**

Long-range couplings can be quite large in acetylenic and allenic systems, although such molecules are not frequently encountered in practice. In allene itself, ^4J is -7 Hz, since the rigid geometry is optimal for σ–π overlap. In 1,1-dimethylallene, ^5J decreases to (+)3 Hz, since σ–π overlap is averaged by methyl rotation. In a rigid system such as **63**, the homoallenic coupling is increased to 4.6 Hz because of optimal overlap. In cumulated systems, e.g., **64**,

64: $(CH_3)_2C{=}C{=}C{=}CH(CHO)$

the ^6J couplings are somewhat smaller. In both methylacetylene (^4J) and dimethylacetylene (^5J), the long-range coupling is about 3 Hz, although the signs are probably opposite. The triple bond has no steric requirements for σ–π overlap, or for conjugation with other triple bonds. As a result, there is only a small diminution for ^5J, ^6J, and ^7J [(+)2.2,(−)1.3, (+)1.3 Hz] in **65–67**. Even over three triple bonds, significant long-range couplings have been observed, $^8J = 0.65$ Hz in **68**, $^9J = 0.4$ in Hz in **69**.

65:	H–C≡C–C≡C–H
66:	CH₃–C≡C–C≡C–H
67:	CH₃–C≡C–C≡C–CH₃
68:	CH₃–C≡C–C≡C–C≡C–H
69:	CH₃–C≡C–C≡C–C≡C–CH₂OH

Conjugation of double bonds is a stereochemically much more complicated situation. For example in butadiene (**70**), there are two four-bond couplings ($^4J_{13} = -0.86$, $^4J_{23} =$

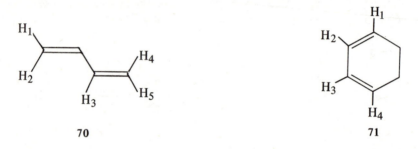

                                    70   **71**

-0.83) and three five-bond couplings ($^5J_{14} = +0.60$ Hz, $^5J_{15} = +1.30$, $^5J_{24} = +0.69$). On the other hand, 1,3-cyclohexadiene (**71**) has only one of each type ($^4J_{13} = +1.06$ Hz, $^5J_{14} = +0.91$). The largest coupling ($^5J_{15}$ in butadiene) undoubtedly has a strong contribution due to the zigzag pathway. Studies of butadiene couplings can give useful conformational information.

If any of the carbon atoms along an H–C–C–C–H path (all single bonds) are sp² hybridized, ^4J is increased over its normal value for three saturated carbon atoms, though rarely to more than 1 Hz in the absence of a zigzag path. The four-bond coupling in acetone (CH₃–CO–CH₃) is 0.54 Hz. Larger values can be obtained in systems in which the coupling protons are parallel to the π lobe on the unsaturated carbon. Zigzag pathways occur frequently in such systems. Thus $^4J_{ee}$ in *trans*-2-bromo-4-*t*-butylcyclohexanone (**72**) is 1.7 Hz, and ^4J in bromo-*p*-benzoquinone (**73**) is 1.9 Hz.

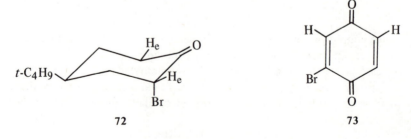

                            72   **73**

Couplings in aromatic systems fall into some of the categories discussed above but nonetheless deserve to be considered separately because of their widespread structural importance. The long-range $^4J_{meta}$ and $^5J_{para}$ may be distinguished from $^3J_{ortho}$ by their smaller magnitudes (see section 4-5). In benzene itself, $^3J_{ortho}$ is 7.54 Hz, $^4J_{meta}$ is 1.37 Hz, and $^5J_{para}$ is 0.69 Hz. The para coupling has a predominantly π mechanism, whereas the ortho and meta couplings are σ. The meta coupling is somewhat larger than normally expected for a ^4J because of the zigzag pathway. These couplings are sensitive to substituent changes. The meta coupling between protons on either side of a halogen atom decreases in the order F (1.37 Hz), Cl (0.88), Br (0.75), I (0.51).

Protons on saturated carbon atoms attached to an aromatic ring couple with all three types of aromatic protons. For toluene, $^4J_{ortho} = -0.75$ Hz, $^5J_{meta} = +0.36$, and $^6J_{para} = -0.62$. These couplings, frequently referred to as *benzylic*, depend on a σ–π interaction between the substituent protons and the π cloud. Thus $^4J_{ortho}$ is a special case of the allylic coupling, and should have a well-defined stereochemical dependence. In addition, $^4J_{ortho}$ is dependent on the mobile bond order of the double bond in the coupling pathway.

We have already mentioned on several occasions the special properties of the four- or five-bond zigzag pathways. Increased couplings over zigzag pathways are observed for $^4J_{H_t-H_a}$ in allylic systems (54a), $^5J_{15}$ in butadienes (70), ^4J in cross-conjugated molecules (72, 73), and $^4J_{meta}$ in aromatics. Other examples include pyrrole-2-carboxaldehyde (74, $^5J = 1.2$ Hz) and quinoline (75, 0.9 Hz). All these couplings have involved pathways containing at least one

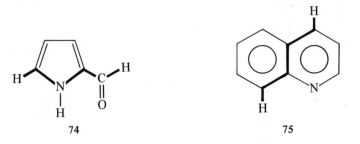

74   75

unsaturated carbon atom. The normal nonzigzag coupling over three saturated carbon atoms is <0.7 Hz. When the four bonds are in the same plane and form a zigzag ("W" or "M") pathway, the coupling can be considerably larger. The bridgehead coupling in norbornanes is 1-1.7 Hz ($^4J_{1-4} = 1.6$ in 76), the anti-endo coupling is 1.5-4 Hz ($^4J_{5-7a} = 2.4$ in 76), the

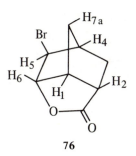

76

exo–exo coupling is 1-2 Hz ($^4J_{2-6} = 1.0$ in 76), and the bridgehead–exo coupling is 1-1.5 Hz ($^4J_{4-6} = 1.0$ in 76). In more highly strained systems, or those with multiple zigzag pathways, the coupling increases, e.g., 7.4 Hz in **77**, 10 Hz in **78**, and (the record) 18 Hz in **15**. The sign

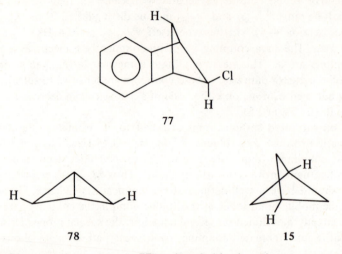

of the coupling is known to be positive in **77**, and probably the others are positive as well. The zigzag couplings decrease with deviations from planarity and with an increase in the distance between the first and third carbon atoms. The large magnitude of these saturated zigzag couplings has been explained in terms of a direct passing of spin information by interaction between the backsides of the C–H orbitals.

The zigzag coupling over a saturated pathway has also been found in several five- and six-bond situations: 2.3 Hz in **79**, 1.7 in **80**, 1.3 in **81**, and 1 Hz in **82**. Although **79**, **81**, and

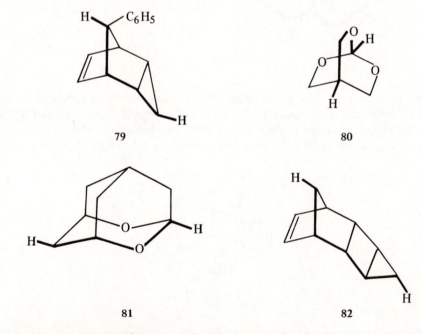

82 offer double zigzag paths, **80** has a triple nonzigzag path. In this last system the evidence certainly favors a direct interaction between the H–C orbitals.

Although coupling information is always passed by electron-mediated pathways, in some cases part of the through-bond pathway may be skipped, as in σ–π allylic overlap (**54b,** **60**) and in **80**. Molecule **83** represents an extreme example of this phenomenon. The six-bond

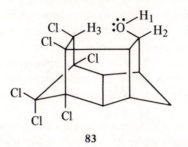

83

(all saturated) coupling between H_1 and H_3 is 1.1 Hz. Since H_3 and the oxygen atom are in van der Waals contact (see section 3–2), the coupling information appears to be transferred from H_3 through the oxygen lone pairs to H_1. Couplings of this type have been termed *through space*, but a more appropriate term would be *lone pair mediated.*

This type of coupling, though in dispute until recently, seems now to be on firm ground. It is found most commonly in H–F and F–F couplings. The six-bond CH_3–F coupling in the structurally similar **84** and **85** are <0.5 Hz and 8.3 Hz, respectively. In **84**, the H–F distance

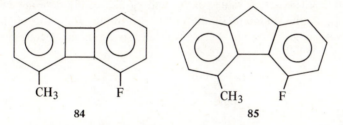

84   **85**

is 2.84 Å, whereas in **85** it is 1.44 Å. In the latter case the nuclei are well within the sum of the van der Waals radii (~2.55 Å), and the coupling information is probably passed from the proton through the lone pair electrons to the fluorine nucleus. The five-bond F–F coupling in **86** is 167–178 Hz, regardless of the nature of X (CH_2Br, CH_2OH, $CO_2C_2H_5$, NH_2). This

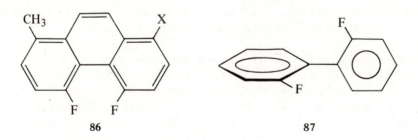

86   **87**

insensitivity to substitution and the larger magnitude in comparison to that in *o,o′*-difluoro-biphenyl (**87**, 16.5 Hz) are strongly evidential for a through space, lone pair mediated coupling. Such a mechanism is probably quite important in the geminal H–C–F coupling also, but the rigid geometry makes the proposition difficult to prove. Long-range couplings between protons and other nuclei such as fluorine are very common, since these couplings are inherently larger than their H–H counterparts.

4-7   Empirical Spectral Analysis

Spectral analysis signifies the derivation of the chemical shift of each nucleus and the coupling constant between each pair of nuclei from an observed spectrum. If all the ν_i and J_{ij} are known, the spectrum may be precisely reproduced by computer calculation. Analysis of first-order spectra is a straightforward process of measuring the distance between certain peaks. Thus an AX spectrum consists of two doublets. The doublet spacing is J_{AX} and the midpoints of the doublets are ν_A and ν_X. In an AMX spectrum, each nucleus produces a quartet resonance. The common spacing in the A and M portions is J_{AM}, etc.; the ν_i are given by the midpoints of the quartets.

If a spectrum is second order, by reason either of magnetic nonequivalence or of a small $\Delta\nu/J$, analysis becomes much more complex. The second-order two-spin system (AB) contains four lines, the inner peaks of which are more intense. Figure 4-5 gives a diagrammatic representation of an AB system. The coupling constant (J_{AB}) is obtained directly from the doublet spacing, but no specific peak positions correspond to the chemical shifts. Because the A peaks are of unequal intensity, ν_A is defined by the weighted average position. The chemical shift difference, $\Delta\nu_{AB}$, may be obtained from eq. 4-20, in which $2C$ is the spacing between alternate

$$\Delta\nu_{AB} = (4C^2 - J^2)^{1/2} \qquad\qquad 4\text{-}20$$

peaks in Figure 4-5. The actual values of ν_A and ν_B are readily determined by adding and subtracting $\frac{1}{2}\Delta\nu_{AB}$ to the midpoint of the entire spectrum. The ratio of intensities of the inner to outer peaks is given by eq. 4-21. Thus all aspects of the spectrum may be derived from knowledge of J and C.

$$\frac{\text{inner}}{\text{outer}} = \frac{1 + J/2C}{1 - J/2C} \qquad\qquad 4\text{-}21$$

Analysis of three-spin systems can range from the trivial to the impossible. The first-order systems AX_2 and AMX may be analyzed by direct inspection. As the chemical shift difference in the AX_2 system grows smaller (Figure 4-6), degeneracies are lifted, intensities are perturbed, and one new peak is sometimes added. In the AB_2 extreme, a total of nine peaks can be observed: four from spin flips of the A proton, four from spin flips of the B protons, and one from simultaneous spin flips of both A and B protons. The ninth peak, called a combination line, is ordinarily forbidden and hence is not observed in the first-order case (AX_2).

When eight peaks are observed (Figure 4-6, numbering from left to right for an AB_2 spectrum with A at lower field, or from right to left with A at higher field), all the spectral parameters may be obtained empirically from the various peak positions. The chemical shift of the A proton corresponds to ν_3, the position of the third peak. The average of ν_5 and ν_7 gives the chemical shift of the B proton (eq. 4-22). The coupling constant is given by a linear

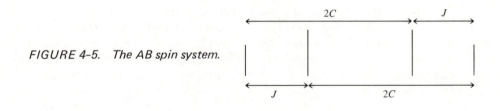

FIGURE 4-5.   The AB spin system.

$$\nu_B = \frac{1}{2}(\nu_5 + \nu_7) \qquad\qquad 4\text{-}22$$

$$J_{AB} = \frac{1}{3}(\nu_1 - \nu_4 + \nu_6 - \nu_8) \qquad\qquad 4\text{-}23$$

combination of four peak positions (eq. 4-23). Thus the entire spectrum may be analyzed without reference to ν_9, the combination peak. If not all eight of the remaining peaks are

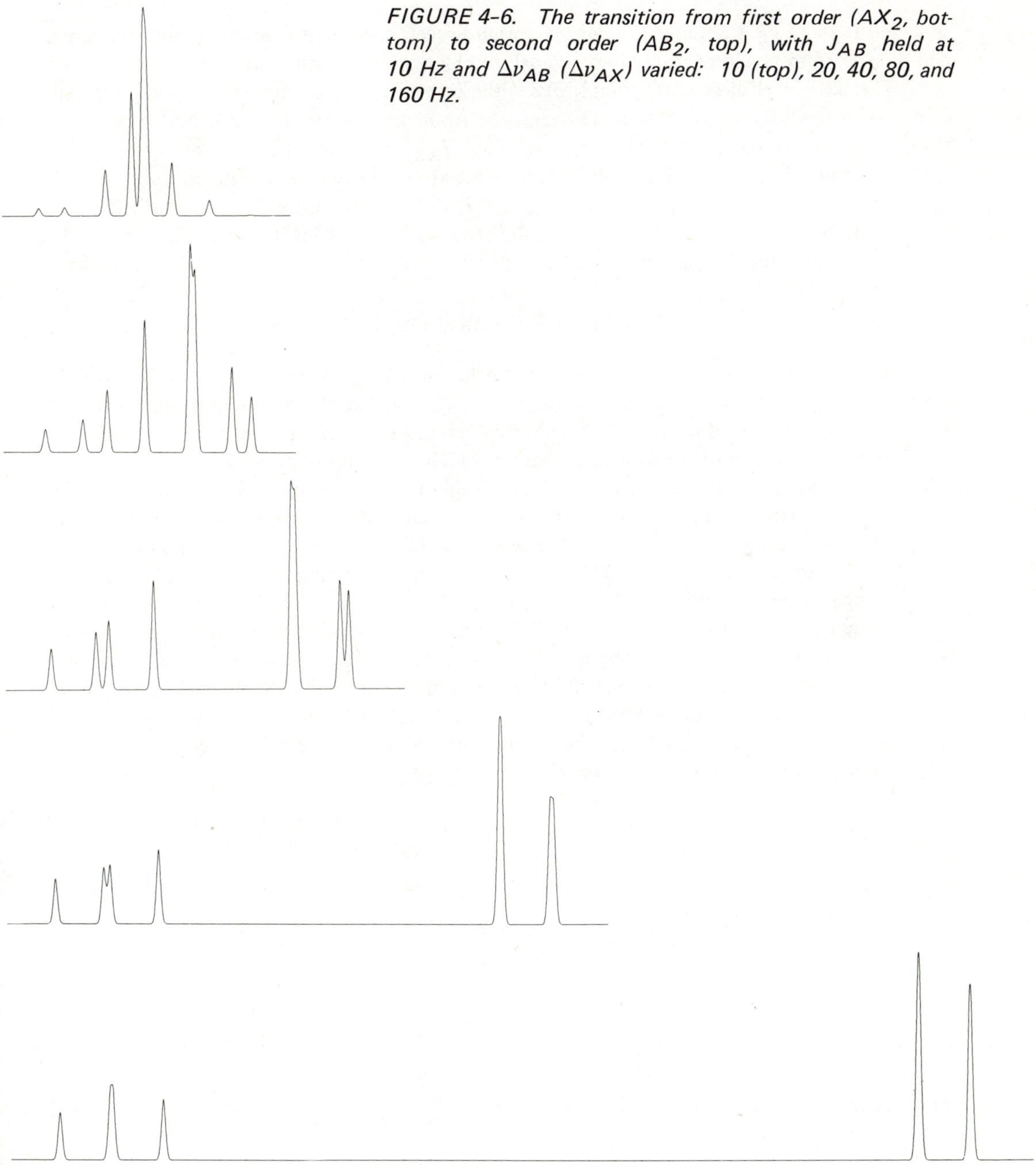

FIGURE 4-6. The transition from first order (AX$_2$, bottom) to second order (AB$_2$, top), with J_{AB} held at 10 Hz and $\Delta\nu_{AB}$ ($\Delta\nu_{AX}$) varied: 10 (top), 20, 40, 80, and 160 Hz.

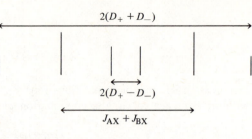

FIGURE 4-7. The X part of an ABX spectrum.

observed (ν_5 and ν_6 are frequently degenerate), however, confusion as to numbering may arise, and the spectrum would probably best be analyzed by computer techniques (next section).

The next level of spectral complexity is the case in which all three nuclei have different chemical shifts, but one of these is well removed from the other two (ABX). This spectrum is determined by six parameters: ν_A, ν_B, ν_X, J_{AB}, J_{AX}, and J_{BX}. The X proton of an ABX spectrum generally has six lines, with the midpoint at ν_X (Figure 4-7). The distance between the tallest peaks is equal to the sum of J_{AX} and J_{BX}. The separation between the outermost lines is twice the sum of D_+ and D_-, and the separation of the innermost peaks is twice the difference of D_+ and D_-, which are defined in eq. 4-24. The value of J_{AB} is obtained from

$$D_{\pm} = \frac{1}{2}\left\{\left[(\nu_A - \nu_B) \pm \frac{1}{2}(J_{AX} - J_{BX})\right]^2 + J^2_{AB}\right\}^{1/2} \qquad 4\text{-}24$$

the AB part of the spectrum (see below). With the knowledge of D_+, D_-, and J_{AB}, eq. 4-24 gives $(\nu_A - \nu_B)$ and $(J_{AX} - J_{BX})$. Since $(J_{AX} + J_{BX})$ is also available, the individual values of J_{AX} and J_{BX} may be obtained. Two solutions are always possible, one in which these two coupling constants have the same sign, another in which they have opposite signs.

The AB portion of the ABX spectrum is made up of two overlapping quartets (Figure 4-8). The doublet separation in both quartets gives four independent measurements of J_{AB} (no relative or absolute sign is determined). The separation of alternate peaks in the quartets gives a redundant determination of D_+ and D_-. The separation of the midpoints of the quartets gives an additional measurement of $\frac{1}{2}(J_{AX} + J_{BX})$.

If the arithmetic for a particular ABX spectrum is followed through, the exact values of ν_X, $(\nu_A - \nu_B)$, and J_{AB} may be obtained, but two sets of J_{AX} and J_{BX} are produced, either with like or unlike signs. Only one of these solutions gives the X portion with the proper intensities. The most straightforward method of differentiating the two sets is therefore to calculate by computer methods the spectrum corresponding to each. The correct set of parameters reproduces the experimental X intensities.

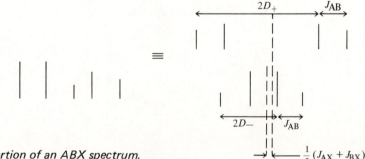

FIGURE 4-8. The AB portion of an ABX spectrum.

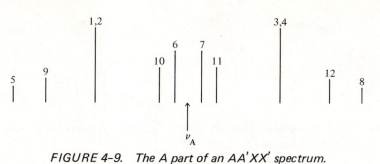

FIGURE 4-9.   The A part of an AA'XX' spectrum.

The ABC spectrum, in which all three nuclei are closely coupled, contains up to fifteen lines (four A-type lines, four B-type, four C-type, and three combination), but analysis by hand is very difficult. The computer methods described in the next section are best adopted for ABC analyses.

"Hand" analysis finds only limited application in four-spin systems. First-order spectra (AX_3, A_2X_2) provide no difficulty; of the second-order spectra (AA'XX', AA'BB', ABXY, ABCX, ABCD, etc.), only the AA'XX' can be readily analyzed. This spectrum, which is second order by reason of magnetic nonequivalence, is determined by six parameters: ν_A, ν_X, $J_{AA'}$, $J_{XX'}$, J_{AX}, $J_{AX'}$. The A and X parts are identical, and the chemical shifts (ν_A, ν_X) are located at their respective midpoints. Figure 4-9 gives a diagrammatic representation of the A part of an AA'XX' spectrum with mirror symmetry about ν_A. As ($\nu_A - \nu_X$) gets smaller, this symmetry is lost, and the spectrum tends toward AA'BB'. The spectrum is sensitive to the sums and differences of coupling constants (eq. 4-25), rather than to the couplings themselves. The

$$K = J_{AA'} + J_{XX'} \quad L = J_{AX} - J_{AX'} \quad M = J_{AA'} - J_{XX'} \quad N = J_{AX} + J_{AX'} \qquad 4\text{-}25$$

distance between the dominant peaks ($\nu_{1,2} - \nu_{3,4}$) gives N, the sum of J_{AX} and $J_{AX'}$. The remainder of the spectrum consists of two quartets (ν_{5-8} and ν_{9-12}). The separation of adjacent outer peaks in a given quartet corresponds to K [($\nu_5 - \nu_6$) or ($\nu_7 - \nu_8$)] and M [($\nu_9 - \nu_{10}$) or ($\nu_{11} - \nu_{12}$)]. Finally, L may be obtained from K, M, and the separation between alternate peaks in a given quartet (eq. 4-26). If all ten peaks are observed, analysis of the AA'XX'

$$\nu_5 - \nu_7 = (K^2 + L^2)^{1/2}$$

$$4\text{-}26$$

$$\nu_9 - \nu_{11} = (M^2 + L^2)^{1/2}$$

spectrum is quite straightforward. The analysis does not distinguish $J_{AA'}$ from $J_{XX'}$ or J_{AX} from $J_{AX'}$.

A common example of the AA'XX' spectrum occurs in the bismethylene fragment ($-CH_2-CH_2-$) of cyclic compounds. The $J_{AA'}$ and $J_{XX'}$ are geminal couplings, and the J_{AX} and $J_{AX'}$ are vicinal couplings. Because geminal couplings are usually about -12 Hz in such situations, K is quite large. As a result, ν_5 and ν_8 are lost in the noise, and ν_6 and ν_7 become degenerate. Because $J_{AA'}$ and $J_{XX'}$ are usually close together in value and sometimes even identical, the separation (M) between ν_9 and ν_{10} (ν_{11} and ν_{12}) is small or zero, so ν_9 and ν_{12} are drawn inside the two large peaks that give the sum of the vicinal couplings (N). If M is zero, then the difference between the vicinal couplings (L) is given by the separation between $\nu_{9,10}$

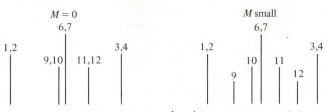

FIGURE 4-10. *The special case of AA'XX' spectra (A part) for a —CH₂CH₂— fragment in a cyclic compound.*

and $\nu_{11,12}$ (Figure 4-10). Otherwise, L is obtained from M and eq. 4-26. This spectrum thus readily gives J_{AX}, $J_{AX'}$ and $(J_{AA'} - J_{XX'})$, but not $(J_{AA'} + J_{XX'})$.

4-8   Computer Techniques for Analysis

When spectra are too complex for hand analysis, recourse is made to computer techniques. Such methods are also used to refine hand analyses. Programs are now available that satisfy most spectral needs. The first step is usually a trial-and-error procedure of guessing the chemical shifts and coupling constants in order to match the observed spectrum. This method is surprisingly successful in the majority of cases with four spins or fewer, since information on most coupling situations is now available. The chemical shifts are varied initially, until the width of the observed and calculated spectra approximately agree. Then the coupling constants or their sums and differences are varied systematically one at a time until a reasonable match is obtained.

Refinements of hand or trial-and-error analyses utilize iterative procedures. The program of Castellano and Bothner-By (LAOCN3, available from the Quantum Chemistry Program Exchange, QCPE 111) iterates on peak positions. The coupling constants and chemical shifts are adjusted for best agreement between observed frequencies and those calculated from the test parameters. The program of Swalen and Reilley (NMREN, NMRIT, QCPE 33, 34, 35, 126, 127) calculates the energies of each spin state from the observed peak positions and then iterates by adjusting the spectral parameters until the energies of the calculated spin states agree with those from the observed peak positions. Both iterative methods require considerable knowledge of the spin system before they are successful. This knowledge is generally gained by the trial-and-error procedure. Analysis is carried as far as possible by this means, and the iteration serves more as a refinement. Iteration is useful, however, because a peak-by-peak analysis of the results can be made to find where the sources of error are. Also, the iterative methods utilize all the spectral data to determine the spectrum, so a more accurate analysis results. These programs find their greatest utility in three- and four-spin systems, but programs have been written for much larger systems.

Programs have also been designed for special situations. Castellano and Waugh's EXAN II gives exact analyses of ABC systems in more than 90% of cases. The size of spin systems can be greatly expanded in some instances by taking magnetic equivalence into account, so that, for example, a methyl group is treated as a single entity. Haight's LAME is a version of LAOCN3 with magnetic equivalence.

The theory of spectral analysis, which is of considerable interest in itself, is beyond the scope of this presentation. It has been dealt with in detail by Roberts (see reference 5 at the end of this chapter) and in the general texts listed at the end of Chapter 1.

4-9   Second-Order Effects

The vast majority of nmr analyses are carried out simply by inspection, without recourse to quantitative techniques. Such procedures are unfortunate because they so frequently lead to incorrect conclusions. This situation led Bothner-By in 1965 to comment that over 90% of reported geminal and vicinal couplings are either incorrect or not experimentally well determined. One of the principal reasons for such widespread failure to obtain correct spectral parameters is erroneous interpretation of second-order phenomenon. In this section we will discuss some qualitative aspects of this problem, since this approach is most useful for day-to-day spectral analyses. The discussion will be divided into three sections:

(a) Spectra with the correct number of lines expected from first-order considerations, but with misleading spacings.
(b) Spectra with apparently too few lines (*deceptive simplicity*).
(c) Spectra with apparently too many lines (*virtual coupling*).

When spectra are second order, peak spacings do not necessarily correspond to coupling constants, even when intuition seems to be clear. For example, the spectrum of protonated thiane-3,3,5,5-d_4 (**88**, Figure 4-11, the *S*-proton is entirely axial) contains a triplet of triplets

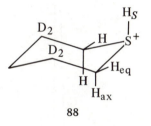

88

for the *S*-proton. The large splitting (12.4 Hz) corresponds to a coupling between H_S and H_{ax}, the small splitting (4.0) to a coupling between H_S and H_{eq}. These spacings, however, are not true coupling constants. In order to calculate a spectrum with these spacings by computer methods, coupling constants of $J_{H_S-H_{ax}} = 14.1$ Hz and $J_{H_S-H_{eq}} = 2.3$ are required. The spectrum gives only the sum of the coupling constants by inspection (16.4 Hz). Computer methods are required to obtain the true values of the coupling constant. This situation is common when a proton is coupled to two (or more) types of nuclei that are themselves closely coupled. This case (**88**) was particularly misleading because the *S*-proton resonance looks first order. Unless the spin system contains no closely coupled nuclei, analysis by inspection must be supplemented by computer simulation.

Frequently, lines can coincide in such a way that the spectrum assumes a simpler appearance than seems consistent with the actual spectral parameters. The ABX spectrum is a frequent perpetrator of such deceptive simplicity. For example, if J_{AB} is much greater than $[(\nu_A - \nu_B) \pm \frac{1}{2}(J_{AX} - J_{BX})]$, then D_+ and D_- both become equal to $\frac{1}{2}J_{AB}$ (see eq. 4-24). Examination of Figures 4-7 and 4-8 shows that the X part becomes a 1/2/1 triplet and the AB part a 1/1 doublet, with the separation between any two adjacent peaks equal to $\frac{1}{2}(J_{AX} + J_{BX})$. (For reasons not apparent from this level of discussion, the outer two X lines and the outer lines in each quartet in the AB part become forbidden.) The resulting spectrum thus resembles a first-order A_2X pattern, even though J_{AX} and J_{BX} are nonzero and unequal. Analysis of the

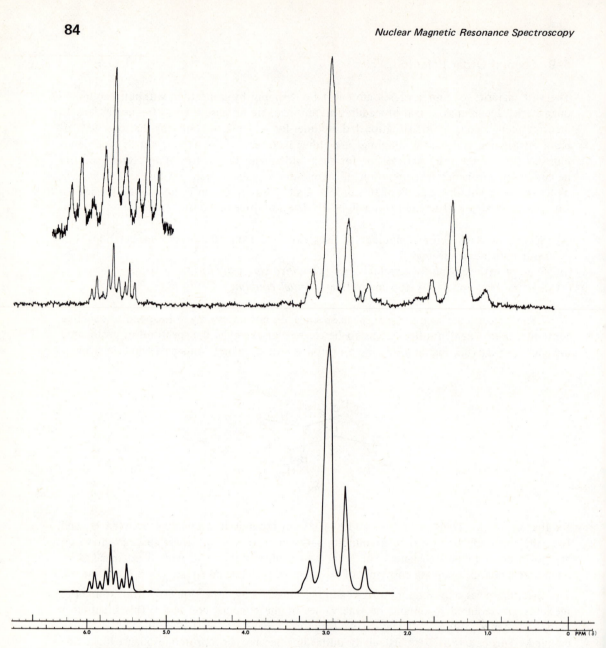

FIGURE 4-11. *The 60 MHz proton spectrum of thiane-3,3,5,5-d₄ in FSO₃H at −30°; observed (upper) and calculated (lower). [Reproduced with the permission of the American Chemical Society from J. B. Lambert, R. G. Keske, and D. K. Weary, J. Amer. Chem. Soc., **89**, 5921 (1967).]*

spectrum is impossible, since only the sum of J_{AX} and J_{BX} is available. The conclusion to be avoided is that J_{AX} and J_{BX} are equal. Figure 4-12 shows a simulated example for such a case. Other patterns of deceptive simplicity can occur in ABX spectra.

The AA′XX′ spectrum also can be deceptively simple. If M and K are both much larger than L (eq. 4-25, Figure 4-9), peaks 6, 7, 10, and 11 become degenerate at the center of the spectrum, and 5, 8, 9, and 12 become too small to be observed. Thus both the A and X positions appear to be first-order triplets, even though J_{AX} and $J_{AX'}$ are unequal.   Again, only the sum of the two coupling constants in question can be obtained from the spectrum.   The 1, 3, 4, and 6 protons of cycloheptatriene (**89**) produce such a pattern.   The 3,4 resonance is a triplet

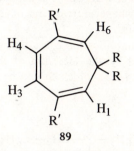

89

FIGURE 4-12.   *(Upper) A deceptively simple ABX spectrum: $\nu_A = 0.0$ Hz, $\nu_B = 3.0$, $\nu_X = 130.0$; $J_{AB} = 15.0$ Hz, $J_{AX} = 5.0$; $J_{BX} = 3.0$.   (Lower) The same parameters, except $\nu_B = 8.0$ Hz.   The larger value of $\Delta\nu_{AB}$ removes the deceptive simplicity and produces a typical ABX spectrum (Figures 4-7 and 4-8).*

even though $J_{13} \neq J_{14}$. Any values given to the coupling constants reproduce the spectrum, so long as $(J_{13} + J_{14})$ is kept equal to 1.6 Hz. The incorrect conclusion to be reached from observation of a pair of triplets is that $J_{AX} = J_{AX'}$. Numerous other varieties of deceptive simplicity can occur in AA′XX′ and AA′BB′ spectra, depending on the relative values of K, L, M, and N. In any spin system, whenever two sets of nuclei have two different coupling constants, the phenomenon is possible.

A different type of spectral simplification can arise by accidental overlap of subspectra. This phenomenon is best illustrated by the spectrum of the nonaromatic protons in benzyl-phenylphosphine oxide (**90**, Figure 4–13), which should be first order (A_2MX), since the P–H

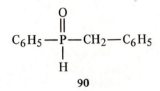

90

and P–CH$_2$ protons are separated by 250 Hz and have a coupling of only 7.5 Hz ($\Delta\nu/J = 33$). The P–H resonance is split into a large doublet by $^1J_{31P–H} = 474$ Hz, and each component should be split into a triplet by $^3J_{H–P–C–H}$. The low field subspectrum is indeed a triplet, but the high field part is a singlet. The large H–P coupling has placed this resonance on top of one of the diastereotopic benzyl proton resonances. The seeming paradox that the low field P–H subspectrum shows coupling to the benzyl protons, but the high field spectrum does not is an artifact of this coincidence. In the spectrum of *n*-butylphenylphosphine oxide, both subspectra of the P–H resonance are clean triplets, since the P–CH$_2$ resonance is shifted to

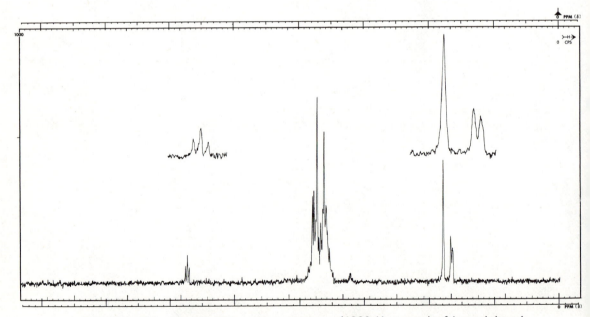

FIGURE 4-13. The 60 MHz proton spectrum (1000 Hz sweep) of benzylphenyl-phosphine oxide (90). [Reproduced with the permission of R. L. Letsinger and T. L. Emmick.]

higher field. (It is to be recalled that isochronous nuclei do not split each other, even if they are chemically nonequivalent and $J \neq 0$. Isochronous subspectra are subject to the same restriction.) Subspectral decoupling of this type is not a widespread phenomenon, but can be very perplexing when it does occur.

Second-order effects can also cause spectra to be apparently too complex. The ABX system furnishes an illustrative example. Consider the case for which A and B are closely coupled ($\Delta\nu_{AB}$ small, J_{AB} large), J_{AX} is large, but J_{BX} is zero. The X spectrum, it would seem, should be a doublet from coupling to A. Since A and B are closely coupled, the spin states of A and B are mixed, and X acts as if it were coupled to both nuclei. The influence of B on X when $J_{BX} = 0$ has been termed *virtual coupling*. This phenomenon can occur in larger spin systems as well. For example, in **91** the methine and methylene protons are closely

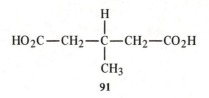

91

coupled. The methyl group is coupled only to the methinyl proton, but its resonance is much more complicated than a simple doublet because of virtual coupling to the methylene protons. In **92**, the carboxyl group shifts the methine proton downfield, so it is no longer closely

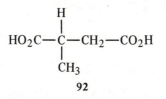

92

coupled to the methylene protons. Consequently, the methyl resonance is a clean doublet because virtual coupling is no longer possible.

The dimethylquinones provide a further example. The spectrum of the 2,5-dimethyl compound (**93**) contains a first-order methyl doublet and an alkene quartet, as expected. The spectrum of the 2,6 compound (**94**, Figure 4–14) is much more complex. Although a given

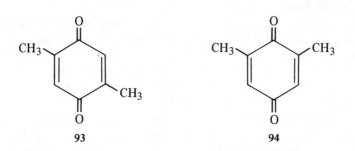

93   **94**

methyl group is coupled only to the adjacent alkenic proton, there is a W coupling between the two equivalent alkenic protons. These protons, which have zero coupling in **93**, are therefore closely coupled in **94** ($\Delta\nu_{AA'} = 0$, $J_{AA'}$ significant), and the methyl group is therefore virtually coupled to the more distant alkenic proton.

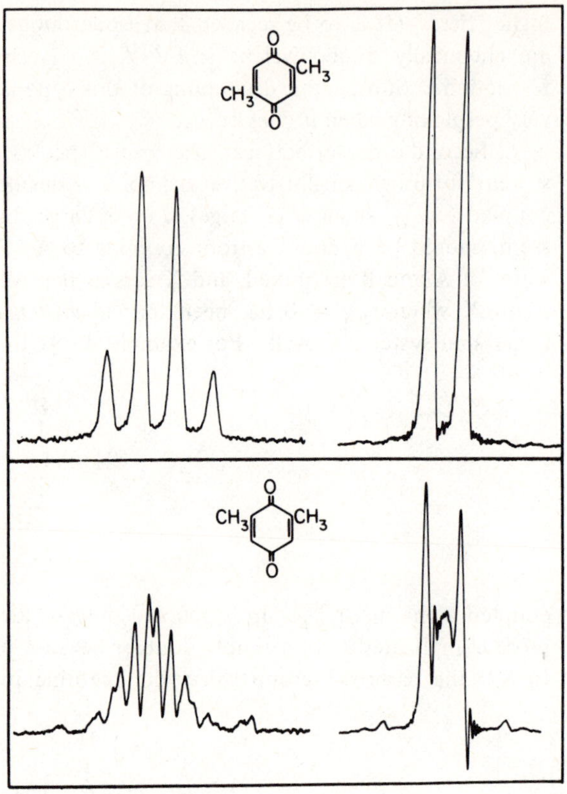

FIGURE 4-14. The 60 MHz proton spectra of the 2,5- and 2,6-dimethyl-quinones (93, 94). [Reproduced with the permission of Academic Press from E. D. Becker, High Resolution NMR, 1969.]

Although special names have been given to specific situations, all these phenomenon derive from the second-order nature of the spectra. Numerous errors in interpretation can be avoided if the widespread and subtle nature of second-order effects is kept in mind.

4-10   Aids in Spectral Analysis

4-10a   Higher Fields

Although a direct, frontal attack on many-spin systems is frequently a successful method of analysis, numerous aids have been developed to make the task easier (or sometimes make it possible). Since the chemical shift is directly proportional to the field strength, the $\Delta v/J$ ratio in a second-order spectrum can be increased by raising the field. This possibility has led to the continuing development of magnets with higher and higher fields. Routine proton spectra today are still taken at 60 MHz, but spectrometers operating at 90 or 100 MHz are usually available for chemical shift clarification. Superconducting magnets operating at 220–360 MHz are less readily available, but can give significant assistance in separating chemical shifts (Figure 4–15). Higher fields serve to decrease second-order effects due to small $\Delta v/J$ ratios, but second-order properties caused by magnetic nonequivalence, as in the AA'XX' spectrum, cannot be altered.

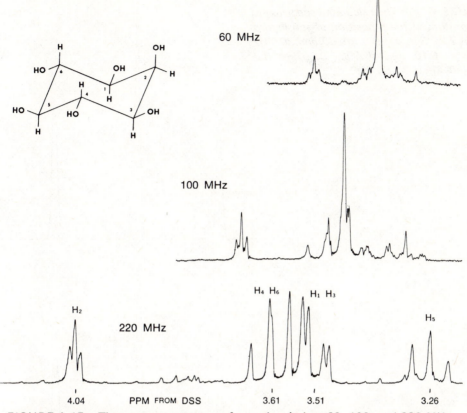

FIGURE 4-15. The proton spectrum of myo-inositol at 60, 100, and 220 MHz.
[Courtesy of Varian Associates.]

4-10b Lower Fields

It is frequently overlooked that a decrease in field strength can also assist in spectral analysis. At lower fields, spectral degeneracies that give rise to deceptive simplicity are sometimes lifted. At 90 MHz the $AA'BB'$ spectrum of tetrahydrofuran-2,2,3,3 -d_4 (**95**, Figure 4–16)

95

consists of two triplets, one broadened by deuterium coupling. Only $(J_{AB} + J_{AB'})$ can be obtained from the spectrum. At 60 MHz, enough fine structure is present to give J_{AB}, $J_{AB'}$, and $(J_{AA'} - J_{BB'})$, as discussed in section 4–9. The 30 MHz spectrum is still easier to analyze. Instruments equipped with a voltage divider can make measurements over the full range of fields. Otherwise, spectra must be taken on separate, fixed-frequency spectrometers.

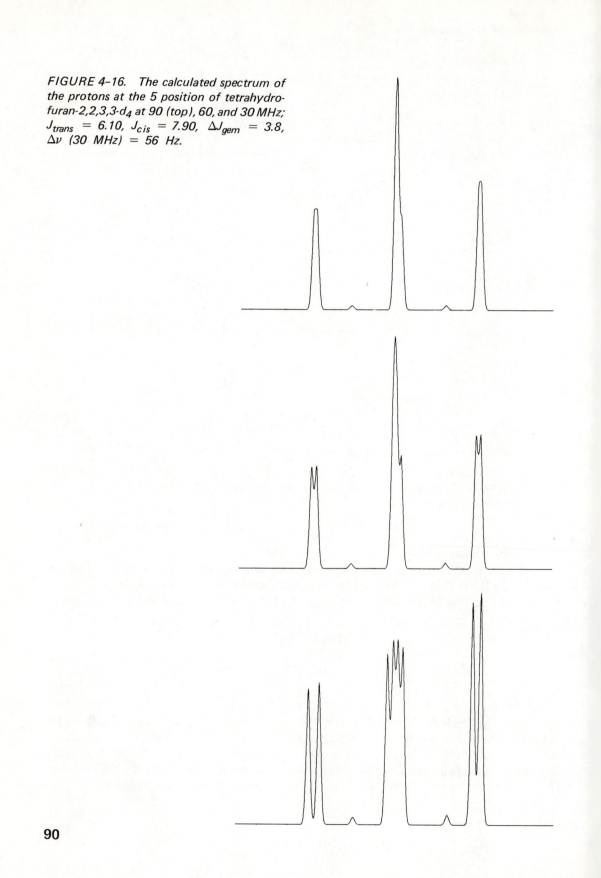

FIGURE 4-16. The calculated spectrum of the protons at the 5 position of tetrahydrofuran-2,2,3,3-d_4 at 90 (top), 60, and 30 MHz; $J_{trans} = 6.10$, $J_{cis} = 7.90$, $\Delta J_{gem} = 3.8$, $\Delta \nu$ (30 MHz) = 56 Hz.

4–10c Double Irradiation

If only certain protons in a complicated spin system are of interest, the coupling properties of the unwanted nuclei may be removed either by double proton irradiation or by deuterium substitution. Therefore, irradiation at the X frequency of an ABX system reduces the spectrum to an AB quartet. This technique is described in full in Chapter 5.

4–10d Deuterium Substitution

Alternatively, the unwanted proton can be replaced synthetically by deuterium. In **96**,

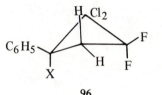

96

the vicinal H–C–C–F couplings were needed to obtain conformational information. When X was H, the resulting five-spin ABCXY spectrum was too complicated for analysis. Because the benzyl proton was too close in frequency to the methylene protons for double irradiation, the molecule was prepared with X = D. The smaller spin system (ABXY) was readily analyzed. Irradiation at the deuterium frequency was necessary to remove small couplings from the benzyl deuterium (Figure 4–17). Tetrahydrofuran has only two α,β vicinal couplings, but they are hidden in an $A_2A_2'B_2B_2'$ spin system. Synthesis of the 2,2,3,3-d_4 derivative (**95**) reduces the problem to the analysis of a four-spin $AA'BB'$ system, from which J_{AB} and $J_{AB'}$ are readily obtained (Figure 4–16).

Analysis of kinetic processes can also be facilitated by double irradiation and deuterium substitution. Analysis of the low temperature spectrum of cyclohexane (see Figure 1–14) was reduced from twelve spins to one by use of the undecadeutero derivative with deuterium irradiation. Similarly, ring reversal in tetrahydropyran (eq. 4–26) was followed by examination

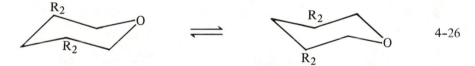

4–26

of the α proton resonance. Two different research groups solved the problem of coupling to the β protons (in order that the problem be only a four-spin system), one group by direct proton irradiation, the other by synthesis with β deuterium atoms (R = D).

4–10e Satellite Spectra

The 1.1% naturally abundant carbon-13 provides a convenient method to measure couplings between chemically equivalent protons. We have previously mentioned the method of measuring the H-D coupling to obtain such H-H couplings. If the equivalent nuclei are on different carbon atoms, the coupling information can be obtained from the ^{13}C satellite. For 1.1% of the molecules (uncorrected for statistical factors), the spin system is $H–^{12}C–^{13}C–H$. The proton on ^{12}C resonates at the same position as the molecules with no ^{13}C, although there is generally a small $H–C–^{13}C$ coupling. The large one-bond $^{13}C–H$ coupling produces multiplets on either side of the centerband and separated from it by about $\frac{1}{2}J_{^{13}C-H}$. The centerband is not precisely at the midpoint between the two sidebands because of a small ^{13}C isotope

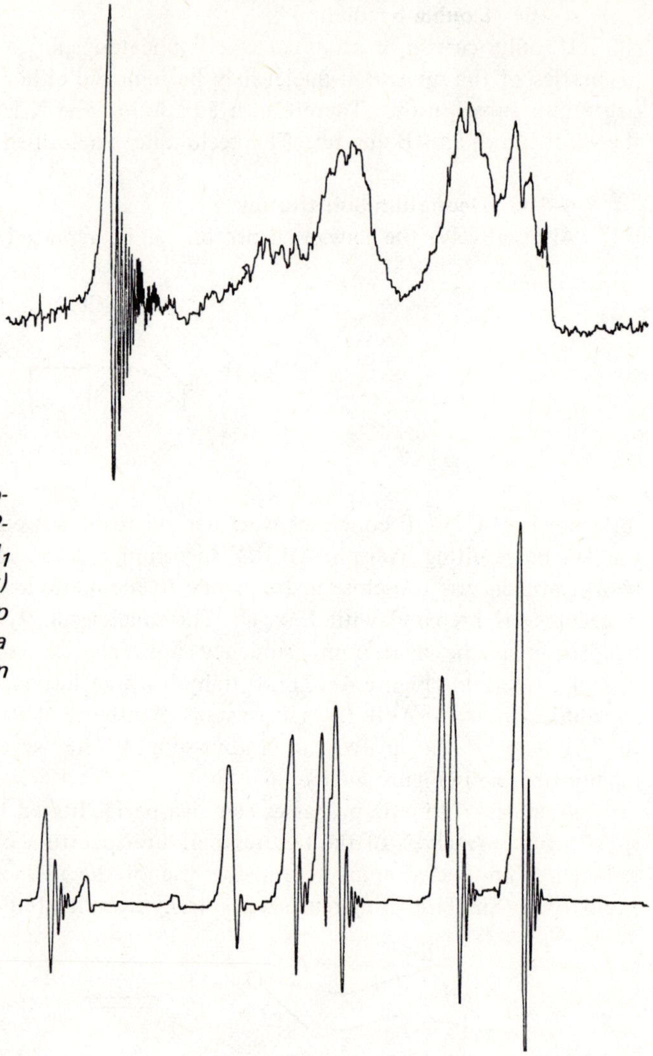

FIGURE 4-17. The 60 MHz proton spectrum of 1,1-difluoro-2,2-dichloro-3-phenylcyclobutane-3-d_1 with (lower) and without (upper) deuterium irradiation. The sharp peak at the left in each case is a sideband of TMS for calibration purposes.

effect on the chemical shift. The separation of each sideband from the centerband serves as an effective chemical shift difference, and coupling between H–^{12}C and H–^{13}C is observed in the satellite. Figure 4-18 shows the ^{13}C satellite of the alkenic protons on cyclopropene. The satellite is a doublet of triplets, since the alkenic proton on ^{13}C is coupled to the other alkenic proton and to the two methylene protons. In this experiment, $^1J_{13C-H}$ was found to be 226 Hz and $^3J_{H-C-C-H}$ to be 1.3. It is fortunate that a ^{13}C satellite experiment has two sites for observation (upfield and downfield of the centerband), since one site is frequently masked by other resonances.

Other satellite spectra can be used for analytical purposes. Nitrogen (^{15}N), silicon (^{29}Si), selenium (^{77}Se), cadmium (^{111}Cd, ^{113}Cd), tin (^{117}Sn, ^{119}Sn), tellurium (^{125}Te), platinum (^{195}Pt), and mercury (^{199}Hg) have reasonably abundant spin-$\frac{1}{2}$ isotopes that give useful satellite resonances in the proton spectrum. Analysis of the proton spectrum of diethyl-mercury furnishes such an example. Mercury-199 has an abundance of 16.8%; the remaining

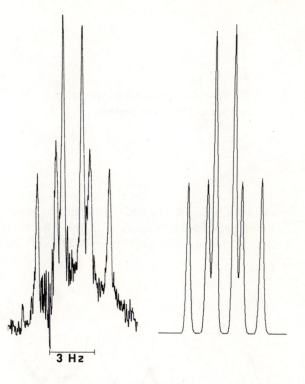

FIGURE 4-18. The 90 MHz spectrum of cyclopropene, showing the downfield ^{13}C satellite of the alkenic protons; observed (left) and calculated (right). [Reproduced with the permission of the American Chemical Society from J. B. Lambert, A. P. Jovanovich, and W. L. Oliver, Jr., J. Phys. Chem., 74, 2221 (1970).]

3 Hz

abundant isotopes either have no spin or exhibit no coupling because of quadrupolar relaxation (Chapter 7). The centerband methylene and methyl resonances are so close together that they form a complicated A_2B_3 pattern. The ^{199}Hg couplings to the methyl and methylene protons, however, are unequal, so that the ^{199}Hg satellite of the methyl resonance is well separated from the methylene satellite. The larger effective chemical shift difference makes the satellite spectra almost first order, and from them, $^3J_{H-C-C-H}$, $^2J_{^{199}Hg-C-H}$, and $^3J_{^{199}Hg-C-C-H}$ can be measured. The midpoint between the ^{199}Hg methylene satellites gives ν for the centerband resonance, and similarly for the methyl resonance. With both $J_{H-C-C-H}$ and $\Delta\nu$, the spectrum can be analyzed completely.

4-10f Subspectral Analysis

The methods of subspectral analysis can sometimes be used to determine spectral parameters very readily from complicated spin systems. The AB_3 case, for example, contains up to sixteen lines (6 A-type, 8 B-type, and 2 combination) and is determined entirely by two parameters, J_{AB} and $\Delta\nu_{AB}$. Among the sixteen lines (usually only fourteen are observed) are four that form an "ab" subspectrum. These lines, properly assigned, give J_{AB} and $\Delta\nu_{AB}$ by use of eq. 4-20. Assignment requires identification of the proper spacings and relative intensities (eq. 4-21). Subspectral analysis can be particularly useful in the analysis of more complicated spin systems, e.g., AB_2X, $AA'A''XX'X''$, AB_2X_2, AB_4, A_2B_3, $AA'X_3X_3'$, $AA'XX'M$, etc.

4-10g Multiple-Quantum Transitions

When very high rf power levels are used, spin systems can sometimes absorb two quanta at the same frequency. The double-quantum transition (DQT) for a two-spin system is illustrated in Figure 4-19. The frequency of ν_{DQ} is at the midpoint of the AB quartet. As the

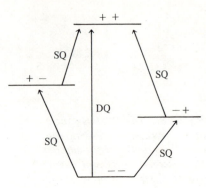

FIGURE 4-19. The double-quantum
transition in a two-spin system.

power level is raised, the single-quantum transitions become saturated, and the sharp double-quantum peak begins to appear.   In the AB part of the ABX spectra, DQT's appear at the centers of the two quartets (Figure 4–20).   Frequently the eight peaks are difficult to sort

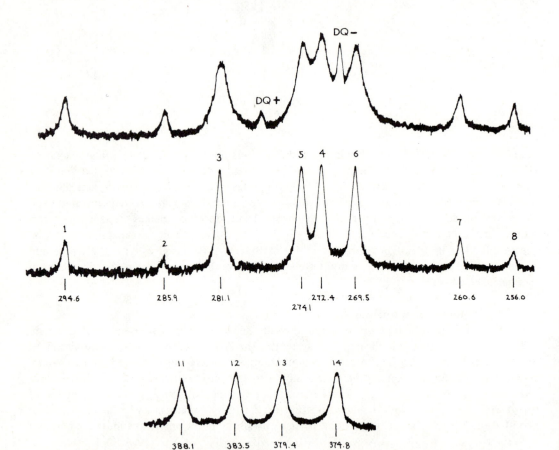

FIGURE 4-20. The 60 MHz proton spectrum of 1-acetoxy-2-nitro-1-phenylethane in CCl_4 with phenyl and acetoxy resonances omitted; X part (bottom) and AB part at high rf power to show the double-quantum transitions. [Reproduced with the permission of the American Chemical Society from E. W. Garbisch, J. Chem. Ed., **45**, 402 (1968).]

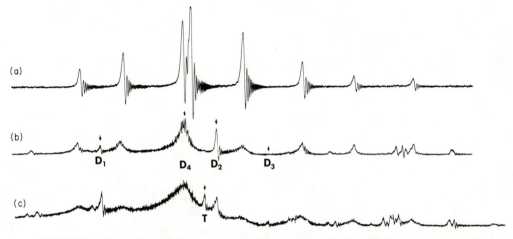

(a)

(b)

D_1 D_4 D_2 D_3

(c)

T

FIGURE 4-21.   The 60 MHz proton spectrum of 1,2,3-trichlorobenzene; (a) single-quantum spectrum; (b) double-quantum spectrum; (c) triple-quantum spectrum. The extra lines are spinning sidebands.   [Reproduced with the permission of Academic Press from P. L. Corio, Structure of High-Resolution NMR Spectra, 1967.]

unambiguously into the two correct quartets. Observation of the DQT's provides an automatic assignment.   In an AB_2 spectrum, DQT's occur at $\frac{1}{2}(\nu_7 + \nu_8)$, $\frac{1}{2}(\nu_2 + \nu_8)$, $\frac{1}{2}(\nu_2 + \nu_6)$, and $\frac{1}{2}(\nu_5 + \nu_6)$, and a triple-quantum transition occurs at $\frac{1}{2}(\nu_A + 2\nu_B)$ (Figure 4-21).   These peaks may be used to confirm peak assignments, and to eliminate extraneous peaks. DQT's may also be used to determine the relative signs of the three coupling constants in ABC spectra.

4-11   Liquid Crystal Solvents: Direct Dipole-Dipole Coupling

Molecules such as p,p'-di-n-hexyloxyazoxybenzene (**97**) exist as nematic liquid crystals in

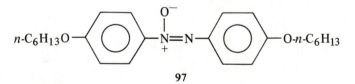

97

certain temperature ranges.   The properties of this mesophase are intermediate between those of solids and liquids.   Large numbers of these rod-like molecules align themselves along their long axis to form a *domain*. A domain can be oriented by a magnetic field so that the long molecular axis is parallel to B_0. Small organic molecules dissolved in a liquid crystal will be aligned also.   The ordering is not so rigid as in the solid phase.   For example, a solute benzene ring is aligned with its plane parallel to the B_0 field, but it freely rotates around the six-fold axis perpendicular to this plane.

Because the solute molecules maintain some degrees of freedom, the nmr spectrum is not the broad, almost featureless mass usually given by solids.   The peaks are relatively sharp,

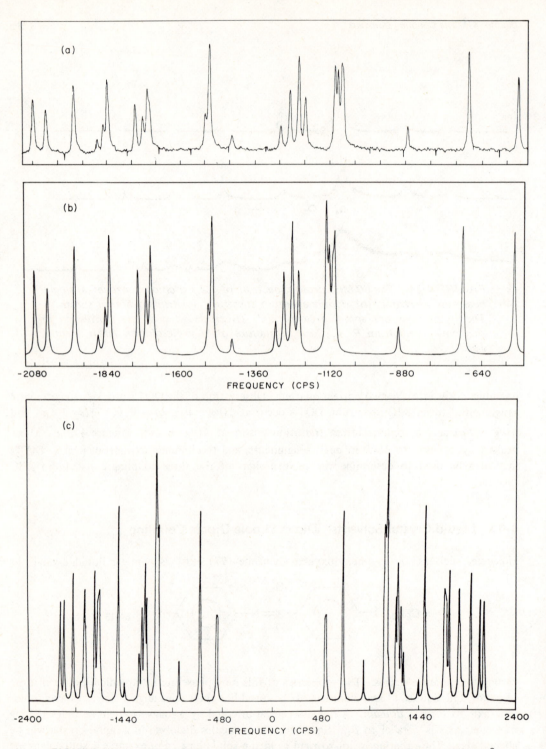

FIGURE 4-22. The 60 MHz proton spectrum of 1,3,5-trioxane (**98**) in **97** at 80°
accumulated over 87 passes; only half the spectrum is shown, the other half being a
mirror image: (a) observed half-spectrum; (b) calculated half-spectrum; (c) calculated full spectrum. [Reproduced with permission from M. Cocivera, J. Chem.
Phys., **47**, 3061 (1967).]

but cover a much larger range of frequencies than do ordinary liquid phase (isotropic) spectra (Figure 4-22). Because of two important distinctions, liquid crystal spectra produce data that are unavailable from other sources. (1) Molecules are no longer tumbling freely, so the spectrum is determined by the direct dipole-dipole couplings (D_{ij}) as well as by the indirect couplings (J_{ij}). Not only do liquid crystal spectra provide a new approach for obtaining D_{ij}, but analysis of the signs of the various couplings gives the absolute signs of J_{ij}. The sign of D_{ij} is known from the mode of orientation, and the spectrum determines the relative signs of D_{ij} and J_{ij}. (2) Coupling constants between chemically equivalent nuclei are observed. Thus an isolated methyl group, as in acetonitrile, gives a triplet resonance, since a given proton exhibits coupling to the other two. Analysis of the fluorine spectrum of hexafluorobenzene dissolved in **97** gave the following parameters: $D_{ortho} = -1452.7$ Hz, $D_{meta} = -271.6$, $D_{para} = -194.2$, $J_{ortho} = -22$, $J_{meta} = -4$, $J_{para} = +6$. Although the signs of the J_{ij} are reliably determined, their magnitudes are accurate only to ± 2 Hz; the D_{ij} magnitudes are quite accurate.

The direct dipolar coupling between nuclei i and j is related to structural parameters by eq. 4-27, in which r_{ij} is the internuclear distance, θ_{ij} is the angle between the field direction (z)

$$D_{ij}^{dir} = (2\pi)^{-1} \gamma_i \gamma_j \hbar \langle (1 - 3\cos^2\theta_{ij})r_{ij}^{-3} \rangle_{av.} \qquad 4\text{-}27$$

and a vector along r_{ij}, and the average is taken over all molecular motions. Because of the dependence of D_{ij} on actual structural parameters, analysis of nmr spectra in liquid crystal solvents can be used to deduce complete structures in terms of bond lengths and valence angles. Unfortunately, the dipolar coupling D_{ij} is not changed by a linear expansion or contraction of the molecular geometry. Therefore, one parameter must be fixed and the others related to it. Furthermore, the molecules must be relatively small and possess certain symmetry elements. Greater molecular complexity has recently been achieved by use of deuterium substitution to simplify the spectrum. The structures of cyclopropane, cyclobutane, 1,3,5-trioxane (**98**), and bicyclobutane (**99**), as well as several smaller molecules, have been determined in this manner.

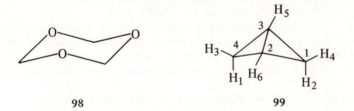

98   99

The nonplanarity of cyclobutane to the extent of about $35°$ is clearly demonstrated by the spectral parameters. Table 4-1 compares the nmr-derived structure of **99** with those obtained from electron diffraction and microwave data. The nmr method gives comparable bond lengths, but there are some differences among the valence angles. Comparable values of the H—C—H angle in simple methane derivatives have been determined by liquid crystal nmr methods and by microwave spectroscopy: CH_3CN ($109°2' \pm 2'$ by nmr; $109°16'$ by microwave); CH_3OH ($110°3' \pm 8'$; $109°2' \pm 45'$); CH_3I ($111°42' \pm 2'$; $111°25'$).

Table 4-1. *The Structure of Bicyclobutane* (**99**) *Determined by Various Techniques*

	NMR	ED	MW
Bond lengths (Å)			
$C_1-C_2{}^a$	$(1.507)^b$	1.507	1.498
C_2-C_3	1.507	1.502	1.497
C_1-H_2	1.167	~1.106	1.093
C_1-H_4	1.194	~1.106	1.093
C_2-H_6	1.142	1.108	1.071
Angles (deg.)			
$H_2-C_1-H_4$	110.2	116.0	115°34′
$H_6-C_2-C_3$	128.0	125.5	~122
$C_2C_3C_4$ plane/$C_1C_2C_3$ plane	120.2	122.8	122°40′
$H_2-C_1-C_1C_2C_3$ plane	126.3	~122	121°34′

^{a}See structure 99 for numbering system.
^{b}The nmr parameters are related to this electron diffraction bond length.

BIBLIOGRAPHY

1. Magnetic Equivalence: K. Mislow and M. Raban, *Top. Stereochem.*, **1**, 1 (1966).
2. Directly Bonded Couplings: W. McFarlane, *Quart. Rev.* (London), **23**, 187 (1969); C. J. Jameson and H. S. Gutowsky, *J. Chem. Phys.*, **51**, 2790 (1969).
3. Geminal, Vicinal, and Long-Range Couplings: S. Sternhell, *Quart. Rev.* (London), **23**, 236 (1969); A. A. Bothner-By, *Advan. Magn. Resonance*, **1**, 195 (1965); V. Bystrov, *Russ. Chem. Rev.*, **41**, 281 (1972).
4. Long-Range Couplings: M. Barfield and B. Charkrabarti, *Chem. Rev.*, **69**, 757 (1969); S. Sternhell, *Rev. Pure Appl. Chem.*, **14**, 15 (1964).
5. Spectral Analysis: J. D. Roberts, *An Introduction to the Analysis of Spin-Spin Splitting in High-Resolution Nuclear Magnetic Resonance Spectra*, W. A. Benjamin, New York, 1961; K. B. Wiberg and B. J. Nist, *The Interpretation of NMR Spectra*, W. A. Benjamin, New York, 1962; R. J. Abraham, *The Analysis of High Resolution NMR Spectra*, Elsevier, Amsterdam, 1971.
6. Computer Programs: C. W. Haigh, *Ann. Reports NMR Spectrosc.*, **4**, 311 (1971); P. Diehl, H. Kellerhals, and E. Lustig, *NMR Basic Prin. Progr.*, **6**, 1 (1972).
7. Subspectral Analysis: P. Diehl, R. K. Harris, and R. G. Jones, *Progr. NMR Spectrosc.*, **3**, 1 (1967).
8. Double-Quantum Transitions: E. W. Garbisch, Jr., *J. Chem. Ed.*, **45**, 311, 402, 480 (1968).
9. Carbon-13 Satellites: J. H. Goldstein, V. S. Watts, and L. S. Rattet, *Progr. NMR Spectrosc.*, **8**, Part 2, 103 (1973).
10. Liquid Crystals: A. D. Buckingham and K. A. McLauchlan, *Progr. NMR Spectrosc.*, **2**, 63 (1967); S. Meiboom and L. C. Snyder, *Science*, **162**, 1337 (1968); G. R. Luckhurst, *Quart. Rev.* (London), **22**, 179 (1968); P. Diehl and W. Niederberger, *Nucl. Magn. Resonance (Spec. Period. Report)*, **3**, 368 (1974).

5

MULTIPLE RESONANCE

5-1 Types of Multiple Resonance

As implied by its name, the technique of multiple resonance utilizes additional rf frequencies to bring about special effects in the nmr spectrum. Consider the 90 MHz proton spectrum of $(C_6H_5)_2PCH_3$ (Figure 5-1). Under ordinary spectral conditions, the methyl resonance is a first-order doublet from coupling to the ^{31}P nucleus. If the ^{31}P resonance frequency (32.1 MHz) is subjected to a second strong rf field ($\nu_2 = \gamma B_2$), the ^{31}P nuclei will rapidly absorb and emit rf energy. The ^{31}P nuclei are shuttled back and forth between the $-\frac{1}{2}$ and $+\frac{1}{2}$ spin states so rapidly that the methyl protons no longer distinguish distinct energy levels. When *decoupled* from ^{31}P in this manner, the proton acts as if ^{31}P were a nonmagnetic nucleus (Figure 5-1).

When the observed nucleus (^1H in the above example) and the irradiated nucleus (^{31}P) are of different types, the experiment is termed *heteronuclear double resonance*, represented by the notation $^1H\{^{31}P\}$. For the general experiment $X\{Y\}$, X represents the observed nucleus and Y, in braces, the irradiated nucleus. *Homonuclear double resonance* involves observation and irradiation of the same type of nucleus. Thus the quartet structure of the methylene resonance in ethylbenzene may be reduced to a singlet by irradiation at the frequency of the methyl resonance. The methylene protons are then said to be decoupled from the methyl protons. Triple and quadruple resonance experiments may be carried out with the use of additional frequencies ($\nu_3 = \gamma B_3$ and $\nu_4 = \gamma B_4$).

Although the terms *double resonance* and *spin decoupling* are sometimes used interchangeably, their meanings are quite distinct. Double resonance is the broader term, denoting any experiment that utilizes a second irradiating field. Spin decoupling only refers to those cases in which double irradiation results in spectral simplification. In section 5-3 we will discuss some instances in which irradiation produces a more complicated spectrum,

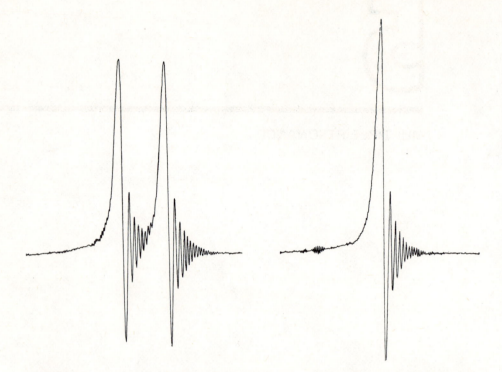

FIGURE 5-1. *The proton spectrum (methyl resonance) of* $(C_6H_5)_2PCH_3$ *at 90 MHz, with (right) and without (left) simultaneous irradiation at the* ^{31}P *resonance frequency (32.1 MHz).*

and in Chapter 7 double resonance experiments are described that only produce intensity changes.

Double resonance experiments may be carried out under either frequency or field sweep conditions. Unlike single resonance operations, the spectra that result from the two methods are not equivalent. Let us consider application of the methods to a homonuclear double resonance experiment. In the frequency sweep mode, the irradiating frequency ν_2 is set at the resonance position of the proton to be decoupled. Then ν_1 is swept in the usual manner to produce a spectrum in which every proton is decoupled from the proton resonating at ν_2. As ν_1 approaches ν_2, the two frequencies produce an interference pattern. Figure 5-2 shows the normal and double resonance spectra of mannosan triacetate (**1**). As the H_5

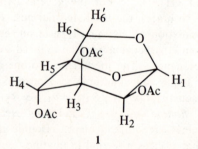

1

resonance is irradiated, spectral simplification is observed in both the H_4 and the H_6

resonances. In addition to the spectral alterations due to removal of couplings, changes in the resonance positions of the decoupled nuclei may be seen in Figure 5-2. This phenomena, called the Bloch–Siegert shift, must be taken into account when carrying out and interpreting double resonance experiments.

In the field sweep mode, homonuclear decoupling experiments are more complicated. If ν_1 and ν_2 both remain constant as the field is swept, decoupling is observed only when $(\nu_1 - \nu_2)$ happens to correspond to the difference between two resonance frequencies. For example, in **1**, if it is desired to decouple H_4 from H_5, $(\nu_1 - \nu_2)$ is set to correspond to about 25 Hz (Figure 5-2). When the spectrum is recorded, the H_4 resonance is decoupled from H_5, but H_6 and H_6' are not. Any other protons that might be separated by 25 Hz would also be decoupled from each other. To decouple H_5 from H_6 and H_6', separate experiments would have to be run with $(\nu_1 - \nu_2)$ set at about 40 and 75 Hz, respectively. Because of the greater amount of information per spectrum and the ease of execution, the frequency sweep technique is much preferred to the field sweep. Most instruments produced today use the frequency sweep method, so that field sweep techniques are becoming obsolete.

Double resonance experiments may also be classified according to the intensity or bandwidth of the irradiating field. A very wide range of frequencies may be irradiated by using either white noise or a broadband oscillator. This technique is commonly used in heteronuclear experiments to remove all the coupling properties of a given type of nucleus. Natural abundance ^{13}C spectra (see Figure 1-4) are *noise-decoupled* from protons as a matter of course (see Section 5-2a).

If the bandwidth of irradiation covers only the peaks of a single resonance, then coupling to this nucleus disappears completely in the resonances of other nuclei. If the irradiation bandwidth covers only part of the resonance, incomplete decoupling is observed elsewhere in the spectrum. This technique, known as *selective spin decoupling*, can give important information about the relative signs of coupling constants. Finally, the bandwidth can be so narrow or the intensity so low that only a single transition is affected. Additional

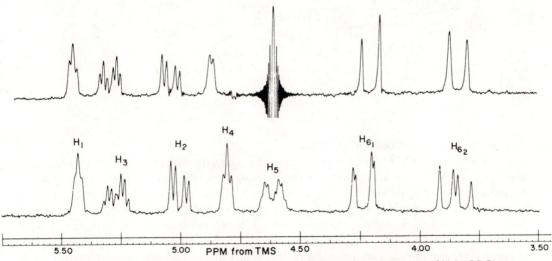

*FIGURE 5-2. The 100 MHz proton spectrum of mannosan triacetate (1) in CDCl₃;
without decoupling (lower) and with frequency sweep double irradiation at δ 4.62
(upper). [Reproduced with the permission of Varian Associates.]*

splittings can actually be observed in resonances of coupled nuclei elsewhere in the spectrum. This *spin tickling* method also gives relative-sign information (section 5-3).

The double resonance experiment can be carried out in an entirely different fashion by sweeping ν_2, the decoupling frequency, rather than ν_1 or B_1, the observing frequency and field. This *internuclear double resonance* technique (indor) is enjoying increased use today as more machines are equipped to sweep ν_2. The indor experiment is basically the reverse of the normal frequency sweep double resonance experiment. The observing frequency (ν_1) is set on the resonance of the nucleus of interest, and changes in the intensity of this peak are monitored. As ν_2 is swept, the pen responds with a peak each time ν_2 passes through a resonance that is coupled to the nucleus at ν_1. This method, discussed in section 5-4, is used for a wide range of purposes, from sign determination to observation of hidden resonances.

5-2 Applications of Spin Decoupling

5-2a Signal Enhancement

Most ^{13}C nuclei in organic molecules are bonded to protons, except in certain functionalities such as nitriles or carbonyls. Because of the large value of the one-bond ^{13}C–H coupling constant, the ^{13}C resonances can be split into a wide-ranging multiplet. Noise decoupling at the ^{1}H frequency, therefore, produces a considerable enhancement in sensitivity by removing all proton couplings. Figure 1-4 shows the ^{13}C spectrum of sucrose (2), with ^{1}H noise decoupling. In addition to the simple collapse of peaks from spin decoupling, enhancement is obtained from the nuclear Overhauser effect (Chapter 7). These two factors derived from noise decoupling have been instrumental in the recent expansion of ^{13}C nmr spectroscopy. This technique may also be used with other nuclei to pinpoint chemical shifts by removal of all ^{1}H couplings.

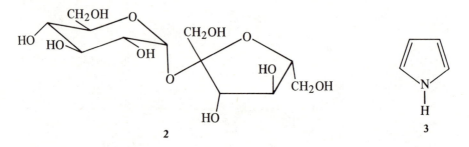

5-2b Removal of Quadrupolar Effects

Quadrupolar nuclei ($I > \frac{1}{2}$) have a special mechanism for relaxation discussed in Chapter 7. Relaxation leads to transitions between the various I_z levels ($-1 \rightleftharpoons 0 \rightleftharpoons +1$ for $I = 1$). At the extreme of slow relaxation, a proton on ^{14}N, for example, is split into a triplet. At fast relaxation, the ^{14}N spin states are averaged by rapid interconversion, and the proton resonance appears as a singlet. Unfortunately, ^{14}N relaxation is frequently between these two extremes. As a result, a proton on or near a ^{14}N nucleus will give a resonance that is broadened by incomplete averaging of the ^{14}N spin states. The spectrum of pyrrole (3) is representative of this effect (Figure 5-3). The proton on nitrogen is broadened almost to the point of invisibility, and the α protons are considerably broadened as well. If the sample is irradiated at the ^{14}N frequency, the effects of incomplete relaxation are nullified by rapid

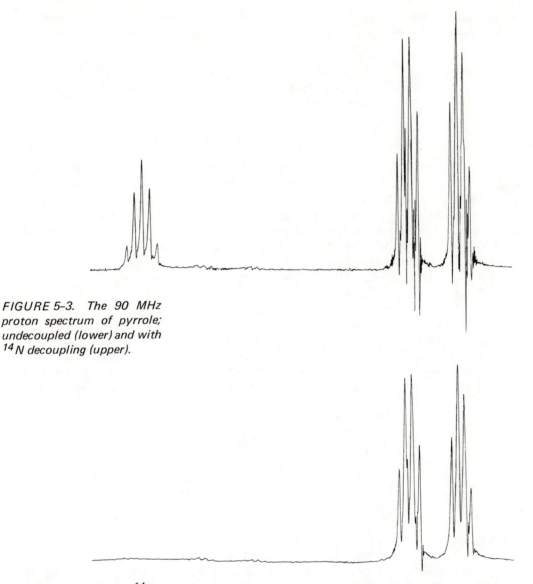

FIGURE 5-3. The 90 MHz proton spectrum of pyrrole; undecoupled (lower) and with ^{14}N decoupling (upper).

interconversion of the ^{14}N spin states. A given proton no longer distinguishes individual ^{14}N spin states, so the spectrum appears as if ^{14}N were a nonmagnetic nucleus. The $^1H\{^{14}N\}$ spectrum of pyrrole, illustrated in Figure 5-3, shows the sharp resonances of the protons close to nitrogen.

5-2c Location of Hidden Resonances

Frequently, proton resonances of interest lie in a region containing an envelope of resonances that mask the properties of any individual nucleus. If the hidden proton of interest (A) is coupled to a proton (X) whose resonance is well separated from the others, some information about A can be salvaged by decoupling techniques. The decoupling frequency (ν_2) is varied through the envelope of resonances. When it reaches the resonance frequency of proton A, the resonance of X loses the A–X coupling. Thus ν_A can be measured by monitoring the

X resonance. (This experiment may also be carried out by the indor method described in section 5-4.)   The spectrum of quinolizidine (**4**) provides an example of this technique.

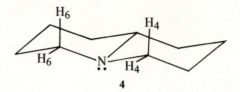

4

As noted in Chapter 3 (section 3–3), the axial–equatorial chemical shift difference in cyclohexanes is about 0.5 ppm. When the methylene group is adjacent to a nitrogen atom with an axial lone pair, however, this quantity increases to about 1.0 ppm. This phenomenon was first established in the rigid *trans*-quinolizidine (**4**). The resonances of the equatorial 4 and 6 protons at δ 2.8 are well separated from those of the remaining protons in the molecule. To determine the location of the hidden axial 4 and 6 protons, the decoupling frequency was swept through the δ 1.0–2.0 envelope. Collapse of the geminal coupling in the δ 2.8 resonance signaled the position of the axial resonances.   By this means, $\Delta\nu_{ae}$ was established to be 0.93 ppm in **4**.

5–2d Spectral Simplification for Analysis

Spin decoupling as a means for simplifying spectra to be analyzed was alluded to in Chapter 4. The complexities of a large spin system can sometimes be reduced by removing the coupling properties of one or more member nuclei. Irradiation of deuterium in molecules **5** and **6** (see Chapter 4) served to remove H–D coupling and to leave an easily analyzable

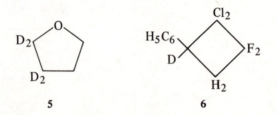

5   **6**

four-spin system. Decoupling is very useful in studying rate processes (see Chapter 6). Analysis of ring reversal in tetrahydropyran (**7**) and of nitrogen inversion in *N*-chlorohomopiperidine (**8**) was made possible by irradiation of the β protons in each case. The α protons then form

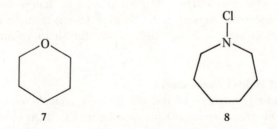

7   **8**

identical two-spin systems $[(A_2)_2]$, which give a singlet when the rate process is fast and an AB spectrum when it is slow. Spectral analysis through double irradiation can sometimes give an immediate differentiation between structural possibilities.   For example, the stereochemistry of the exocyclic double bond in **9** can be determined from the magnitude

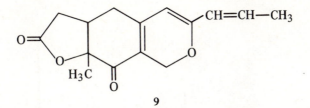

9

of $^3J_{\text{vic}}$. This quantity unfortunately is hidden in the complicated ABX$_3$ spectrum containing the methyl spins. Irradiation at the methyl frequency reduces the spectrum to an AB quartet with $|J_{AB}| = 15$ Hz (trans stereochemistry).

5-2e Structure Elucidation

The structures of large molecules can sometimes be mapped by a sequence of double resonance experiments. Coupled protons are indicated by the collapse of multiplets. The structure of the alkaloid phyllochrysin (**10**) was confirmed by a series of five decoupling

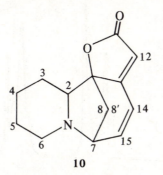

10

experiments (Figure 5-4). The exact chemical shift of only H-12 (δ 5.74) is available directly from the spectrum. The protons next to nitrogen (H-2, H-6, and H-7) are located by their chemical shift behavior in CF_3CO_2H. Irradiation of H$_7$ simplifies the H-14/H-15 resonance to an AB quartet ($J_{14\text{-}15} = 9.3$ Hz, $\Delta\nu_{14\text{-}15} = 0.18$ ppm). The same double irradiation experiment collapses the H-8 quartet to a doublet, so that $J_{8\text{-}8'} = 9.2$ Hz and $\delta(\text{H-8}) = 2.64$ ppm are determined. Irradiation of H-8, in addition to decoupling H-8$'$, transforms the H-7 resonance to a quartet, from which are obtained $\delta(\text{H-7}) = 3.90$ ppm, $J_{7\text{-}14} = 1.1$ Hz, and $J_{7\text{-}15} = 4.5$ Hz. Irradiation with a bandwidth that is broad enough to decouple both H-14 and H-15 reduces the H-7 resonance to a doublet ($J_{7\text{-}8} = 4.5$ Hz). The chemical shifts of H$_2$ (δ 3.67) and H$_6$/H$_6'$ (δ 2.88) are determined by the respective irradiation of H$_3$/H$_3'$ (δ 1.32) and H$_5$/H$_5'$ (δ 1.80). The decoupling experiments, therefore, have been able to locate each chemical shift. In addition, all the coupling constants except those in the piperidine ring are determined.

5-3 Relative Signs of Coupling Constants

In addition to the applications discussed in the previous section, double resonance experiments with narrowband irradiation can provide information about the relative signs of coupling

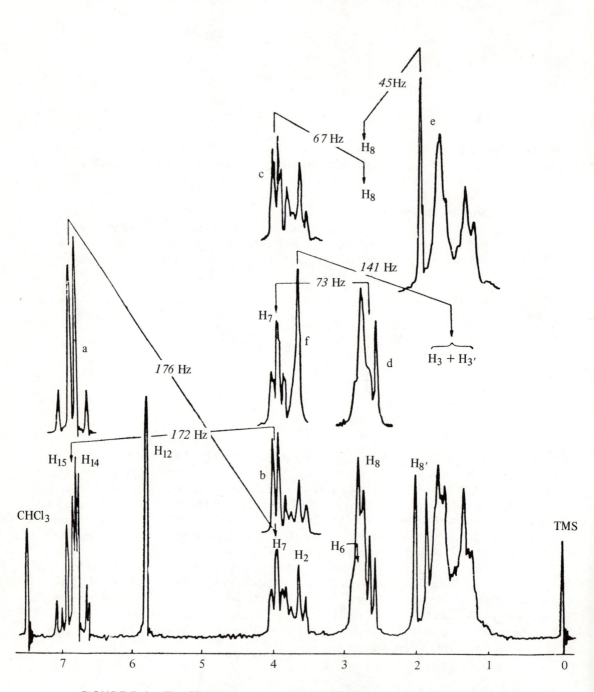

FIGURE 5-4. The 60 MHz proton spectrum of phyllochrysin (**10**) in CDCl₃ with various decoupling experiments. [Reproduced with the permission of Masson & Cie, Paris, from J. Parello, A. Melera, and R. Goutarel, Bull. Soc. Chim. France, 898 (1963).]

constants. In this section we will discuss the methods of selective spin decoupling and of spin tickling.

Selective spin decoupling is best approached by examination of its application to the AMX system. Each nucleus gives a quartet resonance, and the individual peak positions are given by eq. 5-1. A diagrammatic representation is given in Figure 5-5. Thus the frequency of

$$\text{A resonance:} \quad \nu(A_i) = \nu_A + I_z J_{AM} + I_z' J_{AX}$$

$$\text{M resonance:} \quad \nu(M_i) = \nu_M + I_z'' J_{AM} + I_z' J_{MX} \qquad (5\text{-}1)$$

$$\text{X resonance:} \quad \nu(X_i) = \nu_X + I_z J_{MX} + I_z'' J_{AX}$$

$A(I_z = I_z' = +\frac{1}{2})$ is $(\nu_A + \frac{1}{2}J_{AM} + \frac{1}{2}J_{AX})$. Since a positive frequency increment carries the peak to lower field, provided that both J_{AM} and J_{AX} are positive, then $A(+\frac{1}{2},+\frac{1}{2})$ falls at the low field end of the quartet (A_1). Similarly, $A(-\frac{1}{2},-\frac{1}{2})$ is at the high field end (A_4); $A(+\frac{1}{2},-\frac{1}{2})$ is next to low field (A_2) $(J_{AM}$ is arbitrarily taken to be larger than $J_{AX})$; and $A(-\frac{1}{2},+\frac{1}{2})$ is next to high field (A_3). If J_{AX} were larger than J_{AM}, $A(+\frac{1}{2},-\frac{1}{2})$ would be A_3 and $A(-\frac{1}{2},+\frac{1}{2})$ A_2. Similar analysis leads to the labeling of the M and X resonances given in Figure 5-5, assuming that all three coupling constants have a positive sign and that $|J_{AM}| > |J_{MX}| > |J_{AX}|$. If J_{MX} had the same magnitude but a negative sign, the assignment of certain peaks changes. Thus $X(+\frac{1}{2},+\frac{1}{2})$ would be X_3, since I_z is multiplied by a

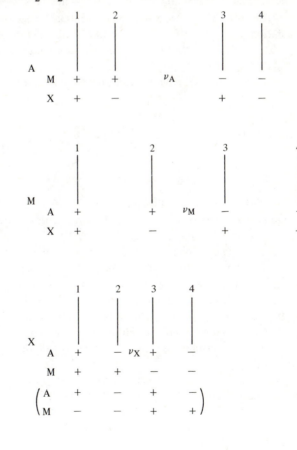

FIGURE 5-5. The AMX spectrum with peaks labeled according to the values of I_z, assuming that all coupling constants have a positive sign and that $|J_{AM}| > |J_{MX}| > |J_{AX}|$.

negative number in the expression for $v(X_i)$ in eq. 5-1. The complete assignment for the X spectrum in which J_{MX} is negative but J_{AM} and J_{AX} are positive, is given at the bottom of Figure 5-5 in parentheses. Changes also occur in the M assignment, but these are not given in the figure.

If the decoupling frequency is centered between peaks A_1 and A_2, only those molecules are irradiated that have $I_z = +\frac{1}{2}$ for the M nucleus. The decoupling power is kept sufficiently low that no irradiation occurs at A_3 and A_4, for which I_z of the M nuclei is $-\frac{1}{2}$. Thus only half the molecules in the sample are actually being irradiated, and only those X nuclei for which $I_z(M) = +\frac{1}{2}$ are decoupled from the A nuclei. When J_{AM} and J_{MX} are both positive, these X nuclei give rise to peaks X_1 and X_2. Irradiation of the A nuclei ($M = +\frac{1}{2}$) causes rapid equilibration of the $+\frac{1}{2}$ and $-\frac{1}{2}$ spin states of A, so that peaks X_1 and X_2 collapse to a singlet at their midpoint. On the other hand when J_{AM} and J_{MX} have opposite signs, the X nuclei for which $M = +\frac{1}{2}$ give rise to X_3 and X_4. Opposite relative signs therefore bring about the collapse of the X_3 and X_4 peaks on irradiation at A_1 and A_2. The selective spin decoupling experiment, therefore, can determine the relative signs of J_{AM} and J_{MX} by irradiation of the A nuclei and examination of the X resonance. Similarly, the relative signs of J_{AM} and J_{AX} could be determined by irradiation of a pair of X peaks and observation of collapse of a pair of M peaks. A third experiment (irradiation of M, observation of A) would confirm the relative signs of J_{AX} and J_{MX}. In this way the complete set of relative signs for all three coupling constants is obtained.

In the general selective decoupling experiment, three types of nuclei are needed to form an $A_iM_jX_k$ system. Irradiation of the A subspectrum for which $I_z(M) = +\frac{1}{2}$ and observation of the X resonance gives the relative signs of the couplings of A and X with M. Irradiation of a downfield subspectrum and collapse of a downfield subspectrum indicate that the couplings to the third nucleus have the same sign; collapse of the upfield subspectrum would have indicated opposite signs.

Figure 5-6 gives the proton spectrum of phosphacyclohexane methiodide (**11**). The

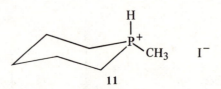

11

H–^{31}P–CH$_3$ portion of the molecule furnishes an AMX$_3$ spectrum suitable for selective decoupling experiments. The methyl resonance appears as a quartet near δ 1.4 from coupling to the ^{1}H and ^{31}P nuclei. The ^{31}P spectrum is neither observed nor irradiated, so only the one experiment that gives the relative signs of $J_{AM} = {}^1J_{H-^{31}P}$ and $J_{MX} = {}^2J_{H-C-^{31}P}$ can be carried out. The A subspectra are clearly observable at δ 3.5 and 9.1, split by the large $^1J = 498$ Hz. Irradiation of the low field A subspectrum (δ 9.1) collapses the high field X (CH$_3$) doublet. Irradiation of the high field A subspectrum (δ 3.5) collapses the low field CH$_3$ doublet. Therefore $^1J_{^{31}P-H}$ and $^2J_{^{31}P-C-H}$ must be opposite in sign. Another experiment, involving either irradiation or observation of the ^{31}P spectrum, would be required to obtain the relative signs of these two couplings with respect to $J_{AX} = {}^3J_{H-^{31}P-C-H}$.

Before examining the use of spin tickling to determine relative signs, we shall explain the general nature of the technique by its application to the AB spectrum. The normal

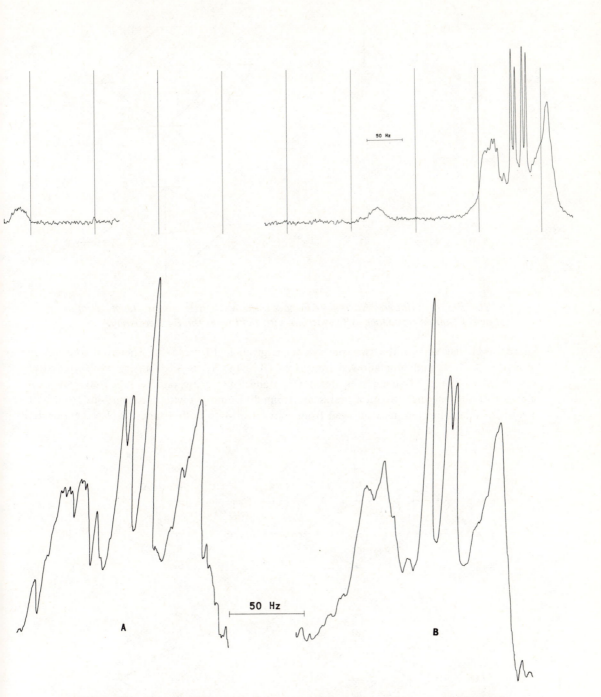

FIGURE 5-6. The 90 MHz proton spectrum of phosphacyclohexane methiodide in CHCl₃; complete spectrum undecoupled (upper) and the region δ 0.8–2.0 with selective spin decoupling (lower). In A (left), irradiation occurs at δ 9.1; in B (right), at δ 3.5. [Reproduced with the permission of Pergamon Press from J. B. Lambert and W. L. Oliver, Jr., Tetrahedron, 27, 4245 (1971).]

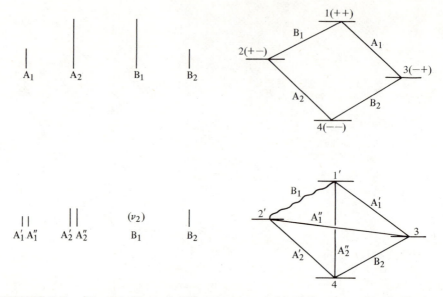

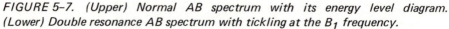

FIGURE 5–7. *(Upper) Normal AB spectrum with its energy level diagram.*
(Lower) Double resonance AB spectrum with tickling at the B_1 frequency.

energy level diagram for the two-spin system is given in Figure 5–7. There are four energy levels (1, 2, 3, 4) and four allowed transitions (A_1, A_2, B_1, B_2). If a very weak irradiation is applied at the B_1 frequency, states 1 (++) and 2 (+−) are rapidly interconverted. As a result of this mixing, two new states are formed ($1'$ and $2'$) with contributions from both 1 and 2. Transitions are now allowed from states 3 or 4 to both $1'$ and $2'$. The net spectral

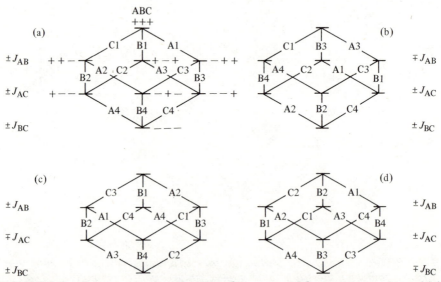

FIGURE 5–8. *Energy level diagram for an ABC spectrum in which*
$|J_{AB}| > |J_{AC}| > |J_{BC}|$: *(a) all signs the same; (b) J_{AB} different in sign; (c) J_{AC}*
different in sign; (d) J_{BC} different in sign. In each case, A_1, B_1, and C_1 are the
lowest field peaks, and A_4, B_4, and C_4 are the highest field peaks.

result is an increase in the total number of possible A lines to four by a splitting of A_1 into A_1' and A_1'', and of A_2 into A_2' and A_2''. Irradiation of B_2 would likewise split A_1 and A_2 into small doublets. In general, irradiation of a given transition will double any lines that are connected to the interconverting energy levels (1 and 2 in Figure 5-7). Connected transitions that have the same values of $F_z = \Sigma I_{zi}$ as the irradiated transition are termed regressive, e.g., A_1 and B_1 in Figure 5-7. Transitions that go on to a higher or lower value of F_z are termed progressive, e.g., A_2 and B_1. An irradiated transition generally splits a regressively connected line into a sharper doublet than a progressively connected line.

This method may be applied to relative sign determinations in the ABC spectrum. For a system in which the nuclei are labeled so that $|J_{AB}| > |J_{AC}| > |J_{BC}|$, four energy level diagrams may be constructed, depending on the relative signs of these coupling constants (Figure 5-8). These diagrams are constructed in the manner used for the AMX spectrum in Figure 5-5.

If line A_1, for example, is irradiated and if all the coupling constant signs are the same (case a), then peaks B_1, C_1, B_3, and C_3 should double. Other sets of relative signs give contrasting expectations (cases b–d). Thus a single spin-tickling experiment gives all the relative signs, whereas at least two selective decoupling experiments are required to obtain the same information. An interesting application of this method has been made to styrene-imine (**12**) and styrene sulfide (**13**). The ring protons on carbon form AKM systems in each

case (Figure 5-9). It is noted that nuclei A, B, and C are assigned so that $|J_{AB}| > |J_{AC}| > |J_{BC}|$.

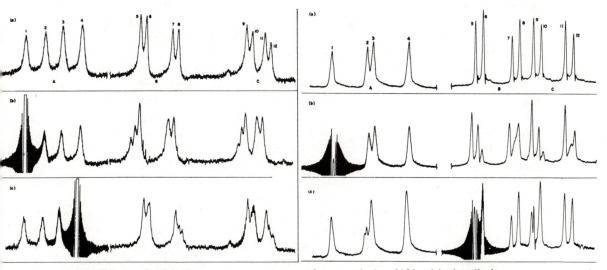

FIGURE 5–9. (Left) The proton spectrum of styreneimine (**12**) with irradiation at A_1 (b) and A_4 (c). (Right) The proton spectrum of styrene sulfide (**13**) with irradiation at A_1 (b) and B_1 (c). The single resonance spectrum is given at the top (a) for each. [Reproduced with the permission of the American Chemical Society from S. L. Manatt, D. D. Elleman, and S. J. Brois, J. Amer. Chem. Soc., **87**, 2220 (1965).]

For **12**, when A_1 is irradiated, B_1, B_3, C_1, and C_3 are perturbed. The molecule, therefore, is assigned to case a, in which all the signs are alike. Irradiation of A_4 confirms this assignment by perturbing B_2, B_4, C_2, and C_4. The first experiment (irradiation of A_1) gives further confirmation because B_1 and C_1 are sharp doublets (regressive) and B_3 and C_3 are only broadened (progressive). Shifts of various peaks are also observed as a result of the Bloch-Siegert effect. Analogous experiments with **13** classify it as case d, with J_{BC} of opposite sign to J_{AB} and J_{AC}. The A nucleus in each molecule is the benzylic proton, and the B, C nuclei are the methylene protons. The vicinal couplings, J_{AB} and J_{BC}, are probably positive (section 4-5), so that the geminal coupling in the imine is positive and in the sulfide negative. The unusual positive geminal coupling in **12** results both from bond angle and electronegativity considerations (see section 4-4).

Spin-tickling methods may be applied to more complicated systems, with the particular view of mapping energy level diagrams. The specific information given by this technique about energy level connections and the regressive or progressive relationships between transitions can be of considerable assistance in the analysis of complex spin systems.

5-4   Internuclear Double Resonance (INDOR) Spectra

The technique of sweeping the irradiating frequency (ν_2) while observing the effects on the intensity of the peak at ν_1 has been used in both heteronuclear and homonuclear contexts. We shall consider each in turn. To illustrate the possibilities of the heteronuclear indor technique, let us examine the effect of sweeping a decoupling frequency through the region of ^{31}P resonances while monitoring the intensity of a peak in the proton resonance spectrum using trimethyl phosphite, $(CH_3O)_3P$, as the example (Figure 5-10). The proton spectrum consists of the methyl doublet, by coupling with ^{31}P. If ν_1 is set at the frequency of one of the methyl peaks, and ν_2 is swept from 24.295720 to 24.295820 MHz, the proton peak is perturbed each time ν_2 passes through a ^{31}P resonance. The recorder is set in operation to monitor these changes. The base line indicates the normal intensity level of the peak at ν_1. A negative peak results if this intensity is decreased, a positive peak if increased. When ν_2 arrives at the frequency of a ^{31}P peak, the ^1H peak is decoupled from those ^{31}P nuclei that resonate at that particular frequency. The recorder therefore produces a negative peak. As each ^{31}P peak is irradiated by ν_2, a peak is recorded, the intensity of which depends on the intensity of the irradiated peak. The final result (Figure 5-10) is a spectrum identical in every respect to the normal ^{31}P spectrum, except that the intensities are negative and observation is made at the ^1H frequency. Thus we have been able to record a ^{31}P spectrum with equipment designed to receive only ^1H signals. Indor spectra can also yield positive peaks, if the irradiating intensity is high enough to bring about nuclear Overhauser effects (section 7-4).

The technique has two advantages. (1) Decoupling equipment is less expensive than the entire package needed to observe a particular nucleus directly. An operator need only purchase the observing equipment (ν_1) for protons and decoupling equipment (ν_2) for the nuclei to be detected by the indor method. In this way, ^{119}Sn, ^{31}P, ^{29}Si, ^{14}N, and particularly ^{13}C spectra, among others, have been obtained. (2) The sensitivity of these nuclei to the direct magnetic resonance experiment is much less than that of ^1H. In the indor experiment, the observed nucleus is ^1H, even though the recorded spectrum is of another nucleus. Thus resonances of these poorly sensitive nuclei can be observed with the sensitivity of the ^1H

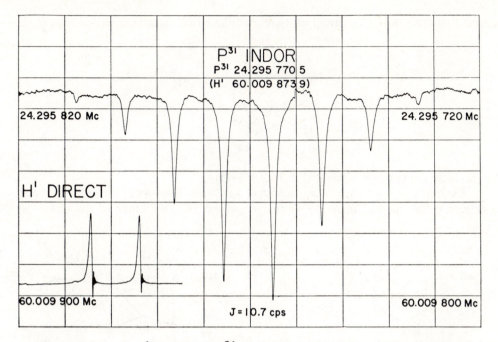

FIGURE 5-10.   The ^1H and indor ^{31}P spectra of $(CH_3O)_3P$.   [Reproduced with the permission of E. B. Baker, L. W. Burd, and C. V. Root, Rev. Sci. Instrum., 34, 243 (1963).]

spectrum.   These advantages have lessened in importance with the advent of multinuclear spectrometers with Fourier transform capabilities.   The indor method has the disadvantage of only being applicable to nuclei that are coupled to a high sensitivity nucleus (^1H or ^{19}F).   Furthermore, spectral simplification by noise decoupling, so useful in $^{13}C\{^1H\}$ spectra, cannot be applied, since the ^1H resonance is needed for observation.   For these reasons, the indor technique for observing nuclei other than protons is becoming less popular.

Although use of the heteronuclear indor technique has been decreasing, that of homonuclear indor is on the increase.   The indor procedure can potentially carry out most of the applications of ordinary decoupling—spin tickling, structure elucidation, location of hidden resonances.   If a single transition in a spectrum serves as the monitor line (ν_1), peaks are observed in the indor spectrum from any other transition that has an energy level in common with those of the observed peak.   A lower level of power (ν_2) is used than even in the spin-tickling experiment, so that intensities are perturbed without splitting of the connected transitions into doublets.   Under these conditions, negative signals are obtained from regressively connected transitions and positive signals from progressively connected transitions.   In the AB case of Figure 5-7, irradiation of peak B_1 produces a negative peak at A_1 and a positive peak at A_2.   Irradiation of peak A_1 for the ABC spectrum of Figure 5-8 produces negative peaks at B_1 and C_1, positive peaks at B_3 and C_3 for case a.   The indor method may be superior to spin tickling for sign determinations and energy level mapping, since a given experiment produces peaks only for connected transitions, and their progressive or regressive nature is graphically demonstrated.   The signs of the coupling constants of the almost first-order ABKL spectrum of 3-acetylpyridine (**14**) have been worked out in this way

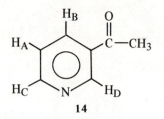

by the indor experiments illustrated in Figure 5–11.

Extensive use of the indor technique has been employed to analyze the proton spectrum of the thermal dimer **15** of the propellane **16**. Only four of the sixteen skeletal

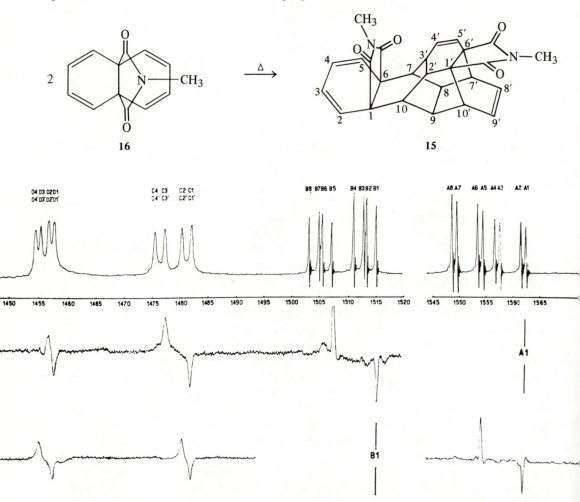

FIGURE 5-11. The proton spectrum of 3-acetylpyridine (**14**): (middle) indor experiment monitoring line A_1; (bottom) indor experiment monitoring line B_1. [Reproduced with the permission of Academic Press from V. G. Kowalewski, D. G. de Kowalewski, and E. C. Ferrá, J. Mol. Spectrosc., **20**, 203 (1966).]

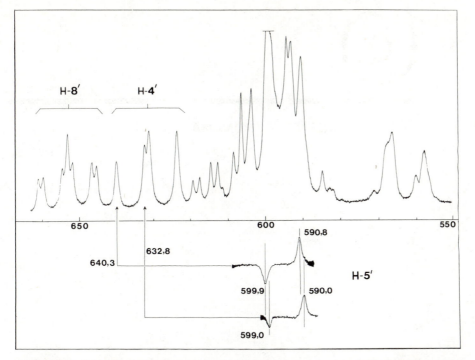

FIGURE 5-12. The alkenic region of the proton spectrum of 15; (middle) indor with the monitor line at the 640.3 Hz line of H-4'; (bottom) indor with the monitor line at the 632.8 Hz line of H-4'. [Reproduced with the permission of Maxwell Scientific, Inc., from O. Sciacovelli, W. von Phillipsborn, C. Amith, and D. Ginsberg, Tetrahedron, 26, 4589 (1970).]

protons give resolved resonances. By monitoring lines from these protons, the obscured resonances of other protons can be defined. Figure 5-12 shows indor experiments with the monitor lines from H-4'. The four lines that make up the H-5' resonance are clearly extracted from the mass of peaks in the 590–610 Hz region. The process can be extended by monitoring frequencies of lines revealed by previous indor experiments. In this way, the chemical shifts and coupling constants of most of the protons in **15** were determined in order to confirm the proposed structure.

BIBLIOGRAPHY

1. W. von Philipsborn, *Angew. Chem., Int. Ed. Engl.*, **10**, 472 (1971).
2. W. McFarlane, *Ann. Rev. NMR Spectrosc.*, **1**, 135 (1968); *Ann. Reports NMR Spectrosc.*, **5A**, 353 (1972).
3. V. J. Kowalewski, *Progr. NMR Spectrosc.*, **5**, 1 (1969).
4. W. McFarlane, *Determ. Org. Struct. Phys. Meth.*, **4**, 139 (1971).
5. J. D. Baldeschwieler and E. W. Randall, *Chem. Rev.*, **63**, 81 (1963).

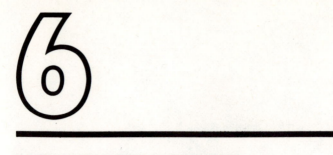

RATE-DEPENDENT PHENOMENA

6-1 The DNMR Method

The use of spectral changes to obtain kinetic information (see section 1-6) has been generally referred to as the *dynamic nuclear magnetic resonance* (dnmr) method. The experimental variable can be the concentration of the reacting species or, for unimolecular reactions, the temperature. An example of a unimolecular rate process is the ring reversal of cyclohexane-d_{11}, described in section 1-6 (see Figure 1-14). In the fast exchange limit, the axial and equatorial protons give a single resonance; in the slow exchange limit, separate resonances. At intermediate rates, the proton resonance passes through a transition from one extreme to the other. The rate of the reaction at each temperature can be obtained by analysis of the line shapes (see below). From a series of rate constants, activation parameters may be calculated. The dnmr method has become a standard and invaluable procedure for determining barriers in the range 5-27 kcal mole^{-1}.

The theory for dnmr processes has been developed by both classical procedures (the Bloch equations) and quantum mechanics (density matrices). One of the most widely used results for the case in which two peaks collapse to a singlet has been the determination of the rate constant at the coalescence temperature (T_c). This quantity is defined as the temperature at which the two peaks just coalesce to a single peak, as the spectrum passes from the slow exchange to the fast exchange limit (see Figure 1-14). At T_c, the unimolecular rate constant for the reaction is given by eq. 6-1, in which $\Delta\nu$ is the distance between the two

$$k_c = \frac{\pi\Delta\nu}{\sqrt{2}} \qquad\qquad 6\text{-}1$$

peaks at slow exchange. This equation is limited to cases in which the nuclei reside in two equally populated sites without mutual coupling. Frequently, however, the exchanging nuclei do couple, as in the ring reversal of 1,3,5-trithiane (eq. 6-2). The diastereotopic methylene

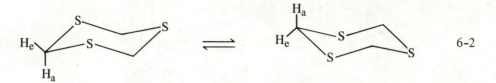

$$6\text{-}2$$

protons give a singlet at room temperature, but an AB quartet below $-80°$ ($\Delta\nu = 0.76$ ppm, $J = 14.5$ Hz). For two-site exchange between coupled nuclei, the rate constant at T_c is given by eq. 6-3.

$$k_c = \frac{\pi}{\sqrt{2}} (\Delta\nu^2 + 6J^2)^{1/2} \qquad\qquad 6\text{-}3$$

The free energy of activation ($\Delta G_c^\ddagger$) at the temperature of coalescence can be calculated from the rate constant by eq. 6-4. Because of its operative simplicity, the T_c method has seen

$$\Delta G_c^\ddagger = 2.3 \, RT_c(10.32 + \log T_c/k_c) \qquad\qquad 6\text{-}4$$

widespread use. Nonetheless, much remains to be desired from one-point kinetics. Because of its dependence on temperature, $\Delta G^\ddagger$ cannot be compared from one system to another unless there is good reason to believe that $\Delta S^\ddagger$ is constant or zero. Frequently, within an homologous series T_c does not vary greatly, and legitimate comparisons can be made.

Equations have been developed by Gutowsky, Alexander, Saunders, and Binsch, among others, that describe the entire spectrum of an exchanging system. Computer programs are available that plot the spectrum calculated from these equations for a particular value of the rate constant. Thus the observed rate constants can be obtained for each temperature either by trial-and-error matching or by direct computer comparison of observed and calculated spectra. A match set of spectra is given by way of example for the α and γ protons in piperidine-3,3,5,5-d_4 (Figure 6-1, eq. 6-5).

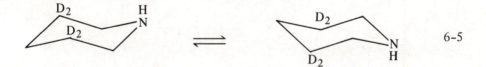

$$6\text{-}5$$

From a plot of log k vs. $1/T$, the Arrhenius activation energy (E_a) and pre-exponential factor (A) can be obtained. Alternatively, a plot of log (k/T) vs. $1/T$ can give $\Delta H^\ddagger$ and $\Delta S^\ddagger$. This *complete lineshape* (cls) method is superior to the T_c method because it utilizes every point on the spectrum and produces rate constants for several temperatures. It is more difficult to apply, however, since it requires the use of a digital computer, and it is more sensitive to systematic errors. Nonetheless, when properly used, the cls method is the more reliable.

Programs are now available to compute spectra for almost any conceivable multisite exchange situation, including many for which there is no possibility of using a T_c method.

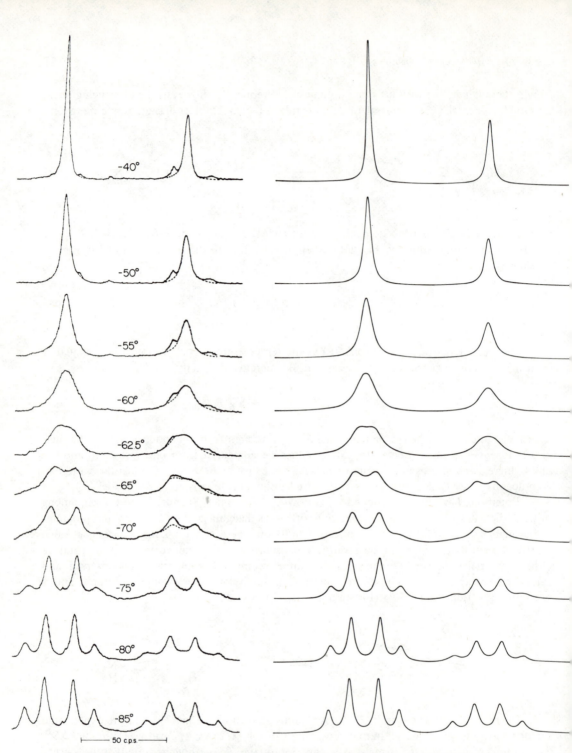

FIGURE 6–1.   The observed and calculated 60 MHz proton spectra of the α and γ protons of piperidine-3,3,5,5-d₄ as a function of temperature.   The impurities come from incompletely deuterated material. Nitrogen inversion and (or) exchange of the N-proton are so rapid that the configuration at nitrogen does not influence the temperature dependence of the spectrum. [Reproduced with the permission of the American Chemical Society from J. B. Lambert, R. G. Keske, R. E. Carhart, and A. P. Jovanovich, J. Amer. Chem. Soc., **89**, 3761 (1967).]

For example, the secondary cyclopropyl protons in the diazanorcaradiene (**1**) are diastereo-topic at slow exchange ($-25°$, Figure 6-2). As the temperature is raised to $+111°$, the A_2BC spectrum becomes A_2B_2. This spectral averaging is explained by the interconversion of **1a** and **1b** through the diazacycloheptatriene intermediate (**2**, eq. 6-6). Figure 6-2 gives

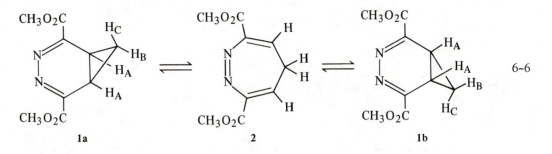

6-6

the observed and calculated spectra for this four-spin, two-site exchange. The observed spectra pass through several points of coalescence, and the spectral richness provides very accurate rate constants over a wide range of temperatures.

The advent of dnmr methods has made available for study many rate processes previously inaccessible by other techniques. The practical range of variable temperature studies is -180 to $+200°$, corresponding to a range in barriers of about 5–27 kcal mole^{-1}. To study a rate process by any dnmr procedure (T_c or cls), the slow exchange spectrum must be observable in order to measure $\Delta\nu$. Failure to observe a chemical shift difference at low temperatures may be attributable either to a process with too low a barrier to freeze out or to a vanishingly small value of $\Delta\nu$. Because of this ambiguity, lower limits to barriers can never be estimated. For cases in which the slow exchange spectrum is unaltered by raising the temperature, upper limits to $\Delta G_c^{\ddagger}$ can be set by the use of eqs. 6-1 and 6-4.

The dnmr method has been applied to a wide variety of molecular processes. In the following sections, we will examine the various classes of reactions that have been studied by this method.

6-2   Hindered Rotation

The barrier to carbon–carbon rotation in ethane is about 3 kcal mole^{-1}, well below the range of the dnmr technique. The barrier to C=C rotation in ethylene is well above the dnmr range. There are many intermediate bonding situations, however, and these have proved to be remarkably amenable to dnmr studies. Amide resonance presents one such case. In dimethylformamide (**3**) the room temperature spectrum contains two methyl resonances,

but as the temperature is raised, the spectrum passes through the familiar transition to a single peak. Lineshape analysis indicates a barrier of about 22 kcal mole^{-1}. Carbamates

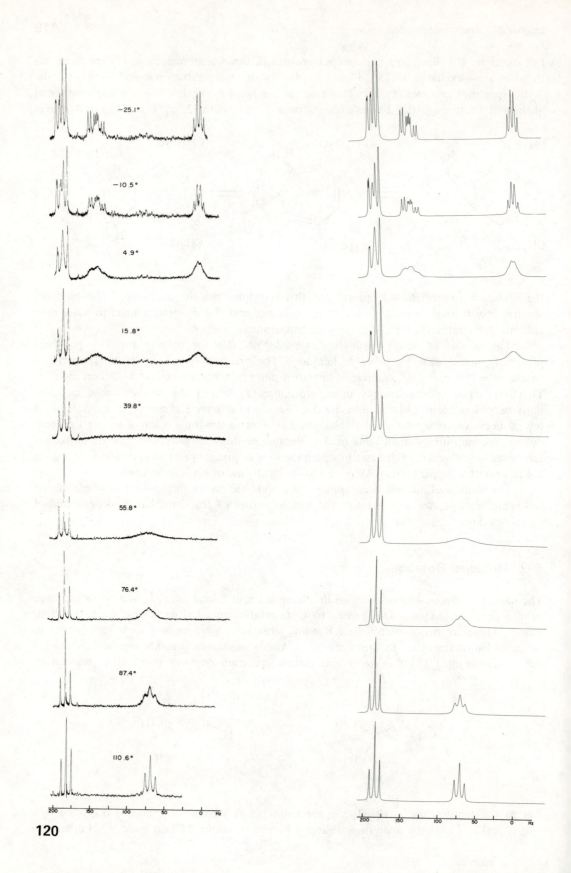

FIGURE 6-2 (opposite). The observed and calculated 60 MHz proton spectra of the ring protons of 2,5-dicarbomethoxy-3,4-diazanorcaradiene (1) as a function of temperature. [Reproduced with the permission of the American Chemical Society from D. A. Kleier, G. Binsch, A. Steigel, and J. Sauer, J. Amer. Chem. Soc., 92, 3787 (1970).]

contain a similar partial double bond (**4**), but the barrier is lower than in amides because

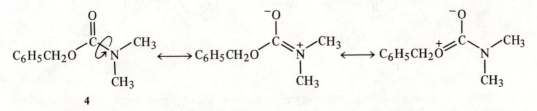

4

the amide-like resonance is partially offset by ester resonance. The barrier in benzyl dimethylcarbamate (**4**) is about 16 kcal mole^{-1}. Numerous other examples of hindered C–N rotation, such as in thioamides, carbamoyl chlorides, vinylamides, and enamines, have been observed.

Partial double bonding can give rise to hindered rotation about the N–N bond in nitrosamines (**5**, dimethylnitrosamine, barrier about 23 kcal mole^{-1}), the N–O bond in alkyl nitrites (**6**, methyl nitrite, 10 kcal mole^{-1}), the C–C bond in diazoketones (**7**, diazoacetone, 15.5 kcal mole^{-1}), the C–C bond in aromatic aldehydes (**8**, benzaldehyde, 8 kcal mole^{-1}), and the B–N bond in aminoboranes (**9**, phenylbis(dimethylamino)borane,

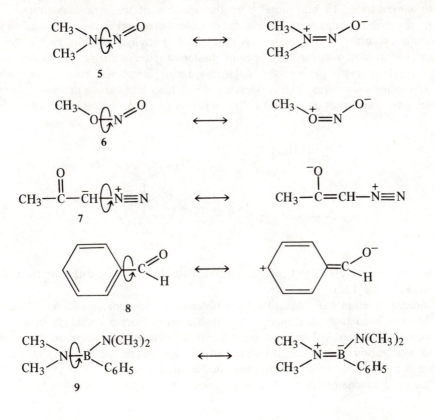

11 kcal mole^{-1}). Formal double bonds can exhibit free rotation when alternative resonance structures suggest partial single bonding. The calicene **10**, for example, has a barrier to rotation about the central bond of only 20 kcal mole^{-1}.

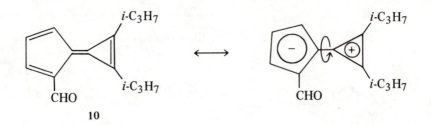

10

Steric congestion can raise the barrier to rotation about single bonds enough to bring it into the dnmr range. Rotation about the sp^2-sp^2 single bond in the biphenyl **11** is

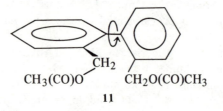

11

raised to a measurable 13 kcal mole^{-1} by the presence of the ortho substituents. In this molecule, the ortho groups serve double duty, since the benzyl protons are used to monitor the rotation process. When rotation is rapid, the methylene protons give a singlet nmr resonance, but at slow rotation they become diastereotopic and hence give an AB resonance. Hindered rotation about an sp^3-sp^3 single bond can sometimes be observed when one of the carbons involved is quaternary. Thus at $-150°$, the *t*-butyl group in *t*-butylcyclopentane (**12**) gives two resonances in the ratio 2/1, whereas at room temperature only a singlet is

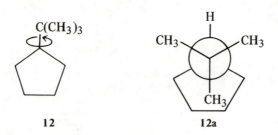

12   **12a**

observed. At slow rotation, two of the *t*-butyl methyl groups are different from the third (Newman projection **12a**).

Hindered rotation has frequently been observed in halogenated alkanes. The increased barrier has not been fully explained, but probably arises from a combination of steric and electrostatic interactions. Hindered rotation in substituted alkanes must be analyzed in terms of the various rotational isomers. In 2,2,3,3-tetrachlorobutane (**13**), the methyl resonance is a singlet at room temperature, but an approximately 2/1 doublet at $-44°$, due to the trans and gauche rotamers. Lineshape analysis gives a barrier of about 15 kcal mole^{-1}.

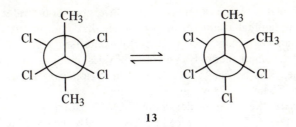

13

This figure is possibly the largest measured for a substituted ethane. In less symmetrical ethanes, there are three distinct rotamers, and the barriers between each pair must be measured. For *dl*-1,2-dibromo-1,2-dichloro-1,2-difluoroethane (**14**), all three rotamers could

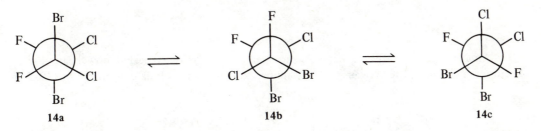

be observed at low temperatures, with **14a** the lowest in energy, **14c** 120 cal mole^{-1} higher, and **14b** another 320 cal mole^{-1} higher. The barrier from **c** to **a** was found to be 9.9 kcal mole^{-1}, but the other barriers could only be established to be greater than 10.1 kcal mole^{-1}.

When both atoms that constitute the single bond in question possess nonbonding electron pairs, the barrier is frequently in the dnmr range. The high barrier has been explained in terms of electrostatic interactions, lone pair–lone pair repulsions, and in some cases (p–d)$_{\pi}$ bonding. Thus hindered rotation has been observed about the S–S bond in disulfides (**15**, dibenzyl disulfide, 7 kcal mole^{-1}), the N–N bond in hydrazines (**16**, tetrafluorohydrazine, 6 kcal mole^{-1}), the N–S bond in sulfenamides (**17**, *N,N*-dibenzyltrichloromethanesulfenamide, 15 kcal mole^{-1}), and the N–P bond in aminophosphines (**18**, chloro(dimethylamino)phenyl-phosphine, 11 kcal mole^{-1}).

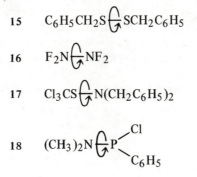

6–3   Ring Reversal

Ring reversal is defined as the interconversion of equivalent ring forms (substitution disregarded) through rotations about single bonds. The structural locations of substituents that are geminal to each other are interchanged, e.g., axial and equatorial. The frequently used term *ring inversion* is somewhat of a misnomer because inversion should refer only to a change in configuration, as in pyramidal atomic inversion, rather than conformation, as in ring reversal. Certainly the observation of ring reversal in cyclohexane (see Figure 1-14) was one of the greatest triumphs of the dnmr method. Very few physical methods are sensitive to degenerate processes of this type. The barrier to ring reversal in cyclohexane is now accepted to be close to 10 kcal mole^{-1}. The use of low temperature nmr spectra in studies of monosubstituted cyclohexanes has led to the accurate determination of axial/equatorial equilibrium constants by direct integration (eq. 6-7). This procedure has proved to be more reliable than the kinetic method.

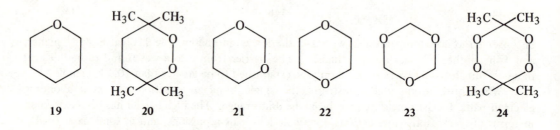

6-7

Almost every conceivable heterocyclic analogue of cyclohexane has now been studied by dnmr methods. The oxygen series is almost complete up to four oxygen atoms (**19-24**).

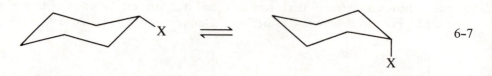

19	**20**	**21**	**22**	**23**	**24**

The barriers in tetrahydropyran (**19**, 10.7 kcal mole^{-1}), 1,3-dioxane (**21**, 9.7 kcal mole^{-1}), 1,4-dioxane (**22**, 9.7 kcal mole^{-1}), and 1,3,5-trioxane (**23**, 10.9 kcal mole^{-1}) are all very similar to that in cyclohexane. Those containing an oxygen–oxygen bond, however, as in the 1,2-dioxane (**20**, 16.1 kcal mole^{-1}) and the 1,2,4,5-tetraoxane (**24**, 14.2 kcal mole^{-1}), have somewhat higher barriers. The barrier to ring reversal is a composite of torsional operations. Because the barrier to rotation about an O–O bond is higher than that about a C–O bond (on account of the lone pair–lone pair interactions mentioned in section 6-2), molecules such as **20** and **24** with two heteroatoms bonded to each other have a higher barrier to ring reversal.

Ring reversal occurs in cyclohexene between two half-chair forms (eq. 6-8). Anet

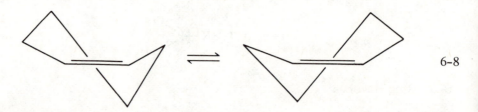

6-8

and Haq utilized the deuterated derivative **25** to measure the barrier by dnmr techniques (5.3 kcal mole^{-1}). Cis-fused decalins can undergo ring reversal (eq. 6-9), although the

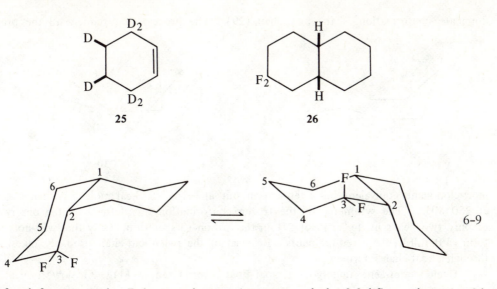

25   26

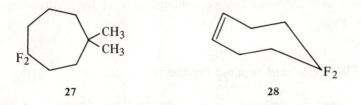

6-9

trans-fused form is rigid.   Roberts and co-workers prepared the 3,3-difluoro derivative **26** (the fluorine atoms serve as a label) in order to determine the barrier (13.4 kcal mole^{-1}) to reversal in *cis*-decalin.

Seven- and eight-membered rings present a much more complex conformational situation. Not only are more conformational forms possible (chair, boat, crown, twist-chair, boat-chair, etc.), but individual conformations can interconvert by the low energy process of pseudorotation as well as by ring reversal. As a result, the nmr spectrum of cycloheptane or 1,1-difluorocyclo-heptane is insensitive to temperature changes down to $-180°$. Introduction of geminal methyl groups (**27**) creates certain steric interactions that permit the observation of a rate process

27   28

with a barrier of 6 kcal mole^{-1}, although the exact nature of the process is uncertain. Pseudo-rotation is blocked when a double bond is present in the ring. What is thought to be a chair-chair reversal has been observed in 5,5-difluorocycloheptene (**28**) by low temperature methods (7.4 kcal mole^{-1}).   1,3,5-Cycloheptatriene undergoes a boat–boat reversal (eq. 6–10) with

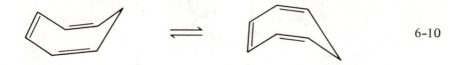

6-10

an activation energy of 7.7 kcal mole^{-1}.

Cyclooctane and its derivatives have been examined extensively. The pentadecadeutero derivative of the parent compound exhibits a rate process below $-100°$ with a free energy of activation of 7.7 kcal mole^{-1}. The concensus of several studies now indicates that the dominate conformation is the boat-chair (**29**). The process that renders all the protons

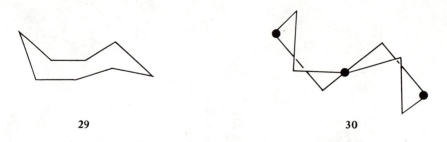

29   30

in cyclononane equivalent has been frozen out at $-165°$ (6 kcal mole^{-1}) when observed at 250 MHz. The symmetry of the frozen conformation was defined by the observation of only two peaks in the ratio of 2/1 in the carbon-13 spectrum. Only the twist-boat-chair form (**30**) fulfills these requirements. Reversal of the twist-boat-chair probably takes place through a boat-chair intermediate.

Cyclooctatetraene undergoes a boat–boat reversal (eq. 6-11). At slow reversal, the

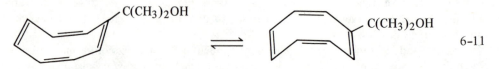

6-11

methyl groups are diastereotopic and give distinct resonances, which coalesce to a singlet above room temperature (14.7 kcal mole^{-1}). The transition state to reversal is thought to be a planar form with alternating single and double bonds. Bond switching may also occur in this molecule, but this subject is considered in section 6-5. Larger annulenes offer a very rich field for dnmr studies, although some of the observed processes are not true ring reversals.

6-4 Atomic Inversion and Related Processes

An atom that possesses a nonbonding pair of electrons and is bonded to three other groups in a pyramidal fashion may undergo unimolecular inversion of configuration. At the transition state to this pyramidal atomic inversion the central atom is trigonally hybridized and the lone pair is in a p orbital (eq. 6-12). In an early study, Bottini and Roberts observed that

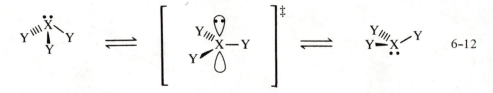

6-12

nitrogen inversion in aziridines is detectable by variable temperature nmr methods. Nitrogen inversion has since been studied in a host of structural situations, and inversion has also been observed about carbon, oxygen, phosphorus, arsenic, and sulfur.

The first and second row elements offer an interesting contrast. Nitrogen and oxygen generally invert too rapidly for detection by the dnmr method, and therefore require some structural modification to raise the barrier to an observable range. Phosphorus, arsenic, and sulfur, on the other hand, invert too slowly, and require a modification that can lower the barrier. The device used by Bottini and Roberts to observe nitrogen inversion was to incorporate the nitrogen atom into a three-membered ring (**31**, eq. 6-13). Angle strain is

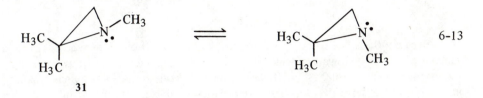

$$6\text{-}13$$

31

increased on passing from the ground state to the transition state, so the barrier is considerably higher (18-20 kcal mole^{-1} in **31**) than in unstrained open chain amines ($\sim$4-7 kcal mole^{-1}). The effect is also observed in azetidines (**32**, 9 kcal mole^{-1}) and bridged-ring systems (**33**,

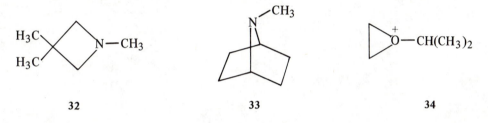

32 **33** **34**

10 kcal mole^{-1}), though to a lesser extent. The only example of inversion about oxygen observed to date (**34**) also utilized the strain of a three-membered ring.

The barrier to inversion may be raised by replacing a carbon substituent with one that is more electronegative. By this means the s character of the ground state lone pair is increased; the transition state lone pair must still be p-hybridized, so the barrier is increased. *N*-Chloropyrrolidine (**35**, $\sim$12 kcal mole^{-1}) and *N*-methyloxazolidine (**36**, 16 kcal mole^{-1})

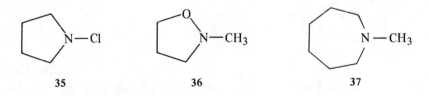

35 **36** **37**

exhibit barriers that are enhanced by this effect. When neither ring strain nor electronegative substituents are present, the barrier to nitrogen inversion is generally below the dnmr range. *N*-Methylhomopiperidine (**37**), whose inversion barrier was measured to be about 7 kcal mole^{-1}, is one example to be observed without these constraints. Systems containing six-membered rings or only open chain fragments are more difficult to analyze because of the possibility that ring reversal or bond rotation may be responsible for the spectral changes.

With second row elements, devices must be developed to reduce the barrier in order to render the process detectable by dnmr techniques. Electropositive substituents may be used to produce the opposite effect to that described above. Thus the barrier to inversion in isopropylphenyltrimethylsilylphosphine (**38**, 19 kcal mole^{-1}) is considerably lower than that of benzylmethylphenylphosphine (**39**, about 32 kcal mole^{-1}), measured by classical kinetic techniques.

$$P(C_6H_5)[CH(CH_3)_2][Si(CH_3)_3] \qquad P(CH_2C_6H_5)(CH_3)(C_6H_5)$$

$$\textbf{38} \qquad\qquad\qquad\qquad \textbf{39}$$

Because the lone pair is p-hybridized in the transition state (eq. 6-12), a substituent capable of direct conjugation will lower the barrier to inversion. It is for this reason that the barrier in 1-phenyl-2,2-dimethylaziridine (11 kcal mole^{-1}) is much smaller than that in the *N*-methyl derivative **31**. This effect is less important in phenyl-substituted phosphines than in amines because $(2p-3p)_\pi$ conjugation is weaker than $(2p-2p)_\pi$ conjugation. A varient of this effect has been recognized as the source of the very low barrier in the phosphole **40**. Although the ground state in **40** is nonplanar, the transition state in the phosphole is

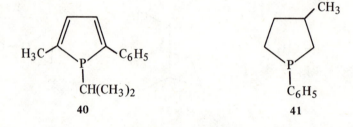

planar and therefore aromatically stabilized. The dnmr-measured barrier in **40** is 16 kcal mole^{-1}, whereas that in the saturated analogue **41** is 36 kcal mole^{-1}.

When nitrogen is trigonally hybridized in the ground state, the process of planar inversion can interconvert syn and anti forms (eq. 6-14). Although the identical result may

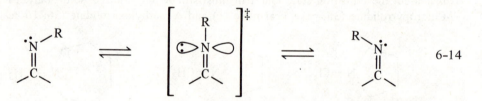

$$6-14$$

be obtained by a rotation about the imine bond, current evidence indicates that the inversion mechanism is favored in most cases. In the *N*-phenylimine of acetone (**42**), the dnmr-measured

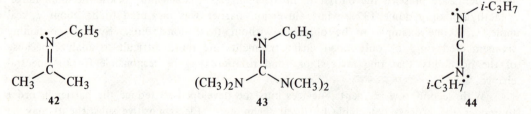

barrier is 20 kcal mole^{-1}. Somewhat lower barriers are observed in the guanidine series (**43**,

13 kcal mole^{-1}). Planar inversion in carbodiimides is extremely rapid. The only measured example to date is the diisopropyl derivative **44**, whose barrier was found to be about 7 kcal mole^{-1}.

Structural reorganizations may take place in certain pentacoordinate bipyramidal molecules. According to the Berry mechanism, the axial and equatorial positions may be interchanged by conversion to an intermediate tetragonal pyramid, followed by reformation of an alternative trigonal bipyramid (eq. 6-15). Variable temperature spectra of sulfur

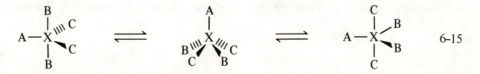

6-15

tetrafluoride may be interpreted in terms of such a mechanism. The room temperature spectrum has only a single ^{19}F resonance. At $-100°$, however, the spectrum contains two distinct triplets, indicating two pairs of fluorine atoms, as in structure **45**. The Berry

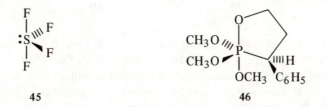

45   46

mechanism adequately explains these spectral changes, although other mechanisms are not excluded. A similar process has been invoked to explain the spectrum of the phosphorane **46**. At room temperature there is only one methoxyl doublet (from coupling to ^{31}P), but at $-70°$ three doublets are present. Structure **46**, with three distinct methoxyl groups, is consistent with these observations, although other structures may also be possible. A sequence of reorganizations of the type outlined in eq. 6-15 renders the three groups equivalent on the average. The barriers to reorganization for both **45** and **46** were found to be about 10 kcal mole^{-1}.

6-5 Valence Tautomerization and Other Chemical Reorganizations

By a chemical reorganization we mean any unimolecular reaction that involves the breakage and reformation of bonds without changes in the atomic constituents. Valence tautomerizations form an important subgroup of this class of reactions. In 1963, Doering and Roth described the preparation and nmr properties of 3,4-homotropilidene (**47**) and thereby

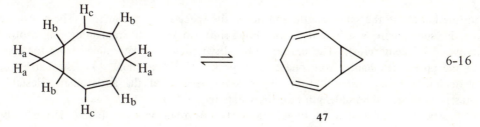

6-16

47

founded a new field of dnmr applications. At low temperatures, the spectrum of **47** has the features expected for the five functionally distinct types of protons (disregarding diastereotopic differences). Above room temperature, however, the Cope rearrangement depicted in eq. 6-16 occurs more rapidly than the nmr time scale. As a result, the spectrum contains only three types of resonances, corresponding to the protons labeled H_a, H_b, and H_c. A kinetic analysis performed several years later on the 1,3,5,7-tetramethyl derivative of **47** yielded an activation energy to rearrangement of about 14 kcal mole^{-1}.

If two nongeminal *a*-type protons in **47** are replaced by a carbon bridge, the Cope rearrangement occurs even more rapidly. The barriers for barbaralone (**48**, 9.6 kcal mole^{-1}),

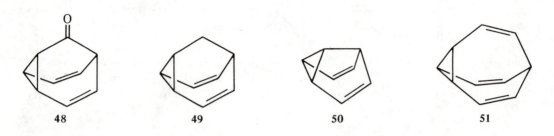

| 48 | 49 | 50 | 51 |

barbaralane (**49**, 7.8 kcal mole^{-1}), and semibullvalene (**50**, 6.4 kcal mole^{-1}, measured for the octamethyl derivative) are considerably lower than that for the parent system **47**. The most satisfying example of this fluxional behavior is bullvalene (**51**, 12.8 kcal mole^{-1}), in which all the C–H groups are rendered equivalent by a sequence of Cope rearrangements. The complex nmr spectrum of the frozen structure at room temperature becomes a singlet above 180°. Although the term *fluxional* is generally reserved for molecules that undergo rapid valence tautomerizations, it could just as accurately apply to most of the other examples given in this chapter.

Cyclooctatetraene offers another example of rapid valence tautomerization detectable by dnmr methods. The process of boat–boat ring reversal has already been discussed (section 6-3, eq. 6-11). In a separate operation, the location of the single and double bonds can become reversed by a process termed *bond switching*. Anet and co-workers determined the barrier for this process with the same molecule used in the ring reversal study (eq. 6-11). The transition state for bond switching (eq. 6–17) differs from that for ring reversal in that

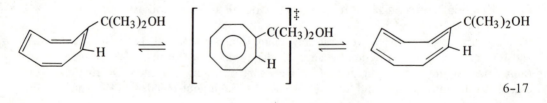

6-17

all the bonds are the same length, and hence the species is antiaromatic. The proton adjacent to the substituent is different in the bond-shift isomers (the remaining ring protons were replaced by deuterium). The barrier to bond switching was determined from the conversion of the proton resonance from one to two peaks (17.1 kcal mole^{-1}). The barrier to bond switching is higher than that to ring reversal because of the antiaromatic destabilization present in the equal-bond-length transition state (eq. 6-17).

The interconversion of oxypin and its norcaradiene form (eq. 6-18) has also been

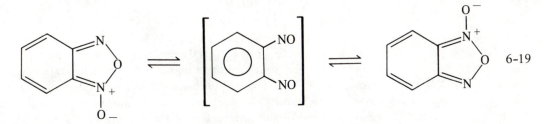

6-18

studied by dnmr techniques. The barrier was found to be only about 9 kcal mole^{-1}. The rapid interconversion of the two equivalent benzofurazan oxides (eq. 6-19) through

6-19

ortho-dinitrosobenzene can be followed by the conversion of the AA′BB′ aromatic spectrum to ABCD (16 kcal mole^{-1}).

Recently, the chemical reorganization of carbonium ions has been opened to study by dnmr methods. One of the first examples studied was the ion produced from *exo*-norbornyl halides at low temperatures. By a sequence of 3,2- and 6,2-hydride shifts and Wagner-Meerwein rearrangements (eq. 6-20), all of the protons are rendered equivalent, so that the room

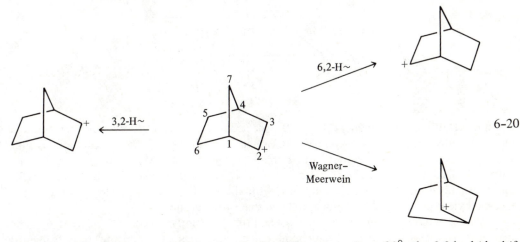

6-20

temperature spectrum contains only a single sharp resonance. By −80°, the 3,2-hydride shift is slowed, so that the 1,2,6 protons give one resonance, the 3,5,7 protons another, and the 4 proton a third. Lineshape analysis gave a barrier of about 11 kcal mole^{-1}.

6-6 Fluxional Organometallic Systems

One of the most fertile areas of activity in the dnmr field today concerns organic molecules that contain metal atoms. The terms *fluxional* or *stereochemically nonrigid* have been used in these areas, though they are no more appropriate than in others. The typical process involves changes in the relationship between the metal and the organic ligand without dissociation. Frequently the process is degenerate in the sense that the product is structurally indistinguishable from the starting material. For example, the dynamic structure of tetramethylalleneiron tetracarbonyl (**52**) was elucidated by temperature-dependent nmr experiments.

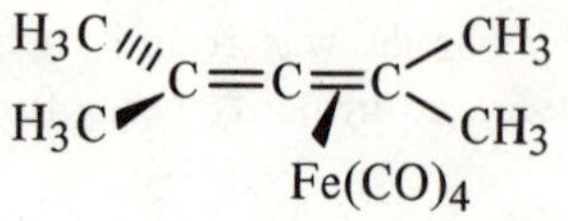

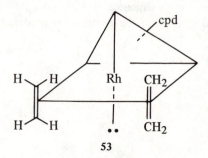

52

At $-60°$, the proton spectrum indicates the presence of three distinct methyl groups, in the ratio of 1/1/2, in agreement with the structure given above. As the temperature is raised to $25°$, the three peaks coalesce to a singlet (the barrier is about 9 kcal mole^{-1}). The $Fe(CO)_4$ unit circulates about the allenic π-electron structure by moving orthogonally from one alkenic unit to the other, then to the reverse side of the first unit, and finally to the opposite side of the second unit. Such a sequence of movements averages the methyl environments.

Alkenes bonded to metals can exhibit dynamic properties simply by rotation about the metal-ligand bond. Bis(ethylene)-π-cyclopentadienylrhodium (**53**) furnished an early

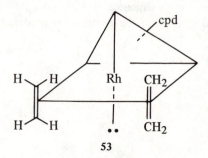

53

example of this phenomenon. The $-20°$ spectrum shows at least two distinct types of alkenic protons, but these are averaged to a single type above $57°$. Model compounds show that the inner and outer protons, rather than the upper and lower protons are chemically shifted at the low temperature extreme. These may be interconverted by a rotation of $180°$ about the metal–ethylene bond. Exchange with the medium was excluded by showing that excess ethylene gives a separate resonance from that of the ligand.

The allyl group bonded to a metal has been frequently observed to exhibit dynamic structural properties. The metal–ligand bond may be either sigma (monohapto) or pi (trihapto). The γ,γ-dimethylallyl Grignard reagent studied by Roberts some time ago provides an example of a σ-bonded situation (eq. 6-21). At $-20°$ the nmr spectrum exhibits two methyl

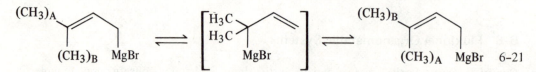

resonances, as expected from the frozen structure. Above $33°$, however, these two resonances coalesce to a sharp singlet. This observation was explained in terms of an isomerization to the α,α isomer (eq. 6-21), in which the methyl groups are equivalent, followed by return to the dominant γ,γ isomer. The details of this interpretation may have to be revised in light of orbital symmetry considerations, since the suprafacial 1,3-sigmatropic shift of eq. 6-21 is thermally forbidden.

The allyl groups in trisallylrhodium are π-bonded. Although the detailed structure is not known, the variable temperature nmr spectrum has given considerable information. At $-74°$, each allyl group gives a distinct AM_2X_2 spectrum; at $10°$, two different allyl resonances are present in a 2/1 ratio; and at $100°$, only one averaged allyl resonance persists. Apparently, one allyl group is less tightly bonded than the other two. Rapid rotation of this group about the allyl–rhodium bond at $10°$ renders the other two groups identical, though still different from the third. Finally, at the highest temperature, metal–ligand rotations average the environments of all three allyl groups. The barrier to rotation for the less tightly bound group was estimated to be about 9.4 kcal mole^{-1}.

Some of the most intricate processes have been observed in metal complexes with cyclic polyenes. Cyclooctatetraeneiron tricarbonyl was the first such to be observed and still serves as an excellent prototype. The spectrum below $-150°$ is consistent with the X-ray data, in that four protons appear to be *bound* and four *unbound*. The iron atom, therefore, is bonded to four π centers. Above $-100°$, the ring protons give only a singlet resonance (barrier of about 7.2 kcal mole^{-1}). The Fe(CO)$_3$ group must shift $45°$ around the ring simultaneously with a bond shift (section 6-5). Eight such operations will bring the iron atom back to its original position and result in complete averaging of the ring protons (eq. 6-22). A similar process has been observed in the cyclooctatetraene complexes of

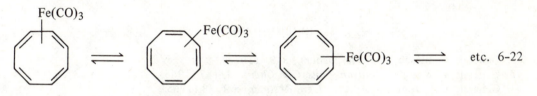

ruthenium, osmium, molybdenum, chromium, and tungsten tricarbonyls, although the number of complexed carbon atoms is not the same in all cases.

A set of 1,5-sigmatropic shifts has been demonstrated to take place in triphenyl-7-cycloheptatrienyltin (eq. 6-23). At $0°$, the spectrum is characteristic of the "frozen"

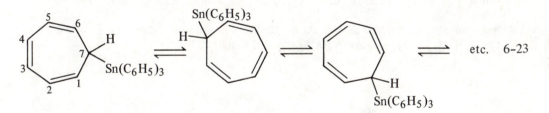

σ complex, but at $100°$, all of the ring protons become equivalent (barrier about 10 kcal mole^{-1}). That the migration is a 1,5 shift to the 3 or 4 positions was demonstrated by a double irradiation experiment. Saturation of the 7-proton resonance at $-10°$ brings about a decrease in intensity of the 3,4-proton resonance. Thus chemical exchange occurs more rapidly than proton relaxation. At slightly higher temperatures, this irradiation additionally decreases the intensity of the 1,6 resonance, so that the second stage of the migration is also a 1,5 shift. In this system at least, the dynamic properties of the molecule are in accord with the predictions of orbital symmetry.

Numerous cyclopentadienyl complexes have also been examined. The compound $Fe(\pi\text{-}C_5H_5)(CO)_2C_5H_5$ (**54**) contains iron that is π-bonded to one cpd and σ-bonded to the

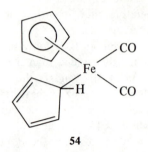

54

other. The low temperature spectrum ($-80°$) contains a singlet for the π-bonded ring and an AA'BB'X spectrum for the σ-bonded ring. Above room temperature, the latter spectrum collapses to a singlet (barrier of 10 kcal mole^{-1}). Detailed analysis of the spectrum favors an interpretation in terms of a sequence of 1,2 rather than 1,3 shifts. The 1,2 shift is equivalent to a 1,5 shift, which is orbital-symmetry-allowed. The 1,2 shift is also the least motion pathway, so that the role of orbital symmetry in this reaction is not clear. Thus the cyclo-heptatrienyl complex of eq. 6-23 is significant because the 1,5 shift is favored even though it is no longer the least motion pathway.

BIBLIOGRAPHY

1. General: G. Binsch, *Top. Stereochem.*, **3**, 97 (1968).
2. General: L. W. Reeves, *Advan. Phys. Org. Chem.*, **3**, 187 (1965).
3. General: C. S. Johnson, *Advan. Magn. Resonance*, **1**, 33 (1965).
4. General: T. H. Siddall and W. E. Stewart, *Progr. NMR Spectrosc.*, **5**, 33 (1969).
5. Hindered Rotation: H. Kessler, *Angew. Chem., Int. Ed. Engl.*, **9**, 219 (1970); W. E. Stewart and T. H. Siddall, *Chem. Rev.*, **70**, 517 (1970).
6. Ring Reversal: J. E. Anderson, *Quart. Rev.* (London), **19**, 426 (1965).
7. Atomic Inversion: J. B. Lambert, *Top. Stereochem.*, **6**, 19 (1971).
8. Atomic Inversion: A. Rauk, L. C. Allen, and K. Mislow, *Angew. Chem., Int. Ed. Engl.*, **9**, 400 (1970).
9. Organometallics: K. Vrieze and P. W. N. M. Vanleeuwen, *Progr. Inorg. Chem.*, **14**, 1 (1971); N. M. Sergeyev, *Progr. NMR Spectrosc.*, **9**, Part 2, 1 (1973).

7

RELAXATION PHENOMENA

7-1 Mechanisms of Relaxation

It will be useful at this point to describe the resonance experiment in a somewhat more rigorous fashion than was necessary in Chapters 1 and 3. Nuclei with a spin (I) of $\frac{1}{2}$ have been said to be lined up either with ($I_z = -\frac{1}{2}$) or against ($I_z = +\frac{1}{2}$) the B_0 field or, more accurately, to be precessing about the $-z$ or $+z$ direction by the action of B_0. A single magnetic nucleus with moment μ_i is thus tilted away from the z axis (Figure 7-1). Alignment with the field is characterized by precession about the $-z$ direction at the natural Larmor angular velocity ($\omega_0 = \gamma B_0$). When the angular velocity (ω) of the B_1 field in the xy plane is identical to that of the nuclear magnet μ_i, the two vectors can remain in phase long enough for an effective interaction to occur. The resulting absorption of energy by the nuclear spin reverses the direction of the z component of the μ_i vector so that precession continues about the $+z$ direction ($I_z = +\frac{1}{2}$). The nmr experiment is sustained because neighboring magnetic nuclei in motion can produce oscillating fields with the same frequency and phase as the excited nucleus. Thus energy may pass from the excited spins to the lattice so that nuclei can return to the lower spin state and be available for another spin excitation by B_1. This mechanism has been termed *spin-lattice relaxation*.

When the collection of nuclei is outside of the B_0 field, there is no net magnetization in the z direction ($M_z = \Sigma I_z = 0$), aside from the very small effect of the earth's magnetic field. After the sample has reached equilibrium in the B_0 field, there is a small excess of nuclei with $I_z = -\frac{1}{2}$, so that M_z attains a small, negative value (M_{z0}). As the magnetic resonance experiment is carried out, the excess nuclei in the $I_z = -\frac{1}{2}$ state are excited to the $+\frac{1}{2}$ state so that the net value of M_z approaches zero. Spin-lattice relaxation is defined as the process whereby the z component of the magnetization (M_z) returns to its equilibrium value (M_{z0}). The rate constant for this first-order decay process is T_1^{-1}, and T_1 is called the *spin-lattice*

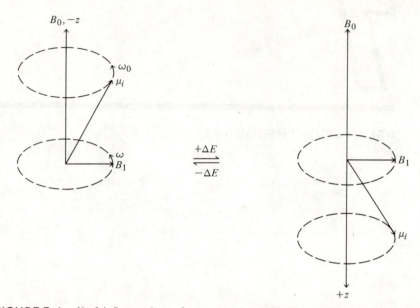

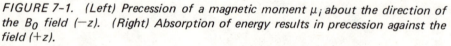

FIGURE 7-1.   (Left) Precession of a magnetic moment μ_i about the direction of the B_0 field $(-z)$.   (Right) Absorption of energy results in precession against the field $(+z)$.

relaxation time. This mechanism is also termed *longitudinal relaxation*, because the decay of magnetization occurs along the longitude (z axis) of the nmr experiment.

The interaction between excited spins and random oscillatory fields produced by the lattice is dependent on viscosity, temperature, and concentration, so T_1 depends on all these factors.   As was mentioned in section 1–5, unpaired electrons in paramagnetic molecules can contribute to this relaxation process.   Because the magnetic moment of the electron is several orders of magnitude greater than that of the proton, it is extremely effective in inducing spin-lattice relaxation.   Paramagnetic materials can be so effective in reducing T_1 that uncertainty broadening (see section 1–5) becomes the dominant factor in determining the linewidth of resonances.   For results at highest resolution, even dissolved oxygen must be removed from a solution before the spectrum is taken.   Longitudinal relaxation in nuclei with a quadrupolar moment $(I > \frac{1}{2})$ can be induced through interaction with the surrounding electron cloud by a special mechanism to be discussed in the next section.

Ordinarily, nuclei with a spin of $\frac{1}{2}$ are relaxed by interactions with surrounding magnetically active nuclei. The closest nuclei will be the most effective.   Thus the relaxation time of a ^{13}C nucleus in a methyl group is much shorter than that of a quaternary ^{13}C nucleus. In the former situation, the three attached protons are available to induce relaxation, whereas in the latter case the only attached nuclei are the magnetically inactive carbons (aside from the 1.1% ^{13}C).   Measurement of ^{13}C relaxation times can assist in structural assignments through considerations such as these.

A second type of relaxation is concerned with magnetization in the xy plane, perpendicular to the z axis. At equilibrium, chemically equivalent magnetic nuclei precess about the z direction with identical angular frequencies, but with an essentially random distribution of phases. This situation may be represented by a set of magnetic moments that are distributed evenly over the surface of a cone whose axis is in the $-z$ direction (Figure 7–2a). Because there is a small excess of $-\frac{1}{2}$ nuclei, the net magnetization M is directed along the $-z$ axis.

FIGURE 7-2. *(a) A random phase distribution of nuclear moments about the —z direction. The small, thick arrow indicates the total magnetization M, which is directed along the —z axis. (b) Action by the applied magnetic field B_1 along the x axis tends to give some phase coherence to the precessing magnetic moments. As a result, the total magnetization M (thick arrow) is no longer directed along the z axis, but possesses a component in the xy plane.*

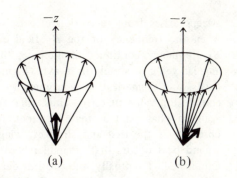

(a) (b)

At equilibrium in the B_0 field, there is no magnetization in the xy plane ($M_x = M_y = 0$). When the applied oscillatory field B_1 attains the Larmor frequency, not only are nuclei excited from the $-\frac{1}{2}$ to the $+\frac{1}{2}$ state, but some phase coherence is conveyed to the precessing nuclei. The B_1 field oscillates in the x direction, 90° out of phase from the direction of nuclear precession. By its action, the net magnetization M is tipped off the z axis, thereby developing a component in the xy plane (Figure 7-2b). It is this induced magnetization in the y direction that is detected by the receiver coil. When the B_1 field passes out of the region of resonance, the x and y components of magnetization slowly decay back to their equilibrium values of zero, and the signal is no longer detected. Because this phenomenon occurs in the plane perpendicular to the applied magnetic field, it is termed *transverse relaxation*. This relaxation is a first-order process with rate constant T_2^{-1}. Also called *spin-spin relaxation*, this phenomenon is defined as the process whereby the xy components of the magnetization (M_x and M_y) return to their equilibrium values (zero).

Any process that tends to destroy phase coherence will contribute to transverse relaxation. A shorter value of T_2 reduces the lifetime of the y magnetization to be detected. The shorter lifetimes will ultimately give rise to uncertainty broadening of the nmr signal. If B_0 is not perfectly homogeneous, each nuclear moment of a given type will not have exactly the same Larmor frequency. Therefore, all nuclei of this type will not precess about the surface of the cone with exactly the same angular velocity. Phase coherence produced by the resonance experiment, as in Figure 7-2b, will be lost as the nuclei precess at different rates about the z axis.

In highly viscous media, two nuclei that are chemically equivalent may reside in slightly different solution environments and hence possess different Larmor frequencies. These differences will also tend to destroy the phase coherence brought about by the resonance experiment. Thus viscosity can contribute to transverse relaxation. A similar phenomenon could be expected to be present in solids. Spin-spin exchanges also contribute to T_2. Two nuclei with opposite spins can simply interchange their I_z spin properties. Such an operation makes no net change in the z magnetization, so it does not contribute to T_1. Because it does limit the lifetime of phase coherence, it contributes to T_2 through uncertainty considerations. Frequently, T_2 is called spin-spin relaxation, although a phrase such as *spin-phase relaxation* might be more descriptive. Through these contributions, T_2 usually dominates lineshape considerations. The theory of Lorentzian lineshapes relates T_2 to the width of a line at half height [$(\text{lw})_{1/2}$ in Hz] by the formula given in eq. 7-1.

$$T_2 = \frac{2}{(\text{lw})_{1/2}} \qquad \qquad 7\text{-}1$$

The familiar ring-out or wiggle-beat pattern that follows sharp peaks recorded at fast sweeps arises from relaxation effects. The sharp maximum of the resonance signal occurs when the frequency of the B_1 field equals the Larmor frequency of the nucleus. As the B_1 frequency continues to increase, the B_1 vector rotates in the xy plane at an increased angular velocity about the z axis (Figure 7-1). Although the resonance condition no longer exists, xy magnetization continues to persist until transverse relaxation returns it to its equilibrium value of zero. Thus when the B_1 vector is $360°$ ahead of the M vector, they are again in phase and a small increment of xy magnetization is detected. This increment is observed as the first and strongest wiggle-beat. When the B_1 vector gains another $360°$, a second beat occurs that is smaller because the residual component of xy magnetization has diminished. Eventually, the ring-out decays to zero. The envelope of the wiggle-beat pattern can thus be used as a measure of T_2, the transverse relaxation time.

7-2 Quadrupolar Nuclei

Quadrupolar nuclei that are surrounded by an unsymmetrical (nontetrahedral, nonspherical) electron cloud have a special mechanism for longitudinal relaxation. A nucleus with a quadrupole moment is ellipsoidal in charge distribution, whereas those with a spin of 0 or $\frac{1}{2}$ are spherical. We have seen in Chapter 1 that a nucleus with a spin of 1 can be oriented with, normal to, or against the applied magnetic field (eq. 7-2). The unsymmetrical surrounding electron

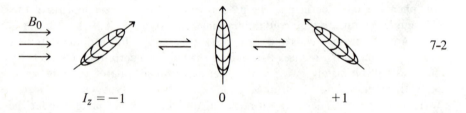

$$I_z = -1 \qquad\qquad 0 \qquad\qquad +1$$

7-2

cloud, tumbling in solution, gives rise to a fluctuating electric field. This field exerts a torque on the unsymmetrical quadrupolar nuclear charge and alters its orientation in the magnetic field. Although the interaction is purely electrostatic, the result is a change in the I_z spin quantum number. Thus the electric field about a tumbling ^{14}N nucleus can bring about transitions among the three spin states ($-1 \rightleftharpoons 0 \rightleftharpoons +1$). If a nucleus has been excited to the $+1$ state by application of the B_1 field, the interaction between the quadrupolar nucleus and the unsymmetrical electron cloud can induce relaxation. Since only the z component of the magnetization is affected, the mechanism is a longitudinal relaxation (T_1).

For this mechanism of relaxation to take place both the nucleus and the electron cloud must have an unsymmetrical charge distribution. If the nucleus is not quadrupolar, the fluctuating electric field will not be able to interact with it. Thus nuclei such as ^1H, ^{13}C, and ^{15}N are unaffected. If the electron cloud is tetrahedral in symmetry, as in the ammonium or borohydride ions, the electric field does not fluctuate as a result of tumbling, and again there is no effect. The most important effect on the quadrupolar nucleus itself is a considerable shortening of T_1, to the point that this quantity is the determining factor in the linewidth. Nuclei such as ^{35}Cl experience very strong quadrupolar relaxation, so that resonances (except for that of the chloride ion) are several kHz in width. Cases such as ^{14}N are termed intermediate, and linewidths may be a few tens of Hz wide. Because ^2H has only s electrons, the electric field is reasonably symmetrical, the quadrupolar effect is small, and linewidths can be on the order of a Hz.

More important to the organic chemist than the resonance linewidths of the quadrupolar nuclei is the effect quadrupolar relaxation has on the resonances of other nuclei in the molecule, particularly protons. If relaxation is extremely rapid, a neighboring nucleus will experience only the average spin environment of the quadrupolar nucleus, which therefore acts as if it were nonmagnetic. The protons in methyl chloride produce a sharp singlet resonance, despite the fact that the ^{35}Cl and ^{37}Cl nuclei have spins of $\frac{3}{2}$. Other nuclei with similar properties are ^{79}Br, ^{81}Br, ^{127}I, and ^{75}As. At the other extreme is the ^{2}H nucleus, which has a weak quadrupolar moment and relatively symmetrical surrounding electric fields. As a result, coupling of protons to neighboring deuterium atoms, with a spin of 1, can be observed. Figure 1-10 shows the proton in $CD_3(CO)CD_2H$ split into a 1/2/3/2/1 quintet by the two neighboring deuterium atoms.

The ^{14}N nucleus (and to some extent ^{11}B) straddles these two extremes. In some structural situations, such as the interior nitrogen in biuret, $NH_2(CO)NH(CO)NH_2$, quadrupolar relaxation is rapid enough to produce a sharp singlet resonance for the attached proton. That the singlet structure is due to quadrupolar relaxation rather than to rapid intermolecular exchange of the NH proton is proved by observation of the ^{15}N–H coupling by means of the ^{15}N sideband. Many examples of singlet N–H resonances, such as in amines, are due to exchange phenomena. In the very symmetric environment of the ammonium ion, the ^{14}N nucleus experiences no appreciable quadrupole relaxation, so the proton spectrum is a sharp triplet (see Figure 1-9). Intermediate rates of quadrupolar relaxation result in extremely broad spectra, exemplified by that of pyrrole (NH) given in Figure 5-4. Incomplete averaging of the ^{14}N spin states produces a proton resonance intermediate in form between the triplet and singlet structures. Sometimes the resonance is so broad that it is effectively invisible. Double irradiation at the frequency of the ^{14}N resonance will bring about rapid equilibration of the ^{14}N spin states and produce a sharp resonance for the attached proton (see section 5-2).

Changes in temperature can produce interesting effects on the spectra of protons attached to quadrupolar nuclei that have intermediate rates of relaxation. The spectrum of pyrrole at $-40°$ contains a broad singlet for the resonance of the proton attached to ^{14}N. At room temperature, the resonance is broadened to the point of invisibility (see Figure 5-4). At $+150°$, a triplet structure has emerged for this resonance. The increase in temperature has brought the relaxation rate from the fast to the slow extreme. As the temperature is raised, the rate of molecular tumbling increases, and the effectiveness of the coupling between the electric field and the quadrupolar nucleus diminishes. These observations provide one of the few examples of an increase in spectral complexity as the temperature is raised. All the examples given in Chapter 6 exhibit spectral simplification with increased temperature.

Nuclei that are anisotropically shielded may undergo relaxation by a mechanism that is formally similar to quadrupolar relaxation. The ^{13}C nucleus in carbon disulfide, for example, experiences a different local field when the axis of the molecule is lined up with the field direction from when it is orthogonal to the field. As the molecule tumbles in solution, the local field experienced by the nucleus therefore fluctuates. Frequency components produced by this fluctuating field at the Larmor frequency of the ^{13}C nucleus, therefore, can induce spin-lattice relaxation. The principal analogy to the quadrupolar mechanism is that in both cases fluctuating fields produced by molecular tumbling lead to longitudinal relaxation. The major difference is that relaxation through chemical shift anisotropy can occur with any nucleus, whereas quadrupolar relaxation only applies to those with a spin greater than $\frac{1}{2}$. It is also of interest that this is the only mechanism of relaxation that is field dependent. This observation is related to the general chemical shift dependence on the magnitude of B_0. Thus relaxation by chemical shift anisotropy can be unambiguously differentiated from other mechanisms by measurements at several fields.

7-3 Contact Shifts and Lanthanide Shift Reagents

Under conditions of very short electron relaxation times, unpaired electron spins can give rise to very large shielding effects. In purely organic free radicals, the primary effect is shortening of the nuclear T_1 and consequent line broadening. When the unpaired electron spin is on a transition metal, however, the chemical shifts of ligand protons can be very severely perturbed. The ethylenediamine complex of Ni^{II} (**1**) gives a proton spectrum

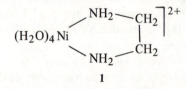

1

containing two resonances separated by about 270 ppm. The protons on carbon resonate at about δ 100 and those on nitrogen at about δ −170. Such shifts are unprecedented from considerations of data derived from diamagnetic materials.

These enormous shifts have been explained in terms of a contact mechanism. An unpaired electron on nickel polarizes the bonding electrons in such a way as to generate a small net unpaired spin density at the hydrogen nuclei. If the favored electron polarization is in the same direction as the B_0 field, it is termed positive; if in the opposite direction, negative. Positive and negative spin densities have opposite effects on the diamagnetic nuclear shielding, so both upfield and downfield shifts are observed. For this reason, the two types of protons in **1** are shifted in opposite directions. The magnitude of the shift is directly proportional to the hyperfine splitting constant a_H between the unpaired spin on nickel and the proton in question. Because the hyperfine splitting is proportional to the electron density ρ_C on the attached carbon (eq. 7-3), observation of the contact shift can serve as a direct measure of ρ_C for cases

$$a_H = Q\rho_C \qquad\qquad 7\text{-}3$$

in which a_H is not measurable from the esr spectrum. In this manner, electron densities have been measured at each of the ten different positions in Ni^{II} N,N'-bis(p-1,3-butadienylphenyl)-aminotroponeiminate (**2**). Because only a small portion of the unpaired spin density is actually

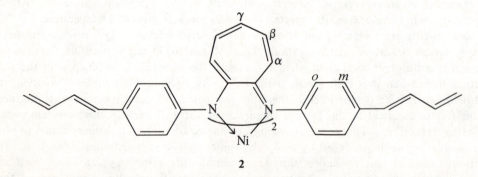

2

conveyed to the ligand nuclei, the measured values of ρ_C are generally on the order of a hundredth of a spin. The measured electron density at the α carbon of **2**, for example, is ±0.041, at the β carbon ∓0.021, at the ortho carbon ±0.0079. By comparison with values calculated from valence bond approaches, it has been estimated that only about a tenth of an electron is delocalized from the nickel atom to the ligand.

Organic chemists have found a far greater utility in the recently discovered lanthanide shift reagents. Hinckley observed that the addition of certain europium complexes to a solution of an organic material bearing a hydroxyl substituent greatly enhances the chemical shifts of the protons in the molecule. It is thought that the shift reagent and the substrate form a Lewis acid–base complex. The chemical shift enhancement is considered to derive from a pseudocontact mechanism, according to which the spin information is passed through space within the complex rather than through the bonding electrons. This mechanism has both a distance and an angular dependence. Because the pseudocontact mechanism drops off very rapidly with distance, the shifts are largest close to the point of complexation. The lanthanide shift reagents have come to be used primarily to minimize second-order spectral effects for ease of analysis. The spectrum of androstan-2β-ol is given in Figure 7-3 with and without tris(dipivalomethanato)europium(III) $\cdot$ 2 pyridine (3). The resonances close to the 2β-hydroxyl

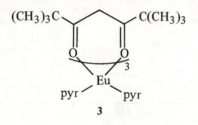

group experience the largest shifts. This reagent, when more commonly used without the two moles of pyridine, is referred to as Eu(dpm)$_3$. 1,1,1,2,2,3,3-Heptafluoro-7,7-dimethyloctane-dionatoeuropium(III), Eu(fod)$_3$ (4), has been found to have superior solubility properties to

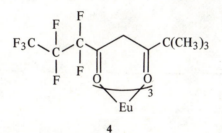

Eu(dpm)$_3$ and to complex with less basic sites.

Because of the dependence of the pseudocontact shift on the distance and angular relationship between the substrate and the shift reagent, quantitative analysis of these shifts can give useful information on structures in solution. The conformations of almost all the isomeric hexanols have been worked out in this manner. Addition of the shift reagent probably does not alter the conformation of the molecule in alcohol complexes, although the question is still open in other cases.

Shift reagents are now available with a wide range of ligands and with several of the rare earths (praseodymium, holmium, dysprosium, ytterbium, etc.). Useful shift information has been obtained with almost all basic organic functional groups: alcohols, amines, ketones, esters, sulfoxides, ethers, thioethers, nitriles, oximes, lactones, and aldehydes. Little or no shift has been found with nitro compounds, halides, and alkenes; decomposition occurs with acids and phenols. Lanthanide shifts also have been observed for nuclei other than ^{1}H, such as ^{19}F and [13]C.

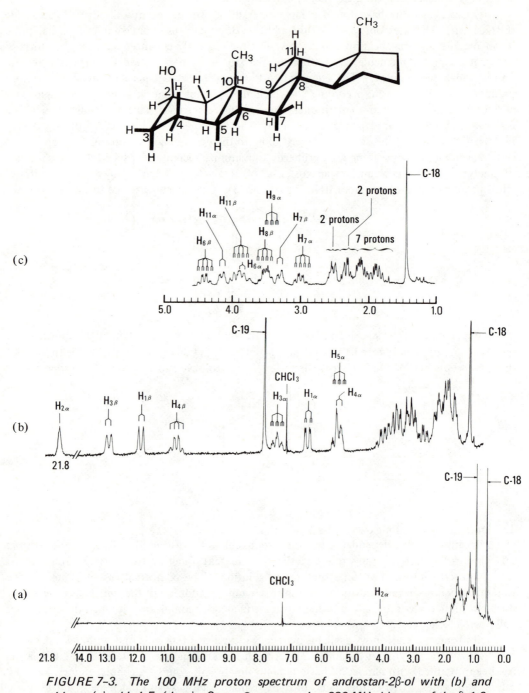

FIGURE 7–3. The 100 MHz proton spectrum of androstan-2β-ol with (b) and without (a) added Eu(dpm)₃·2pyr. Spectrum c is a 220 MHz blowup of the δ 1.0–5.0 region. [Reproduced with the permission of the American Chemical Society from P. V. Demarco, T. K. Elzey, R. B. Lewis, and E. Wenkert, J. Amer. Chem. Soc., **92**, 5737 (1970).]

7-4 Nuclear Overhauser Effect

Frequently two nuclei that are held close to each other by molecular constraints provide the dominant mechanism for mutual relaxation through dipole-dipole interactions. If one of these nuclei is doubly irradiated, the Boltzmann distribution of spins for the other nucleus is altered, and the intensity of its resonance is perturbed. The phenomenon was first observed by Overhauser in cases for which an electron resonance frequency is irradiated and the effects on the resonance of a nucleus are observed. Of more interest to the organic chemist is the case in which both spins are nuclei, hence the term *nuclear Overhauser effect* or nOe. The phenomenon has great structural utility because the dipole-dipole mechanism for relaxation drops off with the inverse of the distance between the two spins. Qualitative conclusions with regard to molecular structure or conformation are readily obtained, and quantitative results are possible.

Figure 7-4 depicts the spin energy states for a pair of uncoupled nuclei of spin $\frac{1}{2}$. Nucleus A is irradiated, nucleus B observed. When the oscillating B_2 field is applied at the resonance frequency of A, states 2 and 4 are rapidly interconverted, as are 1 and 3. The process of cross-relaxation ($++ \rightleftharpoons --; -+ \rightleftharpoons +-$) operates to return a B nucleus that is excited to state 4 from 3, back to state 1; or one excited to state 2 from 1, back to 3. When the new steady state system is attained, the populations of states 1 and 3 and of 2 and 4 are equal (eq. 7-4).

$$\begin{aligned} p_1 &= p_3 \\ p_2 &= p_4 \end{aligned} \qquad \text{7-4}$$

Without irradiation at the A frequency, the ratio of B nuclei with $I_z = +\frac{1}{2}$ (less stable) to those with $I_z = -\frac{1}{2}$ is given by eq. 7-5. When the A frequency is saturated, this ratio

$$\frac{n_B(+\frac{1}{2})}{n_B(-\frac{1}{2})} = \frac{p_2}{p_1} = \frac{p_4}{p_3} = e^{-(E_2 - E_1)/kT} = e^{-(E_4 - E_3)/kT} \qquad \text{7-5}$$

is given by eq. 7-6, which is elaborated by substitution from eq. 7-4. The new population

$$\frac{n_B(+\frac{1}{2})}{n_B(-\frac{1}{2})} = \frac{p_2 + p_4}{p_1 + p_3} = \frac{p_4}{p_1} = e^{-(E_4 - E_1)/kT} \qquad \text{7-6}$$

ratio is determined by an effective energy difference that can be as large as $(E_4 - E_1)$. Because this quantity is potentially larger than $(E_2 - E_1)$ or $(E_4 - E_3)$, the population of the

FIGURE 7-4. Spin states for a system of two uncoupled nuclei with double irradiation at the frequency of nucleus A and observation of B.

lower spin state increases. With a larger number of spins able to resonate, the B nucleus produces a signal with a larger integrated intensity.

The maximum attainable increase in intensity depends on the ratio of the gyromagnetic ratios for the A and B nuclei. If both are protons, only a factor of 1.5 in intensity is obtained (an increase of 50%). If nucleus A is a proton, and B is ^{13}C, the factor is almost 3.0 (200%). Operation of other relaxation processes serves to reduce the observed intensity enhancement. Thus if proton B is in part relaxed by nuclei other than A, the Overhauser enhancement is proportionately decreased. For optimal results, the observed nucleus B should be physically much closer to A than to any other magnetic nuclei, and a solvent without magnetically active nuclei should be used to avoid intermolecular relaxation.

The nuclear Overhauser effect is exploited in ^{13}C nmr experiments simply as a means of signal enhancement. Most ^{13}C nuclei are directly bonded to protons that provide the dominant mode of relaxation. Noise decoupling of all protons thereby produces an Overhauser enhancement of up to a factor of three. This proposition was quantitatively tested by measuring the integrated intensity of the ^{13}C resonance of enriched formic acid with and without double irradiation at the proton frequency. The mean observed enhancement was 2.98 ± 0.15. It should be emphasized that the Overhauser enhancement is completely independent of spectral changes that arise from the collapse of spin multiplets due to spin decoupling. The observation of an nOe does not require that the nuclei A and B be spin-coupled, but only that A relax B through a dipolar mechanism.

The dependence of the nOe on geometry lends itself to spectral, structural, and conformational assignments. In the first such application, Anet and Bourn demonstrated that the lower field methyl resonance in the spectrum of *N,N*-dimethylformamide corresponds to the methyl group cis to the formyl proton. Complicated structural questions may sometimes be answered for materials in the solution phase. The adenosine derivative **6** (2',3'-isopropyl-ideneadenosine) was the subject of a series of Overhauser experiments. The question was

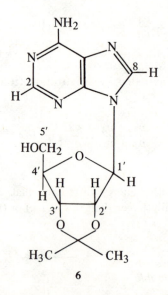

6

whether the purine portion of the molecule lies over the sugar as shown (syn) or in an extended form in which the proton on C_8 lies over the sugar (anti). Saturation of the H-1' resonance brings about a 23% enhancement of the H-8 resonance. Similarly, saturation of H-2' produces a 9% enhancement of H-8, whereas saturation of H-3' or H-5' produces only about a 4% enhancement. The proton at the 8 position, therefore, must be positioned closest to H-1′ but not far from H-2′. In the syn form H-8 is closest to H-1′; in the anti form, to H-5′. The predominant form in solution must then be the syn rotational isomer, as drawn, although some anti may be present to cause the small enhancements at H-3′ and H-5′. In the corresponding guanosine derivative, irradiation at H-3′ or H-5′ gives a 10–12% enhancement, so a significant amount of the anti isomer must be present. Failure to observe an nOe cannot be taken as evidence to reject or accept a particular structure. Enhancements of at least 10% are generally required before a positive assignment can be accepted.

7-5 Chemically Induced Dynamic Nuclear Polarization

Unusual spin polarizations have been observed in product resonances from reactions that pass through radical intermediates. Figure 7–5 shows the reaction mixture from the thermolysis of benzoyl peroxide as a function of time.   The reaction proceeds through the benzoyloxy

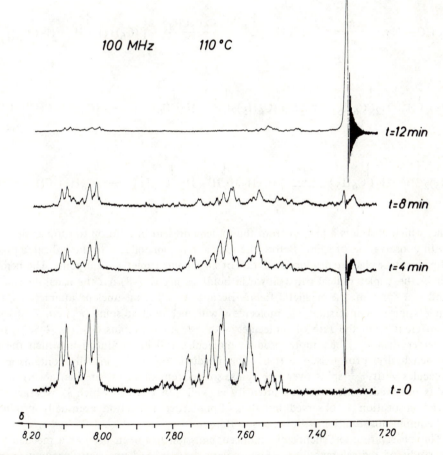

FIGURE 7–5. The 100 MHz proton spectrum of benzoyl peroxide in cyclohexanone at 110° as a function of time. [Reproduced with the permission of Verlag der Zeitschrift für Naturforschung from J. Bargon, H. Fischer, and V. Johnsen, Z. Naturforsch., A, **22**, 1551 (1967).]

and phenyl radicals, the latter of which abstracts a proton to produce benzene (eq. 7-7). At

$$C_6H_5\overset{\overset{\displaystyle O}{\|}}{C}-O-O-\overset{\overset{\displaystyle O}{\|}}{C}C_6H_5 \xrightarrow{\Delta} 2[\,C_6H_5-\overset{\overset{\displaystyle O}{\|}}{C}-O\cdot\,] \xrightarrow{-2CO_2} 2[\,C_6H_5\cdot\,] \longrightarrow C_6H_6$$

7-7

zero time for the reaction, the proton spectrum contains the resonances for benzoyl peroxide. After 4 min, these resonances have diminished in intensity, and a strong benzene resonance has appeared with reverse spin polarization (emission). After 8 min, the peroxide resonance has further diminished, and the benzene resonance has decayed to zero. After 12 min, the peroxide resonances have almost disappeared, and the benzene resonance has grown to its normal state of magnetization.

Observations of this sort, termed *chemically induced dynamic nuclear polarization* (cidnp), have been explained in terms of polarization of spin during singlet–triplet transitions within a correlated radical pair. The decomposition of diphenyldiazomethane in toluene provides a platform for discussion (eqs. 7-8). Sensitized photolysis produces triplet diphenyl-

$$(C_6H_5)_2C-N_2 \xrightarrow[\substack{\text{sens.} \\ -N_2}]{h\nu} (C_6H_5)_2\overset{1}{C}1 \xrightarrow{C_6H_5CH_3} [\,(C_6H_5)_2\overset{1}{C}H \quad H_2\overset{1}{C}C_6H_5\,] \quad 7\text{-}8a$$

$$[(C_6H_5)_2\overset{1}{C}H^\alpha \quad H_2\overset{1}{C}C_6H_5] \underset{}{\overset{k_\alpha}{\rightleftharpoons}} [\,(C_6H_5)_2\overset{\downarrow}{C}H^\alpha \quad H_2\overset{1}{C}C_6H_5\,] \longrightarrow (C_6H_5)_2CH^\alpha-CH_2C_6H_5$$

7-8b

$$[(C_6H_5)_2\overset{1}{C}H^\beta \quad H_2\overset{1}{C}C_6H_5] \underset{}{\overset{k_\beta}{\rightleftharpoons}} [\,(C_6H_5)_2\overset{\downarrow}{C}H^\beta \quad H_2\overset{1}{C}C_6H_5\,] \longrightarrow (C_6H_5)_2CH^\beta-CH_2C_6H_5$$

7-8c

carbene, which abstracts a proton from the toluene present in solution to form a triplet radical pair with conservation of spin. Before the radical pair can collapse to the indicated product, a spin flip must take place to produce a mixture of singlet and triplet radical pairs. The benzhydryl proton in the triplet radical pair (shown in boldface in eqs. 7-8) has the usual distribution of $+\frac{1}{2}$ and $-\frac{1}{2}$ spins in the magnetic field. Because of electron–nucleus interactions, the rate of triplet–singlet conversion of molecules with $+\frac{1}{2}$ nuclear spins (k_α, eq. 7-8b) may be quite different from the rate of molecules with $-\frac{1}{2}$ nuclear spins (k_β, eq. 7-8c). If k_α is much larger than k_β, then more product molecules will be produced in which the I_z spin of the benzhydryl proton is $+\frac{1}{2}$ (α). The product, therefore, initially contains an excess of molecules with $+\frac{1}{2}$ spins over $-\frac{1}{2}$ spins, so that emission, rather than absorption, of rf energy is observed, with enhanced intensity as well. In cases in which k_β is larger than k_α, enhanced absorption is observed initially. Spin-lattice relaxation eventually produces the normal equilibrium situation with net absorption.

The observation of chemically induced emission has been used as a mechanistic tool in the study of radical reactions. Spin polarization in a diamagnetic product is taken as evidence that the nuclear spins resided at some time during the reaction in a radical pair. Halogen–lithium exchange in a mixture of *n*-butyllithium and *sec*-butyl iodide produces polarization of the methylene and methyne protons, respectively, in *n*-butyl iodide and *sec*-butyl iodide. A mechanism has been proposed in which a one-electron transfer gives a radical pair that can either return to polarized *sec*-butyl iodide or go on to polarized *n*-butyl iodide (eq. 7-9). The details of numerous rearrangement reactions have been clarified

$$sec\text{-}C_4H_9\text{--}I + n\text{-}C_4H_9\text{--}Li \rightleftharpoons [sec\text{-}C_4H_9\cdot, \text{I}, \text{Li}, n\text{-}C_4H_9\cdot] \rightleftharpoons sec\text{-}C_4H_9\text{--}Li + n\text{-}C_4H_9\text{--}I \qquad 7\text{-}9$$

by observation of spin polarizations. The Meisenheimer rearrangement of a tertiary amine oxide to an oxygen-substituted hydroxylamine furnishes such an example. The rearrangement of *N,N*-dimethylbenzylamine oxide at 140° produces *N,N*-dimethyl-*O*-benzylhydroxylamine in which both the methyl and the benzyl protons give polarized resonances. The rearrangement probably proceeds by N–C cleavage followed by benzyl migration to the oxygen atom (eq. 7-10). Analogous observations have been made for rearrangements of sulfonium and ammonium ylids.

$$C_6H_5CH_2\text{--}\underset{\overset{+}{N}}{\overset{O^-}{|}}\diagdown\overset{CH_3}{\diagdown CH_3} \rightarrow \begin{bmatrix} C_6H_5\overset{\cdot}{C}H_2 & \cdot\underset{\overset{+}{N}}{\overset{O^-}{|}}\diagdown\overset{CH_3}{\diagdown CH_3} \\[2em] C_6H_5CH_2 & \underset{N}{\overset{\cdot O}{|}}\diagdown\overset{CH_3}{\diagdown CH_3} \end{bmatrix} \rightarrow C_6H_5CH_2\text{--}O\text{--}N\diagdown\overset{CH_3}{\diagdown CH_3}$$

$$7\text{-}10$$

The cidnp technique has the advantage over conventional esr detection of free radicals in increased sensitivity and in the assurance that the observed radical is present in the actual reaction pathway rather than in some extraneous process. Both techniques are limited by the absence of proof that the radicals arise in the major reaction pathway. The polarizations may be associated with a minor pathway that leads to a misconception of the overall reaction mechanism. Polarization only occurs in molecules that recombine within the geminate radical pair. Radicals that escape from the cage before recombination do not contribute to the induced polarization, so that some reactions passing through radical intermediates might go undetected by the cidnp technique.

BIBLIOGRAPHY

1. Contact Shifts: D. R. Eaton and W. D. Philips, *Advan. Magn. Resonance*, **1**, 103 (1965).
2. Lanthanide Shift Reagents: J. R. Campbell, *Aldrichim. Acta*, **4**, 55 (1971); R. v. Ammon and R. D. Fischer, *Angew. Chem., Int. Ed. Engl.*, **11**, 675 (1972); *Nuclear Magnetic Resonance Shift Reagents*, R. E. Sievers, ed., Academic, New York, 1973; J. Reuben, *Progr. NMR Spectrosc.*, **9**, Part 1, 3 (1973); B. C. Mayo, *Chem. Soc. Rev.*, **2**, 49 (1973).
3. Nuclear Overhauser Effect: J. H. Noggle and R. E. Schirmer, *The Nuclear Overhauser Effect*, Academic, New York, 1971.
4. CIDNP: H. R. Ward, *Accounts Chem. Res.*, **5**, 24 (1972); R. G. Lawler, *Accounts Chem. Res.*, **5**, 32 (1972); S. H. Pine, *J. Chem. Educ.*, **49**, 664 (1972); R. G. Lawler, *Progr. NMR Spectrosc.*, **9**, Part 3, 145 (1973); C. Richard and P. Granger, *NMR, Basic Prin. Progr.*, **8**, 1 (1974); G. L. Closs, *Advan. Magn. Resonance*, **7**, 157 (1974).

Part Two

VIBRATIONAL SPECTROSCOPY

1

PRELIMINARY CONSIDERATIONS

1-1 Introduction

Both infrared and Raman spectra give information on molecular vibrations or, more accurately, on transitions between vibrational energy levels in molecules. This information is of considerable help to the organic chemist because it can be directly related to molecular structure.

An infrared spectrum is obtained when a sample *absorbs* radiation in the region of the electromagnetic spectrum known as the infrared. The expression *absorption band* is used to denote a feature that is observed in the spectrum. When the absorption band is quite narrow and sharp, the word *peak* is used.

A Raman spectrum, on the other hand, is produced by a *scattering* process. Monochromatic incident radiation, usually in the visible region of the spectrum and usually produced by a laser, is scattered by the sample. The scattered light is observed instrumentally in a direction perpendicular to the incident radiation. Since early Raman spectra were recorded on photographic plates, the features appeared as *lines* or *bands* on the plates and these words are now used when discussing Raman spectra. The expression *Raman shift* may also be encountered in descriptions of Raman spectra. The significance of this expression will be seen in section 1–3.

The word *frequency* is encountered when reading about infrared and Raman spectra. In both contexts this term can be taken to mean a vibrational frequency of the molecule. In the case of an infrared band, the frequency of vibration corresponds to the frequency of the infrared radiation absorbed. In the case of a Raman frequency, the relationship between the vibrational frequency of the molecule and the frequency of the scattered light is a little more complicated and will be discussed in Chapter 3.

The material for study is usually in the form of a solid, a neat liquid, or a solution. Sometimes, however, a compound in the gas or vapor phase is studied. Under these conditions, in addition to changes in vibrational energy, simultaneous changes in rotational energy can occur and some structure may be observed on the vibrational band. This statement really only applies to infrared spectra, since Raman spectra of gases are difficult to obtain. Several atmospheres pressure of the gas are required, or a complicated multiple reflection cell must be used when such pressures cannot be obtained. Even then, the vapor pressure of the compound must be at least 100 torr. It is likely, therefore, that Raman spectra of gases will only be of importance to the organic chemist in certain special cases.

Infrared and Raman spectra together can provide a variety of information on structure, symmetry, functional groups, purity, structural and geometrical isomers, and hydrogen bonding. Some of this information is obtained from three closely related fields: near infrared, far infrared, and infrared reflectance spectroscopy.

1-2   Notation

For most of this Part on infrared and Raman spectroscopy we shall talk about *fundamental vibrational frequencies* of molecules. By this term we mean one of the classical mechanical fundamental frequencies of a vibrating system comprised of masses (atoms) connected by springs (bonds). In other words, the frequencies of vibration are those of a *classical mechanical model* representing the molecule under consideration.

We know that molecules are usually described using *quantum mechanics*. What we observe as a spectral feature in the infrared or Raman spectrum can be ascribed to a transition between two *energy levels* of the molecule. We will show later that to a good approximation, the classical mechanical model gives the same results as the quantum mechanical treatment. In other words, the difference in energy between the two levels involved in the quantum mechanical transition corresponds approximately to a classical mechanical vibrational frequency. The mathematical relationship is the familiar $E_2 - E_1 = h\nu$. A quantum of electromagnetic radiation $h\nu$ (a photon) is absorbed or released as a result of the transition between energy levels E_1 and E_2. The vibrational frequencies ν_i (vibrations per second) can be calculated using methods of classical mechanics. This procedure is discussed in some detail in Chapter 3.

One other point that must be made is the basic difference between infrared and Raman spectroscopy, which lies in the mechanism by which the electromagnetic radiation interacts with the molecule to produce the spectrum. *Infrared* spectra result from *absorption of photons* by molecules. *Raman* spectra, on the other hand, can be considered as resulting from inelastic *collisions of photons* with molecules. In the case of infrared absorption the photon causes the molecule to increase its vibrational energy by a transition to a higher energy level. In the Raman effect the inelastic collision results in the transfer of some energy either from the photon to the molecule or from the molecule to the photon. In the former case, the molecule will be left in a higher energy level (giving rise to the so-called Stokes lines). In the latter case, the molecule must already be in an excited state so that it can return to a lower state after giving up energy to the photon (giving rise to the so-called anti-Stokes lines). Since most molecules are in their ground vibrational state at normal temperatures, only the Stokes lines are important and it is these that comprise the Raman spectrum of interest.

1-3 Units

A spectrum is recorded graphically with the frequency or wavelength as the abscissa and with the intensity as the ordinate. There are several units used for both ordinate and abscissa scales. The unit of frequency ν is $\sec^{-1}$ (vibrations per second). This quantity is a very large number for molecular vibrations (of the order of $10^{13} \sec^{-1}$) and is inconvenient. Infrared spectroscopists formerly used a unit related to the frequency, namely, the *wavelength* (in microns) of the infrared radiation absorbed. The micron (μ) is 10^{-4} cm. Other related wavelength units are defined as follows:

$$
\begin{aligned}
1 \text{ angstrom (Å)} &= 10^{-8} \text{ cm} \\
1 \text{ millimicron (m}\mu\text{)} &= 10^{-7} \text{ cm} \\
1 \text{ nanometer (nm)} &= 10^{-7} \text{ cm} \\
1 \text{ micron (}\mu\text{)} &= 10^{-4} \text{ cm}
\end{aligned}
$$

In many older texts and papers on infrared spectroscopy, the positions of the infrared absorption bands are given in microns, and most of the earlier instruments produced a spectrum whose abscissa was linear in wavelength.

The positions of Raman lines cannot be expressed in units of wavelength, since the lines are measured as *shifts* from the incident or exciting line. The wavelength of the exciting line depends on the laser used. Since infrared and Raman spectra are used together to give information on molecular structure, it is obviously necessary to have a common unit. One could use the unit of frequency—in fact, one does talk about Raman frequencies. However, as noted above, molecular vibrational frequencies are very large (10^{13}–$10^{14} \sec^{-1}$). A more convenient unit, $\bar{\nu}$, is obtained (eq. 1-1) by dividing the frequency ν by the velocity of light

$$ \bar{\nu} = \frac{\nu}{c} \qquad\qquad 1\text{-}1 $$

(3×10^{10} cm $\sec^{-1}$). Consider, for example, a fundamental vibration of frequency 3×10^{13} $\sec^{-1}$. As eq. 1-1a shows, the result is a convenient unit for vibrational frequencies. This

$$ \frac{3 \times 10^{13} \sec^{-1}}{3 \times 10^{10} \text{ cm} \sec^{-1}} = 10^3 \text{ cm}^{-1} \qquad\qquad 1\text{-}1a $$

1000 cm^{-1} is read as 1000 *centimeters to the minus one, reciprocal centimeters,* or *wavenumbers.* Although this unit is a frequency *divided* by a velocity, it is common practice to refer to 1000 cm^{-1} as a *frequency* of 1000 cm^{-1}, with the division by c understood. Now infrared and Raman spectra can be directly compared, and the unit cm^{-1} will be used exclusively from now on.

There is also strong support on theoretical grounds for using cm^{-1}, since this unit is related directly to energy ($E = h\nu = hc\bar{\nu}$). Wavelength, velocity, and frequency of electromagnetic radiation are related by eq. 1-2, where n is the refractive index of the medium

$$ n\lambda\nu = c \qquad\qquad 1\text{-}2 $$

through which the radiation passes. Now it is the wavelength that is reduced by the factor n, whereas the frequency is unchanged. Thus the frequency is the most important characteristic of the radiation and is the property that should be used to describe it.

Even air has an effect (which is not linear) on the wavelength of the radiation, and corrections must be applied to observed Raman lines measured in terms of wavelength in air in order to obtain the true Raman frequencies. This problem can be circumvented by calibrating Raman spectrometers to give frequency shifts directly in cm^{-1}. The correction is small at infrared wavelengths and can be neglected.

Finally, from eq. 1-2 using λ corrected to vacuum for which $n = 1.0$, we get eq. 1-3.

$$\frac{1}{\lambda} = \frac{\nu}{c} \qquad\qquad\qquad 1\text{-}3$$

If we measure wavelength in cm, we see that $1/\lambda$ is in cm^{-1}, and for this reason the unit cm^{-1} is sometimes called the *wavenumber*.

Several units are used to measure the intensity of an infrared absorption peak. Percent transmission and absorption are most common, but *absorbance* (A) and *transmittance* (T) will also be encountered. The first two units are self-explanatory; absorbance is defined in several ways (eq. 1-4, in which I_0 is the intensity of the incident light and I of transmitted

$$A = \log_{10}\frac{I_0}{I} = \log_{10}\frac{1}{T} = \log_{10}\frac{100}{\%T} \qquad\qquad 1\text{-}4$$

light). The absorbance used to be known as the optical density (O.D.).

The intensity of absorption can be related to the concentration and the path length by the Beer–Lambert law; eq. 1-5, where c is the concentration in moles liter^{-1}, l is the

$$A = \epsilon l c \qquad\qquad\qquad 1\text{-}5$$

pathlength in cm, and ϵ is the molar absorption coefficient, with unit cm^2 mole^{-1}. (Sometimes concentrations are expressed in mole fraction, a dimensionless quantity. In this case, ϵ will have the unit of cm^{-1}. Older names for ϵ include extinction coefficient and molar absorptivity.) A strong infrared band would have an ϵ of the order of 100 cm^2 mole^{-1} or an A of the order of 1.0, while a weak band would have an ϵ of the order of 10 cm^2 mole^{-1} or an A of about 0.1 (A of course depends on pathlength).

Intensities in Raman spectra are much less quantitative, since the height of a peak depends on many factors, such as the power of the laser, the wavelength, the slit width, the sensitivity of the photomultiplier detector, and the amplification system used. Thus quantitative results can only be obtained if an *internal standard* is used to determine the amount of sample actually in the laser beam and giving rise to scattering.

One final important point. The intensity of a Raman line is a *linear* function of concentration, whereas the intensity of an infrared absorption band is a logarithmic function of concentration. Doubling the concentration of a solution should double the intensities of all Raman lines for identical instrumental settings, whereas the apparent effect on the infrared peak heights will depend on the peak. For example, a weak infrared band will appear to be affected much more than a strong absorption, since doubling the concentration will almost double the intensity of a weak band, while changing that of a strong band by only about 10%. It is clear that caution must be exercised in discussing both infrared and Raman band intensities.

1-4 Infrared Spectroscopy

Infrared spectroscopy is a well-established technique and commercial instruments have been available since the late 1940s. Over the years a very large number of infrared spectra have accumulated in the literature, and collections of reference spectra are commercially available. All this makes infrared spectroscopy a very useful tool for determination of molecular structure. Direct information about the presence of functional groups is immediately available from an infrared spectrum, and comparison of the infrared spectrum of an unknown material with a reference spectrum, or with the spectrum of a known compound, can give an empirical proof of the identity of the unknown substance.

For the present purposes the normal infrared range will be assumed to be 4000–650 cm^{-1}. Many infrared spectrometers, however, cover a wider range, overlapping the far infrared to 400 or even 200 cm^{-1}.

1-5 Raman Spectroscopy

Standard "bench top" laser Raman spectrometers, covering the range from 20 to 3500 cm^{-1} are now commercially available, and sampling and recording of spectra are as simple as in the infrared technique. Samples in the form of liquids, solutions, powders, and single crystals can be handled by a standard instrument. Gases, for which the Raman intensity is usually several orders of magnitude weaker, require a more sophisticated arrangement involving a higher power laser, multiple reflection cells, and often a specially cooled photomultiplier.

Raman spectra complement infrared spectra, and the two techniques taken together can provide a powerful tool in organic structure determination. An important difference between the two techniques is the cost, which for comparable resolution and accuracy is about two to three times higher for Raman than for infrared. The main reason for this difference is the fact that a more sophisticated monochromator is needed for a Raman spectrometer than for an infrared instrument (see section 2-2).

Experienced users of Raman spectroscopy could obtain the same kind of structural information as the infrared spectroscopist, but they would use a different set of group frequencies. Vibrations that give rise to strong characteristic infrared absorption are often very weak in the Raman spectrum. The converse is also true.

There are certain advantages of Raman spectroscopy over infrared. One is the simpler spectra usually observed in the Raman because of the absence of overtone or combination bands. The latter are an order of magnitude weaker in the Raman than in the infrared and are usually too weak to be observed. A second advantage is the wider choice of solvents for solution spectra. Water in particular can be used and other solvents have more clear regions in the Raman than in the infrared. Information on the lower frequency region 50–200 cm^{-1}, corresponding to the far infrared, is easily obtained.

There are, however, certain disadvantages of Raman spectroscopy. One is the inherent weakness of Raman spectra; often they are masked by the background and scattering from dirty samples. Fluorescence can sometimes be a nuisance, but this problem is virtually eliminated when the He/Ne red laser is used. It is often hard to get good spectra from solids unless they are crystalline. As mentioned earlier, a disadvantage of Raman spectroscopy is the high cost of the instruments. The cost of the lowest priced Raman instruments is in the price range of a good infrared spectrometer.

1-6 Far Infrared Spectroscopy

The region of the electromagnetic spectrum known as the far infrared is usually taken as $200\text{-}10$ cm^{-1}. However, the range of the small routine instruments generally available in organic laboratories often ends at 650 or 400 cm^{-1}. For the present purposes, the "no man's land" between 650 and 200 cm^{-1} will be included with the far infrared range.

Many fundamental vibrational frequencies, such as bending and stretching modes involving heavy atoms and torsional vibrations, occur below 650 cm^{-1} and are important for identification and structure determination. Stretching and bending of the $X\text{–}H\cdots X$ group in hydrogen-bonded compounds also have frequencies in the far infrared. Considerable experimental data have been gathered on infrared spectra in the region 650–300 cm^{-1}. The very far infrared region, below 100 cm^{-1}, is not too important for organic chemists.

1-7 Near Infrared Spectroscopy

Although limited in use, near infrared spectroscopy can be important for the study of compounds containing $O\text{–}H$, $N\text{–}H$, and $C\text{–}H$ groups, particularly alcohols, phenols, amines, and hydrocarbons. The near infrared region is also useful for the study of hydrogen bonding.

The spectral range known as the near infrared extends from the red end of the visible spectrum $(12{,}500$ $cm^{-1})$ to 4000 cm^{-1}, and several commercial instruments are available for this region. Although it is not so useful as the mid-infrared for qualitative analysis, it is particularly useful for quantitative work. Association studies involving hydrogen-bonded species, the determination of water as a contaminant, and the quantitative determination of methyl and methylene groups in aliphatic compounds are examples of important applications of near infrared spectroscopy.

1-8 Internal Reflection Spectroscopy

When a ray of light strikes an interface between two *nonabsorbing* materials of different refractive index (n_1 and n_2), the light is partially transmitted and partially reflected. Suppose the direction of the light is from material 1 to 2, with n_1 less than n_2 (for example, air to glass). This arrangement is known as *external* reflection. For external reflection the reflectivity can never be 100% and is usually much less.

When n_1 is greater than n_2, we have *internal* reflection. In this case the reflection is total (i.e., the reflectivity is 100%) when the angle of incidence is between the critical angle and $90°$ (grazing incidence). In order to understand how internal reflection can be used to obtain spectra we shall study the phenomenon a little further.

Total internal reflection can be observed in a glass of water. When the inside of the glass is viewed through the water surface, it appears to be completely silvered and opaque. When the outside of the glass is touched with a finger, details of the ridges and whorls on the skin are clearly seen, but the silvered effect remains between these features. The total reflection is destroyed where the skin actually makes contact with the glass. This result can be explained by a penetration of the electromagnetic field of the light into the rarer medium (smaller refractive index) by a fraction of a wavelength, as illustrated in Figure 1–1.

If the light is in the infrared region and if the rarer medium is a compound that absorbs infrared radiation, then the penetrating radiation field can interact by means

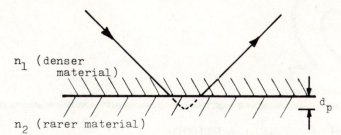

n_1 (denser material)

n_2 (rarer material)

FIGURE 1-1. The path of a ray of light in total internal reflection. The ray penetrates a fraction of a wavelength (d_p) beyond the reflecting surface into the rarer medium of refractive index n_2.

of an absorbing mechanism to produce an attenuation of the total reflection (atr). This interaction can be described in terms of an *effective thickness*, which corresponds to a sample thickness in a normal absorption process.

The technique of internal reflection spectroscopy consists of recording through the usual infrared region, the wavelength dependence of the *reflectivity*, of the interface between a sample compound and a material that has a higher refractive index, but is transparent to the infrared radiation. The radiation approaches and leaves the interface through the denser medium. In the simplest experimental arrangement, the material of higher refractive index is in the form of a prism, as shown in Figure 1-2. The spectra recorded are very similar to those obtained by normal transmission techniques.

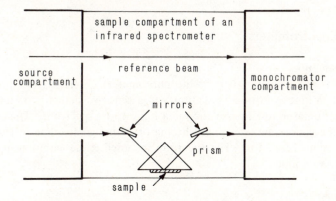

FIGURE 1-2. A diagram of a simple experimental arrangement for obtaining an internal reflection spectrum.

2

EXPERIMENTAL METHODS

2-1 Infrared Spectrometers

2-1a Introduction

An infrared spectrometer (or spectrophotometer), consists of three basic parts: (1) a source of continuous infrared radiation, (2) a sensitive detector of infrared radiation, and (3) a monochromator to disperse the radiation into its spectrum. The sample is usually placed between the source and the monochromator, although in some instruments it can be placed after the monochromator. The spectrum recorded is the variation in transmission by the sample with wavenumber.

2-1b Sources

The source of infrared radiation is usually a coil of wire, such as Nichrome with high resistance, or a rod of partially conductive material such as silicon carbide, or rare earth oxides heated by an electrical current. Temperatures of about $1200°C$ ($1500°K$) give the optimum yield of energy in the infrared. Figure 2-1 shows the energy distribution of a typical *black body* (nonreflecting) radiator at several temperatures. It can be seen that at $1800°K$ there is only a small increase in intensity at low wavenumbers (long wavelengths) but a large increase at high wavenumbers (short wavelengths).

2-1c Detectors

The most commonly used detector for the infrared region is the thermocouple. The radiation from the monochromator is focused by means of a mirror onto a junction of a very sensitive thermocouple maintained under a vacuum (Figure 2-2). The emf from this thermocouple is amplified and further treated by electronic devices in order to produce the recorded spectrum. Bolometers have also been used for both the infrared and far infrared.

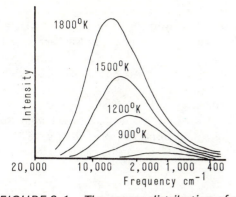

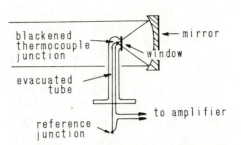

FIGURE 2-1. *The energy distribution of a black body at several temperatures.*

FIGURE 2-2. *Diagram of a thermocouple detector.*

For the near infrared, a more sensitive device, the lead sulfide photoconductive detector (similar to that used in a light meter) is used. For the far infrared, a detector known as a Golay cell (after its inventor) can be employed. The Golay cell is illustrated in Figure 2-3. A gas chamber has a thin film of metal at one end and a flexible mirror at the other. Infrared radiation striking the absorbing surface causes heating of the gas and deformation of the mirror. The deformation is measured by an auxiliary optical system using visible light and a photocell. A grid is placed between the light source and the pneumatic cell in such a way that only when the flexible mirror is distorted can light pass to the photocell. This rather complicated device is more sensitive than a thermocouple in the far infrared.

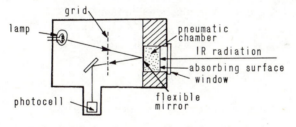

FIGURE 2-3. *Diagram of a Golay detector.*

2-1d The Monochromator

As the name implies, the monochromator disperses the continuous radiation into its spectrum of monochromatic components. These component frequencies are passed sequentially and continuously by a mechanical scanning device to the detector. In this way, the detector can sense which frequencies have been absorbed or partially absorbed by the sample and which frequencies have been unaffected. The radiation enters the monochromator through a slit and after dispersion leaves through another slit. The width of the entrance slit determines how much energy enters the monochromator and the width of the exit slit determines the width of the narrow band of frequencies simultaneously reaching the detector.

The ability of the instrument to distinguish between absorptions at closely similar frequencies is known as the resolution. The resolution depends on the width of the slits and, more importantly, on the dispersion element in the monochromator. Two types of dispersion element are used in spectrometers, prisms and gratings.

2-1e Prisms

The dispersion element in older spectrometers was invariably a prism of a material transparent to the infrared. The most commonly used prism material was NaCl (rock salt), and the region of the spectrum covered was from 4000 to 650 cm^{-1}. Below 650 cm^{-1} NaCl itself absorbs, and at higher frequencies the dispersion is poor. All prism materials exhibit the phenomenon of good dispersion at frequencies approaching the absorption band of the material itself. Thus, to cover a wide range of frequencies with good dispersion, say from 4000 to 200 cm^{-1}, several prisms would have to be used. For example, the combination of LiF, CaF$_2$, NaCl, and KBr prisms would provide good dispersion over the whole range from 4000 to 400 cm^{-1}. For further information on prisms as dispersion elements the reader is referred to Chapter 2 of reference (1).

2-1f Gratings

Gratings disperse the radiation into its constituent wavelengths by means of diffraction. A grating consists of a large number of equally spaced grooves ruled on a metal surface (Figure 2-4). In order for the rays that are diffracted from successive grooves to be in phase, the relationship given in eq. 2-1 must be obeyed, where a is the distance between

$$a(\sin i + \sin \theta) = n\lambda \qquad\qquad 2\text{-}1$$

grooves, i is the angle of incidence, θ is the angle of reflection from the groove, n is the diffraction order (1, 2, 3, etc.) and λ is the wavelength of the radiation. Equation 2-1 is the fundamental equation for the diffraction grating. In most infrared spectrometers the angle 2α between the incident and diffracted rays is fixed. If the angle between the normal to the grating and the bisector of the angle between the incident and diffracted rays is ϕ, we have eqs. 2-2. This situation is illustrated in Figure 2-5.

$$i = \phi - \alpha \quad\text{and}\quad \theta = \phi + \alpha \qquad\qquad 2\text{-}2$$

Substituting for i and θ in eq. 2-1, we have

$$a\left[\sin(\phi - \alpha) + \sin(\phi + \alpha)\right] = n\lambda \qquad\qquad 2\text{-}3$$

Using the trigonometric relation

$$\sin(A + B) + \sin(A - B) = 2\sin A \cos B$$

we obtain another useful form of the grating equation (eq. 2-4). If we consider radiation of

$$2a \sin\phi \cos\alpha = n\lambda \qquad\qquad 2\text{-}4$$

normal incidence to the grating, then eq. 2-1 becomes eq. 2-5.

$$a \sin\theta = n\lambda \qquad\qquad 2\text{-}5$$

Equations 2-1 and 2-5 express the condition for maximum intensity in the diffracted beam, namely that the path difference between rays striking adjacent grooves should be an integral number of wavelengths (Figure 2-6).

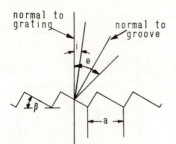

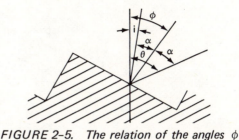

FIGURE 2-4.   Characteristics of a diffraction grating.

FIGURE 2-5.   The relation of the angles ϕ
and α to the angles of incidence and diffraction from a groove of a grating.

We are usually only interested in the $n = 1$ or possibly the $n = 2$ diffraction (the
first and second orders). Higher orders, i.e., when $\lambda = \frac{1}{3}, \frac{1}{4}$, etc., can be eliminated by means
of filters.

A grating gives high efficiency at angles of incidence near to the blaze angle β, but
the efficiency falls off as i becomes greatly different from β. The wavelength of the greatest
efficiency is obtained from eq. 2-6. The angle α is usually small, so that $\cos \alpha \cong 1$; hence,
eq. 2-7.

$$\lambda_\beta = \frac{2a}{n} \sin \beta \cos \alpha \qquad\qquad 2\text{-}6$$

$$\lambda_\beta = \frac{2a}{n} \sin \beta \qquad\qquad 2\text{-}7$$

Gratings are often used in two orders in modern infrared spectrometers. As an example,
consider a grating with 100 lines mm^{-1} (i.e., spacing $a = 10^{-3}$ cm), blazed for maximum
efficiency in the first order at 10 μ (1000 cm^{-1}). From eq. 2-7, the blaze angle β is
calculated to be $30°$. Such a grating will have another efficiency maximum at $\lambda_\beta/2$ (i.e.,
5 μ or 2000 cm^{-1}). Thus this grating used in first and second orders could be used to cover
the normal spectral range of an infrared spectrometer, from 4000 to 650 cm^{-1}.

To see how a spectrum is scanned we go back to eq. 2-5. It is seen that if $\sin \theta$ is
varied in a linear manner, the spectrum will be scanned linearly in wavelength. However,
taking the inverse of eq. 2-5, we get eq. 2-8, where $\bar{\nu} = 1/\lambda$ (in cm^{-1}). Hence, by varying

$$\frac{1}{a} \operatorname{cosec} \theta = \frac{1}{n} \bar{\nu} \qquad\qquad 2\text{-}8$$

FIGURE 2-6.   Diagram illustrating the path difference
$BC = a \sin \theta$ between rays diffracted from adjacent grooves
of a grating.

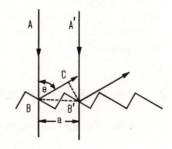

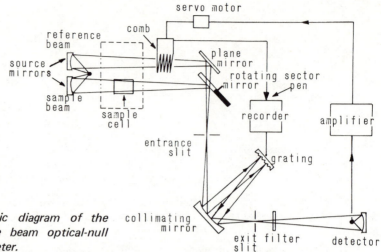

FIGURE 2-7. Schematic diagram of the operation of a double beam optical-null infrared spectrophotometer.

cosec θ in a linear manner, the spectrum will be linear in wavenumber. This method is used in many infrared spectrometers. A more detailed discussion of diffraction gratings has been given in a recent text on the theory and operation of infrared spectrophotometers (reference 2).

2-1g Recording of an Infrared Spectrum

A diagram of the path of the radiation through a typical double beam infrared spectrometer is shown in Figure 2-7. The operation is based on the optical-null principle.

Radiation from the source is divided into two beams, a reference beam and a sample beam. After passing through the sample compartment the beams are transmitted alternately by means of a semicircular rotating mirror into the monochromator. After dispersion by the grating, the radiation arrives at the detector. When the beams are balanced, a steady dc signal is produced by the detector. When the sample absorbs energy, there will be a stronger signal from the reference beam. Thus an alternating current will be produced by the detector. After amplification, this ac signal is used to drive a servomotor, which pushes a comb into the reference beam, and energy is blocked off from the reference beam. When sufficient energy has been removed from the reference beam, the two beams will again be balanced and there will no longer be an ac signal produced by the detector. The movement of the comb is linked to a pen that draws a trace, usually on a moving chart. (In some instruments, the pen carriage moves while the chart remains stationary.)

The rate of movement of the chart (or the pen holder) is coupled to the rate of rotation of the grating so that the spectrum is recorded as the grating rotates. Optical filters are automatically changed at the appropriate places in the scan. Often, two or more gratings, mounted back to back or on a turret, are used to cover the whole range of frequencies. The appropriate grating is rotated into place as required during the course of the scan.

2-2 Raman Spectrometers

2-2a Comparisons with Infrared Spectrometers

Some of the instrumentation of a Raman spectrometer is similar to that of an infrared spectrometer. The source of radiation in this case is the sample itself. The Raman scattering is excited by the intense monochromatic radiation from a laser. The scattered light is focused by a lens onto the entrance slit of a monochromator, where it is dispersed into its spectrum. Several basic differences exist between Raman and infrared spectra. The infrared spectrum consists of a continuum with a few parts missing where the sample has absorbed radiation. The Raman spectrum, on the other hand, consists of mostly nothing, with a few narrow regions of radiation emitted from the sample. There is also very strong scattering at the frequency of the exciting line, but this can be reduced by an optical filter.

The infrared continuum is relatively strong while the Raman lines in the visible are inherently very weak. Infrared detectors, however, have low sensitivity, whereas very sensitive detectors (photomultipliers) can be obtained for radiation in the visible. Slits of several hundred microns give reasonable resolution in the infrared, but much smaller slits are needed in the visible. Another very important point is stray light, which, if not effectively removed, can produce such a high background that the Raman spectrum is lost. A Raman spectrum is usually recorded as the variation of photomultiplier response with distance (in cm^{-1}) from the frequency of the exciting line.

2-2b   Raman Monochromators

In order to remove as much stray light as possible, a multiple (double or triple), or at least a double-passed single, monochromator must be used. Either of these will automatically give better stray light rejection and higher resolution than single-passed single monochromators. A double monochromator, as the name implies, is simply two monochromators used in series, with a slit in between. We shall only discuss grating monochromators here. One of the main problems with a double monochromator is that the two gratings must rotate simultaneously. If the gratings are not rotated exactly in phase, then there will be a reduction in the amount of radiation reaching the exit slit. The gratings are either mounted in tandem with a common drive mechanism or mounted vertically over each other on a single common rotating shaft.

A typical arrangement of a double monochromator is shown in Figure 2-8. Light entering slit S1 is collimated by a concave mirror M1 onto the first grating G1, then diffracted

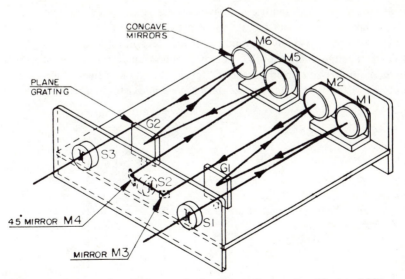

FIGURE 2-8. Schematic diagram of the Spex Industries Model 1401 monochromator. [Courtesy of Spex Industries, Inc.]

to M2. The radiation then passes through slit S2 into the identical second monochromator. The two gratings, G1 and G2, are turned simultaneously by means of a parallelogram linkage. A linear wavenumber counter can be set to zero at any excitation frequency. Thus Raman frequencies can be read directly. The gratings in this particular instrument have 1200 grooves mm^{-1} and are used in the first order.

2–2c Lasers

The laser is one of the outstanding technological developments in recent years. The state of the art is advancing rapidly, and the cost of commercial lasers for use in Raman spectroscopy has been steadily diminishing since their initial appearance on the market. The word *laser* stands for "light amplification by stimulated emission of radiation." The feature of the laser that makes it so valuable as a source for Raman spectroscopy is its ability to produce an intense beam of coherent radiation with a small diameter at a single frequency.

In order for a system to emit radiation, the atoms (or molecules) must first be raised to an excited state. This condition can be achieved by heating, as in the incandescent lamp, or by electrical discharge, as in neon lighting. Normal absorption of ultraviolet or visible radiation is accompanied by excitation of atoms or molecules to excited electronic states. Getting the atoms into excited states is known as *pumping.* Normally, the atoms do not stay in the excited state for very long. In a laser system, the excited atoms go from the initial excited state, by means of collisions or other radiationless transitions, to a second excited state of lower energy from which they are stimulated to go to another lower state (not necessarily the ground state) by radiation of the appropriate wavelength. There is, therefore, an amplification at the stimulating frequency by the emission from atoms undergoing the transition. When pumping is continuous, as, for example, in the helium–neon and the argon ion lasers, then a buildup of the intensity of the original stimulating radiation is achieved. The portion of this intense monochromatic radiation that is then allowed to escape constitutes the laser beam. The operation of a typical laser system will now be described in a little more detail. For further information on the subject of lasers, see references (3) and (4).

2–2d The Helium–Neon Gas Laser

Several companies manufacture helium–neon gas lasers, which give continuous operation at 6328 Å in the red at powers from 1 milliwatt or less up to 100 milliwatts. An electrical discharge through a helium–neon gas mixture at low pressure in a tube excites electrons of the helium atoms to excited energy levels. Energy is then transferred to neon atoms by collisions, so that the neon atoms end up in an excited level. When more neon atoms are present in the excited state than in a lower state, a *population inversion* exists (Figure 2-9).

FIGURE 2-9. The energy levels in the He and Ne atoms involved in the He–Ne gas laser. See reference (3) for further information on the notation of these levels.

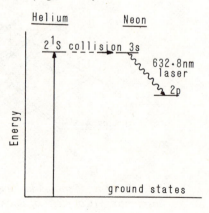

When radiation of energy equal to the separation of the two states passes through the gas mixture, some of the neon atoms are stimulated into making the transition to the lower state. This process is accompanied by emission of radiation of the same energy (wavelength) as the initial radiation. The probability that a photon will stimulate the emission of another photon in one pass through the gas is about 20%, so the protons must pass several times up and down the tube in order to produce amplification. To achieve this condition, reflectors are placed at each end of the tube containing the gas. These reflectors are partially transmitting, and about 3% of the radiation passes out of the laser *cavity*. This output beam is the coherent, monochromatic radiation characteristic of the laser.

By having a continuous discharge and by keeping most of the radiation traveling back and forth inside the laser cavity, continuous operation of the laser is achieved. A diagram of a helium-neon laser is given in Figure 2-10. This system can also give laser action at wavelengths of 6118, 10,840, and 11,520 Å, and others. These are rejected by means of a prism reflector.

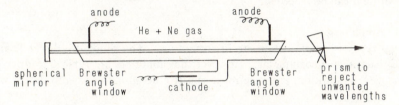

FIGURE 2-10. *Schematic diagram of a helium-neon gas laser.*

A list of the lasers most commonly used for Raman spectroscopy is given in Table 2-1.

Table 2-1. *Lasers Used in Raman Spectroscopy*

Laser	Wavelength (Å)^a	Typical Power (milliwatts)
He–Ne	6328	15 and 50
Argon ion	4880	100 and 1000
	5145	100 and 1000
Krypton ion	5682	200
	6471	400
He–cadmium vapor	4416	50

^{a}All of these lasers can give several other wavelengths, but with lower output powers.

2-2e Photomultipliers

All laser Raman spectrometers use photomultiplier (pm) tubes as detectors. The operation of a photomultiplier or electron multiplier is shown in Figure 2-11. A photon from the

FIGURE 2-11. *Diagram of a photomultiplier.*

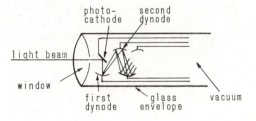

light beam causes an electron to leave the cathode. This electron is accelerated by a positive potential and strikes a second sensitive surface called a dynode. The impact of the fast moving electron causes several secondary electrons to be released. These in turn are accelerated and strike the second dynode. The process is repeated ten to fifteen times so that an amplification of several million is obtained.

For Raman spectroscopy, a tube is chosen with the best possible response in the red end of the spectrum. Photomultiplier tubes have high sensitivity at the blue end of the spectrum, but the response falls off rapidly in the red. For this reason, when using He–Ne laser excitation, Raman lines in the 3000 cm^{-1} region and beyond often appear to be weak.

Even when no light is actually reaching the photomultiplier, a signal due to the so-called *dark current* is observed. The dark current can be reduced by cooling with a current of cold nitrogen gas from evaporating liquid nitrogen passed through a jacket surrounding the pm tube. Thermoelectric cooling devices are also available. The signal from the photomultiplier must be amplified before it can be recorded.

2-2f Amplifiers

Several types of amplification are in current use. The simplest is direct dc amplification using a very sensitive picoammeter. However, dc amplifiers are subject to drift, causing the zero to change. In addition, even the most sensitive dc amplifier fails to detect weak Raman lines.

The system in most common use is known as photon counting. This method measures the number of pulses from the pm that fall within a certain range of pulse heights during a fixed counting interval. An electronic discriminator is used to reject pulses of height outside the chosen range. The pulses are then further processed before reaching the recorder. In this way, very few of the dark pulses are counted and this method gives a very good signal/noise ratio for weak signals.

A third method of amplification, known as phase sensitive detection, was used in the first commercial laser Raman instrument, the Perkin–Elmer LR 1. This method is also known as synchronous detection. The light is chopped before it reaches the photomultiplier tube. The amplifier (known as a lock-in amplifier) is designed to process only those signals from the pm tube that have the chopper frequency. In this way, the unchopped dark current and any stray light reaching the pm tube after the chopper are eliminated. One disadvantage is that only one half of the Raman radiation reaches the pm tube.

2-3 Sampling Methods for Infrared Transmission Spectra

2-3a Introduction

The very simplest sampling techniques will often give quite satisfactory results. However, in many cases potentially important features of the spectrum can be lost, distorted, or obscured by careless sampling methods. Sampling of pure liquids is relatively simple. Solutions and solid samples, on the other hand, require more care. We shall discuss some sampling techniques in the next few paragraphs. In all transmission studies, except the pressed-pellet method, windows transparent to the infrared are needed. It is appropriate, then, to start with a few words about windows and polishing methods.

2-3b Window Materials

There are several materials in common use for infrared spectroscopy. A summary of the

Table 2-2. *Properties of Infrared Window Materials*

Material	IR Transmission Limit (cm^{-1})	Refractive Index at $4000\ cm^{-1}$	Solubility in Water $(g/100\ ml)$ at $20°C$
NaCl	650	1.5	36
KCl	500	1.5	35
KBr	400	1.5	53
CsI	150	1.7	80
AgCl	400	2.0	insoluble
CaF_2	1100	1.4	0.002
BaF_2	850	1.5	0.1
KRS-5	200	2.4	0.02
Irtran-2	800	2.3	insoluble

more important of these is given in Table 2-2. Most of these crystals can be purchased as sawn blanks ready for polishing.

NaCl. This is perhaps the most commonly used material. The rock salt region is the traditional range (2–16 μ) studied by earlier workers. The least expensive window material, NaCl, is easy to polish but breaks easily and is only transparent as far as 650 cm^{-1}.

KCl. This is as cheap as NaCl, easy to polish, but fractures when subject to stress. It is transparent to 500 cm^{-1}.

KBr. This is another traditional material. It costs about 50% more than NaCl or KCl. It is easy to polish, but is fragile. KBr is transparent to 400 cm^{-1}.

CsI. This is a very useful material; it does not fracture under mechanical or thermal stress. However, it is very water soluble, very soft, and difficult to polish. It's cost is quite high (about ten times that of NaCl), but it is transparent to 150 cm^{-1}.

AgCl. Polished windows of AgCl are only slightly more expensive than polished NaCl windows. AgCl does not cleave. In fact, in the form of rolled sheets, it can be bent. It is resistant to attack by corrosive liquids, including HF, and is insoluble in water. The rolled sheets have ready polished surfaces and can be obtained in very thin sections. The material darkens on exposure to uv radiation. AgCl transmits to about 400 cm^{-1}.

CaF_2. This is only very slightly soluble in water, so it may be used for aqueous or D_2O solutions. The cost is about five times that of polished NaCl. CaF_2 windows should be purchased already polished, since this material is very difficult to polish. Fortunately, it is very hard and normally does not need repolishing. A disadvantage of CaF_2 is that it is only transparent to 1100 cm^{-1}.

BaF_2. This material is somewhat more soluble in water than CaF_2, but has an extended transmission range to about 850 cm^{-1}. Both CaF_2 and BaF_2 will fracture with thermal or mechanical shock.

KRS-5. This material is a mixed thallium bromide–iodide compound. It forms bright red crystals that are sparingly soluble in water, does not cleave, and transmits to 200 cm^{-1}. The main disadvantages of KRS-5 are the high price and the high refractive index, which may cause losses of transmitted energy by scattering. It is also toxic and may be attacked by some compounds in alkaline solution.

Irtran-2. This material is made by hot-pressing powdered ZnS. It has high strength and resistance to shock. It is virtually insoluble in water and is resistent to attack by a large variety of organic and inorganic materials. It has a high refractive index and transmits to about 800 cm^{-1}. It is very expensive.

2-3c Window Polishing

Soft materials such as NaCl, KBr, and CsI may be cleaned and repolished very simply and satisfactorily for qualitative work by the following method. A piece of plate glass, about 4″ square, is obtained. The sharp edges are removed and a chamois cloth is stretched over the glass and fixed by means of staples, stitching, or adhesive tape, etc. The cloth is wetted with distilled water and very fine Linde type-A aluminum oxide polishing powder is sprinkled on the cloth. This powder may be obtained from the Union Carbide Company. The powder is rubbed into the surface of the wet cloth with a finger. This is the polishing surface. It should be neither too wet nor entirely dry. Only by experience can the user find the best combination of amount of powder and moisture to give the best results. If the crystal is badly etched or scratched, or if a sawn blank is to be polished for the first time, then preliminary polishing with successively finer emery papers is needed.

To produce a clean, transparent surface on the window, the procedure is as follows. The crystal is placed on the moist chamois cloth and polished with a rotatory motion for three or four rotations *only*; then the crystal is quickly dried by rubbing on the ball of the thumb of the other hand. The procedure is repeated a few times until a satisfactory finish is obtained. Excess wet polishing powder should be removed from the edges of the window with tissue. The crystal is then turned over and the procedure repeated for the other face.

This very simple method produces entirely satisfactory results for qualitative work. It will not, of course, produce a perfectly flat, homogeneous surface. For such a finish, a polishing machine is usually required. Occasionally some trouble with fogging will be experienced at the edge of the window where it was held by the fingers. This is particularly troublesome for the very soluble materials such as CsI. In these cases, a rubber or polyethylene glove may be worn on the hand holding the window. A glove should be worn when polishing KRS-5 because of its toxicity.

Polishing kits are available from the manufacturers of spectrometer accessories. These may or may not give better results than the method described here. One supplier (International Crystal Laboratories, Irvington, N.J. 07111) will repolish windows. Further information on crystal polishing may be found in Chapter 3 of reference (5).

2-3d Liquids and Solutions

Probably the easiest method to obtain a qualitative infrared spectrum of a liquid is to place one drop of the liquid onto a disc of NaCl, KBr, etc., cover the drop with a second disc and mount the pair in a holder such as the one illustrated in Figure 2-12. Teflon spacers may be used to give various pathlengths. This method may lead to total absorption in some regions and very weak spectra in others. The former problem may be overcome by using a thinner spacer. If the absorption is still too strong, a drop of solvent such as CS_2 or CCl_4 could be added. The absorption of the solvent must then be taken into account.

In addition to the demountable liquid cell illustrated in Figure 2-12, fixed-pathlength, sealed cells may be purchased. These usually have amalgamated silver or lead spacers. Such a cell is illustrated in Figure 2-13. The cells are filled, emptied, or flushed by means of a syringe through conventional stainless steel Luer ports. Teflon stoppers are used to close the ports.[1]

[1] Another type of fixed-pathlength cell is the cavity cell, which is made from a small block of NaCl or KBr with polished faces and a slot-like cavity drilled in the center.

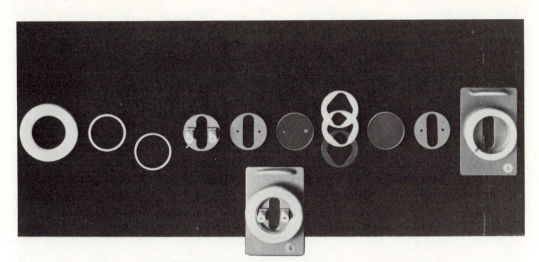

FIGURE 2-12.   A demountable liquid cell.   [Courtesy of Barnes Engineering Company.]

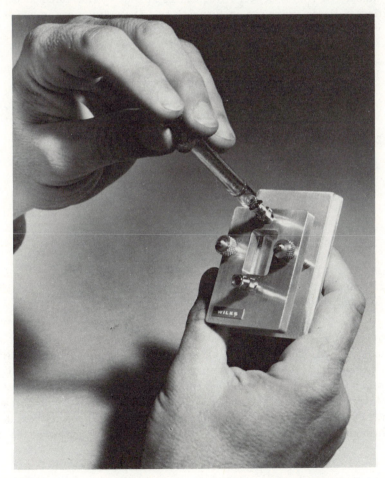

FIGURE 2-13.   A fixed-pathlength cell.   [Courtesy of Wilks Scientific Corporation.]

For the far infrared spectrum, polyethylene cells are available with various pathlengths. These become easily contaminated, but are of low enough cost to be disposable.

2-3e Determination of Cell Pathlength

Manufacturers of accessories for infrared spectroscopy supply Teflon or lead spacers of thicknesses varying from 0.01 mm up to 1 mm or more. These spacers give a good idea of the pathlength. However, spacers can be compressed and windows might be thicker in the center than at the edges unless they have been very carefully polished. For quantitative work one needs to know the pathlength accurately, and two methods are available for calibrating cells.

(i) *The fringe method.* The empty cell is placed in the sample beam and the percent transmission adjusted to approximately 80%. A portion of the spectrum is then scanned until 20 fringes are produced. The cell thickness (l) is calculated using eq. 2-9, where

$$l\,(\text{cm}) = \frac{n}{2(\bar{\nu}_1 - \bar{\nu}_2)} \qquad\qquad 2\text{-}9$$

n is the number of fringes between frequencies $\bar{\nu}_1$ and $\bar{\nu}_2$ (measured in cm^{-1}). Three typical fringe patterns are shown in Figure 2-14. The useful range in which thicknesses can be obtained by this method is from 0.5 to 0.01 mm.

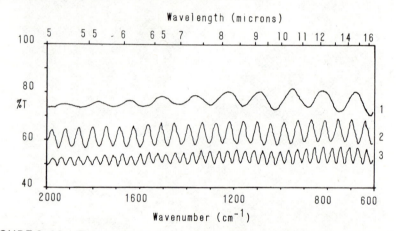

FIGURE 2-14. *Three typical fringe patterns from small pathlength liquid cells. The reader may care to verify that the pathlengths are (1) 0.037, (2) 0.086, and (3) 0.134 mm, making use of eq. 2-9.*

(ii) *The standard absorber method.* If the absorbance of an infrared band of a substance is measured for an accurately known pathlength, then this absorbance may be used to determine the pathlength of other liquid sample cells. Benzene can be used for this purpose by the following procedure. A band at $1960\ \text{cm}^{-1}$ in the infrared spectrum of benzene shows an absorbance of 1.0 when the sample is in a cell of exactly 0.1 mm pathlength. This band, therefore, is suitable for calibrating cells whose pathlengths are less than 0.1 mm. The pathlength in mm is given by eq. 2-10. A second weaker band in the spectrum of

$$l\,(\text{mm}) = \frac{0.1}{1.0} \times \text{absorbance} = 0.1 \times \text{absorbance} \qquad\qquad 2\text{-}10$$

benzene at $845\ \text{cm}^{-1}$ can be used to calibrate cells of pathlengths between 0.1 and 4 mm.

In this case, the absorbance of a 0.1 mm sample is 0.24, so that the pathlength in mm is given by eq. 2-11.

$$l \text{ (mm)} = \frac{0.1}{0.24} \times \text{absorbance} = 0.42 \times \text{absorbance} \qquad \qquad 2\text{-}11$$

2-3f Solvents for Infrared Spectroscopy

It is often convenient to record the infrared spectrum of a compound in solution. Unfortunately, some of the best solvents have very strong infrared absorption bands that obscure parts of the spectrum of the compound. Water, for example, absorbs strongly throughout the spectrum and is rarely used in infrared work. However, H_2O and D_2O can be useful solvents for the infrared spectroscopy of compounds such as sugars, amino acids, and compounds of biochemical interest. Special window materials, such as CaF_2, BaF_2, AgCl, Irtran-2, or KRS-5, must be used. Discussion of the use of aqueous solution infrared spectroscopy may be found in Chapter 11 of reference (5) and in reference (6).

The answer to the problem of solvent absorption is to use a pair of matched cells with the solvent in the reference beam and the solution in the sample beam. Only regions in which the solvent does not absorb strongly can be used, so that a series of solvents will be needed for a complete spectrum. It should be noted that during the recording of a spectrum using matched cells, when a region in which the solvent absorbs strongly is scanned, essentially no energy passes in either the sample or reference beam. The instrument will be comparing "nothing" with "nothing" and the pen will be observed to be "dead" in such a region. The more sophisticated infrared spectrometers can produce a spectrum from surprisingly little energy (for example, the Perkin–Elmer Model 180). However, for the low-priced instruments the solvents used must be chosen carefully. A partial list of solvents is given in Table 2-3 with useful regions indicated.

Table 2-3. *Useful Solvents for Infrared Solution Spectra*

Solvent	Useful Regions (cm^{-1})	Typical Pathlength (nm)
CS_2	all except 2200–2100 and 1600–1400	0.5
CCl_4	all except 850–700	0.5
$CHCl_3$	all except 1250–1175 and below 820	0.25
$CHBr_3$	all above 700 except 1175–1100 and 3050–3000	0.5
C_2Cl_4	all except 950–750	0.5
Benzene	all above 750 except 3100–3000	0.1
CH_2Cl_2	all above 820 except 1300–1200	0.2
Acetone	2800–1850 and below 1100	0.1
Acetonitrile	all except 2300–2200 and 1600–1300	0.1
$CBrCl_3$	all above 820	0.5
Cyclohexane	below 2600	0.1
N,N-Dimethylacetamide	2750–1750 and below 950	0.05
N,N-Dimethylformamide	2750–1750 and below 1050	0.05
Ethyl acetate	2850–1800 and below 1000	0.05
Ethyl ether	all except 3000–2700 and 1200–1050	0.05
Heptane and hexane	all except 3000–2800 and 1500–1400	0.2
Dimethyl sulfoxide	all except 1100–900	0.05
Toluene	2800–750	0.05
H_2O^a	2800–1800	0.05
D_2O^a	above 2800 and	0.1
	2200–1300	0.05

^{a}Special windows needed (see section 2-3a).

2-3g Mulling Techniques

A small amount of the sample is ground in an agate or mullite mortar. Then a drop of a paraffin oil, usually Nujol, is added and the grinding continued. Nujol is available at very low cost in drugstores. The mixture should have the consistency of a thin paste. It is transferred to an infrared window, and a second window lowered onto it. A thick film free of air bubbles should be produced. Any of the windows described at the beginning of this section may be used.

The two plates with the mull between are placed in a cell holder and the spectrum is recorded. There will be strong bands at 2900, 1470, and 1370 cm^{-1} and a weak band at 720 cm^{-1} due to Nujol. If the Nujol bands are stronger than the peaks from the sample, then more sample and less Nujol must be ground. If the sample peaks are too strong the two windows can be squeezed together or a small drop of Nujol can be added. The user will have to experiment with the mull to obtain the best results.

When the region near 2900 cm^{-1} is important, other mulling materials must be used. The usual second compound is Fluorolube, a chlorofluorocarbon oil, which is available from chemical suppliers. This material is completely opaque below 1400 cm^{-1} in the pathlengths normally used in mulls. It also has a band at 1650 cm^{-1}. Another compound useful for mulls, when the 2900 cm^{-1} region is to be studied, is hexachlorobutadiene. This compound has no absorptions above 1650 cm^{-1}. It also has a useful "window" between 1500 and 1250 cm^{-1}.

Scattering encountered in mulls can result in low transmission, a problem that is often serious at the high frequency end of the spectrum. A reference beam attenuator of the type illustrated in Figure 2-15 can be used to bring the baseline back to near 100% transmission,

FIGURE 2-15. Reference beam attenuator. [Courtesy of Barnes Engineering Company.]

with some sacrifice in performance of the instrument. More grinding often helps, since the scattering is proportional to the particle size. Scattering is discussed further in the following paragraphs on pressed pellets.

2-3h The Pressed-Pellet Technique

For solid compounds that are insoluble in the usual solvents, a convenient sampling method is the pressed-pellet technique. A few milligrams of the sample are ground together in an agate or mullite mortar with about 100 times the quantity of a material (the matrix) that is transparent in the infrared. The usual material is KBr, although other compounds such as CsI, TlBr, and polyethylene are used in special circumstances. The finely ground powder is introduced into a stainless steel die, usually 13 mm in diameter, which is then evacuated for a few minutes with a rotary vacuum pump. The powder is then pressed in the die between polished stainless steel anvils at a pressure of about 30 tons per square inch on the disc.[2]

A well-made KBr pellet may be transparent, but the real criterion is whether the transmittance is sufficiently high (80-90%) in regions where the sample does not absorb (below 3000 cm^{-1}). At the high frequency end of the spectrum the transmission will often be low because of scattering effects. The amount of scattering by mulls and pellets depends on the refractive indices of the sample and the matrix or mulling material, according to formula 2-12, where n_1 and n_2 are the refractive indices of the sample and matrix, respec-

$$\text{Scattering} \propto \left(\frac{n_1 - n_2}{n_1 + n_2}\right)^2 \left(\frac{d}{\lambda}\right)^4 \qquad \text{2-12}$$

tively, d is the particle size, and λ is the wavelength. It can be seen that for relatively similar refractive indices, large particles at short wavelengths will cause serious scattering. At longer wavelengths, the particle size becomes less important. However, d should be less than about 20 μ for good pellets. This dimension can usually be achieved by hand grinding in a hard mortar or by mechanical shaking of the sample in a tube containing stainless steel balls. When scattering reduces the transmission, a reference beam attenuator can be used to restore the baseline to near 100%. As the spectrum is scanned, however, less and less attenuation will be needed because the scattering will become less serious at longer wavelengths.

An interesting phenomenon, known as the Cristiansen effect, can sometimes be observed in mulls or pellets of samples that are highly scattering as a result of a large difference between the two refractive indices, n_1 and n_2. Near an absorption band of frequency v, the refractive index of the sample n_2 changes as in Figure 2-16a. The refractive index of the mulling or matrix material n_1 does not change, and if n_2 becomes approximately equal to n_1, the scattering is greatly reduced. This condition occurs at frequencies slightly greater than v. Then, as the

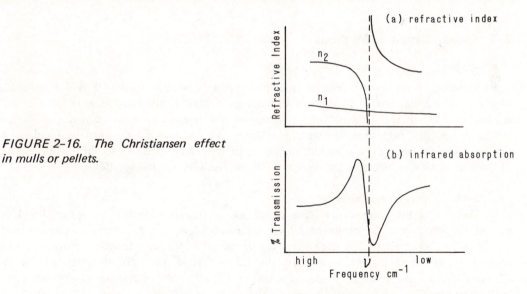

FIGURE 2-16. *The Christiansen effect in mulls or pellets.*

[2] A very inexpensive minipress can be made from two half-inch diameter stainless steel bolts and a stainless steel nut. The ends of the bolts must be polished flat and parallel. One bolt is inserted about half way into the nut, and the KBr + sample mixture added. The second bolt is then screwed into the nut and pressure applied by tightening the bolts together. Usually a vice is needed for this operation. When the bolts are carefully withdrawn, a pellet suitable for infrared transmission work remains. The pellet is not removed from the nut, which acts as a holder in the spectrometer. Minipresses are available from the manufacturers of infrared sampling accessories.

absorption maximum is passed, n_1 becomes different from n_2 and scattering again occurs. The resulting bandshape is shown in Figure 2-16b. Thus the true frequency and intensity of the absorption band cannot be determined.

KBr pellets can be prepared from thin layer chromatography spots. The spot is extracted with a suitable solvent, and the resulting solution placed onto a small quantity of KBr powder. The solvent is evaporated, and the KBr made into a pellet. For very small quantities of sample a micro pellet may be necessary (see section 2-5c). A somewhat more sophisticated technique is available from the Harshaw Chemical Company, Cleveland, Ohio. The chromatogram spot is removed and mixed with a small quantity of solvent in a small vial. A triangular section of pressed KBr (known as a wick-stick) is placed with the base in the solution. As the solution evaporates, the compound migrates to the tip of the KBr triangle. The tip is broken off and made into a micro pellet in the usual way.

Matrices other than KBr may be used. CsI is useful when spectra down to 150 cm^{-1} are required. One disadvantage of CsI is that it is a highly water-soluble material. Lower pressures are required for CsI pellet formation than for KBr. TlBr is used when materials of high refractive index are studied. It has a refractive index of 2.3 in the infrared region and is transparent out to 230 cm^{-1}. It has low water solubility and can be used in high humidity conditions. It is also useful in the near infrared. Powdered polyethylene has been used for making pellets for far infrared spectra because its spectrum between 400 and 10 cm^{-1} is essentially free of absorption.

In addition to the scattering problems discussed above, the pressed-pellet method has other disadvantages. Changes may occur in the sample during the grinding and pressing process, and the sample may react with absorbed water or even with the matrix material.

2-4 Raman Sampling Methods

2-4a Introduction

Sampling methods for Raman spectroscopy are usually simpler than for infrared. In principle, all that has to be done is to place the sample in the laser beam in front of the entrance slit of the monochromator. Focusing of the laser beam may improve the Raman intensity, but this may also damage the sample. The scattered light must be focused on the entrance slit, and a polarization analyzer can be placed before the slit. Multipassing of the laser beam also increases the Raman scattering from liquids and solutions.

2-4b Sample Compartments

There is really no necessity for a sample compartment as such. An optical bench is needed on which can be mounted lenses, sample holders, the polarization analyzer, and a polarization scrambler. The laser beam can impinge on the sample from below or above (known as 90° illumination), or horizontally (known as 0° or 180° illumination), but in all cases it is simple enough to mount a lens, an interference filter, and a half-wave plate in the laser beam before it reaches the sample.

All the commercial instruments are sold complete with a sample compartment containing the items mentioned above. A schematic diagram of a typical sample compartment is shown in Figure 2-17. Starting from the laser itself, the components have the following uses. The interference filter is of a type that passes only the laser frequency and a narrow band on either side so that nonlasing plasma lines are eliminated. It is often useful to take this filter out

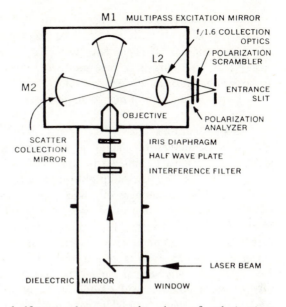

FIGURE 2-17. The arrangement of the optical components in a typical sample compartment. [Courtesy of Spex Industries, Inc.]

when accurate calibration is required. The half-wave plate turns the plane of polarization through 90°. The iris diaphragm also reduces interference from the laser plasma. The microscope objective focuses the light to a spot in the sample.

The scattered light is focused by means of the lens (L2) onto the entrance slit of the monochromator. A spherical mirror (M1) causes multipassing of the beam, and a mirror (M2) reflects 180° Raman scattering back toward the monochromator. The polarization analyzer is usually a piece of polaroid, and the polarization scrambler makes sure that no polarized light gets into the monochromator. This component eliminates any effects produced by the polarizing properties of the spectrometer itself in measurements of depolarization ratios.

2-4c Liquids

A small volume quartz cell, such as that supplied for uv absorption work, could be used as a Raman liquid sample container. This type of cell will have polished windows. The ends of the cell also should be polished. Although quartz is preferable because it has low fluorescence, ordinary Pyrex glass is usually satisfactory.

Raman spectroscopy is ideally suited for microsampling. The laser beam can be focused to a very small area, and samples contained in melting point capillaries yield virtually the same spectra as much larger samples. No special accessories are needed as in the case of infrared microsampling. It is important, however, to align the sample exactly with the slit of the monochromator. This operation may be done easily with the aid of a low-powered auxiliary laser. Figure 2-18 illustrates a typical arrangement.

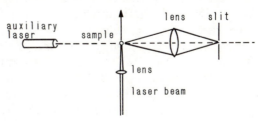

FIGURE 2-18. An arrangement for aligning a small sample with the monochromator slit.

The auxiliary laser is aligned with the slit, and the sample adjusted so that it is exactly in line with both the exciting laser beam and the auxiliary laser beam. The Cary 82 instrument has an alignment lamp that projects an image of the slit into the sample area. This arrangement is very useful for positioning both the sample and the laser beam at the focal point of the collection optics.

Further discussion of microsampling techniques is given in references (7) and (8).

2-4d Solutions

The techniques for handling solutions are essentially the same as for pure liquids. However, care must be taken with the solvents. They should be clean, pure, and sometimes filtered and redistilled to avoid problems from fluorescence.

One of the advantages of Raman spectroscopy over infrared is in the area of solution spectra. As we saw in section 2-3c, most of the common solvents have extensive regions of absorption in the infrared. Although this is not the case for the Raman, there is still the drawback that the intensities obtained from dilute solutions cannot be increased by having longer pathlength. Thus there is a lower limit to the useful concentrations that can be used. This limit is usually several percent by weight, which is a considerably higher concentration than can be used successfully for infrared absorption measurements. Only with a very high laser power and a very good monochromator and detection system can dilute solutions be studied.

Of course, all solvents have a Raman spectrum, but Raman lines are much narrower than infrared absorption bands. In addition, overtones and combinations are very much weaker in the Raman. Hence, there are many completely clear regions in the Raman spectra of all the usual solvents. Water is an excellent solvent for Raman solution studies, a fact that is extremely important in the examination of compounds of biological interest where the natural environment is usually aqueous solution.

Common solvents are listed in Table 2-4. The most important Raman lines are given in the second column. The stronger lines extend for approximately 20-40 cm^{-1} on either side of the frequency quoted, and the weaker lines obscure a region of about 10-20 cm^{-1}. When Raman spectra of dilute solutions are recorded, the instrument is used at its highest sensitivity. Under these conditions even the very weakest lines of the solvent are observed. These lines are listed in the third column of Table 2-4. Of course, a spectrum of the solvent must be recorded in all cases. As far as possible, the same conditions should be used to observe the spectra of the solution and the solvent.

2-4e Depolarization Measurements

Any Raman line can be classified as polarized or depolarized. The light from a laser is polarized and the Raman scattering is also polarized so that when an analyzer is placed before the entrance slit of the monochromator there is a difference in intensity between the light transmitted through the analyzer in the parallel and perpendicular orientations. The ratio of the perpendicular intensity to the parallel is known as the depolarization ratio ρ. For laser-excited Raman spectra, ρ has a maxiumum value of $\frac{3}{4}$. For spectra excited by unpolarized light from the mercury arc source, ρ has a maximum value of $\frac{6}{7}$. The depolarization ratio is usually measured by comparing peak heights. A band with ρ less than $\frac{3}{4}$ is said to be polarized and a band with ρ exactly equal to $\frac{3}{4}$ is said to be depolarized. The origin of this number is discussed in section 3-11.

Depolarization measurements are useful for assigning frequencies to the totally symmetric vibrations of the molecule (see section 3-10). Another use is in separation of over-

Table 2-4. *Common Raman Solvents*

Solvent	Raman Frequencies (rounded to the nearest 5 cm^{-1})	
	Most Important Lines	*Weak Lines*
Carbon disulfide	650(vs), 795(m)	400
Carbon tetrachloride	220(s), 315(s), 460(vs), 760–790(m)	1540
Chloroform	260(s), 365(vs), 670(vs), 760(m), 1220(m), 3025(w)	–
Methanol	1035(m), 1460(w), 2840(m), 2900–2950(w)	1110, 1165, 3405
Acetonitrile	380(s), 920(s), 1375(w), 2250(s), 2950(s)	1440, 3000
Methylene chloride	284(s), 700(vs), 740(w), 2985(w)	1155, 1425, 3045
Nitromethane	480(m), 655(s), 920(vs), 1370–1410(m), 2965(m)	610, 1105, 1560, 2765, 3040, 3060
Acetone	530(m), 790(vs), 1070(m), 1225(m), 1430(m), 1710(m), 2920(s)	390, 490, 900, 1365, 2845, 2965
Benzene	605(m), 990(vs), 1180(m), 1580–1610(w), 3040–3070(m)	405, 780, 825, 850, 2620, 2950
Cyclohexane	385(w), 430(w), 800(s), 1030(m), 1160(w), 1265(m), 1440–1470(m), 2855(m), 2920–2950(m),	1345, 2630, 2670, 2700, 2905
Ethanol	880(s), 1450–1490(m)	430, 1050, 1095, 1275, 2875, 2930, 2975
Dimethylformamide	320(w), 360(m), 410(m), 660(s), 870(s), 1095(m), 1405(s), 1440(m), 1660(m)	1065, 2800, 2860, 2930
Dimethyl sulfoxide	300–350(s), 385(m), 650–710(vs), 955(w), 1045(s), 1420(m), 2915(m), 3000(w)	2885

lapping bands. When one band is due to a totally symmetric vibration, its intensity is often drastically reduced upon rotating the analyzer through 90°. For nontotally symmetric vibrations, the reduction in intensity will only be 0.75. An example of the separation of overlapping bands of different polarizations is given in Figure 2-19. A band in the Raman spectrum of quinoxaline (1,4-diazanaphthalene) was recorded with both parallel and perpendicular orientations of the analyzer. Two lines are clearly present in this band, and the frequencies of the polarized and depolarized lines can be accurately determined from the parallel and perpendicular spectra, respectively.

To make depolarization measurements the spectrum is scanned twice with identical monochromator and amplifier settings. The only difference is the orientation of the analyzer

FIGURE 2-19. Part of the Raman spectrum of quinoxaline run with both parallel and perpendicular orientations of the analyzer.

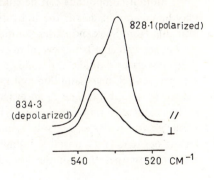

(parallel and perpendicular). Small regions of the spectrum are scanned sequentially, peak heights and baselines are noted for each orientation, and depolarization ratios are calculated from eq. 2-13.

$$\rho = \frac{(I - I_0)_\perp}{(I - I_0)_\parallel} \qquad\qquad 2\text{-}13$$

2-4f Solids

Solids may be examined as powders, lumps, pellets, or crystals. Powders are best handled in capillary tubes or tamped into a conical depression in the end of a metal rod. Irregular pieces of solid materials may be glued to a glass rod and held in the laser beam. Powders may be pressed into pellets and examined in the same way. Multipassing of the laser beam is not possible for solids, and high background scatter may be often observed. In such cases, a crystal of the compound is desirable.

A crystal 1 or 2 mm long can be mounted on a goniometer head in the same way that crystals are mounted for X-ray examination. Single crystals can be carefully positioned so that the laser beam passes along one of the axes of the crystal. In this case, when a polarization analyzer is used, important information concerning the symmetry of the normal vibrations of the molecule can be obtained. A discussion of single crystal Raman spectroscopy is given in a book by Gilson and Hendra (reference 9).

Water-sensitive solid samples can be loaded into capillary tubes in a dry box and sealed. Larger bore tubes (nmr tubes) can be connected by a ground joint to a stopcock and evacuated on a vacuum line. The sensitive compound is transferred to the tube in the glove box and the stopcock rejoined. After evacuation the tube can be sealed.

Polymers present some problems, but often a piece of the material can be held in the laser beam and a Raman spectrum obtained directly. Colored or dark materials may absorb energy from the laser beam, and fluorescent impurities can obscure the spectrum. A recent article (reference 10) gives further information on the handling of polymer samples for Raman spectroscopy.

2-4g Special Raman Sampling Methods

Raman spectra of samples at very high and very low temperatures require special consideration. These and other techniques for absorbed species and colored samples are discussed briefly here. A very important sampling method, which has not yet been widely used in connection with Raman spectroscopy, is gas chromatography. The ways in which this method have been applied to infrared absorption spectroscopy are discussed in section 2-5e. Some applications to Raman spectroscopy are apparent, and undoubtedly new techniques will be developed in the future.

Molten compounds can be studied in quartz or Pyrex tubes heated in a small furnace that may be custom-made for the individual instrument. A variable temperature sampling device is available from Spex Industries, Metuchen, N.J., and liquid helium cryostats designed for Raman studies are available from several manufacturers.

For gases, liquids, and solids having a vapor pressure of a few torr at room temperature, a convenient Raman sampling method involves condensing the gas or vapor on a copper block cooled by liquid nitrogen in an evacuated Dewar. This method has been used successfully for such widely differing samples as C_2F_5CN (normally a gas), $TiCl_4$ (normally a fuming liquid), s-triazine (normally a wet solid), and the rare valence isomers of benzene: benzvalene and Dewar benzene (very unstable liquids). At liquid nitrogen temperatures, Raman lines of solids are much sharper than in the liquid phase. Underlying hot bands are also eliminated. One must be aware of the possibility of splitting of lines due to crystal effects, but this problem is usually important only for molecules having a threefold or higher axis of symmetry.

Colored samples often absorb laser energy and boil or vaporize with decomposition. If the sample is rotated rapidly this problem can be avoided. The method was developed by Kiefer and Bernstein (reference 11) and can be used for both liquid and solid samples. The laser beam illuminates a continuously moving sample, which, if homogenous, gives a constant level of Raman scattering.

Laser Raman spectroscopy lends itself to novel sampling techniques, since the laser beam can be brought in at any angle (even at 180° to the direction of the monochromator slit). Pyrex glass can be used to fabricate cells, Dewars, etc., and special situations often can be handled on an ad hoc basis with homemade sampling devices.

2-5 Special Techniques in Infrared Spectroscopy

2-5a Introduction

In this section some special techniques, not covered previously in the sections on infrared transmission sampling methods, will be discussed. Several areas will not be discussed in this book since they are of less interest to the organic chemist: gas phase Raman spectra, spectra of adsorbed species, infrared specular reflectance methods, etc. These are dealt with in more specialized texts on infrared and Raman spectroscopy.

2-5b Gas Phase Infrared Spectroscopy

The gas or vapor phase spectrum of an organic molecule recorded under low resolution will give essentially the same information as the liquid or solution spectrum. Complications due to rotation of the molecules will at most be observed as splitting of some peaks or broadening. Gases are usually studied in a 10 cm glass cell with alkali halide windows at pressures of between 10 and 100 torr. A vacuum line is needed to handle the gas. Volatile liquids or solids can be studied in the vapor phase if the cell is warmed with a heating tape until sufficient pressure of vapor is produced. In some cases this procedure may be a useful alternative to the other methods.

2-5c Microsampling

For infrared microsampling, special small volume cells are available. Very small KBr discs can also be prepared. Two types of microcell are in common use, the cavity cell and a miniaturized version of the standard liquid cell. The former consists of a small block of material such as NaCl, KBr, etc., with parallel polished faces and a microcavity drilled ultrasonically in the center. An example of such a cell is shown in Figure 2-20. Various

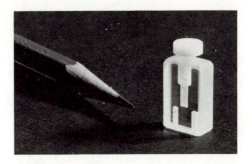

FIGURE 2-20. A microcavity infrared cell.
[Courtesy of Barnes Engineering Company.]

FIGURE 2-21. A reflection-type beam condenser. [Courtesy of Barnes Engineering Company.]

volumes are available from a fraction of a microliter up to 0.5 ml. Standard pathlengths in the range 0.05–5 mm can be obtained. The second kind of microcell is usually purchased completely assembled and sealed and with a fixed pathlength.

Because of the small area exposed to the infrared beam, some kind of beam-condensing device is necessary. A typical reflection type of beam condenser is illustrated in Figure 2-21. This instrument produces a three-to-one image reduction of the sample. To compensate for intensity losses, a second beam condenser or an attenuator is used in the reference beam. Refracting beam condensers are also available. The Perkin-Elmer accessory shown in Figure 2-22 utilizes three KBr lenses. The frequency range of this type of beam condenser is limited by the material used to make the lenses. Other disadvantages are the lower transmission and the danger of fogging of the lenses.

FIGURE 2-22. A refraction-type beam condenser. [Courtesy of The Perkin–Elmer
Corporation.]

FIGURE 2-23. *Die for making micro KBr pellets.*
[Courtesy of The Perkin-Elmer Corporation.]

Microgram quantities of sample can be pressed into KBr pellets using the special die shown in Figure 2-23. The pellets are pressed into openings centered in stainless steel discs. Various sizes of pellet from 0.5 mm up to 13 mm diameter can be made. The 0.5 mm pellet requires about 1 mg of KBr mixed with 1 μg of sample. A standard KBr die could also be used with paper inserts, but the smallest practical diameter is somewhat larger.

2-5d Attenuated Total Reflectance (ATR)

In section 1-8 the principles of atr spectroscopy were outlined. Some of the available instrumentation and applications will be described here. In addition to atr, the letters matr, mir, or fmir may be encountered. These stand for multiple atr, multiple internal reflectance, and frustrated multiple internal reflectance. They all mean that multiple internal reflections are produced by a special crystal of the form shown in Figure 2-24. The word *frustrated* is synonymous with attenuated in the present context.[3]

FIGURE 2-24. *A multiple internal reflection element.*

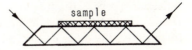

sample

The simple atr unit has the advantage that the angle of incidence can be varied and thus different levels of penetrations of the sample can be achieved. The internal reflection element in this case is usually in the form of a hemicylinder. The disadvantage of this method is that a single internal reflection may give only a very weak spectrum.

Multiple reflections can produce enhancement of the spectra of weakly absorbing samples. The angle of incidence is fixed in this method, and several plates will be needed to study a variety of compounds. The thinner the plate, the more internal reflections there will be. A typical plate is 2 mm thick, 5 cm long, and 2 cm high with the ends cut at 45°.

[3]The use of internal reflectance spectroscopy in the infrared was first suggested independently by both Fahrenfort and Harrick. Fahrenfort called the phenomenon frustrated total reflectance (ftr), while Harrick called it attenuated total reflectance (atr). Thus the two names are synonymous for the present purposes. Unfortunately, the expression frustrated total reflectance has been used by Harrick in his book (reference 22) when the internal reflection is reduced by a nonabsorbing film. We are interested, of course, only in the attenuation of the internal reflection by an absorbing layer.

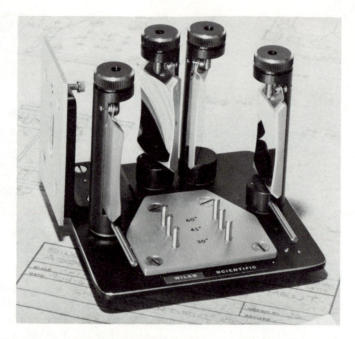

FIGURE 2-25. An internal reflection accessory. The sample is held in contact with an internal reflection element of the type shown in Figure 2-24 by means of a holder, which is then positioned on the appropriate pins. [Courtesy of Wilks Scientific Corporation.]

This arrangement gives about 24 reflections. The ends of the plates can be cut at various angles, with 30, 45, or 60° most commonly used. A large energy loss of 50% or more occurs in the sample beam, so that an attenuator or a second internal reflection unit should be used in the reference beam. A typical unit is shown in Figure 2-25.

An interesting device originally suggested by Harrick (reference 12) is the internal reflection probe. In this accessory, available from Harrick Scientific Corp., Ossining N.Y. 10562, the light beam enters and leaves from the same end of an internal reflection plate. In this way, the other end is free for immersion into sample material, and liquids or powders can be examined. The only other apparatus needed is a beaker to hold the sample.

The depth of penetration of the infrared radiation into the sample depends on the wavelength, the refractive indices of sample and crystal, and angle of incidence of the reflected ray at the interface. The most commonly used crystal materials are AgCl, KRS-5, Irtran-2, and germanium. The properties of the first three compounds are discussed in section 2-3a. Germanium has a very high index of refraction (4.0 at 1000 cm^{-1}) and transmits to about 700 cm^{-1}.

Good spectra, comparable to infrared transmission spectra, often can be obtained with atr methods, if proper choice is made of crystal material and angle of incidence. Differences will be observed between transmission and atr spectra. In general, bands are shifted slightly to lower frequencies in atr, and the high frequency bands are weaker in atr than in transmission. Although the refractive index of the internal reflection element must be greater than that of the sample, the best spectra are obtained when the difference in refractive indices is small. The method of atr is extremely useful for difficult samples and for studying films or coatings. However, whenever transmission methods can be applied they are usually preferable to atr methods.

Some applications of atr are discussed in references (6) (p. 50) and (13).

2-5e Gas Chromatography and Infrared Spectroscopy

In many practical situations, the samples to be studied are in the form of mixtures. In larger scale preparations, mixtures are separated into their components by standard methods such as distillation and crystallization. For very small samples, some form of chromatography is invariably used. If the material is sufficiently volatile and relatively stable, gas chromatography (gc) is an ideal method for separation. Subsequent study by infrared can be conveniently carried out using special techniques, some of which will now be briefly outlined. Further information is available in references (5) and (6).

A simple and versatile technique involves collecting a gc fraction on KBr powder and subsequently pressing the powder into a pellet. Very small pellets can be made by using paper inserts placed on the lower anvil of the die.

Gas chromatography effluents can be collected in small traps cooled with liquid nitrogen or dry ice-acetone mixtures. The trapped material is transferred to an infrared cell. Various fraction collection devices that are available commerically usually eliminate the transfer step by collecting the material directly in an infrared cell. Because of the very small volumes involved, it is necessary to use microcells in conjunction with a beam-condensing device (see section 2-5c).

A thermoelectrically cooled fraction collection apparatus, available from Wilks Scientific Corporation (Figure 2-26), collects and analyzes the fraction in the same cell. A sample is condensed as a thin layer between two internal reflection plates. A diagram of this cell is given

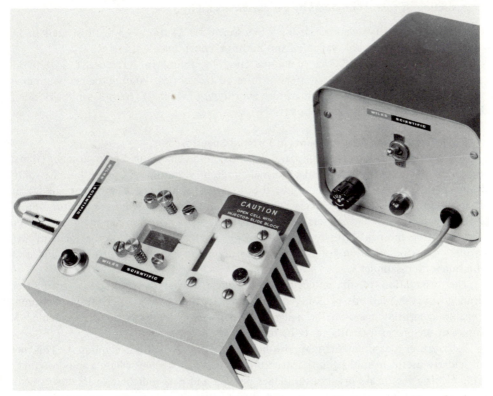

FIGURE 2-26. Thermoelectrically cooled device for collecting gc fractions for ir spectroscopy. [Courtesy of Wilks Scientific Corporation.]

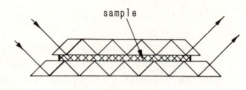

FIGURE 2-27.   *A capillary internal reflection cell.*

in Figure 2-27.   The spectrum is recorded by the multiple internal reflection technique described in section 2–5d.

2–5f   Pyrolysis Techniques

Indirect evidence on the composition of a substance can be obtained from examination of the infrared spectra of pyrolysis products.   This technique is useful for such intractable substances as hard polymers, elastomers, wood, coal, etc.   Even the atr technique probably would not be able to handle such substances.

Devices are commercially available from companies such as Barnes Engineering Company and Wilks Scientific Corporation for pyrolyzing materials and studying the pyrolysis products. The following sampling techniques are available:

(i)   Collection of the condensable part of the pyrolysis products on a transmission window (NaCl, KBr, etc.).

(ii) Trapping of the condensable material on an atr crystal.

(iii) Examination of the vapor phase pyrolysis products directly by transmission.

(iv) Separation of volatile products by gas chromatography.

The pyrolysis accessory available from Wilks Scientific Corporation is illustrated in Figure 2-28. Similar accessories are available from Barnes Engineering Co.

Further description of the technique may be found in reference (14) and in the literature supplied by the manufacturers of infrared accessories. Among the many applications of the method to be found in the literature are rubber (15), coal (16), polyvinyl chloride (17), and many complex organic compounds (6, 18).

2–5g   Polarization Measurements in Infrared Spectroscopy

The intensity of an absorption band depends on the change in dipole moment (transition moment) associated with a particular normal mode of vibration. In gases, liquids, solutions, and amorphous or polycrystalline solids, the direction of the dipole moment change in the molecule is immaterial. However, in crystalline solids and certain polymers and natural products, such as polysaccharides and proteins, the intensity of the absorption depends on the direction of the transition moment with respect to the plane of polarization of the incident infrared radiation. Thus polarization (or dichroic) measurements may be used to study and evaluate orientation within samples.

The orientation of the molecules in a single crystal of a compound can often be determined by the methods of X-ray diffraction. In these cases, polarization measurements can enable absorption bands to be assigned to particular vibrations of the molecule. Many examples of such measurements are to be found in the literature.

Polarization measurements may also be made using atr or fmir techniques. This method is particularly useful in studies on the structures of polymer samples where a thin polymer strip and the electric vector are oriented in planes parallel and perpendicular to the instrument slit. Certain vibrations have drastically reduced intensity when the sample is turned through 90°.

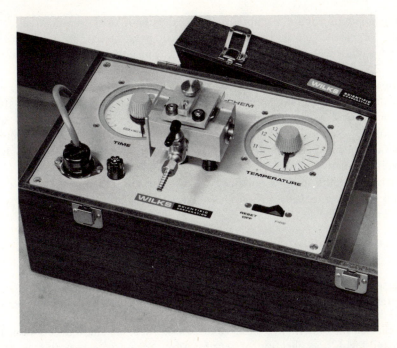

FIGURE 2-28.   An apparatus for pyrolysis of compounds for direct recording of
the infrared transmission of gaseous products and ATR spectra of condensable
products.   An internal reflection element is held over a port at the top of the cell.
The cell can be located directly onto the pins of the internal reflection apparatus
shown in Figure 2-25.   [Courtesy of Wilks Scientific Corporation.]

If these vibrations have been previously assigned to specific groups in the polymer, then the
orientation of these groups can be established.

Polarizers are relatively simple devices and are available from the instrument manufacturers. Two commonly used types of polarizer are the wire grid polarizer and the stacked
silver chloride plate polarizer.   The former, which is proprietary to the Perkin-Elmer
Corporation, is made by vapor depositing a grid of closely spaced parallel gold wires on a
suitable material transparent in the infrared. The substrate used depends on the region to be
scanned. AgBr is used for the range 5000–300 cm^{-1} and polyethylene from 500 cm^{-1} down
into the far infrared.   This polarizer is mounted very simply in the infrared spectrometer,
either in the common beam of a double beam instrument or in the sample space of the
instrument, so that even the simplest infrared spectrometers can make use of the wire grid
polarizer.   The second type of polarizer consists of two stacks of silver chloride plates,
illustrated in Figure 2-29.   Each stack is arranged with the plates parallel to each other
and at the Brewster angle of 26.5° to the incident beam.   [The theory of this kind of
polarizer is discussed in the book by Potts (reference 19, pp. 245–48).]   The two stacks
are placed as shown in Figure 2-29 in order to bring the diffracted rays back onto the
axis of the incident beam.

FIGURE 2-29.   Diagram showing the path
of the light through a parallel plate polarizer.

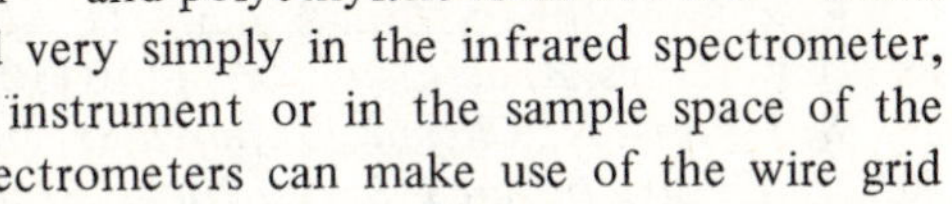

Two final points should be noted. First, the theoretical transmission of either the perpendicular or parallel electric vector of a perfect polarizer is 50%. An actual polarizer will transmit about 80% of theoretical, or 40%. Thus the usable energy is reduced by a factor of 2.5, and either a reference beam attenuator (see Figure 2-15) or a second polarizer should be used in the reference beam to enable the full ordinate range of the chart to be used. This loss in energy can be compensated by either increasing the amplifier gain by a factor of 2.5 (noisy spectra) or by opening the slits by a factor of $\sqrt{2.5} = 1.6$ (lower resolution). Second, a grating spectrometer itself has polarization characteristics that may vary with frequency. It is preferable, therefore, to rotate the sample through 90° rather than to rotate the polarizer.

A discussion of the uses of polarized radiation in infrared studies of polymers is given in Chapter 4 of reference (20). A specialized text on polarized light has been written by Shurcliff (reference 21).

BIBLIOGRAPHY

1. N. L. Alpert, W. E. Keiser, and H. A. Szymanski, *IR – Theory and Practice of Infrared Spectroscopy,* 2nd ed., Plenum, New York, 1970; (paper) Plenum/Rosetta, 1973.
2. J. E. Stewart, *Infrared Spectroscopy,* Marcel Dekker, New York, 1970.
3. T. P. Melia, *An Introduction to Masers and Lasers,* Chapman and Hall, London, 1967.
4. "Lasers and Light," a collection of reprints from *Scientific American,* W. H. Freeman, San Francisco, 1969.
5. R. G. J. Miller and B. C. Stace, eds., *Laboratory Methods in Infrared Spectroscopy,* 2nd ed., Heyden and Son, 1972.
6. H. A. Szymanski, ed., *Progress in Infrared Spectroscopy,* Vol. 3, Plenum, New York, 1967.
7. S. K. Freeman and D. O. Landon, *The Spex Speaker,* Vol. VII, December (1968). Available from Spex Industries, Metuchen, N.J. 08840.
8. G. F. Bailey, S. Kint, and J. R. Scherer, *Anal. Chem.,* **39**, 1040 (1967).
9. T. R. Gilson and P. J. Hendra, *Laser Raman Spectroscopy,* Wiley, London, 1970.
10. M. J. Gall, P. J. Hendra, D. S. Watson, and C. J. Peacock, *Appl. Spectrosc.,* **25**, 423 (1971).
11. W. Kiefer and H. J. Bernstein, *Appl. Spectrosc.,* **25**, 500 (1971).
12. N. J. Harrick, *Anal. Chem.,* **36**, 188 (1964).
13. B. Katlafsky and R. E. Keller, *Anal. Chem.,* **35**, 1665 (1963).
14. D. A. Vassallo, *Anal. Chem.,* **33**, 1823 (1961).
15. M. Tryon, E. Horowitz, and J. Mandel, *J. Res. Nat. Bur. Stand.,* **55**, 219 (1955).
16. J. D. Brooks, R. A. Durie, and S. Sternhell, *Aust. J. Appl. Sci.,* **9**, 303 (1958).
17. R. R. Stromberg, S. Straus, and B. G. Achhammer, *J. Res. Nat. Bur. Stand.,* **60**, 147 (1958).
18. D. L. Harms, *Anal. Chem.,* **25**, 1140 (1953).
19. W. J. Potts, Jr., *Chemical Infrared Spectroscopy,* Wiley, New York, 1963.
20. A. Elliott, *Infrared Spectra and Structure of Organic Long-Chain Polymers,* Edward Arnold, London, 1969.
21. W. A. Shurcliff, *Polarized Light,* Harvard University Press, Cambridge, Mass., 1962.
22. N. J. Harrick, *Internal Reflection Spectroscopy,* Wiley, New York, 1967.
23. R. T. Conley, *Infrared Spectroscopy,* 2nd ed., Allyn and Bacon, Boston, 1972.

3

THEORY OF INFRARED ABSORPTION AND RAMAN SCATTERING

3-1 Introduction

In this chapter we shall discuss in a simplified way how molecules absorb or scatter electromagnetic radiation and how such processes are related to molecular vibrations and the symmetry of molecules. We will need to know something about the properties of electromagnetic radiation and how the interaction of molecules with this radiation results in absorption or scattering. We are interested particularly in molecular vibrations, or changes in vibrational energy. Molecular vibrations can be treated by classical mechanics using a simple ball and spring model, whereas vibrational energy levels and transitions between them are concepts taken from quantum mechanics. Both approaches are useful and will be treated here.

3-2 Electromagnetic Radiation

3-2a The Wave Description
Visible light and infrared radiation are two forms of electromagnetic radiation, which consists of continuous electric and magnetic waves. The waves are transverse; that is, the displacements are perpendicular to the direction of propagation, with the electric and magnetic displacements perpendicular to each other. Only the electric waves interact significantly with molecules and it is those that are of importance in explaining Raman scattering and infrared absorption.

The waves are characterized by a wavelength λ (m) and a frequency ν (sec^{-1}) related by eq. 3-1, where c is the velocity of light (3×10^8 m sec^{-1}).

$$\lambda = \frac{c}{\nu} \qquad\qquad 3\text{-}1$$

The transverse displacement of the waves, i.e., the electric field strength, varies in a regular periodic manner, which can be described by a sine or a cosine function such as eq. 3-2,

$$E = E_0 \cos(2\pi\nu t) \qquad\qquad 3\text{-}2$$

where E is the magnitude of the displacement at any time t and E_0 is the amplitude of the wave (maximum displacement).

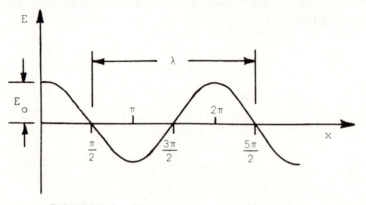

FIGURE 3-1.   *Characteristics of an electric wave.*

An electromagnetic wave can be represented by the curve shown in Figure 3-1. We have no information on the magnitude of the displacement, but we assume that as waves spread out from a source, their amplitudes decrease steadily. The waves carry energy, which can be transferred to molecules. The rate of flow of energy (the power) is proportional to the square of the amplitude. When electromagnetic radiation passes into a sample, the molecules are exposed to an oscillating electric field, and energy is available to excite molecular vibrations either directly by absorption or indirectly through scattering.

3-2b   The Particle (Photon) Description

Although light behaves as if it were a train of waves, it also behaves as if it were a stream of tiny particles called *photons*. One photon of light of frequency ν has energy $h\nu$, where h is Planck's constant (6.63×10^{-34} J $\cdot$ sec). As we shall see, collisions of photons with molecules provide a convenient way of explaining Raman scattering. Absorption of a photon of the appropriate frequency ν in the infrared can cause a molecule to undergo a transition from an initial vibrational energy state E_1 to another state E_2 as in eq. 3-3.

$$E_2 - E_1 = h\nu \qquad\qquad 3\text{-}3$$

Only a few of the many possible vibrational transitions are actually "allowed," and even those that are allowed may not give rise to strong absorption or scattering. The reasons for this lie in the mechanism of absorption and scattering, to be discussed next. It is possible to predict which transitions are allowed by means of the so-called quantum mechanical *selection rules* described in section 3-7.

3-3 Infrared Absorption

3-3a Introduction

A molecule can absorb energy from the oscillating electric field of a beam of infrared radiation if it possesses an oscillating dipole. The frequency of oscillation of the dipole must be exactly the same as that of the applied electric field. All molecules except homonuclear diatomics have some vibrations that produce a change of dipole moment. Even symmetrical molecules with no permanent dipole moment, such as CO_2 have a transient dipole when they vibrate in an unsymmetrical way. The exact matching of frequencies of the oscillating dipole and the electric field of the radiation is ensured by the use of a source that gives continuous infrared radiation (see section 2-1a) so that all appropriate frequencies are present.

3-3b Transition Probability

The extent to which radiation is absorbed (known as intensity of absorption) depends on the probability that a transition between two energy levels of energy E_a and E_b can be made to occur by an external perturbation. The electric field of a beam of electromagnetic radiation provides the necessary perturbation and the energy $(E_b - E_a)$ needed for the transition.

To calculate the probability of a transition occurring in a certain time interval, the methods of quantum mechanics must be used. The starting point for the calculation is the time-dependent Schrödinger equation (eq. 3-4). In this equation Ψ is a time-dependent wave

$$\left(\frac{ih}{2\pi}\right)\left(\frac{\partial \Psi}{\partial t}\right) = \mathcal{H}\,\Psi \qquad \text{3-4}$$

function and $\mathcal{H}$ is the Hamiltonian of the time-independent Schrödinger equation $(\mathcal{H}\psi = E\psi)$ used to calculate the energy levels of the molecule (see section 3-6).

The perturbation produced by the electric field of the radiation can be taken into account by adding a term $\mathcal{H}'$ to the Hamiltonian of eq. 3-4. The wave function is also changed by this perturbation, so that the time-dependent Schrödinger equation is now given by eq. 3-5.

$$\left(\frac{ih}{2\pi}\right)\left(\frac{\partial \Psi'}{\partial t}\right) = (\mathcal{H} + \mathcal{H}')\,\Psi' \qquad \text{3-5}$$

For the case of infrared absorption, the perturbation is the interaction between an oscillating dipole moment in the molecule and the electric field of the radiation. This interaction can be expressed mathematically by eq. 3-6, where μ is the dipole moment and E is the electric field vector from eq. 3-2.

$$\mathcal{H}' = \mu \cdot \mathbf{E} = \mu E_0 \cos 2\pi\nu t \qquad \text{3-6}$$

Use of eq. 3-6 in eq. 3-5 leads to the transition probability (eq. 3-7), where E_0 is the

$$\text{probability} = \left(\frac{4\pi^2}{h^2}\right)|\mu_{ab}|^2 E_0^2 t \qquad \text{3-7}$$

electric field amplitude (see eq. 3-2) and $|\mu_{ab}|$ is known as the *transition dipole moment* (eq. 3-8).

$$|\mu_{ab}| = \int_{-\infty}^{+\infty} \psi_b |\mu| \psi_a d\tau \qquad\qquad 3\text{-}8$$

Details of the derivation of eq. 3-7 are given in most standard texts on quantum mechanics, some of which are listed at the end of this chapter. In eq. 3-8 ψ_a and ψ_b are the wave functions for the lower and upper states, μ is a dipole moment function, or operator, $d\tau$ is an infinitesimal volume of space ($dxdydz$), and the integral is evaluated over all space. It should be noted that μ is *not* the permanent dipole moment μ_0 of the molecule, but a function that takes into account the changes in dipole moment associated with vibrations.

3-3c The Dipole Moment
The dipole moment function can be written as a Taylor series (eq. 3-9), where $(d\mu/dq)_0$

$$\mu = \mu_0 + \left(\frac{d\mu}{dq}\right)_0 q + \left(\frac{d^2\mu}{dq^2}\right)_0 q^2 + \cdots \qquad\qquad 3\text{-}9$$

is the change in the dipole moment per unit change in the vibrational coordinate q.

For a diatomic molecule, q would be simply the change in bond length. For polyatomic molecules, q would be in general a combination of bond extensions and changes in valence angles, known as a normal coordinate. This quantity is discussed in section 3-9g. The subscript zero in eq. 3-9 means that $d\mu/dq$ is evaluated at $q = 0$ (that is, as the vibration passes through the equilibrium position of the molecule). Equation 3-9 is substituted into eq. 3-8 in order to evaluate $|\mu_{ab}|$. When this is done, it is found that there need not be any permanent dipole moment μ_0 in order for a molecule to absorb infrared radiation, because μ_0 occurs in the vanishing integral

$$\int_{-\infty}^{+\infty} \psi_b \mu_0 \psi_a d\tau = \mu_0 \int_{-\infty}^{+\infty} \psi_b \psi_a d\tau = 0$$

It is also found that terms involving q^2 and higher terms are not important,[1] so that $|\mu_{ab}|$ can be written as eq. 3-10.

$$|\mu_{ab}| = \int_{-\infty}^{+\infty} \psi_b \left(\frac{d\mu}{dq}\right)_0 q \, \psi_a dq \qquad\qquad 3\text{-}10$$

3-3d The Change in Dipole Moment Associated with a Vibration
The quantity $(d\mu/dq)_0$ can be given the following physical meaning. Two charges of magnitude e, one positive and one negative, separated by a distance r, constitute a dipole with moment $\mu_0 = re$. In a molecule the charges would be regions of high and low electron density, as for example, in the heteropolar bond of HCl. When the bond stretches in a vibration, the separation of the charges increases and there is an increase in dipole moment. For HCl, $(d\mu/dq)_0 = 0.88$ Debye Å^{-1}. For a homonuclear diatomic molecule, both μ_0 and $(d\mu/dq)_0$ are zero and the molecule cannot absorb in the infrared. We shall need eq. 3-10

---

[1] Higher terms in the dipole moment function account for electrical anharmonicity.

again later when selection rules are discussed.   From eq. 3-7 and 3-10 it should now be clear that *there must be a change in dipole moment accompanying a vibration in order for a molecule to absorb in the infrared.*

3-3e Experimental Measurement of Absorption—The Beer–Lambert Law

What we actually record in a spectrum (Figure 3-2) is the percent transmittance T, which can be converted to a quantity known as absorbance A by the relationship of eq. 3-11,

$$A = \log_{10} \frac{T_0}{T} \qquad\qquad 3\text{-}11$$

in which T_0 is the incident (baseline) transmittance. In the example of toluene (Figure 3-2), the absorbance of the band at 3050 cm^{-1} is $\log_{10} (93/15) = 0.79$. It is possible to obtain chart paper with logarithmic absorbance as the ordinate scale.   The peak maximum at 3050 cm^{-1} would be read as $\log 100/15 = 0.82$ and the baseline as $\log 100/93 = 0.03$, so that $A = 0.82 - 0.03 = 0.79$ as before. The quantity $100/T$ is known as the intensity I, so that eq. 3-11 could be written as eq. 3-12. Another quantity that is sometimes used for

$$A = \log_{10} \frac{I}{I_0} \qquad\qquad 3\text{-}12$$

intensity is percent absorption, which is simply $(100 - T)$.

In solutions it is found experimentally that absorbance depends on concentration c and pathlength l according to the Beer-Lambert Law (eq. 3-13).   The experimental molar

$$A = \epsilon c l \qquad\qquad 3\text{-}13$$

absorption coefficient ϵ is related to the transition moment of eq. 3-8 by a rather complicated equation (eq. 3-14, where N is the Avogadro number). The second quantity in eq. 3-14 is

$$\epsilon = \left(\frac{h\nu_{ab}N}{1000}\right)\left(\frac{8\pi^3}{3h^2}\,|\mu_{ab}|^2\right) \qquad\qquad 3\text{-}14$$

known as the Einstein coefficient of absorption B_{ab}. Thus, in principle at least, infrared

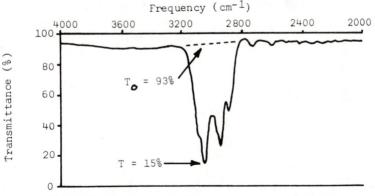

FIGURE 3-2. Part of the infrared spectrum of toluene.

intensities can be predicted from a knowledge of the dipole derivative and the wave functions of the vibrational states involved. In practice, however, $(d\mu/dq)_0$ is not known, and experimental intensities are sometimes used to estimate the dipole derivative.

The Beer-Lambert law accounts for the observation that halving the concentration (or pathlength) of a sample does *not* lead to a halving of peak height. To illustrate this fact, consider a very strong peak with 98% absorption (2% transmittance). Assume that the baseline transmittance is $T_0 = 100\%$. Then $T_0/T = 50$ and $A = \log T_0/T = 1.70$. Now, to halve the peak height means to reduce the absorption to 49.0%; T then becomes 51%, $T_0/T = 1.96$, and $A = 0.29$. This means that the concentration (or the pathlength) would have to be reduced by a factor of $1.70/0.29 = 5.9$ in order to reduce the intensity of this particular peak by a factor of two.

Practical use can be made of this theory when preparing samples for infrared transmission work. If the first pellet, mull, or solution prepared produces very strong peaks (near 100% absorption), then make the next sample thinner or more dilute by a factor of *five*, not two.

> *Problem 3-1* A very strong band in the infrared spectrum of a solution of a certain compound has a transmittance of 1% (99% absorption). By what factor must the concentration (or pathlength) of the solution be reduced in order to increase the transmittance to 25%.

3-4 Raman Scattering

3-4a Introduction

When a laser beam passes through a sample, light is scattered in all directions. The Raman spectrometer usually collects the light scattered at right angles to the direction of the laser beam. The scattered light consists mainly of radiation of frequency unchanged from that of the laser (Rayleigh scattering). In addition, there is some much weaker scattering consisting of frequencies greater and less than the laser frequency. This is the Raman scattering, the intensity of which is usually of the order of 10^{-4} times weaker than the intensity of the Rayleigh scattering. The Raman lines of lower frequency than the laser frequency are known as the Stokes lines and the higher frequency lines the anti-Stokes.

3-4b Pictorial Description of Scattering

A rather simple explanation of Raman scattering, based on collisions of photons with a molecule is illustrated in Figure 3-3. The incident photon of frequency ν collides with a

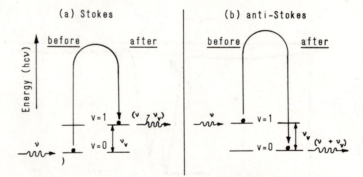

FIGURE 3-3.   The mechanism of Raman scattering: (a) Stokes and (b) anti-Stokes.

molecule that is in either the ground or an excited vibrational state. If no interaction occurs (a perfectly elastic collision), then there is no exchange of energy and the scattered photon has exactly the same frequency as before. This is the Rayleigh scattering. If exchange of energy does occur, then the scattered photon can have either less energy (lower frequency and longer wavelength) or more energy (higher frequency, shorter wavelength). In the former case, as illustrated in Figure 3–3a, the molecule that was originally in the ground vibrational state ends up in an excited state. This process gives rise to the Stokes Raman lines, which are the ones normally recorded as the Raman spectrum.

3–4c Relative Intensities of Stokes and Anti-Stokes Lines

The number of molecules N_v present in an excited vibrational state of energy E_v compared to the number N_0 in the ground state at temperature T is given by the Boltzmann equation (3–15), where k is the Boltzmann constant (1.38×10^{-13} J K^{-1}). The number

$$N_v = N_0 e^{-E_v/kT} \qquad \text{3–15}$$

N_v is quite small compared to N_0 at room temperature, unless E_v is small (a low frequency mode of vibration). As a result, only the low frequency vibrations give rise to anti-Stokes lines of significant intensity.

Problem 3-2 Calculate N_v/N_0, using eq. 3–15, for energy levels of 100 and 1000 cm^{-1} and $T = 300$K. Note that E_v must be converted to Joules by multiplying by hc. Use $hc/k = 1.439$ cm · K.

3–4d Polarizability and Scattering Theory

We have as yet said nothing about the actual mechanism of scattering. According to classical electromagnetic theory, an oscillating dipole emits radiation with the same frequency as the oscillation. A dipole μ^{ind} is induced in a molecule by the electric field **E** of the incident radiation according to eq. 3–16, where α is the polarizability of the molecule. Since both μ^{ind}

$$\mu^{\text{ind}} = \alpha\mathbf{E} \qquad \text{3–16}$$

and **E** are vector quantities, the polarizability α is a tensor and eq. 3–16 should really be written as three equations (3–17a) or in matrix notation (3–17b). The tensor α is symmetric so that

$$\mu_x^{\text{ind}} = \alpha_{xx}E_x + \alpha_{xy}E_y + \alpha_{xz}E_z$$
$$\mu_y^{\text{ind}} = \alpha_{yx}E_x + \alpha_{yy}E_y + \alpha_{yz}E_z \qquad \text{3–17a}$$
$$\mu_z^{\text{ind}} = \alpha_{zx}E_x + \alpha_{zy}E_y + \alpha_{zz}E_z$$

$$\begin{bmatrix} \mu_x^{\text{ind}} \\ \mu_y^{\text{ind}} \\ \mu_z^{\text{ind}} \end{bmatrix} = \begin{bmatrix} \alpha_{xx} \alpha_{xy} \alpha_{xz} \\ \alpha_{yx} \alpha_{yy} \alpha_{yz} \\ \alpha_{zx} \alpha_{zy} \alpha_{zz} \end{bmatrix} \begin{bmatrix} E_x \\ E_y \\ E_z \end{bmatrix} \qquad \text{3–17b}$$

$\alpha_{xy} = \alpha_{yx}$, $\alpha_{xz} = \alpha_{zx}$, and $\alpha_{yz} = \alpha_{zy}$. Thus there are usually six components of the polarizability tensor.

Most of the time the induced dipole oscillates with the frequency of the incident radiation and the emission has the same frequency. This is the Rayleigh scattering. However, if one or more components of α should change in a periodic way, as, for example, in a vibration of the molecule, then the frequency of the oscillating dipole would be modified and the scattered light could have a frequency different from that of the incident radiation. This is the Raman scattering. The frequency difference corresponds to a vibrational frequency of the molecule. This simple theory predicts that the Raman spectrum should consist of lines of frequencies $(\nu \pm \nu_v)$, where ν is the frequency of the exciting laser line and ν_v is one of the $(3N - 6)$ vibrational frequencies.[2]

The magnitude of the oscillating electric field of the incident radiation can be described as before by eq. 3-2. Now μ^{ind} will only have its frequency modified if α changes during a vibra-

$$E = E_0 \, \cos \, (2\pi\nu t) \tag{3-2}$$

tion of the molecule (or more properly if at least one component of α changes during a vibration). To see how the polarizability varies with a vibration, we consider a diatomic molecule and write α as a Taylor series in the vibrational coordinate q (eq. 3-18),[3] where the subscript zero

$$\alpha = \alpha_0 + \left(\frac{d\alpha}{dq}\right)_0 q + \cdots \tag{3-18}$$

on α_0 and $(d\alpha/dq)_0$ means that these quantities are evaluated at the equilibrium position of the vibration. Since q varies periodically with time, if the vibrational frequency is ν_v, we can write eq. 3-19. Substituting in eq. 3-18 gives eq. 3-20, and this expression substituted in

$$q = A \, \cos \, 2\pi\nu_v t \tag{3-19}$$

$$\alpha = \alpha_0 + \left(\frac{d\alpha}{dq}\right)_0 A \, \cos \, 2\pi\nu_v t + \cdots \tag{3-20}$$

eq. 3-16, together with E from eq. 3-2, gives eq. 3-21a, which on multiplying out gives

$$\mu^{\text{ind}} = \left[\alpha_0 + \left(\frac{d\alpha}{dq}\right)_0 A \, \cos \, 2\pi\nu_v t\right] [E_0 \, \cos \, 2\pi\nu t] \tag{3-21a}$$

eq. 3-21b. Now using the trigonometric relation, $\cos A \, \cos B = \frac{1}{2} \, [\cos(A + B) + \cos(A - B)]$,

$$\mu^{\text{ind}} = \alpha_0 E_0 \, \cos \, 2\pi\nu t + AE_0\left(\frac{d\alpha}{dq}\right)_0 (\cos \, 2\pi\nu t \, \cos \, 2\pi\nu_v t) \tag{3-21b}$$

eq. 3-21 becomes eq. 3-22.

$$\mu^{\text{ind}} = \alpha_0 E_0 \, \cos \, 2\pi\nu t + \frac{AE_0}{2}\left(\frac{d\alpha}{dq}\right)_0 [\cos \, 2\pi(\nu + \nu_v)t + \cos \, 2\pi(\nu - \nu_v)t] \tag{3-22}$$

[2] $(3N - 5)$ for a linear molecule.

[3] Terms in q^2 and higher powers are very small and can be neglected.

We can draw several conclusions from eq. 3-22. First, when $(d\alpha/dq)_0 = 0$, only the first term remains. This is the term that accounts for the Rayleigh scattering. The second and third terms account for lines observed at frequencies higher and lower ($\nu \pm \nu_v$) than the exciting frequency (ν). These are the anti-Stokes and Stokes Raman lines, respectively. We can also see that there will only be Raman scattering if at least one component of $(d\alpha/dq)_0$ is not zero.

3-4e Intensity of Scattered Light

Classical electromagnetic theory predicts that the intensity of scattered light is given by eq. 3-23. Thus the intensities are proportional to the fourth power of the frequency ν and to

$$I = \left(\frac{16\pi^4\nu^4}{3c^3}\right)(\mu^{\text{ind}})^2 \qquad \qquad 3\text{-}23$$

the square of the polarizability derivative $(d\alpha/dq)_0$. The fourth power relationship accounts for the blue color of the sky, since the blue part of the solar spectrum is scattered more strongly than the red. This relationship also implies that a stronger Raman spectrum will be obtained using the 4880 or 5145 Å line from an argon ion laser than from the 6328 or 6471 Å line (of equal power) obtained from a helium–neon or krypton ion laser, respectively.

Molecular vibrations producing a large change in polarizability will give strong Raman lines. These are usually highly symmetrical modes or modes involving very polarizable bonds. Some examples of vibrations giving strong Raman bands are the symmetric stretching modes of highly symmetrical molecules such as CCl_4, SF_6, etc.; stretching of carbon–carbon and carbon–nitrogen double and triple bonds; and stretching of carbon–sulfur and sulfur–sulfur bonds.

3-5 Molecular Vibrations—Classical Mechanics

3-5a Vibration of a Single Particle

The vibrations of an organic molecule involving stretching of X–H bonds, where X is C, N, O, etc., can be described by the very simple model illustrated in Figure 3-4. In this model the particle (a hydrogen atom) is connected by a spring (the bond) to a large mass (the rest of the molecule). We only consider motion in the x direction.

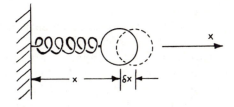

FIGURE 3-4. The vibration of a single particle.

The spring, which represents the X–H bond, is assumed to obey Hooke's law. This law states that the restoring force F in the spring is proportional to the displacement δx from the equilibrium position. This relationship can be written as eq. 3-24, where k is the force

$$F = -k\delta x \qquad \qquad 3\text{-}24$$

constant and the minus sign implies that the force acts in a direction opposite to that in which δx increases. One of Newton's laws of motion states that a force F acting on a body of mass m will produce an acceleration a in the body. In the present case the mass starts from an instantaneous stationary position (at the extended or the compressed position of the spring) and moves with acceleration toward the equilibrium position.

The acceleration can be written as $d^2(\delta x)/dt^2$, and we then have a second expression (eq. 3-25) for the force acting in the spring. Equations 3-24 and 3-25 are equal, so that we

$$F = m\left(\frac{d^2(\delta x)}{dt^2}\right) \qquad\qquad 3\text{-}25$$

have eq. 3-26. Now the motion of the particle is sinusoidal in nature so the displacement

$$-k\delta x = m\left(\frac{d^2(\delta x)}{dt^2}\right) \qquad\qquad 3\text{-}26$$

can be written as eq. 3-27. When eq. 3-27 is differentiated twice with respect to t, we have

$$\delta x = A \cos 2\pi \nu t \qquad\qquad 3\text{-}27$$

eq. 3-28. Substitution in eq. 3-26 and solution for ν gives eq. 3-29.

$$\frac{d^2(\delta x)}{dt^2} = -4\pi^2\nu^2 A \cos 2\pi\nu t = -4\pi^2\nu^2\delta x \qquad\qquad 3\text{-}28$$

$$\nu = \left(\frac{1}{2\pi}\right)\sqrt{\frac{k}{m}} \qquad\qquad 3\text{-}29$$

To test this simple theory consider a C–H bond in a molecule. The force constant is approximately 5.0×10^5 dyne cm^{-1} (5.0×10^2 N m^{-1}) and the mass of the hydrogen atom is 1.66×10^{-27} kg. Putting these numbers into eq. 3-29, gives $\nu = 8.77 \times 10^{14}$ sec^{-1} or $\underline{2920\ cm^{-1}}$. This frequency is just about the center of the region in which C–H stretching vibrations are observed in infrared and Raman spectra. The simple model will not work for bonds involving two heavy atoms. However, such bonds can be thought of as diatomic molecules.

3.5b Vibration of a Diatomic Molecule

For the classical treatment of the vibration of a diatomic molecule we consider the molecule as two balls, with masses equal to the masses of the two atoms, connected by a spring, which represents the bond. During a vibration each atom is displaced from its equilibrium position as shown in Figure 3-5.

It is clear from Figure 3-5 that the total stretching of the bond is given by $(r - r_e)$.

FIGURE 3-5. The displacements of the atoms of a diatomic molecule in a vibration.

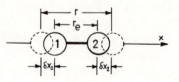

This quantity is known as the *vibrational coordinate q*. It can also be seen that $(r - r_e) = (\delta x_2 - \delta x_1)$ in an algebraic sense, since atom 1 moves in a direction opposite to atom 2. We can therefore write the Hooke's law equation in several ways (eq. 3-30), and there will

$$F = -k(\delta x_2 - \delta x_1) = -k(r - r_e) = -kq \qquad \text{3-30}$$

be *two* Newton's law equations (eqs. 3-31 and 3-32).

$$F = m_1 \left(\frac{d^2(\delta x_1)}{dt^2} \right) \qquad \text{(for atom 1)} \qquad \text{3-31}$$

$$F = m_2 \left(\frac{d^2(\delta x_2)}{dt^2} \right) \qquad \text{(for atom 2)} \qquad \text{3-32}$$

We could proceed by solving two simultaneous differential equations obtained from eq. 3-30 with 3-31 and 3-32. This approach leads to two solutions, one of which is a translation in the x direction (zero frequency) and the other is the vibration. However, we use here a simpler method, involving the single coordinate q, that automatically eliminates the translation in the x direction.

Going back to the definition of q, we note that the change in bond length $(r - r_e)$ is the sum of δx_2 and $-\delta x_1$. Therefore, from eqs. 3-33 and 3-34 we get eq. 3-35.

$$-\delta x_1 = \left(\frac{m_2}{m_1 + m_2} \right)(r - r_e) \qquad \text{3-33}$$

$$\delta x_2 = \left(\frac{m_1}{m_1 + m_2} \right)(r - r_e) \qquad \text{3-34}$$

$$\delta x_2 - \delta x_1 = (r - r_e) = q \qquad \text{3-35}$$

We can now write *either* of the two equations obtainable from 3-31 and 3-32 with 3-30 as eq. 3-36. Now $[m_1 m_2/(m_1 + m_2)]$ is known as the reduced mass and has the symbol μ,

$$\left(\frac{m_1 m_2}{m_1 + m_2} \right)\left(\frac{d^2(r - r_e)}{dt^2} \right) = -k(r - r_e) \qquad \text{3-36}$$

so that eq. 3-36 can be rewritten as eq. 3-37. The vibrational coordinate q varies with time

$$\frac{d^2 q}{dt^2} = -\frac{k}{\mu} q \qquad \text{3-37}$$

in a periodic manner. Expressing this relationship by eq. 3-38, which is analogous to eq. 3-37,

$$q = A \cos 2\pi \nu t \qquad \text{3-38}$$

we can differentiate twice with respect to t, substitute in eq. 3-37, and solve for ν to get eq. 3-39. This result is the same as eq. 3-29 except for the replacement of m by μ; it gives

$$v = \frac{1}{2\pi} \sqrt{\frac{k}{\mu}} \qquad\qquad 3\text{-}39$$

the vibrational frequency (in $\sec^{-1}$) of any diatomic molecule or group. Examples of the usefulness of eq. 3-39 are given in section 4-1.

The vibrational coordinate q is known as an internal coordinate because it does not depend on any external cartesian coordinate system. This kind of coordinate is very useful in calculations of vibrational frequencies of polyatomic molecules because the six zero frequency modes (3 rotations + 3 translations) are automatically omitted.

Problem 3-3 Calculate the vibrational frequency of the C=O group using equation 3-39. Take the force constant to be 12.0×10^2 N m^{-1} and the masses of the C and O atoms to be 12.0 and 16.0, respectively (each should be multiplied by the mass of the H atom, 1.66×10^{-27} kg).

3-5c Stretching Vibrations of a Linear Triatomic Molecule

The single particle model of section 3-5a is satisfactory for vibrations involving a light atom bonded only to a heavy atom, such as in X—H groups. The diatomic molecule model of section 3-5b is useful for groups well separated, either physically by intervening atoms or in frequency, from other vibrating groups. These models, however, are of no use in predicting the frequencies of vibration of adjacent groups that have similar "diatomic" frequencies because such vibrations will be mechanically coupled. In the case of two such adjacent groups, two new vibrations will result, which can be described as in-phase and out-of-phase combinations of the original two "diatomic" vibrations. To illustrate this situation, we will calculate the stretching frequencies of the linear triatomic molecule ABA.

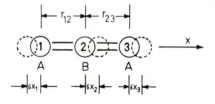

FIGURE 3-6. The linear triatomic molecule ABA.

The three atoms are arranged along the x axis as shown in Figure 3-6. The equilibrium internuclear distances are $r_{12} = r_{23} = r$. The displacement of an atom i from its equilibrium position is denoted by δx_i. Expressions for the linear motion of the three atoms can be written by equating the Hooke's law equation ($F = -k\delta x$) for each atom with the Newton's law equation ($F = m_i \delta \ddot{x}_i$, where $\delta \ddot{x}_i$ means $d^2(\delta x_i)/dt^2$ and is the acceleration of the ith atom). Assuming the force constant to be the same for both bonds we obtain three equations (3-40, 3-41, 3-42).[4]

$$-k(\delta x_1 - \delta x_2) = m_1 \delta \ddot{x}_1 \quad \text{(for atom 1)} \qquad\qquad 3\text{-}40$$

---

[4]It should be noted that the left-hand sides of eqs. 3-40, 3-41, and 3-42 are not simply $F = -k\delta x_i$ because, in calculating the restoring force, it is the motion of an atom *relative* to its neighbors that determines F.

$$-k(2\delta x_2 - \delta x_1 - \delta x_3) = m_2\, \delta\, \ddot{x}_2 \quad \text{(for atom 2)} \qquad 3\text{-}41$$

$$-k(\delta x_3 - \delta x_2) = m_3 \delta \ddot{x}_3 \quad \text{(for atom 3)} \qquad 3\text{-}42$$

These three second-order differential equations have solutions (3-43) of the form used

$$\delta x_i = A_i \cos 2\pi\nu t = A_i \cos \theta \qquad 3\text{-}43$$

previously in section 3-5a, where $\theta = 2\pi\nu t$. Differentiating twice gives eq. 3-44 or 3-45, where $\lambda = 4\pi^2\nu^2$.

$$\delta \ddot{x}_i = -4\pi^2\nu^2 A_i \cos 2\pi\nu t \qquad 3\text{-}44$$

$$\delta \ddot{x}_i = -A_i \lambda \cos \theta \qquad 3\text{-}45$$

After substitution of eqs. 3-43 and 3-45 in 3-40, 3-41, and 3-42, canceling the $-\cos\theta$ factor, and using $m_3 = m_1$, we obtain 3-46, which, on rearranging, give 3-47.

$$\begin{aligned} k(A_1 - A_2) &= m_1 \lambda A_1 \\ k(2A_2 - A_1 - A_3) &= m_2 \lambda A_2 \\ k(A_3 - A_2) &= m_1 \lambda A_3 \end{aligned} \qquad 3\text{-}46$$

$$\begin{aligned} (k - m_1\lambda)A_1 - kA_2 &= 0 \\ -kA_1 + (2k - m_2\lambda)A_2 - kA_3 &= 0 \\ -kA_2 + (k - m_1\lambda)A_3 &= 0 \end{aligned} \qquad 3\text{-}47$$

These three homogeneous equations in A_1, A_2, and A_3 will have a nontrivial solution only when the determinant of the coefficients of A_1, A_2, and A_3 is zero, that is, when eq. 3-48 is satisfied. Multiplying out the determinant of eq. 3-48 gives eq. 3-49, so

$$\begin{vmatrix} (k - m_1\lambda) & -k & 0 \\ -k & (2k - m_2\lambda) & -k \\ 0 & -k & (k - m_1\lambda) \end{vmatrix} = 0 \qquad 3\text{-}48$$

$$(k - m_1\lambda)^2 (2k - m_2\lambda) - 2k^2 (k - m_1\lambda) = 0 \qquad 3\text{-}49$$

$(k - m_1\lambda) = 0$ is a solution (N.B., λ is not wavelength; see eq. 3-45), which gives eq. 3-50.

$$\lambda_1 = \frac{k}{m_1} \quad \text{or} \quad \bar{\nu}_1 = \left(\frac{1}{2\pi c}\right)\sqrt{\frac{k}{m_1}} \qquad 3\text{-}50$$

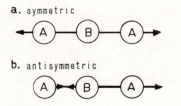

FIGURE 3-7. The stretching modes
of a linear triatomic molecule.

This frequency corresponds to the symmetric (in-phase) stretching of the two A—B groups because it involves no motion of atom B. This mode of vibration, which we call ν_1, is illustrated in Figure 3-7a.

Taking out the factor $(k - m_1\lambda)$ from eq. 3-49 leaves eq. 3-51, which gives eq. 3-52.

$$(k - m_1\lambda)(2k - m_2\lambda) - 2k^2 = 0 \qquad\qquad 3\text{-}51$$

$$m_1 m_2 \lambda^2 - (2m_1 + m_2)k\lambda = 0 \qquad\qquad 3\text{-}52$$

Equation 3-52 has two solutions, one of which is trivial ($\lambda = 0$, corresponding to translation in the x direction). The second solution is shown by eq. 3-53. This vibration is

$$\lambda_2 = \frac{k(2m_1 + m_2)}{m_1 m_2} \quad \text{or} \quad \overline{\nu}_2 = \left(\frac{1}{2\pi c}\right)\sqrt{\frac{k(2m_1 + m_2)}{m_1 m_2}} \qquad\qquad 3\text{-}53$$

the unsymmetrical (out-of-phase) stretching mode and is illustrated in Figure 3-7b. If the masses were all equal, then the frequencies ($\overline{\nu}_1$ and $\overline{\nu}_2$) would be $(1/2\pi c)\sqrt{k/m}$ and $(1/2\pi c)\sqrt{3k/m}$, compared with the diatomic model frequency from eq. 3-39 of $(1/2\pi c)\sqrt{2k/m}$. We note that the calculated frequencies for the coupled oscillators are found above and below the uncoupled frequency. Examples and further discussion of mechanical coupling will be given in section 4-2b.

Problem 3-4 For CO_2 the observed antisymmetric stretching frequency is 2350 cm^{-1}. Use eqs. 3-53 and 3-50 to calculate the frequency of the symmetric stretching mode.

3-6   Molecular Vibrations—Quantum Mechanics

3-6a Introduction

We saw in the previous section how successfully simple classical mechanical models can predict vibrational frequencies for diatomic molecules or groups. However, the concepts of vibrational energy levels and transitions between them are necessary to account for infrared and Raman intensities and selection rules.

To calculate energy levels we use the time-independent Schrödinger equation (3-54),

$$\mathcal{H}\psi = E\psi \qquad\qquad 3\text{-}54$$

where ψ is the wave function corresponding to a level of energy E, and $\mathcal{H}$ is the Hamiltonian operator. The vibrational Hamiltonian for a diatomic molecule is eq. 3-55, where μ is the

$$\mathcal{H} = \left(\frac{-h^2}{8\pi^2\mu}\right)\left(\frac{d^2}{dq^2}\right) + V \qquad\qquad 3\text{-}55$$

reduced mass, q is the vibrational coordinate, h is Planck's constant, and V is a potential energy function. Using the simple harmonic oscillator approximation,[5] the potential energy is given by eq. 3-56, which is related to the Hooke's law force of the previous section by eq. 3-57.

$$V = \frac{1}{2}kq^2 \qquad\qquad\qquad 3\text{-}56$$

$$F = -\frac{dV}{dq} = -kq \qquad\qquad\qquad 3\text{-}57$$

We can now write the Schrödinger equation for the vibrational energy levels of a diatomic molecule as eq. 3-58.

$$\left(\frac{-h^2}{8\pi^2\mu}\right)\left(\frac{d^2\psi}{dq^2}\right) + \frac{1}{2}kq^2\psi = E\psi \qquad\qquad 3\text{-}58$$

The wave functions ψ_v of eq. 3-58 are the harmonic oscillator functions, the first few of which are

$v = 0$ (the ground state): $\qquad \psi_0 = \left(\frac{\alpha}{\pi}\right)^{1/4} e^{-\alpha q^2/2}$

$v = 1$ (the first excited state): $\qquad \psi_1 = \left(\frac{4\alpha^3}{\pi}\right)^{1/4} q e^{-\alpha q^2/2}$

$v = 2$ (the second excited state): $\psi_2 = \left(\frac{\alpha}{4\pi}\right)^{1/4}(1 - 2\alpha q^2)\, e^{-\alpha q^2/2}$

where $\alpha = (2\pi/h)\sqrt{\mu k}$.

3-6b Energy Levels of Diatomic Molecules

The energy levels of diatomic molecules are obtained by substituting the preceding values of ψ_v in eq. 3-58, which gives eq. 3-59. The quantity $(1/2\pi)\sqrt{k/\mu}$ will be recognized

$$E_v = (v + \tfrac{1}{2})\frac{h}{2\pi}\sqrt{\frac{k}{\mu}} \quad (v = 0, 1, 2, \ldots) \qquad\qquad 3\text{-}59$$

as the classical vibrational frequency ν, so that the energy levels are given by eq. 3-60. Thus

$$E_v = (v + \tfrac{1}{2})h\nu \quad (v = 0, 1, 2, \ldots) \qquad\qquad 3\text{-}60$$

the quantum mechanical treatment predicts a set of equally spaced energy levels with a separation between any two levels of $h\nu$.

Problem 3-5 By substitution of the wave functions for $v = 0$ and $v = 1$ in eq. 3-58, verify that the energies are those given by eq. 3-59.

---

[5]To allow for anharmonicity, cubic and quartic force constants l and m could be added to $V = \frac{1}{2}kq^2$, so that V would become $V = \frac{1}{2}kq^2 + \frac{1}{6}lq^3 + \frac{1}{24}mq^4$.

What we observe in infrared or Raman spectra are *transitions* between energy levels. The observed frequencies, however, are essentially the same as the vibrational frequencies of classical mechanical models.

All molecular vibrations are in fact anharmonic, although the anharmonicity may be very small in many cases. This means in practice that the energy levels are not equally spaced, but actually get closer and closer together until the spacing becomes zero at the dissociation limit. We are only concerned, however, with the lowest few levels.

3-6c Polyatomic Molecules

The harmonic oscillator treatment can be extended to polyatomic molecules by considering the molecule to be a set of independent harmonic oscillators. The energy levels are found to be, as in eq. 3-61. The quantum numbers $v_1, v_2, \ldots$ can have integral values

$$E(v_1, v_2, v_3, \ldots) = \sum_{i=1}^{3N-6} (v_i + \tfrac{1}{2})h\nu_i \quad (v_1, v_2, v_3, \ldots = 0, 1, 2, \ldots) \qquad 3\text{-}61$$

$0, 1, 2, \ldots$, so we see that for a large molecule there are a very large number of energy levels.

3-6d Transitions—Fundamentals, Overtones, and Combinations

Transitions between energy levels give rise to the observed infrared and Raman spectra. Fortunately most molecules are in the ground state (all $v_i = 0$) at room temperature (see the Boltzmann equation, 3-15, section 3-4) and only those transitions for which $\Delta v = \pm 1$ are "allowed" by the harmonic oscillator selection rules. This means that only a limited number of absorption bands or Raman lines are observed. The transitions described as *fundamentals* are those between the ground state and levels for which *only one* of the $v_i = 1$. Transitions from the ground state to a state in which only one of the $v_i = 2, 3, \ldots$ and all others are zero, are known as *overtones*. Transitions from the ground state to a state for which $v_i = 1$ *and* $v_j = 1$ simultaneously are known as binary *combinations*. Other combinations, such as $v_i = 1, v_j = 1, v_k = 1$, or $v_i = 2, v_j = 1$, etc., are also possible. Some of these transitions are illustrated in Figure 3-8.

Overtones and combinations are not allowed, but may appear weakly because of anharmonicity or Fermi resonance. These possibilities are discussed in sections 3-7c and 3-8.

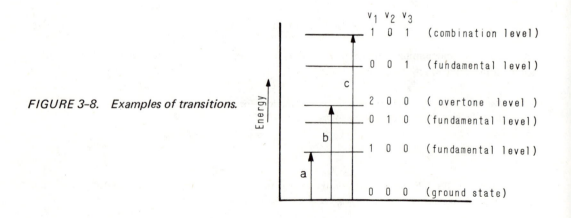

FIGURE 3-8. *Examples of transitions.*

3-7   Quantum Mechanical Selection Rules

3-7a   Introduction

From section 3-3 we know that the intensity of the infrared absorption associated with a transition between two energy levels a and b depends on the transition moment integral (eq. 3-8). This integral arises in the quantum mechanical treatment of absorption of

$$|\mu_{ab}| = \int_{-\infty}^{+\infty} \psi_b \, |\mu| \, \psi_a d\tau \qquad\qquad 3\text{-}8$$

electromagnetic radiation by molecules. A similar treatment of Raman scattering shows that the intensity depends on an analogous integral, as shown in eq. 3-62.

$$|\alpha_{ab}| = \int_{-\infty}^{+\infty} \psi_b \, |\alpha| \, \psi_a d\tau \qquad\qquad 3\text{-}62$$

We have already examined the properties of the dipole moment and polarizability functions μ and α (see eqs. 3-9 and 3-18) and found that for nonzero infrared or Raman intensities $(d\mu/dq)_0$ or $(d\alpha/dq)_0$ must be nonzero. Symmetry properties can be used to decide which molecular vibrations give rise to changes in dipole moment and polarizability (see section 3-10). However, even if $(d\mu/dq)_0$ or $(d\alpha/dq)_0$ is nonzero, the transition may still be forbidden if the overall integrals of eqs. 3-8 and 3-62 are zero. Such considerations lead to the quantum mechanical selection rules.

3-7b   Harmonic Oscillator Selection Rule

The values of the integrals of eqs. 3-8 and 3-62 depend on the wave functions ψ_a and ψ_b. For the harmonic oscillator model, we will show that only transitions for which $\Delta v = \pm 1$ are allowed. In other words, only *fundamentals* (as defined in section 3-6d) are allowed by the quantum mechanical harmonic oscillator selection rule. All overtones and combinations are forbidden in this approximation.

The selection rule $\Delta v = \pm 1$ can be proved for a diatomic molecule using the wave functions listed in section 3-6. The calculation can be carried out for either μ or α, since these functions have identical forms when expanded as Taylor series (eqs. 3-9 and 3-18). Substitution of eq. 3-9 for the dipole moment function in eq. 3-8 gives eq. 3-63. The

$$|\mu_{ab}| = \mu_0 \int_{-\infty}^{+\infty} \psi_b \psi_a dq + \left(\frac{d\mu}{dq}\right)_0 \int_{-\infty}^{+\infty} \psi_b q \psi_a dq \qquad\qquad 3\text{-}63$$

first term of this equation always vanishes because the wave functions are orthogonal in the sense that $\int_{-\infty}^{+\infty}\psi_b\psi_a d\tau = 0$. This result shows that the value of μ_0 is not important. The molecule need not have a permanent dipole moment for infrared absorption to occur.

Equation 3-63 can then be written as eq. 3-64. Note that there are three components

$$|\mu_{ab}| = \left(\frac{d\mu}{dq}\right)_0 \int_{-\infty}^{+\infty} \psi_b q \psi_a dq \qquad\qquad 3\text{-}64$$

of $|\mu_{ab}|$. These are $|\mu_{ab}|_x$, $|\mu_{ab}|_y$, and $|\mu_{ab}|_z$. For the polarizability there are six different integrals $|\alpha_{ab}|_{ij}$, where i and j can be x, y, or z.

Now we consider two cases in detail:

(i) a is the ground state $v = 0$ and b is the first excited state $v = 1$. This is the *fundamental* ($\Delta v = 1$).

(ii) a is again the ground state, but b is the second excited state $v = 2$. This is the *first overtone* ($\Delta v = 2$).

Using the harmonic oscillator functions for case (i), we obtain eq. 6-65.[6] The integral

$$|\mu_{01}| = \left(\frac{d\mu}{dq}\right)_0 \sqrt{\frac{2}{\pi}}\,\alpha \int_{-\infty}^{+\infty} q^2 e^{-\alpha q^2}\,dq \qquad\qquad 3\text{-}65$$

of this equation can be found in tables of standard integrals and the transition moment becomes

$$|\mu_{01}| = \left(\frac{1}{\sqrt{2\alpha}}\right)\left(\frac{d\mu}{dq}\right)_0$$

This quantity is clearly never zero unless $(d\mu/dq)_0$ is zero. Hence the $0 \rightarrow 1$ transition is allowed. Using the wave functions for the $v = 0$ and $v = 2$ levels, for case (ii) we obtain

$$|\mu_{02}| = \sqrt{\frac{\alpha}{2\pi}}\left(\frac{d\mu}{dq}\right)_0 \int_{-\infty}^{+\infty} (q - 2\alpha q^3)e^{-\alpha q^2}\,dq$$

This integral involves two odd (antisymmetric) functions integrated from $-\infty$ to $+\infty$. Hence the integral is always zero, and we say that the $0 \rightarrow 2$ transition is *forbidden* in the harmonic oscillator approximation.

Consideration of other transitions leads to the general harmonic oscillator selection rule $\Delta v = \pm 1$.[7]

3-7c Anharmonicity

In a more exact treatment of molecular vibrations, higher terms are included in the potential function V of eq. 3-56 to allow for anharmonicity (see footnote 5). When this is done, solution of the Schrödinger equation gives the energy levels (eq. 3-66), where

$$E_v = (v + \tfrac{1}{2})h\omega_e - (v + \tfrac{1}{2})^2 \omega_e x_e + \text{higher terms} \qquad\qquad 3\text{-}66$$

$\omega_e = (1/2\pi)\sqrt{k_e/\mu_{\text{red}}}$ is a vibrational frequency and $\omega_e x_e$ is the anharmonicity constant (N.B.: ω_e is not an angular frequency here). The force constant k_e is the harmonic force constant. The symbol ω_e has been used to distinguish this frequency from the harmonic oscillator frequency ν of eq. 3-60.

---

[6] Note that α $(= (2\pi/h)\sqrt{k\mu_{\text{red}}}$, where μ_{red} is the reduced mass defined in section 3-5b) is not related in any way to the polarizability.

[7] See, for example, the book by Eyring, Walter, and Kimball (reference B3).

The solutions of the Schrödinger equation are the anharmonic oscillator wave functions, which may be used to evaluate the integrals of eqs. 3–8 and 3–62 for situations when $\Delta v = 2, 3,$ ... (overtones). In this case the integrals do not vanish and the transitions become allowed. However, the intensities of overtones and combinations in polyatomic molecules are usually very small because the values of the integrals are small compared with the integrals for the fundamentals.

Near infrared spectroscopy is concerned with absorptions due to overtones and combinations. The fundamentals involved are usually stretching vibrations of X–H groups or multiple bonds. Other fundamentals have their overtone and combination frequencies in the normal ir region.

Approximate frequencies can be calculated for purposes of identification and assignment of observed bands by taking multiples or sums of fundamental frequencies. The exact positions and intensities of the observed bands, however, depend on the anharmonicity of the fundamentals involved. Without anharmonicity there would be no observable intensity, because there is anharmonicity the observed frequencies are not simply multiples or sums of the fundamentals. In most cases the observed frequencies will be a few percent lower than calculated.

3–8 Fermi Resonance

3–8a Introduction

For a polyatomic molecule there are $(3N - 6)$ energy levels, for which only one vibrational quantum number is 1 when all the rest are zero. We have seen that these are called the fundamental levels, and a transition from the ground state to one of these levels is known as a *fundamental*. In addition, there are the levels for which one v_i is 2, 3, ... (overtones) or for which more than one v_i is nonzero (combinations). There are, therefore, a very large number of vibrational energy levels, and it quite often happens that two of these have very nearly the same energy. This situation is termed an *accidental degeneracy*. In some cases an interaction known as Fermi resonance can occur between two such levels, with the restriction that the *symmetries* of the levels must be the same for interaction to occur. When this happens, additional features are observed in the infrared or Raman spectra, and an erroneous interpretation of the spectra could be made if Fermi resonance is not taken into account.

3–8b Theory

When interaction occurs, the unperturbed energy levels are subjected to a *repulsion*—one is moved up and the other down. There is also a sharing of intensity between the two transitions involved. From a quantum mechanical first-order perturbation treatment, the energy levels resulting from the interaction of an exactly (accidentally) degenerate pair of levels are given by eq. 3–67, where E^0 is the energy of each unperturbed level and W_{ij} is the perturbation energy.

$$E = E^0 \pm W_{ij} \qquad\qquad 3\text{–}67$$

For the case where the unperturbed levels are not exactly degenerate, the resulting energy levels are given by eq. 3–68a, where E^0_{av} is the average of the unperturbed energies and ΔE is the separation of the unperturbed levels. We can see that when $W_{ij} = 0$, the result

$$E = E^0_{av} \pm \sqrt{W_{ij}^2 + \left(\frac{\Delta E}{2}\right)^2}$$ 3-68a

is eq. 3-68b. In other words, there is no perturbation. This happens when two would-be

$$E = \frac{E_i + E_j}{2} \pm \frac{E_i - E_j}{2} = E_i \text{ or } E_j$$ 3-68b

interacting energy levels are of different symmetry. The magnitude of W_{ij} depends on the anharmonicity associated with the interacting levels. For many vibrations the anharmonicity is small, and in such cases the Fermi resonance interaction will also be small.

3-8c An Example in the CO_2 Molecule

The interaction is greatest for the case of exact (accidental) degeneracy and becomes less as ΔE increases. It can still be noticeable for unperturbed levels as far apart as 50 cm^{-1} or more. The best-known example of Fermi resonance is found in the CO_2 molecule. The O–C–O bending mode ($\bar{\nu}_2$) of CO_2 is observed at 667 cm^{-1} in the infrared spectrum. The overtone of this fundamental is expected near 1330 cm^{-1} (or a little lower depending on the anharmonicity) in the Raman spectrum. However, the symmetric C–O stretching mode ($\bar{\nu}_1$) happens to have about the same frequency (or it would have in the absence of the Fermi resonance perturbation). Now one of the components of $2\bar{\nu}_2$ has the same symmetry as $\bar{\nu}_1$ and a strong interaction occurs. The result is that the observed Raman spectrum contains two strong lines at 1388 and 1286 cm^{-1} instead of the expected one strong and one very weak line near 1340 and 1330 cm^{-1}, respectively (Figure 3-9). It will be noticed that there are

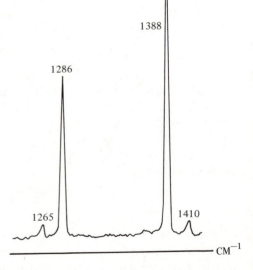

FIGURE 3-9. The Raman spectrum of CO_2 gas in the region of the ν_1 fundamental. Two Fermi doublets are observed: $\bar{\nu}_1/2\bar{\nu}_2$ and the hot bands $(\bar{\nu}_1 + \bar{\nu}_2 - \bar{\nu}_2)/(3\bar{\nu}_2 - \bar{\nu}_2)$.

two additional weak satellite bands at 1265 and 1410 cm^{-1}. These are due to a second Fermi doublet involving the hot bands[8] $(\bar{\nu}_1 + \bar{\nu}_2 - \bar{\nu}_2)$ and $(3\bar{\nu}_2 - \bar{\nu}_2)$.

Numerous other well-characterized examples are known, some of which are discussed in section 4-2c. In large unsymmetrical molecules where the symmetry requirement for the wave function is always met (if there is no symmetry, all fundamentals, overtones, and combinations are of the same symmetry), Fermi resonance is likely to be very common, although in most cases it will be unrecognized.

3-9   Vibrations of Polyatomic Molecules

3-9a   Introduction

The simple equations of section 3-5 become very complex for larger molecules, so that a systematic treatment has been developed for calculating vibrational frequencies of polyatomic molecules. The calculations involve the use of matrices and an equation of classical mechanics. Because the determinantal equations corresponding to eq. 3-48 of section 3-5c are very large, computers have to be used to solve them. Programs are available for the calculations, and approximate vibrational frequencies of quite large molecules can be calculated. The interested reader is referred to the list of texts on the subject at the end of this chapter for further details of the procedure; only an outline is presented here.

Molecular vibrations are associated with continuous changes in kinetic and potential energy within the molecule. The total vibrational energy at any instant is the sum of the kinetic energy associated with the moving atoms and the potential energy associated with stretched or deformed bonds in the molecule. To handle these different kinds of energy, two convenient coordinate systems are set up. *Cartesian displacements* (δx, δy, and δz) are used to describe the kinetic energy, and *internal coordinates* are appropriate for the potential energy.

3-9b   Internal Coordinates (Group Vibrations)

The most effective use of infrared and Raman spectroscopy for the organic chemist is the identification of functional groups from characteristic frequencies in the spectra. Stretching of bonds and deformation of valence angles are examples of vibrations that give rise to useful *group frequencies*. Also it is helpful in the calculation of vibrational frequencies to consider the vibrations as arising from changes in bond lengths and valence angles rather than displacements in x, y, and z directions of the atoms of the molecule. Changes in bond lengths and valence angles are examples of *internal coordinates*.

We have already seen in eq. 3-30 of section 3-5b how the use of the internal coordinate q simplifies the Hooke's law equation for the restoring force produced during the stretching of a bond. In general, $(3N - 6)$ internal coordinates can describe the vibrations of a molecule with N atoms, whereas $3N$ cartesian displacement coordinates are necessary.

In addition to bond stretching and angle bending, other displacements such as wagging or twisting of small groups could be used as internal coordinates. In what follows, we use the symbol R_i for internal coordinates and the general symbol ξ_i to denote a cartesian displacement coordinate such as δx_i, δy_i, or δz_i.

---

[8]Hot bands are due to transitions in which the lower state is not the ground state.

3-9c   Vibrational Energy of a Molecule

The kinetic energy T is a sum of contributions of the form $\frac{1}{2}m_i(\dot{\xi}_i)^2$, where $\dot{\xi}_i$ means $(d\xi_i/dt)$ and represents the velocity of the ith atom in one of the directions x, y, or z. There will be three such contributions for each atom, and the total kinetic energy will include three translations and three rotations of the whole molecule. For the total kinetic energy we can write eq. 3-69.

$$2T = \sum_{i=1}^{3N} m_i\dot{\xi}_i^2 \qquad\qquad 3\text{-}69$$

The potential energy V is the sum of the contributions from the Hooke's law forces produced during stretching of bonds, deformation of valence angles, etc. If we assume that these forces are independent of each other, we can write the vibrational potential energy as eq. 3-70, where f_i is the *force constant* associated with the internal coordinate R_i. It

$$2V = \sum_{i=1}^{3N-6} f_i R_i^2 \qquad\qquad 3\text{-}70$$

can be seen that eq. 3-70 reduces to eq. 3-30 of section 3-5b when there is only one internal coordinate. It should also be noted that in many cases, the restoring forces in a distorted molecule are *not* independent of each other. In such cases, interaction force constants have to be included, and terms such as $(f_{ij}R_iR_j)$ must be added to eq. 3-70 to describe the potential energy of the molecule correctly.[9]

3-9d   Calculation of Vibrational Frequencies

During each cycle of a vibration, an atom starts from rest at the extreme position of the displacement, then accelerates to a maximum velocity as it passes through the equilibrium position and finally slows down until it again comes to rest for an instant at its other extreme displacement.

In general, a vibration of a polyatomic molecule involves some movement of most or all of the atoms. The forces associated with the accelerations of the atoms (Newton's law forces) can be obtained from the kinetic energy and combined with the Hooke's law forces obtained from the potential energy. This operation is accomplished using an equation of classical mechanics known as Lagrange's equation (3-71). This equation cannot be used as it stands

$$\frac{d}{dt}\left(\frac{dT}{d\dot{\xi}}\right) + \frac{dV}{dR} = 0 \qquad\qquad 3\text{-}71$$

$\quad$(Newton's law force)$\qquad$(Hooke's law force)

because two different coordinate systems are involved (ξ and R). The kinetic energy T must be expressed in terms of internal coordinates. This transformation, which forms the basis

---

[9]In matrix notation the kinetic and potential energies can be written as $2T = \dot{\xi}'\mathbf{M}\dot{\xi}$ and $2V = \mathbf{R}'\mathbf{FR}$, where $\dot{\xi}$ and $\mathbf{R}$ are now columns; $\dot{\xi}'$ and $\mathbf{R}'$, rows; and $\mathbf{M}$ and $\mathbf{F}$, diagonal matrices whose elements are the masses and the force constants, respectively. When interaction terms are included in the potential energy, $\mathbf{F}$ is no longer diagonal.

for the Wilson FG matrix method, can be accomplished by standard procedures. (Complete details are given in reference A4).

For a polyatomic molecule with N atoms, there will be many forces acting when the atoms move during a vibration. In general, there will then be $(3N - 6)$ equations like eq. 3-71, with the kinetic energy expressed in terms of the internal coordinates (eq. 3-72):

$$\frac{d}{dt}\left(\frac{dT}{d\dot{R}_i}\right) + \frac{dV}{dR_i} = 0 \qquad\qquad 3\text{-}72$$

When T and V are substituted in 3-72, a set of equations is obtained similar to those of section 3-5c (eqs. 3-40, 3-41, and 3-42). The solution follows the procedure of section 3-5c. Each internal coordinate will vary in a periodic manner during a vibration, and this variation can be described by a cosine function of the form $R_i = A_i \cos 2\pi v t$.

Further algebra leads to a determinantal equation that is like eq. 3-48 but with $(3N - 6)$ rows. This equation is solved numerically using a computer and the vibrational frequencies are obtained. Programs for the cartesian-to-internal-coordinate transformation and the complete vibrational frequency calculation are available.

3-9e   Normal Modes of Vibration

The $(3N - 6)$ frequencies obtained from the calculation outlined above are the normal or characteristic vibrational frequencies of a classical ball and spring model of the molecule. In general, several or even all internal coordinates are involved in a normal vibration, which can be expressed algebraically as a linear combination of internal coordinates (eq. 3-73).

$$Q_i = a_{i1} R_1 + a_{i2} R_2 + \cdots \qquad\qquad 3\text{-}73$$

The quantities Q are known as *normal coordinates*. Each normal vibration is associated with a single normal coordinate, although, in general, all the internal coordinates may contribute something to the vibration. However, certain vibrations in molecules containing hydrogen atoms or multiple bonds are in fact localized mainly in a single internal coordinate or group of similar coordinates. Some examples are X–H stretching (X = C, N, O, S, etc.), X≡Y stretching (X and Y = C or N), X=Y stretching (X = C or N; Y = C, O, or N), and H–X–H bending (X = C or N). These are examples of the well-known *group frequencies* that make vibrational spectroscopy so valuable as a tool for identification of molecules and in the determination of structure.

3-9f   A Classification of Normal Modes

The majority of the normal vibrations of a complex organic molecule cannot be assigned to any specific internal coordinate or group of internal coordinates. These frequencies only give an indication of the complexity of the molecule, or go to make up a complex pattern sometimes known as a *fingerprint* of the molecule.

R. N. Jones has suggested a classification of vibrational frequencies, as follows:

(i)   A frequency should be called a *group frequency* only when 66% or more of the potential energy is located in *one* internal coordinate.

(ii) A frequency should be called a *zone frequency* if 66% or more of the potential energy is located in *two* internal coordinates.

(iii) A frequency should be called a *delocalized frequency* if less than 33% of the potential energy is localized in one coordinate.

The majority of normal modes of large molecules are of the *delocalized* type. Some examples are skeletal modes of organic molecules, vibrations of fluorinated organic compounds, and ring deformation modes in cyclic molecules.

3-10 Molecular Symmetry

3-10a Introduction
A study of the symmetry of a molecule can be of considerable help in the discussion of vibrational spectra. Although the majority of organic molecules have little or no overall symmetry, they can often be considered to be made up of groups such as $-CH_3$, $-CH_2$, and C_6H_5- attached to a skeleton. The skeleton, as well as the attached groups, may have some symmetry. In such cases the *local* symmetry can be useful in the interpretation of the infrared and Raman spectra of the molecule.

Molecular symmetry is defined in terms of the *symmetry elements* that the molecule possesses. Four kinds of symmetry elements are found in molecules: the *n*-fold axis (C_n), the plane of symmetry (σ), the center of symmetry (i) and the rotation-reflection axis (S_n). All molecules also have the identity element (E). These symmetry elements will be explained in more detail shortly.

The collection of all symmetry elements that a molecule possesses is known as a *point group*. Every molecule "belongs" to a point group, which a provides way of classifying the symmetry of the molecule. Symmetry is also discussed in terms of *symmetry operations* such as reflections or rotations, with respect to the corresponding *symmetry elements*. After carrying out an operation on a molecule in its equilibrium configuration, an equivalent configuration, indistinguishable from the initial one is obtained.

When a molecule is associated with a point group, it is found that there are only a certain number of different types of symmetry that describe the vibrations, rotations, translations, and other properties of the molecule under the operations of the group. These symmetry types are described by expressions such as totally symmetric, antisymmetric with respect to a certain symmetry element, and doubly degenerate. They are also given symbols such as A_1, A_2'', B_{2u}, and E. These expressions and symbols of group theory will be explained presently. We will also discuss how molecular vibrations can be classified into symmetry types. Finally, we will show how considerations of symmetry can lead to a prediction of the number of frequencies active in the infrared and to the number and polarization properties of lines expected in the Raman spectrum of a molecule.

3-10b Degenerate and Nondegenerate Vibrations
Two vibrations are said to be doubly degenerate if they have identical frequencies as a consequence of symmetry. Molecules such as CO_2, $CHCl_3$, cyclopropane, benzene, and 1,3,5-trisubstituted benzene derivatives (identical substituents) all have doubly degenerate vibrations. Occasionally, the frequency of an overtone or combination (see section 3-6d) will be the same as a fundamental. This situation is known as *accidental* degeneracy, and if the energy levels involved have the same symmetry, Fermi resonance can occur (see section 3-8). It is also possible to have three vibrations with identical frequencies in tetrahedral (e.g., CH_4) and octahedral (e.g., SF_6) molecules. These examples are known as triply degenerate vibrations.

A simple example of a doubly degenerate pair of vibrations is found in linear ABA molecules such as CO_2. In Figure 3-10 the bending modes of the ABA molecule are

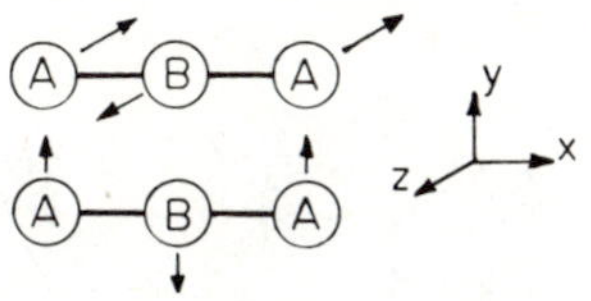

FIGURE 3–10.   The degenerate bending vibrations of
a linear ABA molecule.

illustrated. In this example it is easy to see that the two vibrations have identical frequencies.
In many cases, however, it is not so obvious. Any molecule having a threefold or higher
axis of symmetry [see section 3–10c (i)] will have degenerate vibrations.

Single vibrations with individual frequencies are known as nondegenerate vibrations.
The displacements of the atoms in nondegenerate vibrations remain unchanged or simply
change direction when an operation of the group is performed. Degenerate modes, on the
other hand, have to be considered in pairs (or threes) when a symmetry operation is carried
out.

3–10c   Symmetry Elements and Operations

(i) The axis of symmetry. An axis of symmetry C_n can be classified according to the
angle $(360°/n)$ through which the molecule is rotated about the axis to bring it into an
indistinguishable position. The act of rotating the molecule through $360°/n$ is the symmetry
operation. Performing the rotation n times will bring the molecule back into its original
position, so that the axis is termed n-fold. The direction of the axis is usually chosen as the
z axis. If there are two or more n-fold axes, then the one with the highest n is chosen as the z
axis. Some examples of n-fold axes are shown in Figure 3–11.

There are $(n - 1)$ operations associated with an n-fold axis. The general symbol is C_n^k,
which means rotation through $2\pi k/n$ about the axis. For example, in the chloroform molecule
of Figure 3–11, rotation by $120°$ $(k = 1)$ or $240°$ $(k = 2)$ will produce configurations
indistinguishable from the initial one.

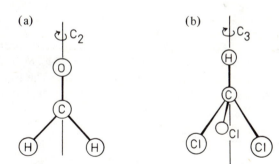

(a)

(b)

FIGURE 3–11.   Examples   of   two-fold
and three-fold rotation axes: (a) the C_2
axis in formaldehyde and (b) the C_3 axis
in chloroform.

(ii) The plane of symmetry. The plane of symmetry σ is designated σ_v, σ_h, or σ_d when
the plane is vertical, horizontal, or diagonal, respectively; in cases where this notation is
ambiguous, the notation $\sigma(xy)$, $\sigma(xz)$ or $\sigma(yz)$ is used. A plane of symmetry bisects a molecule
into two equivalent parts, the one being the mirror image of the other. Chloroform has three
σ_v planes, and cyclopropane has a σ_h plane; in p-dibromobenzene, there are three σ planes
denoted by $\sigma(xy)$, $\sigma(xz)$, and $\sigma(yz)$. Two planes of symmetry are illustrated in Figure 3–12.

The benzene molecule has three σ_d planes which, respectively, bisect the C_1-C_2 and
C_4-C_5 bonds, the C_2-C_3 and C_5-C_6 bonds, and the C_3-C_4 and C_6-C_1 bonds. The notation σ_d
is used to distinguish these planes from the three σ_v planes passing respectively through C atoms
1 and 4, 2 and 5, and 3 and 6. One of the σ_d planes of benzene is shown in Figure 3–13.

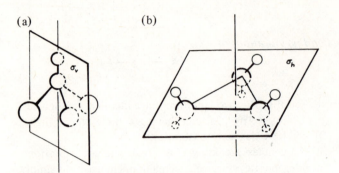

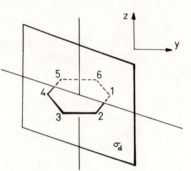

FIGURE 3-12. Planes of symmetry: (a) one of the three σ_v planes of an XYZ_3 molecule and (b) the σ_h plane of cyclopropane.

FIGURE 3-13. One of the σ_h planes in benzene.

(iii) Center of symmetry. A molecule has a center of symmetry i if a line drawn from any atom of the molecule through the center and continued an equal distance beyond arrives at an identical atom, as in *trans*-dichloroethylene and *p*-dibromobenzene (Figure 3-14).

FIGURE 3-14. Centers of symmetry: (a) the center of symmetry in trans-dichloroethylene and (b) the center of symmetry in p-dibromobenzene.

(iv) The rotation-reflection axis. The most difficult symmetry element to recognize is the rotation-reflection axis S_n. This element is a combination of an n-fold axis with a plane of reflection perpendicular to the axis. The molecule need not possess either the C_n axis or the plane of symmetry as *independent* elements. Two S_n axes are illustrated in Figure 3-15.

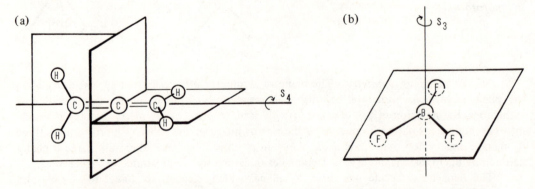

FIGURE 3-15. Examples of S_4 and S_3 axes: (a) the S_4 axis of allene and (b) the S_3 axis of boron trifluoride.

Allene has an S_4 axis because rotation through $90°$ about an axis through the three carbon atoms, followed by reflection in the plane perpendicular to the axis and passing through the central carbon atom, produces an orientation indistinguishable from the original molecule. Note that allene does not possess independently a C_4 axis, nor the perpendicular plane involved in the S_4 symmetry element. Boron trifluoride has an S_3 axis coincident with the C_3 axis. The plane involved in the S_3 symmetry element is also a plane of symmetry σ_h of the molecule.

The general symbol for the rotation-reflection operation is $S_n{}^k$, which means k successive rotations of $2\pi/n$ about the axis, each rotation followed by a reflection in the plane perpendicular to the axis. Many of these operations are equivalent to other operations; for example, $S_6{}^2 = C_3$ and $S_4{}^2 = C_2$. There are two special cases, $S_n{}^n = \sigma$ and $S_2{}^1 = i$.

3-10d   Point Groups

A point group is the collection of symmetry elements that a molecule possesses. The term *point group* is used because for every symmetry operation there is at least one point in the molecule (not necessarily at an atom or on a bond) that remains unmoved during the operation.

There are only a few point groups of importance and the symbols used by spectroscopists for them are due to Schönflies. These symbols are entirely different from those used by crystallographers. The groups of interest have the symbols C_n (where $n = i, s, 1, 2, 3, 4, 5, 6, 7, 8$), D_n, C_{nv}, C_{nh}, D_{nh}, D_{nd} (where $n = 2, 3, 4, 5, 6$), and S_n (where $n = 4, 6, 8$). Only a few of these are common as far as organic molecules are concerned. There are also the special groups T_d, O_h, $C_{\infty v}$ and $D_{\infty h}$.

The symbol C_1 means no symmetry at all, while C_s means that the molecule has only a plane of symmetry. A molecule belonging to one of the C_n groups has one or more (common) n-fold axes only, while the D_n groups have n twofold axes perpendicular to the (common) C_n axes. The C_{nv} groups have a plane (or planes) of symmetry containing the n-fold axis. The C_{nh} and D_{nh} groups have a plane (or planes) perpendicular to the n-fold axis [known as the horizontal plane(s)]. The D_{nd} groups have vertical planes that bisect the angles between adjacent pairs of C_2 axes. These planes are sometimes known as diagonal planes. The S_n groups contain S_n axes, as expected.

Tetrahedral or octahedral molecules belong to point groups T_d and O_h, respectively. For linear molecules two possibilities exist. The molecule either does or does not possess a center of symmetry. The respective groups are $D_{\infty h}$ or $C_{\infty v}$.

A list of point groups, with their symmetry elements, and examples of molecules belonging to them is given in Table 3-1. A systematic procedure for finding the point group of a molecule is given in Table 3-2.

Molecules having threefold or higher axes of symmetry will have degenerate vibrations. The point groups to which these molecules belong are known as *degenerate* point groups. The most commonly encountered degenerate point groups are C_{3v}, D_{3h}, and T_d. Molecules having only twofold axes of symmetry or no symmetry axes belong to *nondegenerate* point groups. Common examples of nondegenerate groups in organic molecules are C_s and C_{2v}.

Problem 3-6   Identify the symmetry elements of the benzene molecule (point group D_{6h}).

Problem 3-7   To which point groups do the following molecules belong: CH_3OH; CH_2F_2; *t*-butyl cyanide $((CH_3)_3CCN)$; 1,3,5-trinitrobenzene; cyclopentadiene; anti-1,1,2,2-tetrabromoethane; CH_3NO_2; C_2Cl_4?

Table 3–1. *Point Groups and Symmetry Elements*

Group	Symmetry Elements	Examples
C_1	None	sugars, steroids, amino acids
C_2	C_2 only	H_2O_2, cyclohexene
C_s	σ only	C_2H_5CN, *ortho*–C_6H_4BrCl, $(CH_3)_2AsI$
C_{2v}	C_2, $\sigma(xz)$, $\sigma(yz)$	H_2O, HCHO, CH_2Cl_2, *cis*-$C_2H_2Cl_2$, CH_2CCl_2, ethylene oxide, urea, furan, C_6H_5X, pyridine
C_{3v}	C_3, $3\sigma_v$	$CHCl_3$, $CH_3C\equiv CCl$, CF_3CN, $(CH_3)_3CC\equiv CH$
C_{2h}	C_2, σ_h, i	*trans*-$C_2H_2Cl_2$, *trans*-$C_4H_4Cl_2$ (2,3-dichloro-1,3-butadiene), 1,4-dioxane
D_{2d}	S_4, $2C_2$, C_2', $2\sigma_d$	allene, C_5H_8 (spiropentane)
D_{2h}	$C_2(z)$, $C_2(y)$, $C_2(x)$, i, $\sigma(xy)$, $\sigma(xz)$, $\sigma(yz)$	ethylene, *p*-$C_6H_4X_2$, naphthalene
D_{3d}	S_6, C_3, $3C_2$, $3\sigma_d$, i	ethane (staggered), $CH_3C\equiv CCH_3$, cyclohexane (chair)
D_{3h}	C_3, $3C_2$, σ_h, S_3, $3\sigma_v$	CO_3^{2-}, 1,3,5-$C_6H_3X_3$, *s*-triazine, ethane (eclipsed)
D_{6h}	C_6, C_3, $3C_2$, $3C_2'$, C_2'', σ_h, $3\sigma_v$, $3\sigma_d$, S_6, S_3, i	benzene
T_d	$4C_3$, $6\sigma_d$, $2S_4$, $3C_2$	CH_4, $(CH_3)_4M$ (M = C, Si, Ge, Sn, Pb)
O_h	$8C_3$, $6C_2$, $6C_2'$, $6C_4$, $6S_4$, $8S_6$, $3\sigma_h$, $6\sigma_d$, i	SF_6
$C_{\infty v}$	C_∞, $\infty\sigma_v$	HCN, OCS, $HC\equiv CCN$
$D_{\infty h}$	C_∞, ∞C_2, σ_h, $\infty\sigma_v$, S_∞, i	CO_2, C_2H_2, C_2N_2, $NCC\equiv CCN$

3–10e Symmetric and Antisymmetric Vibrations

The symmetry of a vibration of a molecule can be defined in terms of the behavior of the displacements of the atoms under the operations of the group. If the displacements of the atoms are represented by arrows, there is the question whether any of these arrows point in a different direction after a symmetry operation has been performed. If the answer is no, then the vibration is *symmetric* with respect to the symmetry operation. If the answer is yes, then the vibration is *antisymmetric* with respect to that operation.

Examples of symmetric and antisymmetric vibrations with respect to a C_2 axis are given for the formaldehyde molecule in Figure 3–16. It can be seen that the vibration illustrated in Figure 3–16a is also symmetric with respect to the planes of symmetry that the molecule possesses, whereas the vibration in Figure 3–16b is antisymmetric with respect to the vertical plane through the C—O bond, but symmetric with respect to the plane of the molecule.

If a vibration is symmetric with respect to all symmetry elements of the group, it is said to be a *totally symmetric* vibration. The vibration in Figure 3–16a is an example of a totally symmetric mode. This type of vibration is always allowed in the Raman spectrum

Table 3-2. *A Systematic Method for Finding the Point Group of a Molecule*[a]

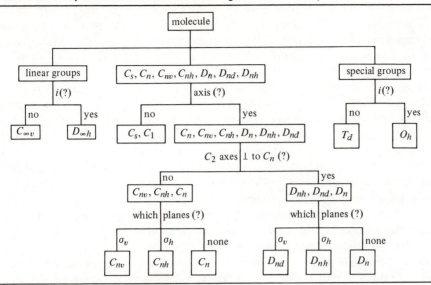

[a] Only the important groups are included here.

and usually gives rise to a strong, highly polarized line (see section 3-10i). Only totally symmetric vibrations have polarized Raman lines.

All vibrations of a molecule that are not totally symmetric are *antisymmetric* with respect to at least one symmetry element. Therefore the word antisymmetric should be qualified by stating with respect to which symmetry element(s) the vibration is antisymmetric. A symmetry type symbol is assigned to a vibration to denote such properties, as we shall see in the following section.

It should be noted that no vibration can be totally antisymmetric. Therefore the word asymmetric should never be used to describe a molecular vibration. All nontotally symmetric vibrations give rise to depolarized Raman lines (see section 3-11), when they are allowed (see section 3-10j). Degenerate modes can never be totally symmetric; hence they always give depolarized lines if they are allowed in the Raman spectrum.

FIGURE 3-16. The behavior of two vibrations of the formaldehyde molecule under a symmetry operation.

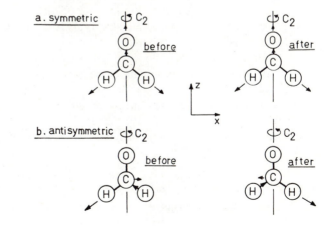

3-10f   A Notation for Symmetry Types

Symmetry types could be described by phrases such as (i) symmetric with respect to all symmetry operations or (ii) symmetric with respect to rotation about the axis of symmetry but antisymmetric with respect to reflection in the planes of symmetry. On the other hand, we could use a convenient short hand such as A_1 for example (i) above and A_2 for example (ii). These symbols are clearly very convenient.

The symbols are based on four letters[10] A, B, E, and F (sometimes T): A means symmetric with respect to the principal axis, B means antisymmetric with respect to the principal axis, E means doubly degenerate, and F (or T) means triply degenerate.

Subscripts 1 and 2 are added to A to distinguish between the totally symmetric species A_1 and the species A_2, which is antisymmetric with respect to one or more planes or other axes of symmetry. Subscripts 1 and 2 (occasionally 3) are added to B to distinguish the B symmetry types among themselves. The actual subscript depends on the convention adopted for naming planes and axes and conveys no further information, unless the convention is stated. For example, the formaldehyde vibration in Figure 3-16b is of species B_1 in a convention that defines the plane of the molecule as the xz plane. The same vibration, however, would be denoted B_2 if we chose the yz plane as the plane of the molecule.

Superscripts single-prime and double-prime are added to distinguish species that are symmetric or antisymmetric with respect to a σ_h plane of symmetry. As an example consider the BF_3 molecule of Figure 3-15b. Rotation of the three F atoms around the C_3 axis (the z axis) is of symmetry type A_2', while translation of the whole molecule in the z direction is of species A_2''. Rotations about the x and y axes form a degenerate pair of species E''.

Finally, subscripts g and u are added to distinguish species that are symmetric or antisymmetric with respect to inversion through a center of symmetry. In *trans*-dichloro-ethylene (Figure 3-14a), a vibration involving the symmetric stretching of the C—Cl bonds is of species A_g, whereas the antisymmetric C—Cl stretch is of symmetry B_u. See section 5-18e for diagrams and further discussion of these vibrations.

Problem 3-8   What do the following symbols reveal concerning the symmetry type: A_1''; B_{2g}; E'; E_{1u}; T_2?

3-10g   Symmetry Types

For every point group there are a certain number of symmetry types or species. This number is often equal to the number of different symmetry operations of the group.[11] The group theory name for symmetry type is irreducible representation, but we will not go into the reasons for this term here.

Symmetry types describe how certain properties of a molecule behave when the symmetry operations of the group are carried out. The properties of interest here are molecular vibrations, rotations and translations of the whole molecule, the components of the dipole moment vector, and the components of the polarizability tensor.

We shall see in section 3-10j that in order for a vibration to be *active* in the infrared, it must belong to the same symmetry type (behave in the same way) as a component of the dipole

---

[10] For the groups $C_{\infty v}$ and $D_{\infty h}$ to which linear molecules belong, the symbol Σ is often used for nondegenerate species, while Π and Δ are used for doubly degenerate species.

[11] The number of symmetry species equals the number of symmetry operations for nondegenerate point groups.

Table 3-3. *Symmetry Types and Molecular Properties for Common Point Groups*

Point Group	Symmetry Types	Rotations	Translations and Components of μ	Components of α	Infrared Activity	Raman Activity
C_s	A'	R_z	T_x, T_y, μ_x, μ_y	xx, yy, zz, xy	yes	pol
	A''	R_x, R_y	T_z, μ_z	yz, xz	yes	depol
C_2	A	R_z	T_z, μ_z	xx, yy, zz, xy	yes	pol
	B	R_x, R_y	T_x, T_y, μ_x, μ_y	yz, xz	yes	depol
C_{2v}	A_1	—	T_z, μ_z	xx, yy, zz	yes	pol
	A_2	R_z	—	xy	no	depol
	B_1	R_y	T_x, μ_x	xz	yes	depol
	B_2	R_x	T_y, μ_y	yz	yes	depol
C_{3v}	A_1	—	T_z, μ_z	$(xx + yy), zz$	yes	pol
	A_2	R_z	—	—	no	no
	E	$(R_x; R_y)$	$(T_x; T_y)$ $(\mu_x; \mu_y)$	$((xx - yy); xy)$ $(xz; yz)$	yes	depol
C_{2h}	A_g	R_z	—	xx, yy, zz, xy	no	pol
	B_g	R_x, R_y	—	xz, yz	no	depol
	A_u	—	T_z, μ_z	—	yes	no
	B_u	—	T_x, T_y, μ_x, μ_y	—	yes	no
D_{2h}	A_g	—	—	xx, yy, zz	no	pol
	B_{1g}	R_z	—	xy	no	depol
	B_{2g}	R_y	—	xz	no	depol
	B_{3g}	R_x	—	yz	no	depol
	A_u	—	—	—	no	no
	B_{1u}	—	T_z, μ_z	—	yes	no
	B_{2u}	—	T_y, μ_y	—	yes	no
	B_{3u}	—	T_x, μ_x	—	yes	no
D_{3h}	A_1'	—	—	$(xx + yy), zz$	no	pol
	A_2'	R_z	—	—	no	no
	E'	—	$(T_x; T_y), (\mu_x; \mu_y)$	$((xx - yy); xy)$	yes	depol
	A_1''	—	—	—	no	no
	A_2''	—	T_z, μ_z	—	yes	no
	E''	$(R_x; R_y)$	—	$(xz; yz)$	no	depol

moment vector, which, in turn, behaves in the same way as a pure translation. On the other hand, for a normal mode to give rise to Raman scattering, it must have the same symmetry properties as a component of the polarizability tensor.

It is clear, then, that for a molecule with some symmetry, it is useful to know the number of vibrations belonging to each symmetry type of the point group of the molecule. The total number of infrared and Raman active fundamentals can then be predicted, together with the number of polarized lines expected in the Raman spectrum. The symmetry types and the associated molecular properties for some common point groups are listed in Table 3-3.

3-10h   The Number of Vibrations of Each Symmetry Type

A classification of the vibrations of a molecule according to the various symmetry types can be made from a knowledge of the point group of the molecule and an examination of how the $3N$ cartesian displacements of the atoms behave under the operations of the group. Space limitation prevents the presentation of full details of the method here. The reader is urged to consult the texts on applications of group theory listed at the end of this chapter for further information. We will, however, discuss some simple cases to illustrate how knowledge of the symmetry types of the normal modes of vibration of a molecule can be useful.

As a first example consider the six normal modes of vibration of the formaldehyde molecule, H_2CO, which belongs to the point group C_{2v}. It can be shown that the normal modes are of three symmetry types: $3A_1 + 2B_1 + B_2$. There are no vibrations of species A_2. All modes are both infrared and Raman active, and the A_1 modes give rise to polarized lines in the Raman spectrum.

In a second example we encounter degenerate vibrations. A tetrahedral XYZ_3 molecule such as chloroform, $CHCl_3$, belongs to the point group C_{3v}. The six normal vibrations can be classified into $3A_1 + 3E$ modes. All are both infrared and Raman active. The A_1 modes give rise to polarized Raman lines, whereas the E modes give depolarized Raman lines.

There are several shortcut methods for finding the number of vibrations of each symmetry type for a molecule. Perhaps the most straightforward of these methods is the one described by G. Herzberg (reference A2). A second procedure, known as the "correlation method," has been described by W. G. Fateley et al. (reference D4); and L. L. Boyle has described a method called "ascent in symmetry" (reference D5). The latter two methods are useful for the analysis of vibrations in crystals.

These shortcut methods all depend on recognizing sets of symmetrically equivalent atoms in a molecule. The sets consist of identical nuclei that are transformed into one another by the symmetry operations of the point group of the molecule. For example, in the formaldehyde molecule, H_2CO (point group C_{2v}), the two H atoms form a set, whereas the C and O atoms each form one-atom sets.

To use the method described by Herzberg, one must divide the atoms into equivalent sets and note the location of the atoms in each set with respect to the various symmetry elements. Using the formaldehyde molecule as an example again, we note that the two H atoms lie on the $\sigma(xz)$ plane, whereas the C and O atoms both lie on all symmetry elements. In a larger molecule of symmetry C_{2v}, it may happen that sets of symmetrically related nuclei lie on no symmetry element, or on the other plane of symmetry $\sigma(yz)$.

Herzberg (reference A2) gives tables from which the number of vibrations of each species can be determined for practically any molecule. As an illustration of the use of the tables we will discuss two examples of molecules belonging to the C_{2v} point group (see Table 3-4).

Table 3-4. *The Number of Vibrations of Each Symmetry Type for the Point Group* C_{2v}

Symmetry Type	Number of Vibrations[a]
A_1	$3m + 2m_{xz} + 2m_{yz} + m_0 - 1$
A_2	$3m + m_{xz} + m_{yz} - 1$
B_1	$3m + 3m_{xz} + m_{yz} + m_0 - 2$
B_2	$3m + m_{xz} + 2m_{yz} + m_0 - 2$

Excerpted from G. Herzberg, *Infrared and Raman Spectra*, Van Nostrand, New York, 1972, Table 39, p. 134.

[a] m is the number of equivalent sets with nuclei on *no* symmetry elements; m_{xz} and m_{yz} are the numbers of sets on the σ_{xz} and σ_{yz} planes respectively; m_0 is the number of nuclei on *all* symmetry elements. The -1 and -2 at the ends of the rows take account of translations and rotations.

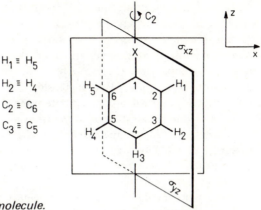

$H_1 \equiv H_5$

$H_2 \equiv H_4$

$C_2 \equiv C_6$

$C_3 \equiv C_5$

FIGURE 3-17.   A monosubstituted benzene molecule.

For formaldehyde $m = 0$, $m_{xz} = 1$, $m_{xy} = 0$, and $m_0 = 2$. Hence the number of A_1 modes is $2 + 2 - 1 = 3$, the number of A_2 modes is $1 - 1 = 0$, the number of B_1 modes is $2 + 2 - 2 = 2$, and the number of B_2 modes is $1 + 2 - 1 = 1$. In summary, there are $3A_1 + 2B_1 + 1B_2$ modes of vibration in the formaldehyde molecule.

For a second example we consider the monosubstituted benzene molecule C_6H_5X illustrated in Figure 3-17. Hydrogen atoms 1,5 and 2,4 as well as carbon atoms 2,6 and 3,5 form four sets. These atoms lie on the $\sigma(xz)$ plane, so that $m_{xz} = 4$. The two carbon atoms on the axis, H atom 3, and the X atom each lie on all symmetry elements, so that $m_0 = 4$. Again, there are no atoms in the $\sigma(yz)$ plane only, and there are no atoms on no symmetry element. Hence $m = 0$ and $m_{yz} = 0$. Using Table 3-4, we find

$$8 + 4 - 1 = 11 \quad A_1 \text{ modes}$$

$$4 - 1 \qquad = \ \ 3 \quad A_2 \text{ modes}$$

$$8 + 4 - 2 = 10 \quad B_1 \text{ modes}$$

$$4 + 4 - 2 = \ \ 6 \quad B_2 \text{ modes}$$

Reference to Table 3-3 for the C_{2v} point group gives us the infrared and Raman activities of the normal modes together with the polarizations expected in the Raman spectrum. In summary, the normal modes of vibration of a monosubstituted benzene derivative can be classified as follows: $11\,A_1$ modes, ir & R (pol); $3A_2$ modes, R only (depol); $10\,B_1$ modes, ir & R (depol); and $6\,B_2$ modes, ir & R (depol).

Problem 3-9 Use Table 3-4 to determine the number of vibrations of each symmetry type for CH_2Br_2.

3-10i Local Symmetry

It very often happens that the vibrations of certain groups of atoms have more or less the same frequencies in any molecule. This is the well-known group frequency concept that is the basis of the qualitative interpretation of infrared and Raman spectra. Not only do these groups have a fairly constant characteristic set of frequencies, but they also behave as if they have approximately the symmetry of the free group. Three well known examples will be discussed here, the methyl group, CH_3 (local symmetry C_{3v}), the methylene group, CH_2, and the phenyl group, C_6H_5 (local symmetry C_{2v}).

A *free* methyl group has $(3N - 6)$ normal modes of vibration $(2A_1 + 2E)$ comprising symmetric and antisymmetric stretching and bending modes. When the methyl group is attached to a molecule, three new modes appear: a torsional mode of symmetry type A_2 and a degenerate pair of rocking vibrations of species E. These motions would be rotations in the free methyl group. Thus there are four regions of the spectrum where we expect to find methyl group vibrations. This conclusion is amply supported experimentally. The presence of the methyl group will also contribute three skeletal modes to the vibrations of the molecule. These correspond to translations of the free methyl group.

When the methyl group is part of a molecule with lower symmetry, the degeneracies are removed and doublets may be observed in the regions of the spectrum where the methyl group frequencies are expected. The A_2 torsional mode is actually inactive under the C_{3v} point group selection rules. It may be allowed, however, by the symmetry of the whole molecule; in fact, methyl torsions are frequently observed in the far infrared.

The vibrations of the methylene group can be described in terms of the local C_{2v} symmetry of the group. The six modes of vibration associated with a CH_2 group are shown in Figure 3-18. The usual descriptions of these modes are given on the figure. It should be noted, however, that mixing with other modes in the molecule may occur and give rise to variations in the location and intensities of the vibrations of the CH_2 group.

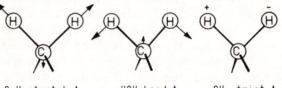

C-H stretch A_1 HCH bend A_1 CH_2 twist A_2

FIGURE 3-18.   *Vibrations of a CH_2 group.*

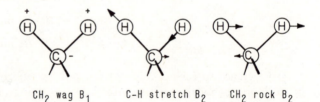

CH_2 wag B_1 C-H stretch B_2 CH_2 rock B_2

The phenyl group, C_6H_5, can be treated under the C_{2v} point group and many of the vibrations of a monosubstituted benzene derivative can be recognized in the spectra of compounds containing one or more phenyl groups. Using the methods of section 3-10h, we find that the vibrations of the phenyl group consist of $10A_1 + 3A_2 + 9B_1 + 5B_2$ modes. The A_2 modes are not active in the infrared. All are active in the Raman and the 11 A_1 modes will be polarized. If the symmetry of the phenyl group is lowered in the molecule, then some of the A_2 modes might appear as weak bands in the infrared and more than 11 polarized Raman lines would be observed. In large unsymmetrical molecules, there will be a large number of infrared and Raman lines. Nevertheless, many characteristic phenyl group vibrations can be observed.

Problem 3-10   Verify that the vibrations of the phenyl group, C_6H_5, comprise $10A_1 + 3A_2 + 9B_1 + 5B_2$ modes.

3-10j   Symmetry Selection Rules

In section 3-3d it was shown that a change in dipole moment is a necessary condition for infrared absorption, and in section 3-4d it was shown that a change in polarizability is necessary for Raman scattering. The quantum mechanical selection rules discussed in section 3-7 predict which transitions are allowed, regardless of the actual value of $(d\mu/dq)$ or $(d\alpha/dq)$. We will now show how symmetry considerations can predict the vibrations for which there will be a change in dipole moment or polarizability.

Symmetry selection rules are quite general. Since we are considering transitions between energy levels, we examine the symmetry of the wave functions that describe the levels. The ground state wave functions are always totally symmetric with respect to all operations of the point group of the molecule. A singly excited level has the symmetry of one of the symmetry types of the point group.

From eq. 3-8 of section 3-3b we see that a transition will give rise to absorption only when the integral $\int_{-\infty}^{+\infty} \psi_b \mu \psi_a d\tau$ is not zero where ψ_a and ψ_b are the ground and excited state wave functions and μ is the electric dipole function. In order for this integral to be non-zero, it must be *totally symmetric* with respect to any symmetry operation of the point group of the molecule. (Obviously, intensity does not change under a symmetry operation.) Now the only way that the product $(\psi_b \mu \psi_a)$ can be totally symmetric is for ψ_b to have the *same symmetry* as one or more of the 3 components of μ. In other words, ψ_b must belong to the same symmetry type as one of the translations of the molecule because μ behaves like a vector with x, y, and z components.

A similar selection rule can be stated for Raman scattering. In this case, the criterion for nonzero scattering intensity is that the integral $\int_{-\infty}^{+\infty} \psi_b \alpha \psi_a d\tau$ is not zero, where α is the polarizability function of eq. 3-18 (section 3-4d). Now the components of α (see eq. 3-17) have the same symmetries as the six products, xx, yy, zz, xy, yz, and zx. Thus again we can see that unless ψ_b has the same symmetry as one or more of these products the integral will be zero. The last column of Table 3-3 summarizes the infrared and Raman activities of the various symmetry types.

3-10k   Infrared-Raman Exclusion Rule

It can be noted from the last column of Table 3-3 that for the point groups C_{2h} and D_{2h} there are no vibrations active in *both* infrared and Raman. This *exclusion rule* always holds when the molecule has a *center of symmetry*. An example of the rule in the *trans*-dichloroethylene molecule is given in section 4-2a. Even for molecules with no center of symmetry, it is frequently observed that strong Raman bands correspond to very weak absorptions in the infrared, and vice versa, because vibrations that give rise to large changes in polarizability often involve small changes in dipole moment.

3-11   Polarization of Raman Lines

3-11a   Introduction

Consider a sample illuminated by a laser beam traveling in the z direction, with the scattered light observed in the x direction, *i.e.*, at $90°$ to the laser beam. The laser beam will be plane-polarized and usually the plane of polarization is the yz plane. So that the electric vector of the laser beam is in the y direction. This standard geometry for recording Raman spectra is illustrated in Figure 3-19.

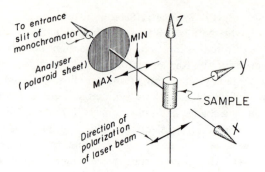

FIGURE 3-19. *Measurement of the depolarization ratio of scattered light.*

Now for liquid or gas samples it is found that the scattered light shows different intensities for different orientations of a polarization analyzer (usually a sheet of polaroid) placed in front of the entrance slit of the monochromator.[12]   For any Raman line, the maximum intensity is observed when the analyzer is adjusted to pass light polarized in the y direction, i.e., parallel to the polarization of the laser beam. The minimum intensity is found when the analyzer is turned through $90°$ to pass light polarized in the z direction.

3-11b The Depolarization Ratio

The significant point here is that in nearly every case some of the scattered light is polarized with its electric vector at $90°$ to the direction of polarization of the incident laser beam.   The ratio of the intensity of light polarized in the z direction ($I_\perp$) to that polarized in the y direction ($I_\parallel$) is known as the depolarization ratio (eq. 3-74).

$$\rho = \frac{I_\perp}{I_\parallel}$$

3-74

When $I_\perp$ is zero, ρ will be zero and the line is said to be completely polarized. This situation only occurs for totally symmetric stretching modes of spherical molecules such as CCl_4 (point group T_d) and SF_6 (point group O_h). In all other cases ρ will be found to be greater than zero up to a maximum value of 0.75 for plane-polarized incident light. Lines for which $\rho = 0.75$ are said to be *depolarized.* Other lines for which ρ lies between zero and 0.75 are said to be partially polarized or simply *polarized.*   Totally symmetric vibrations always give polarized lines. Nontotally symmetric and degenerate vibrations give depolarized lines.

3-11c A Simplified Derivation of the Depolarization Ratio

When the expanded form of the basic equation for scattering (eq. 3-17) is applied to one molecule, it is seen that a dipole is induced in the molecule by E_y only, since the polarized laser beam has its electric vector entirely confined to the yz plane. The situation is further simplified because only the light scattered in the x direction is observed.   Thus only μ_y^{ind} and μ_z^{ind} contribute to the observed spectrum. These components are given by

$$\mu_y^{\mathrm{ind}} = \alpha_{yy}E_y \quad \text{and} \quad \mu_z^{\mathrm{ind}} = \alpha_{zy}E_y$$

---

[12]It should be noted that light scattered from polycrystalline or powdered samples does not exhibit this behavior.

The components of the polarizability tensor α_{yz} and α_{yy} are responsible for the observed intensities $I_{\parallel}$ and $I_{\perp}$, respectively. In a liquid or gaseous sample the molecules have completely random orientations, and an average over all orientations for N molecules must be taken. The results, which are detailed in reference A1, are given by eqs. 3–75 and

$$I_{\parallel} \propto \gamma^2 + \frac{4\beta^2}{45} \qquad\qquad 3\text{–}75$$

3–76 where $\gamma = \frac{1}{3}(\alpha'_{xx} + \alpha'_{yy} + \alpha'_{zz})$ and $\beta^2 = \frac{1}{2}[(\alpha'_{xx} - \alpha'_{yy})^2 + (\alpha'_{yy} - \alpha'_{zz})^2 +$

$$I_{\perp} \propto \frac{3\beta^2}{45} \qquad\qquad 3\text{–}76$$

$(\alpha'_{zz} - \alpha'_{xx})^2 + 6(\alpha'_{xy} + \alpha'_{yz} + \alpha'_{zx})]$ (the primes on the α_{ij} denote derivatives). In eqs. 3–75 and 3–76, γ is known as the mean value or isotropic part of the polarizability and β^2 is known as the anisotropy.

From the definition of the depolarization ratio we have eq. 3–77, from which it can

$$\rho = \frac{I_{\perp}}{I_{\parallel}} = \frac{3\beta^2}{45\gamma^2 + 4\beta^2} \cdot \qquad\qquad 3\text{–}77$$

be seen that only when β^2 is zero will ρ be zero, and the Raman line is completely polarized. Complete polarization only happens for totally symmetric vibrations of spherical molecules. On the other hand, γ^2 is zero for all nontotally symmetric vibrations and this leads to the upper limit value of 0.75 for depolarized lines. For totally symmetric vibrations of most molecules, both γ^2 and β^2 are nonzero, so that ρ lies between zero and 0.75. The corresponding Raman lines are partially polarized.

BIBLIOGRAPHY

A. Theory of Vibrational Spectroscopy
1. L. A. Woodward, *Introduction to the Theory of Molecular Vibrations and Vibrational Spectroscopy*, Oxford University Press, London, 1972.
2. G. Herzberg, *Infrared and Raman Spectra of Polyatomic Molecules*, Van Nostrand, New York, 1945.
3. N. B. Colthup, L. H. Daly, and S. E. Wiberly, *Introduction to Infrared and Raman Spectroscopy*, Academic, New York, 1964.
4. E. B. Wilson, Jr., J. C. Decius, and P. C. Cross, *Molecular Vibrations*, McGraw-Hill, New York, 1955.
5. G. M. Barrow, *Molecular Spectroscopy*, Mc-GrawHill, New York, 1962.

B. Quantum Mechanics
1. H. L. Strauss, *Quantum Mechanics*, Prentice-Hall, Englewood Cliffs, N. J., 1968.
2. W. G. Laidlaw, *Introduction to Quantum Concepts in Spectroscopy*, McGraw-Hill, New York, 1970.
3. H. Eyring, J. Walter, and G. E. Kimball, *Quantum Chemistry*, Wiley, New York, 1944.

C. Raman Spectroscopy

1. T. R. Gilson and P. J. Hendra, *Laser Raman Spectroscopy*, Wiley–Interscience, London, 1970.
2. M. C. Tobin, *Laser Raman Spectroscopy*, Wiley–Interscience, New York, 1971.
3. H. A. Szymanski, ed., *Raman Spectroscopy*, Plenum, New York, 1967, two volumes.

D. Group Theory

1. F. A. Cotton, *Chemical Applications of Group Theory*, Wiley–Interscience, New York, 1971.
2. G. Davidson, *Introductory Group Theory for Chemists*, Elsevier, London, 1971.
3. J. R. Ferraro and J. S. Ziomek, *Introductory Group Theory*, Plenum, New York, 1969.
4. W. G. Fateley, F. R. Dollish, N. T. McDevitt, and F. F. Bentley, "Infrared and Raman Selection Rules for Molecular and Lattice Vibrations: The Correlation Method, "*Appl. Spectros.*, **25**, 155 (1972). Also published under the same title in book form by Wiley–Interscience, New York, 1972.
5. L. L. Boyle, "The Method of Ascent in Symmetry," I. *Acta Crystallogr.*, **A28**, 172 (1972); II. *Spectrochim. Acta*, **28A**, 1347 (1972).

GROUP FREQUENCIES: INFRARED AND RAMAN

4-1 The Origin of Group Frequencies

The subject of group frequencies is essentially empirical in nature, but there is a sound theoretical basis for it. Infrared and Raman spectra of a large number of compounds containing a certain functional group, such as carbonyl, amino, phenyl, nitro, etc., are found to have certain features which appear, more or less, at the same frequency for every compound containing the group. It is reasonable, then, to associate these features with the functional group, provided a sufficiently large number of different compounds containing the group have been studied. For example, the infrared spectrum of any compound that contains a C=O group has a strong band between 1800 and 1650 cm^{-1}. Compounds containing $-NH_2$ groups have two infrared bands between 3375 and 3300. The Raman spectrum of a compound containing the C_6H_5- group has a strong polarized line near 1000 cm^{-1}, and nitro groups are characterized by infrared and Raman bands near 1550 and 1350 cm^{-1}. These are just four examples of the many characteristic frequencies of chemical groups observed in infrared or Raman spectra.

To illustrate the theoretical justification for the group frequency concept, we shall make use of the theory developed in sections 3–5a and 3–5b. Two types of diatomic group will be considered, the C–C group where the bond is triple, double, or single, and the X–H group where X is O, N, or C.

In a series of acetylene compounds, a vibrational frequency is observed near 2100 cm^{-1} in either the infrared or Raman spectrum, or both, for each compound. Similarly, for a series of molecules containing an ethylenic double bond, a frequency near 1650 cm^{-1} is observed. These frequencies are characteristic of the groups and enable their presence in a molecule

225

to be detected. Carbon–carbon single bonds, on the other hand, do not have a characteristic group frequency because the vibration of this group is invariably coupled with vibrations of other parts of the molecule.

In section 3–5b it is shown that the vibrational frequency (in cm^{-1}) of a diatomic molecule in the harmonic oscillator approximation is given by eq. 4-1. If k is expressed in the

$$\bar{\nu} = \frac{1}{2\pi c} \sqrt{\frac{k}{\mu}}$$ 4-1

usual units of 10^5 dynes cm^{-1} (10^2 Newton m^{-1} or millidyne Å^{-1}) and μ is in atomic mass units, then eq. 4-1 becomes eq. 4-2, where $1303 = (1/2\pi c)\sqrt{N \times 10^5}$ (N = Avogadro's

$$\bar{\nu} \text{ (cm}^{-1}) = 1303 \sqrt{\frac{k}{\mu}}$$ 4-2

number, 6.02×10^{23}). The reduced mass for a diatomic molecule consisting of two carbon atoms is $\mu = M_1 M_2/(M_1 + M_2) = (12)(12)/(12 + 12) = 6.0$ amu, so that the frequency of vibration of this hypothetical diatomic molecule is given by eq. 4–3.

$$\bar{\nu} \text{ (cm}^{-1}) = 532\sqrt{k}$$ 4-3

The force constant k depends on the type of bond. For triple, double, and single carbon–carbon bonds, k has the approximate values 16, 10, and 5 mdyne Å^{-1}, respectively, so that the vibrational frequencies of the *diatomic molecule* are approximately 2100, 1700, and 1200 cm^{-1} for triple, double, and single bonds, respectively.

Thus we see that this crude model predicts vibrational frequencies close to those observed in acetylenic and ethylenic compounds. Organic compounds usually have several vibrational frequencies in the 1200-800 cm^{-1} region. The stretching of the C–C single bond certainly contributes to these bands, but it is rarely possible to associate a specific frequency with C–C stretching.

For the second example, consider the series O–H, N–H, and C–H. The force constants are approximately 7, 6, and 5 mdyne Å^{-1}, and the reduced masses are 0.94, 0.93, and 0.92 amu. Using eq. 4-2,[1] we obtain the frequencies 3600, 3300, and 3000 cm^{-1}, respectively, for the vibrations of O–H, N–H, and C–H. When these diatomic groups form part of a larger molecule, there is always a vibration of the molecule with a frequency close to that calculated for the "free" diatomic group. This rule does not hold when there is an interaction between the diatomic group and some other part of the molecule. Such interactions can take several forms and are discussed in section 4–2b.

When two or more identical groups are present in a molecule, there will be two or more similar frequencies in the spectrum which may or may not be resolved. The CH groups in ethylene (C_2H_4) provide such an example. Four CH stretching frequencies can be identified near 3270, 3105, 3020, and 2990 cm^{-1}. One would have to observe both infrared and Raman spectra to see all four frequencies, since C_2H_4 has a center of symmetry and only the 3105 and 2990 cm^{-1} unsymmetrical stretching frequencies are seen in the infrared.

---

[1]It may be seen that using eq. 3-29 in place of eq. 4-2 is equivalent to putting $m_H = \mu = 1.0$ amu, the atomic weight of hydrogen. This approach would give calculated frequencies for OH, NH, and CH of approximately 3500, 3200, and 2900 cm^{-1}, respectively.

The explanation for these characteristic diatomic group frequencies lies in the approximately constant values of the stretching force constant of a group in different molecules. Polyatomic groups also have characteristic frequencies that involve both stretching and bending vibrations or combinations of these. No simple relationship such as eq. 4-2 can be found for polyatomic groups, and the best way to establish whether or not a certain group, $-CH_2$, $-CH_3$, $-NH_2$, $-C_6H_5$, etc., has characteristic frequencies is to examine the vibrational spectra of a large number of compounds containing these groups. Such studies have been carried out for infrared spectra, and many tabulations of group frequencies in the infrared have been made. With the recent revitalization of Raman spectroscopy by the laser, it is to be expected that similar extensive tabulations of group frequencies will soon be available for Raman spectra. Since group frequencies in the Raman generally are not the same as in the infrared, increasing use of Raman spectra for structure determination is inevitable.

Tabulations of the most important infrared and Raman group frequencies are given in sections 4-3 through 4-6. First, some discussion of the factors affecting the positions and intensities of group frequencies is necessary.

4-2 Factors Affecting Group Frequencies

4-2a Symmetry

Occasionally a group frequency may not be observed in the infrared or Raman spectrum because the vibration does not produce a change in dipole moment or polarizability. Symmetry arguments concerning such cases are discussed in Chapter 3. As an example, Figure 4-1 shows part of the infrared spectrum of *trans*-dichloroethylene. The C=C stretching frequency is known to be 1580 cm^{-1} in this molecule, but no band is observed at this frequency in the infrared spectrum. The C=C stretching mode is symmetric with respect to the center of symmetry and is not allowed in the infrared. The vibration, however, is active in the Raman spectrum, as can be seen in Figure 4-2. (The weak band at 1665 cm^{-1} in the infrared spectrum of Figure 4-1 is assigned to a combination of fundamentals at 845 and 820 cm^{-1}.)

Another example is the $-C\equiv C-$ stretching mode. In methylacetylene ($CH_3C\equiv CH$), the vibration is both infrared and Raman active, and a strong infrared band is observed at 2150 cm^{-1}, whereas in dimethylacetylene ($CH_3C\equiv CCH_3$), no band is observed in the infrared

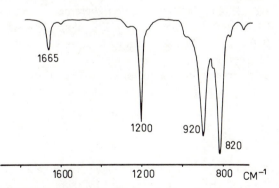

FIGURE 4-1. *Part of the infrared spectrum of trans-dichloroethylene.*

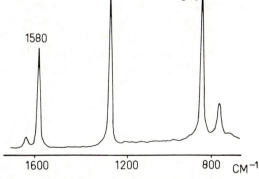
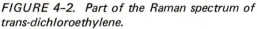

FIGURE 4-2. *Part of the Raman spectrum of trans-dichloroethylene.*

near 2150 cm^{-1}. In a larger, more complicated molecule, however, it is not always so easy to understand why a group frequency is absent or weak in the spectrum. In such cases, a Raman spectrum will often confirm the presence (or absence) of a functional group. In any case, modern organic chemists have several other methods at their disposal to check deductions from the infrared or Raman spectra.

4–2b Mechanical Coupling of Vibrations

Two completely free, identical diatomic molecules will, of course, vibrate with identical frequencies. When the two diatomic groups are part of a molecule, however, they can no longer vibrate independently of each other because the vibration of one group causes displacements of the other atoms in the molecule. These displacements are transmitted through the molecule and interact with the vibration of the second group. The resulting vibrations appear as in-phase and out-of-phase combinations of the two diatomic vibrations. When the groups are widely separated in the molecule, the coupling is very small and the two frequencies may not be resolved.

Consider the two CH stretching modes in acetylene, which are observed at 3375 cm^{-1} in the Raman and 3280 cm^{-1} in the infrared. In diacetylene, however, the two C–H stretching vibrations have closer frequencies, near 3330 and 3295 cm^{-1}. The vibrations of two *different* diatomic groups will *not* be coupled unless the uncoupled frequencies are similar as the result of a combination of mass and force constant effects. For example, in thioamides and xanthates the C=S group has a force constant of about 6.5 mdyne Å^{-1} and the reduced mass is 8.72 amu, so that the vibrational frequency calculated using eq. 4-2 is approximately 1120 cm^{-1}. The C–N and C–O groups have force constants of about 4.8 and 5.1 mdyne Å^{-1}, respectively, and the reduced masses are 6.46 and 6.86 amu. The calculated frequencies are both approximately 1120 cm^{-1}. Consequently, in any compound containing a C=S group adjacent to a C–O or a C–N group, there may be an interaction between the stretching vibrations of the groups. In compounds such as thioamides and xanthates, where the carbon atom is common to both groups, the coupling will be large and the two vibrations will interact with each other to produce two new frequencies, neither of which is in the expected region of the spectrum.

The way in which such mechanical coupling occurs can be illustrated for the case of two C=C groups coupled through a common carbon atom, as in the allene molecule, $CH_2=C=CH_2$ (see Figure 3-15). In the absence of strong coupling one might expect to observe a band in the infrared spectrum near 1600 cm^{-1} due to the out-of-phase (unsymmetrical) vibrations of the C=C groups and a line in the Raman spectrum from the in-phase (symmetrical) modes at a similar frequency. For the 1, 3-butadiene molecule ($CH_2=CH-CH=CH_2$), these bands are, in fact, observed (1640 cm^{-1} in the infrared and 1600 cm^{-1} in the Raman). For allene, however, the observed frequencies are near 1960 and 1070 cm^{-1}. This result can be understood in terms of the simple linear triatomic model discussed in section 3–5c.

If we put $m_1 = m_2$ in eq. 3–51 and use the conversion factor of eq. 4–2, then the two frequencies predicted for this model are given by eqs. 4–4. In the present case m is the mass

$$\bar{\nu}_1 = 1303 \sqrt{\frac{k}{m}} \quad \text{and} \quad \bar{\nu}_2 = 1303 \sqrt{\frac{3k}{m}} \qquad \text{4-4}$$

of the C atom (12.0 amu) and k is the force constant for the C=C bond (about 8.5 mdyne Å^{-1}). Using these numbers in eq. 4-4, we calculate the frequencies to be 1100 and 1900 cm^{-1}.

These calculated frequencies are in excellent agreement with the experimental frequencies, considering the simple model used, and they illustrate the effect of the mechanical coupling of the two vibrations. It was shown in section 3-5c that the high frequency corresponds to the unsymmetrical stretching mode of the two C=C groups, while the low frequency corresponds to the symmetrical case.

The results of this simple calculation can be extended to include the cases of nonlinear groups such as CH_2. The two CH stretching modes interact, but because of the angle between them, the coupling is weaker. This result is mathematically reasonable since two vibrations at right angles would have zero coupling. In the propylene molecule ($CH_3CH=CH_2$), for example, the two CH stretching frequencies of the methylene group are 3080 cm^{-1} for the nonsymmetric vibration and 2980 cm^{-1} for the symmetric mode. It is usually found that the nonsymmetric vibration has a higher frequency than the symmetric mode.

It is also possible for coupling to occur between dissimilar modes, such as stretching and bending vibrations, when the frequencies of the vibrations are similar and the two groups involved are adjacent in the molecule. An example is found in secondary amides, where the C–N stretching vibration is of the same frequency as the NH bending mode. Interaction of these two vibrations gives rise to two bands in the spectrum, one at a higher and one at a lower frequency than the uncoupled frequencies.

Singly bonded carbon atom chains, of course, are not linear, so that the simple model used for the allene molecule would have to be modified. In addition, we were able to ignore the bending of the C=C=C group in allene. This mode is not of the same symmetry type as the stretching modes because it takes place at right angles to the stretching vibrations. For nonlinear chains, the calculations become very much more complex. Nevertheless, one can see in a qualitative way that mechanical coupling will always occur between C–C single bonds in an organic molecule, so that there is no simple C–C group stretching frequency. One can expect that there will always be several bands in the infrared and Raman spectra in the 1200-800 cm^{-1} range in compounds containing saturated carbon chains. Certain branched chain structures, such as the *tert*-butyl group, $(CH_3)_3C-$ and the isopropyl group, $(CH_3)_2CH-$, do have characteristic group frequencies involving the coupled C–C stretching vibrations.

A special case of mechanical coupling, known as Fermi resonance, is described in section 3-8. This phenomenon can shift group frequencies and introduce extra bands. Care must be exercised in cases where Fermi resonance can occur because the result is sometimes the same as that produced by two similar groups in the molecule. As an example, two peaks are observed in the carbonyl stretching region of benzoyl chloride, at 1784 and 1740 cm^{-1}. If this were an unknown compound, one might be tempted to suggest that there are two nonadjacent carbonyl groups in the molecule. Cyclopentanone also has two bands in the same region at 1750 and 1730 cm^{-1}, and *p*-benzoquinone provides another example of a Fermi doublet in the carbonyl stretching region, with two peaks at 1656 and 1671 cm^{-1}.

Perhaps the two most important cases of Fermi resonance are the doublet at 3300 and 3050 cm^{-1} in the infrared spectra of polyamides (nylons), peptides, and proteins and the doublet at 2900 and 2700 cm^{-1} in the infrared spectra of aldehydes. The former doublet is due to an interaction between the N–H stretching mode of the (–CO–NH–) group and the overtone of the N–H deformation mode. In the second example the C–H stretching mode of the –CHO group interacts with the overtone of the CH bending vibration. In both cases there is considerable sharing of intensity and neither member of the doublet can be assigned specifically to the fundamental.

In many molecules, mechanical coupling of the group vibrations is so widespread that there are few, if any, frequencies assignable to functional groups. Such an example is found in aliphatic fluorine compounds where the CF and CC stretching modes are coupled with each other and with FCF and CCF bending vibrations. The presence of fluorine can be deduced from several very strong infrared bands in the region between 1400 and 900 cm^{-1}. These vibrations give very weak Raman lines.

4-2c Chemical and Environmental Effects

Hydrogen bonding, electronic and steric effects, physical state, solvent, and temperature effects all contribute to the position, intensity, and appearance of the bands in the infrared and Raman spectra of a compound. Lowering the temperature usually makes the bands sharper, and better resolution can be achieved especially in solids at very low temperatures. However, there is a possibility of splittings due to crystal effects that must be considered when examining the spectra of solids under high resolution. Solvents can often produce significant shifts in certain group frequencies when hydrogen bonding or molecular association occurs. An inert solvent, on the other hand, can cause the breakdown of hydrogen bonding in dilute solutions and permit a compound to exhibit its normal group frequencies.

Hydrogen bonding not only affects the X–H group frequencies, but it also changes the frequencies of the group with which the H bond is formed. Thus in carboxylic acids both the OH and CO stretching frequencies are lowered by hydrogen bonding. Dimers and polymers are formed, the spectra become complex in the carbonyl stretching region, and a broad unresolved band is observed far below the free OH stretching region. Usually the stretching and bending vibrations of the OH or NH groups involved in hydrogen bonding are shifted much more than the vibrations of the acceptor group. A great deal of experimental data has been collected on the effects of hydrogen bonding on infrared spectra and much of this is to be found in the books by Pimentel and McClellan, by Bellamy, and by Jones and Sandorfy (references A6, A2, and A4). We shall mention hydrogen-bonding effects as they occur in our subsequent discussions of group frequencies. It should be mentioned that O-H stretching and bending vibrations give rise to weak Raman lines. Much less work has been done on Raman spectra of hydrogen-bonded systems. Frequency shifts are the same as for infrared spectra, but intensity changes are not necessarily similar because of the different mechanism of Raman scattering.

Steric effects on group frequencies are quite interesting and useful for diagnostic purposes. As an example, consider the series of alicyclic ketones: cyclohexanone, cyclopentanone, and cyclobutanone. The observed carbonyl stretching frequencies are 1720, 1740, and 1780 cm^{-1}. This increase in frequency with increasing angle strain is generally observed for double bonds directly attached (exocyclic) to rings. Similar frequency changes are observed in the series of compounds methylenecyclohexane, methylenecyclopentane, and methylenecyclobutane where the observed C=C stretching frequencies are 1650, 1660, and 1680 cm^{-1}. When the double bond is a part of the ring (endocyclic), a decrease in the ring angle is accompanied by a lowering of the C=C stretching frequency. The observed frequencies for cyclohexene, cyclopentene, and cyclobutene are 1645, 1610, and 1565 cm^{-1}, respectively. Steric effects can give rise to rotational isomers and certain group frequencies may be somewhat different in the different isomers. Two bands in the spectrum of 1,2-dichloroethane at 1290 and 1235 cm^{-1} have been identified with the anti and gauche forms of this compound. Substitution in the ortho position of phenols can change the position of the OH group vibration. It has been shown that the OH stretching frequency in cyclohexanols is slightly higher for axial isomers than for the equatorial isomers.

Effects due to the change in the distribution of electrons in a molecule by a substituent atom or group can often be detected in the vibrational spectrum. There are several mechanisms, such as inductive and resonance effects, which can be used to explain observed shifts and intensity changes in a qualitative way. These effects involve changes in electron distribution in a molecule and cause changes in the force constants that are, in turn, responsible for changes in group frequencies. Inductive and resonance effects have been used successfully to explain the shifts observed in C=O stretching frequencies produced by various substituent groups in compounds such as acid chlorides, amides, etc. High C=O stretching frequencies are usually attributed to inductive effects, and low frequencies arise when delocalized structures are possible. For example, in acetyl chloride the C=O frequency is 1800 cm^{-1}, which is high compared to a normal C=O frequency such as that observed for acetone (1740 cm^{-1}). On the other hand, in acetamide the carbonyl frequency is lower (near 1670 cm^{-1}). In the first case the electronegative group adjacent to the carbonyl group causes an increase in frequency, whereas in the second case the delocalized electronic structure of amides (1 ↔ 2) lowers the C=O stretching frequency.

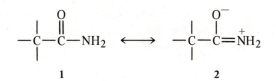

For aromatic compounds similar effects are encountered. The C=O stretching frequency in benzoic acid is 1705 cm^{-1}. Substitution of electronegative groups, such as $-NO_2$ and $-Cl$, in the para position causes an increase in the frequency of this vibration. Electron-releasing groups, such as $-OCH_3$ and $-NH_2$, cause a lowering of the frequency.

Conjugation of double bonds tends to lower the double bond character and increase the bond order of the intervening single bond. In compounds where a carbonyl group could be conjugated with an ethylenic double bond, the C=O stretching frequency is lowered by 20-30 cm^{-1}. This frequency lowering is often accompanied by an increase in intensity in both the infrared and Raman spectra.

4-3 Infrared Group Frequencies

We shall define as a "good" group frequency one that falls within a fairly restricted range regardless of the compound in which the group is found. Extreme cases of mechanical coupling or other effects discussed in the previous section may occasionally cause even a good group frequency to misbehave, so one should be aware of this possibility.

It is unlikely that many more good infrared group frequencies will be discovered; therefore the technique of group frequency assignment is considered to be well established as an aid in analysis and structure determination. Some of the vibrations of chemical groups giving rise to characteristic bands in the infrared spectrum between 4000 and 650 cm^{-1} are presented in this section: an alphabetical list of groups with frequency ranges and intensities in Table 4-1 and a list of frequency ranges from 4000 down to 650 cm^{-1}, with possible groups that could absorb within a given range, in Table 4-2. Further details of frequencies of the vibrations of certain groups in various types of compounds appear in the tables in Chapter 5. These tables are by no means comprehensive. To make full use of the group frequency method for structure determination, the references cited at the end of this chapter must be consulted.

4-4   Raman Group Frequencies

The vibrational motions of a molecule may give rise to infrared absorption or Raman scattering or both. In the last case, group frequencies will be numerically the same in Raman and infrared, but the intensities will often be quite different. In some cases information obtained from the Raman spectrum duplicates that obtained from the infrared, but in many cases the Raman spectrum provides additional information, especially in the low frequency region where far infrared spectra may not be available.

Since infrared absorption depends on change of dipole moment, we expect polar bonds or groups to give strong infrared bands. On the other hand, a change in polarizability is necessary for Raman scattering so that bonds or groups with symmetrical charge distributions are expected to give rise to strong Raman lines. Some of the most important Raman group frequencies are given by $-C=C-$, $-N=N-$, $-C\equiv C-$, and $-S-S-$ stretching modes.

A tabulation of some Raman group frequencies is given in Table 4-3. A bibliography of sources of Raman group frequencies is listed at the end of this chapter. It is not nearly so extensive as for the infrared, but the most recent addition to the list (reference B7) will do much to remedy the situation. Weak Raman bands are included in Table 4-3 if they are characteristic of a group.

4-5   Far Infrared Group Frequencies

All organic compounds have several vibrational frequencies due to skeletal deformation modes below 650 cm^{-1}. Coupling always occurs among these modes, unless the molecule has high symmetry, in which case vibrations with similar frequencies can belong to different symmetry types and therefore do not interact. Good group frequencies are not usually to be expected in the far infrared. Nevertheless, some correlations have been noted among certain classes of compounds, and these could be useful for identification purposes when the information obtained from the regular infrared region is ambiguous.

Before the 1960's, the far infrared region, especially that below 400 cm^{-1}, was not readily available to the organic chemist. There was a paucity of data on group frequencies in this region until the publication of reference (C1) in 1968, which, contains a comprehensive coverage of the region 700-300 cm^{-1}. However, several important group frequencies also occur in the 300-100 cm^{-1} region. These include metal-carbon vibrations in organometallic compounds, vibrations involving bonds between carbon and other heavy elements such as arsenic, torsional vibrations of methyl or CF_3 groups, ring puckering modes, and hydrogen-bond vibrations.

In Table 4-4, some groups or classes of compounds are listed with frequency ranges in which infrared absorption usually occurs. These frequency ranges are often much wider than those for the regular infrared. For this reason, no attempt has been made to compile a table similar to Table 4-2 for these far infrared group frequencies. Further details can be found in the references at the end of this chapter.

4-6   Near Infrared Group Frequencies

The spectra observed in the near infrared are due to vibrational transitions from the ground state to overtone or combination levels. These transitions are "forbidden" in the harmonic oscillator approximation. The fact that absorption is observed in this region means that the vibrations involved are anharmonic. Conversely, we might predict that only those vibrations that have considerable anharmonicity will give prominent peaks in the near infrared. In addition, since the near infrared region ends (arbitrarily) at 4000 cm^{-1}, vibrations with fundamental frequencies greater than 2000 cm^{-1} usually are involved. All these circumstances serve to limit the number of chemical groups giving rise to the absorptions and hence the number of peaks.

Table 4-5 lists some of the more important near-infrared group frequencies. Further details, together with correlation charts, are given in references (D1) and (D2). Some analytical applications of near infrared spectroscopy are described in reference (D3). It will be noted in Table 4-5 that the group frequency ranges are much broader than those of the regular infrared. It should also be remembered that because the intensities of overtones and combinations are low, long pathlengths (up to 10 cm) may be needed for solutions in CCl$_4$ and other solvents. Most of the correlations given in Table 4-5 are for spectra run in CCl$_4$ solution.

4-7 Vibrational Spectra–Structure Correlation Tables

There are several sources of correlation tables that aid in the analysis of a compound in terms of its constituent functional groups. Discussion and examples of how this analysis is carried out is given in Chapter 5. It should be pointed out, however, that a complete structure determination from the vibrational spectra alone is rarely possible except in the case of small, simple molecules. Nevertheless, comparison of the spectra of a compound with published spectra of compounds of known structure can be very helpful, sources of such spectra are given in the appendix. Here we list and describe some sources of special spectra-structure correlation data that are not included in the references at the end of this chapter.

1. IRSCOT, *Infrared Structural Correlation Tables and Data Cards*, H. A. Willis and R. G. J. Miller, eds., Heyden and Son Ltd., London, 1969. There are eleven tables covering hydrocarbons, halogen compounds, oxygen compounds, carboxylic acids and derivatives, nitrogen compounds, compounds containing N—O groups, heterocycles, and sulfur, silicon, boron, and phosphorus compounds. Each table refers to data cards, kept in binders, that give details of absorption bands of structural groups, with frequencies and wavelengths, literature references, and examples.

2. H. A. Szymanski, *Interpreted Infrared Spectra*, 3 vols., Plenum Press Data Division, New York, 1964, 1966, and 1967. Spectra–structure correlations are discussed in terms of group frequencies for alkanes, alkenes, benzene derivatives, alicyclic compounds, alkynes, alcohols, phenols, ethers, and peroxides.

3. H. A. Szymanski and R. F. Erickson, *Infrared Band Handbook*, 2 vols., IFI/Plenum, New York, 1970. Nearly 28,000 entries of absorption bands are arranged in 1 cm^{-1} intervals from 4240 to 29 cm^{-1}. There is also an alphabetic listing of compounds with references to published spectra.

4. IR correlation chart, Barnes Engineering Co., 30 Commerce Road, Stamford, Ct. 06902. This circular chart correlates a given wavelength or wavenumber with compounds that produce ir absorption in that region. It also presents wavelengths and intensities of the main absorption bands of 61 classes of functional groups. The correlations unfortunately are all given in wavelengths.

Table 4-1. *Alphabetic Listing of Functional Groups and Classes of Compounds with Their Absorption Frequencies in the Infrared*

Group or Class	Range (cm^{-1}) and Intensity[a]	Assignment and Remarks
Acetylenes $RC{\equiv}C-$	3300–3250 (m–s)	CH stretch for R = H (3320–3300 in CCl_4 soln)
	2250–2100 (w)	$C{\equiv}C$ stretch; frequency raised by conjugation
Acid halides $\begin{smallmatrix}&O\\&\|\\&C\\R&\ \ X\end{smallmatrix}$		
aliphatic	1810–1790 (s)	C=O stretch
	965–920 (m)	C–C= stretch
aromatic	1785–1765 (s)	C=O stretch; weaker band (1750–1735 cm^{-1}) due to Fermi resonance
	890–850 (s)	C–C stretch
Alcohols		
primary $-CH_2OH$	3640–3630 (s)	OH stretch; in dil CCl_4 soln
	1060–1030 (s)	C–O stretch; lowered by unsaturation
secondary $-CHROH$	3630–3620 (s)	OH stretch; in dil CCl_4 soln
	1120–1080 (s)	C–O stretch; lower when R is a branched chain or cyclic
tertiary $-CR_2OH$	3620–3610 (s)	OH stretch; in dil CCl_4 soln
	1160–1120 (s)	C–O stretch; lower when R is branched
general $-OH$	3350–3250 (s)	OH stretch; broad band in pure solids or liquids
	1440–1320 (m–s)	C–O–H bend; broad
	680–620 (m–s)	C–O–H twist; broad
Aldehydes $\begin{smallmatrix}&O\\&\|\\&C\\R&\ \ H\end{smallmatrix}$	2830–2810 (m) ⎫ 2740–2720 (m) ⎭	CH stretch with overtone of C–H bend; Fermi doublet
	1725–1695 (vs)	C=O stretch; slightly higher in CCl_4 soln
	1440–1320 (s)	H–C=O bend; aliphatic aldehydes
Alkyl R–	2980–2850 (m)	CH stretch, several bands
	1470–1450 (m)	CH_2 deformation
	1400–1360 (m)	CH_3 deformation
	740–720 (w)	CH_2 rocking
Amides		
primary $-CONH_2$	3540–3520 (m) ⎫ 3400–3380 (m) ⎭	NH stretch; dil soln; bands shift to 3360–3340 and 3200–3180 in solid
	1680–1660 (vs)	C=O stretch (amide I)
	1650–1610 (m)	NH_2 deformation; sometimes appears as a shoulder (amide II)
	1420–1400 (m–s)	C–N stretch (amide III)
secondary $-CONHR$	3440–3420 (m)	NH stretch; dil soln; shifts to 3280–3260 in pure liquid or solid
	1680–1640 (vs)	C=O stretch; amide I band
	1560–1530 (vs)	NH stretch; amide II band
	1310–1290 (m)	assignment uncertain
	710–690 (m)	assignment uncertain
tertiary $-CONR_2$	1670–1640 (vs)	C=O stretch
Amines		
primary	3460–3280 (m)	NH stretch; broad band may have some structure
	2830–2810 (m)	CH stretch
	1650–1590 (s)	NH_2 deformation

[a]s = strong, m = medium, w = weak, v = very.

Table 4-1. *(cont.)*

Group or Class	Range (cm^{-1}) and Intensity	Assignment and Remarks
secondary $-NHR$	3350–3300 (vw)	NH stretch
	1190–1130 (m)	C–N stretch
	740–700 (m)	NH deformation
Amine hydrohalides		
$RNH_3^+ X^-$	2800–2300 (m–s)	NH stretch, several peaks
$R'NH_2R^+ X^-$	1600–1500 (m)	NH deformation, one or two bands
α-Amino acid $\begin{array}{c} NH_2 \\ | \\ -C-COOH \\ | \end{array}$	3200–3000 (s)	H-bonded NH_2 and OH stretch; v broad band in solid state
	1600–1590 (s)	COO^- antisym. stretch
(or $-CNH_3^+ COO^-$)	1550–1480 (m–s)	NH_3^+ deformation
	1400 (w–m)	COO^- sym. stretch
Ammonium NH_4^+	3200 (vs)	NH stretch; broad band
	1430–1390 (s)	NH_2 deformation; sharp peak
Anhydrides $\begin{array}{c} -CO \\ \backslash \\ O \\ / \\ -CO \end{array}$	1850–1780 (variable)	antisym. C=O stretch
	1770–1710 (m–s)	sym. C=O stretch
	1220–1180 (vs)	C–O–C stretch (higher in cyclic anhydrides)
Aromatic compounds	3100–3000 (m)	CH stretch; several peaks
	2000–1660 (w)	overtones and combination bands
	1630–1590 (m) and 1520–1480 (m)	C=C stretch, intensity varies
	900–650 (s)	CH out-of-plane deformation, one or two bands, depending on substitution
Azides $-\overset{-}{N}-\overset{+}{N}{\equiv}N$	2160–2120 (s)	N≡N stretch
Bromo $-\overset{|}{\underset{|}{C}}-Br$	700–550 (m)	C–Br stretch
t-Butyl $(CH_3)_3C-$	2980–2850 (m)	CH stretch; several bands
	1400–1390 (m) 1380–1360 (s) $\Big\}$	CH_3 deformation
Carbodiimides $-N{=}C{=}N-$	2150–2100 (vs)	N=C=N antisym. stretch
Carbonyl C=O	1870–1650 (vs, broad)	C=O stretch
Carboxylic acids $\begin{array}{c} O \\ \| \\ C \\ R^{\diagup}\,\diagdown OH \end{array}$	3550 (m)	OH stretch (monomer, dil soln)
	3000–2400 (s, v broad)	OH stretch (solid and liquid states)
	1760 (vs)	C=O stretch (monomer, dil soln)
	1710–1680 (vs)	C=O stretch (solid and liquid states)
	1440–1400 (m)	C–O stretch/OH deformation
	960–910 (s)	C–O–H deformation
Chloro $-\overset{|}{\underset{|}{C}}-Cl$	850–650 (m)	C–Cl stretch
Diazonium salts $-N{\equiv}N^+$	2300–2240 (s)	N≡N stretch
Esters $\begin{array}{c} O \\ \| \\ C \\ R^{\diagup}\,\diagdown OR' \end{array}$	1765–1720 (vs)	C=O stretch
	1290–1180 (vs)	antisym. C–O–C stretch

(table continues)

Table 4-1. *(cont.)*

Group or Class	Range (cm^{-1}) and Intensity	Assignment and Remarks
Ethers $-C-O-C-$	1285–1240 (s)	C–O–C stretch; alkyl aryl ethers
	1140–1110 (vs)	C–O–C stretch; dialkyl ethers
	1275–1200 (vs) and	C–O–C stretches; vinyl ethers
	1050–1020 (s)	
	1250–1170 (s)	C–O–C stretch; cyclic ethers
Fluoroalkyl $-CF_3$, $-CF_2-$, etc.	1400–1000 (vs)	C–F stretch
Isocyanates $-N=C=O$	2280–2260 (vs)	N=C=O antisym. stretch
Isothiocyanates $-N=C=S$	2140–2040 (vs, broad)	C=N=S antisym. stretch
Ketone $C=O$	1725–1705 (vs)	C=O stretch; saturated ketones
	1705–1665 (s) and	C=O and C=C stretches in α,β-unsaturated
	1650–1580 (m)	ketones
	1700–1650 (vs)	C=O stretch; aromatic ketones
	1750–1730 (vs)	C=O stretch cyclopentanones
	1725–1705 (vs)	C=O stretch; cyclohexanones
Lactams $\begin{matrix} CH_2-NH \\ CH_2-C=O \end{matrix}$	695–655 (m–s)	N–C=O bend
Lactones $\begin{matrix} CH_2-O \\ CH_2-C=O \end{matrix}$	1850–1830 (s)	C=O stretch; β-lactones
	1780–1770 (s)	C=O stretch; γ-lactones
	1750–1730 (s)	C=O stretch; δ-lactones
Methyl $-CH_3$	2970–2850 (s)	CH stretch in C–CH_3 compounds
	2835–2815 (s)	CH stretch in methyl ethers O–CH_3
	2820–2780 (s)	CH stretch in N–CH_3 compounds
	1385–1375 (m)	CH_3 deformation in C–CH_3 compounds
	1400–1380 (ms) and	CH_3 deformations when more than one CH_3
	1375–1365 (m)	group on a single C atom
Methylene $-CH_2-$	2940–2920 (m) and	CH stretches in alkanes
	2860–2850 (m)	
	3090–3070 (m) and	CH stretches in alkenes
	3020–2980 (m)	
	1470–1450 (m)	CH_2 deformation
Nitriles $-C{\equiv}N$	2260–2240 (w)	C≡N stretch; aliphatic nitriles
	2240–2220 (m)	C≡N stretch; aromatic nitriles
Nitro $-NO_2$	1570–1550 (vs) and	N=O stretches; aliphatic nitro compounds
	1380–1360 (vs)	
	1480–1460 (vs) and	N=O stretches; aromatic nitro compounds
	1360–1320 (vs)	
	920–830 (m)	C–N stretch
Oximes $=NOH$	3600–3590 (vs)	OH stretch (dil soln)
	3260–3240 (vs)	OH stretch; solids
	1680–1620 (w)	C=N stretch; strong in Raman
Phenyl C_6H_5-	3100–3000 (w–m)	CH stretch
	2000–1700 (w)	four distinct bands in thicker samples; overtones and combinations
	1250–1025 (vs)	CH in-plane bending (five bands)
	770–730 (vs)	CH out-of-plane bend
	710–690 (vs)	ring bending mode

Table 4-1. (cont.)

Group or Class	Range (cm^{-1}) and Intensity	Assignment and Remarks
Phosphines $-PH_2$	2290–2260 (m)	P–H stretch
$-PH-$	1100–1040 (m)	P–H deformation
Pyridyl $-C_5H_4N$	3080–3020 (m)	CH stretch
	1620–1580 (vs) and	C=C and C=N stretches
	1590–1560 (vs)	
	840–720 (s)	CH out-of-plane deformation; one or two bands, depending on substitution
Silanes $-SiH_3$	2160–2110 (m)	Si–H stretch
$-SiH_2-$	950–800 (s)	Si–H deformation
Silanes (fully substituted)	1280–1250 (m–s)	Si–C stretch
	1110–1050 (vs)	Si–O–C stretch (aliphatic)
	840–800 (m)	Si–O–C deformation
Sulfates $R-O-SO_2-O-R$	1440–1350 (s) and	S=O stretches in covalent sulfates
	1230–1150 (s)	
$R-O-SO_3-M$	1260–1210 (vs)	S=O stretch in alkyl sulfate salts
($M = Na^+, K^+$, etc.)	810–770 (s)	C–O–S stretch
Sulfonic acids $-SO_2OH$	1250–1150 (vs, broad)	S=O stretch
Sulfoxides S=O	1060–1030 (s, broad)	S=O stretch
Thiocyanates $-S-C≡N$	2175–2160 (m)	C≡N stretch
Thiols $-S-H$	2590–2560 (w)	S–H stretch; strong in Raman
	700–550 (w)	C–S stretch; strong in Raman
Triazines $C_3N_3Y_3-$	1550–1510 (vs)⎫	ring stretching
(1,3,5 trisubst.)	1380–1340 (vs)⎭	
	820–800 (s)	out-of-plane CH deformation
Vinyl $CH_2=CH-$	3095–3080 (m) and	CH stretches
	3010–2980 (w)	
	1645–1605 (m–s)	C=C stretch
	1000–900 (s)	CH deformation; two bands

Table 4-2. *Numerical Listing of Wavenumber Ranges in Which Some Functional Groups and Classes of Compounds Absorb in the Infrared*

Range (cm^{-1})	Group or Class	Assignment and Remarks
3700–3600	$-OH$ alcohols (s), phenols (s)	dil soln
3520–3320	$-NH_2$ aromatic amines (s), primary amines (m) amides (m)	dil soln
3420–3250	$-OH$ alcohols (s), phenols (s)	liquids and solids
3370–3320	primary amides	solids
3320–3250	$-NOH$ (oximes) (m); C≡C–H (m)	sharp peak for C≡C–H
3300–3280	$-NHR$ (secondary amides) (s)	NH stretch; also in polypeptides and proteins
3260–3150	NH_4^+ (ammonium ion) (s)	broad band
3210–3150	$-NH_2$ (primary amides) (s)	solid
3200–3000	$-NH_3^+$ (amino acids) (m)	v broad band

(table continues)

Table 4-2. *(cont.)*

Range (cm^{-1})	Group or Class	Assignment and Remarks
3100–2400	—COOH carboxylic acids (v broad)	obscures the region
3110–3000	aromatic C–H, =CH$_2$, and $\underset{}{\overset{H}{\diagdown}}C=C\overset{H}{\diagup}$	all bands of medium intensity
2990–2850	C–CH$_3$ (m); –CH$_2$– (s)	two bands for –CH$_2$– group
2850–2700	O–CH$_3$ (m); N–CH$_3$ (m); aldehyde (m)	two bands for aldehyde
2750–2350	–NH$_3^+$ X$^-$ amine hydrohalides (s)	broad band
2720–2560	$\overset{O}{\overset{\|}{-P}}$–O–H phosphorus oxyacid (m)	associated OH stretch
2600–2540	S–H alkyl mercaptan (w)	strong in Raman
2410–2280	P–H phosphines (m)	sharp peak
2300–2240	diazonium salts (m)	aq soln
2280–2220	–O–C≡N (s); –C≡N (variable)	lower frequency when conjugated
2260–2190	–C≡C– (w)	conjugated, or nonterminal
2190–2130	–CNS (m); –NC (m)	C=N stretch
2180–2100	Si–H (s); –N=$\overset{+}{N}$=$\overset{-}{N}$ (m)	silanes, azides
2160–2100	R–C≡C–H (w–m)	may not be seen
2150–2100	N=C=N (vs)	carbodiimides
2000–1650	phenyl (w)	several bands due to overtones and combinations
1980–1950	–C=C=C– (s)	allene derivatives
1870–1650	C=O	carbonyl compounds
1870–1830	β-lactones (s)	C=O stretch
1870–1790	anhydrides (vs)	antisym. C=O stretch
1820–1800	acid halides R–CO–X (s)	frequency lower when R is aromatic
1780–1760	γ-lactones (s)	C=O stretch
1765–1725	anhydrides (vs)	sym. C=O stretch
1750–1730	δ-lactones (s)	C=O stretch
1750–1740	esters (vs)	saturated (20 cm^{-1} lower if unsaturated)
1740–1720	aldehydes (s)	saturated (30 cm^{-1} lower if unsaturated)
1720–1700	ketones (s)	saturated (20 cm^{-1} lower if unsaturated)
1710–1690	carboxylic acids (s)	fairly broad
1690–1640	C=N– (variable)	oximes and imines
1680–1620	primary amides (s)	two bands
1680–1650	nitrite esters	–N=O stretch
1680–1655	$\overset{}{\diagdown}C=C\overset{H}{\diagup}$ (s)	trisubstituted
1680–1660	$\overset{}{\diagdown}C=N–$ (m–s)	aliphatic Schiff bases

Table 4–2. *(cont.)*

Range (cm^{-1})	Group or Class	Assignment and Remarks
1670–1655	secondary amides (vs)	aromatic
1670–1650	$C_6H_5-\overset{\underset{\|}{O}}{C}-$ (s)	benzophenone derivatives
1670–1640	tertiary amides (s)	C=O stretch
1670–1630	$\diagdown$C=C$\diagup$ (m–s)	mono or disubstituted
1650–1590	urea derivatives (s)	two bands
1640–1620	$\diagdown$C=N– (m–s)	aromatic Schiff bases
1640–1610	nitrites R–O–N=O	also nitrates (R–ONO$_2$)
1640–1580	–NH$_3{}^+$ (s)	amino acid zwitterion
1640–1530	β-diketones, β-keto esters (vs, broad)	chelate compounds
1620–1595	primary amines (s)	–NH$_2$ deformation
1615–1605	vinyl ethers (s)	C=C stretch
1615–1590	phenyl (m)	sharp peak, sometimes weak, occasionally a doublet
1615–1565	pyridines (s)	sharp doublet
1610–1580	amino acids (broad)	NH$_2$ deformation
1610–1560	salt of carboxylic acids (vs)	$-C\overset{\diagup O}{\underset{\diagdown O}{-}}$ antisym. stretch
1590–1580	–NH$_2$ primary alkyl amides (m)	amide II band; dil soln
1575–1545	–NO$_2$ (vs)	aliphatic nitro compounds
1565–1475	secondary amides (vs)	NH deformation; amide II band
1560–1510	triazines (s, sharp)	ring stretch
1550–1490	–NO$_2$ (s)	aromatic nitro compounds
1530–1490	–NH$_3{}^+$ (s)	amino acids or hydrochlorides
1515–1485	phenyl (m)	sharp peak; sometimes weak
1475–1450	–CH$_2$– (vs); CH$_3$ (vs)	CH$_2$ scissor vibration; –CH$_3$ antisym. deformation
1440–1400	carboxylic acids (m)	in-plane OH bending and C–O stretch of dimers
1430–1395	NH$_4{}^+$ ion (m–s)	NH deformation
1420–1400	–CO–NH$_2$	primary amides
1400–1370	*t*-butyl (m)	two bands
1400–1310	salts of carboxylic acids (broad)	$-C\overset{\diagup O}{\underset{\diagdown O}{-}}$ sym. stretch
1390–1360	–SO$_2$Cl (s)	$\diagdown S\overset{\diagup O}{\underset{\diagdown O}{}}$ antisym. stretch
1380–1370	C–CH$_3$ (s)	CH$_3$ deformation
1380–1360	C–(CH$_3$)$_2$ (m)	two bands
1375–1350	–NO$_2$ (s)	aliphatic nitro compounds
1360–1335	–SO$_2$NH$_2$	sulfonamides

(table continues)

Table 4-2. *(cont.)*

Range (cm^{-1})	Group or Class	Assignment and Remarks
1360–1320	$-NO_2$ (vs)	aromatic nitro compounds
1335–1295	$>S{<}^O_O$ (vs)	sulfones
1330–1310	$-CH_3$ (vs)	attached to a benzene ring
1310–1250	$-N=N(O)-$ (s)	azoxy compounds
1300–1200	$>N{\rightarrow}O$ (vs)	pyridine *N*-oxides
1300–1175	P=O (vs)	phosphorus oxyacids and phosphates
1300–1000	C–F (vs)	aliphatic fluoro compounds
1285–1240	Ar–O (vs)	alkyl aryl ethers
1280–1250	$Si-CH_3$ (vs)	silanes
1280–1240	$>C{-}C{<}$ with O	epoxides
1280–1180	$-C-N-$ (s)	aromatic amines
1280–1150	$-C-O-C-$ (vs)	esters, lactones
1255–1240	*t*-butyl (m)	also at 1210–1200 cm^{-1}
1245–1155	$-SO_3H$ (vs)	sulfonic acids
1240–1070	$-C-O-C-$ (s–vs)	alicyclic compounds
1230–1100	$-C-N-$ (s)	amines
1200–1165	$-SO_2Cl$ (s)	$-SO_2-$ sym. stretch
1200–1015	C–OH (vs)	alcohols
1190–1140	Si–O–C (s)	silicones, silanes
1170–1145	$-SO_2NH_2$	sulfonamide
1170–1140	$-SO_2-$	sulfone
1170–1130	$Ar-CF_3$ (s)	two bands
1160–1100	$>C=S$ (m)	thiocarbonyl compounds
1150–1070	C–O–C (vs)	aliphatic ethers
1140–1090	$-C-O-H$ (s)	secondary or tertiary alcohols
1120–1030	$-C-NH_2$ (s)	primary aliphatic amines
1095–1015	Si–O–Si (vs); Si–O–C (vs)	silicones, silanes
1080–1040	$-SO_3H$ (s)	sulfonic acids
1075–1020	$-C-O-C-$ (s)	vinyl ethers
1065–1015	$>CH-O-H$ (s)	cyclic alcohols
1060–1025	$-CH_2-O-H$ (vs)	primary alcohols
1060–1045	$>S=O$ (vs)	alkyl sulfoxides
1055–915	P–O–C (vs)	strongest bands and highest frequencies for aliphatic compounds

Table 4–2. *(cont.)*

Range (cm^{-1})	Group or Class	Assignment and Remarks
1030–950	ring vibration (w)	many cyclic compounds
1000–970	$-CH=CH_2$ (vs)	antisym. CH out-of-plane deformation
980–960	$-CH=CH-$ (vs)	$=C-H$ out-of-plane bend (trans isomers)
960–910	$-C-OH$ (variable)	out-of-plane OH bending in carboxylic acid dimers
920–910	$-CH=CH_2$ (vs)	CH out-of-plane deformation
900–875	$CH_2=C\overset{R}{\underset{R'}{\diagdown}}$ (vs)	CH out-of-plane deformation
890–805	1,2,4-trisubst. benzenes (vs)	CH out-of-plane deformation; two bands
860–760	$R-NH_2$ (vs, broad)	NH_2 wag, primary amines
860–720	$-Si-C-$ (vs)	silicon compounds
850–830	1,3,5-trisubst. benzenes (vs)	CH out-of-plane deformation
850–810	$Si-CH_3$ (vs)	$Si-C$ stretch
835–800	$-CH=C\diagup\diagdown$ (m)	CH out-of-plane deformation
830–810	*p*-disubst. benzenes (vs)	CH out-of-plane deformation
825–805	1,2,4-trisubst. benzenes (vs)	CH out-of-plane deformation
820–800	triazines (s)	CH out-of-plane deformation
810–790	1,2,3,4-tetrasubst. benzenes (vs)	CH out-of-plane deformation
800–690	*m*-disubst. benzenes (vs)	two bands
785–680	1,2,3-trisubst. benzenes (vs)	two bands
770–690	monosubst. benzenes (vs)	two bands
760–740	*o*-disubst. benzenes (s)	CH out-of-plane deformation
760–510	$C-Cl$ (s)	$C-Cl$ stretch
740–720	$-(CH_2)_n-$ (w, intensity depends on n)	CH_2 rocking in methylene chains
730–675	$-CH=CH-$ (s)	cis isomers

Table 4–3. *Characteristic Frequencies of Functional Groups in the Raman Spectra of Complex Molecules*

Group or Class	Range (cm⁻¹) and Intensity	Assignment and Remarks
Acetylenes C≡CH	3340–3270 (s)	C≡C stretch
R–C≡C–R	2300–2190 (s)	C≡C stretch; disubst. acetylene; sometimes two bands (Fermi doublet)
R–C≡CH	2140–2100 (s)	C≡C stretch; monoalkyl acetylenes
	650–600 (m)	–C≡CH deformation
Acid chlorides $\overset{\displaystyle O}{\underset{R\quad Cl}{\underset{\diagdown}{\overset{\|}{\underset{\diagup}{C}}}}}$	1800–1790 (s)	C=O stretch
Alcohols R–OH	3650–3250 (w)	–OH stretch ⎫
	1440–1320 (w)	–OH in-plane bend ⎬ usually v weak or
	1160–1020 (w)	–C–OH stretch ⎭ absent
	950–850 (s)	–C–C– stretch
Aldehydes $\overset{\displaystyle O}{\underset{R\quad H}{\underset{\diagdown}{\overset{\|}{\underset{\diagup}{C}}}}}$	1730–1700 (m)	C=O stretch
n-Alkanes (general)	2980–2800 (vs)	CH stretch
	1475–1450 (s)	CH_3 antisym. deformation
	1350–1300 (m–s)	CH_2 bend
	340–230 (s)	–C–C–C– bend; lowest frequency for long-chain molecules
Alkenes (general)	3090–3010 (m)	CH stretch
	1675–1600 (m–s)	C=C stretch; stronger than ir
	1450–1200 (vs)	C–H in-plane deformation
cis-alkenes R′CH=CHR	590–570 (m) ⎫	
	420–400 (m) ⎬	skeletal deformations
	310–290 (m) ⎭	
trans-alkenes R′CH=CHR	500–480 (m) ⎫	skeletal deformations
	220–200 (m) ⎭	
terminal alkenes $RCH=CH_2$	500–480 (m)	skeletal deformations
$RR'C=CH_2$	440–390 (m)	skeletal deformations; two bands
	270–250 (m)	skeletal deformations
Allenes C=C=C	2000–1960 (s)	–C=C=C– antisym. stretch
	1080–1060 (vs)	–C=C=C– sym. stretch
Amides	3540–3520 (w)	antisym. NH_2 stretch, dil soln
primary $-CONH_2$	3400–3380 (w)	sym. NH_2 stretch, dil soln
	1680–1660 (m)	C=O stretch; weaker than ir
	1420–1400 (s)	C–N stretch
secondary –CONHR	3440–3420 (s)	NH stretch, dil soln
	1680–1640 (w)	amide I band
	1310–1280 (s)	amide III band
tertiary $-CONR_2$	1670–1640 (m)	amide I band
Amines		
primary RNH_2	3550–3330 (m)	antisym. NH_2 stretch
	3450–3250 (m)	sym. NH_2 stretch
	1090–1070 (m)	C–N stretch
secondary R′NHR	3350–3300 (w)	NH stretch
	1190–1130 (m)	C–N stretch

Table 4-3. *(cont.)*

Group or Class	Range (cm^{-1}) and Intensity	Assignment and Remarks
aromatic	1380–1250 (s)	C—N stretch
Amino acids —CNH$_2$COOH or CNH$_3$$^+COO^-$	1600–1590 (w)	$C \overset{O}{\underset{O}{\diagup}}$ antisym. stretch
	1400–1350 (vs)	$C \overset{O}{\underset{O}{\diagup}}$ sym. stretch
	900–850 (vs)	C—C—N sym. stretch
Anhydrides 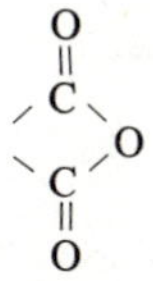	1850–1780 (w–m)	antisym. C=O stretch
	1770–1710 (m)	sym. C=O stretch
Aromatic compounds	3070–3020 (s)	CH stretch
	1620–1580 (m–s)	C=C stretch; may be weak in ir
	1045–1015 (m)	CH in-plane bend
	1010–990 (vs)	ring breathing (absent in *o*- and *p*-disubst. compounds)
	900–650 (m)	CH out-of-plane deformation; one or two bands
Arsenic–carbon (organoarsenic compounds) As—C	570–550 (vs)	C—As stretch
	240–220 (vs)	C—As—C deformation
Azides —N—N≡N	2170–2080 (s)	antisym. NNN stretch
	1345–1175 (s)	sym. NNN stretch
Azo —N=N—	1580–1570 (vs)	N=N stretch; nonconjugated compounds
	1420–1410 (vs)	N=N stretch; conjugated to aromatic ring, e.g., azobenzene
	1060–1030 (vs)	C—N stretch; aromatic azo compounds
Benzenes		
monosubst.	630–610 (s)	out-of-plane CH deformation; absent in ir
1,2- and 1,2,4-trisubst.	750–700 (s)	out-of-plane CH deformation; absent in ir
1,3-disubst.	750–700 (s)	out-of-plane CH deformation; absent in ir
	480–450 (m)	out-of-plane ring deformation
1,2,3-trisubst.	655–645 (s)	out-of-plane CH deformation; absent in ir
1,3,5-trisubst.	570–550 (s)	out-of-plane CH deformation; absent in ir
Bromo C—Br	650–490 (vs)	C—Br stretch
	310–270 (s)	C—C—Br bend
t-Butyl	1250–1200 (m–s)	CH$_3$ deformations; two bands
	940–920 (s)	CH$_3$ deformations; weak in ir
Carbonyl C=O	1870–1650 (s)	C=O stretch; weaker than in ir
Carboxylic acids	1680–1640 (s)	sym. C=O stretch of dimer
Chloromethyl	740–720 (m) ⎫ 670–660 (vs) ⎭	CH$_2$ wag and C—Cl stretch; CH$_2$Cl— group attached to an aliphatic hydrocarbon chain
Chloro C—Cl	850–650 (s)	C—Cl stretch
	340–290 (s)	C—C—Cl bend

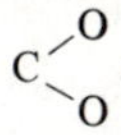

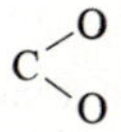

(table continues)

Table 4–3. *(cont.)*

Group or Class	Range (cm^{-1}) and Intensity	Assignment and Remarks
Cumulenes	2070–2030 (vs)	$>C=C=C=C<$ stretch
Cyanamides	1150–1140 (vs)	$-N=C=N-$ sym. stretch
Cyclobutanes	1000–960 (vs)	ring breathing
	700–680 (s)	ring deformation
	180–150 (s)	ring puckering
Cyclohexanes	1460–1440 (s)	CH_2 scissoring
	825–815 (s)	ring vibration (boat)
	810–795 (s)	ring vibration (chair)
Cyclopentanes	1450–1430 (s)	CH_2 scissoring
	900–880 (s)	ring breathing
Cyclopropanes	1210–1180 (s)	ring breathing
	830–810 (s)	ring deformation
Disulfides C–S–S–C	550–430 (vs)	S–S stretch
Epoxides	1280–1260 (s)	sym. ring stretch
Esters R'COOR	1100–1025 (s)	sym. C–O–C stretch
Ethers		
cyclic	820–800 (s)	ring stretching; six-membered ring
	920–900 (s)	ring stretching; five-membered ring
	1040–1010 (s)	ring stretching; four-membered ring
aromatic	1310–1210 (m) and 1050–1010 (m)	C–O–C stretches
aliphatic saturated	1140–1110 (m)	C–O–C stretch
aliphatic unsaturated	1275–1200 (m)	C–O–C antisym. stretch
–C=C–O–C–	1075–1020 (s)	C–O–C sym. stretch
Isocyanates –N=C=O	1440–1400 (vs)	N=C=O sym. stretch
Isopropyl $(CH_3)_2CH-$	1180–1160 (m)	CH_3 deformation
	835–795 (ms)	CH_3 deformations; weak in ir
Ketenes $>C=C=O$	2060–2040 (vs)	$>C=C=O$ stretch
Ketones $\begin{smallmatrix}R\\R'\end{smallmatrix}C=O$	1725–1705 (m)	C=O stretch; saturated compounds
	1700–1650 (m)	C=O stretch; aromatic compounds
	1750–1705 (m)	C=O stretch; alicyclic ketones
Lactams R$\langle$CH$_2$–C=O / CH$_2$–NH$\rangle$	1750–1700 (m)	amide I band
Lactones R$\langle$CH$_2$–C=O / CH$_2$–O$\rangle$	1850–1730 (s)	C=O stretch
Lead–carbon (organolead compounds) Pb–C	480–420 (s)	C–Pb stretch
Mercaptans C–SH	850–820 (vs)	S–H in-plane deformation
	700–600 (vs)	C–SH stretch; weak in ir
Mercury–carbon (organomercury compounds) Hg–C	570–510 (vvs)	C–Hg stretch
Methyl $-CH_3$	2980–2800 (vs)	CH stretch
	1470–1460 (s)	CH_3 deformation

Table 4–3. *(cont.)*

Group or Class	Range (cm^{-1}) and Intensity	Assignment and Remarks
Methylene =CH$_2$ –CH$_2$–	3090–3070 (s) and 3020–2980 (s) 2940–2920 (s) and 2860–2850 (s) 1350–1150 (m–s)	=CH$_2$ stretches –CH$_2$– stretches } antisym. and sym. CH$_2$ wag and twist; weak or absent in ir
Nitrates –ONO$_2$	1285–1260 (vs)	ONO sym. stretch
Nitriles –C≡N	2260–2240 (s) 2230–2220 (s) 1080–1025 (s–vs) 840–800 (s–vs) 380–280 (s–vs)	C≡N stretch; nonconjugated C≡N stretch; conjugated C–C–C stretch C–C–CN sym. stretch C–C≡N bend
Nitrites –ONO	1660–1620 (s)	N=O stretch; alkyl nitrites
Nitro –NO$_2$	1570–1550 (w) and 1380–1360 (s) 920–830 (s) 650–520 (m)	N=O stretches C–N stretch NO$_2$ bend
Oximes C=NOH	1680–1620 (vs)	C=N stretch; may not be seen in ir
Peroxides –C–O–O–C–	900–850 (variable)	O–O stretch; weak in ir
Phosphines –PH–	2350–2240 (m)	P–H stretch
Pinanes 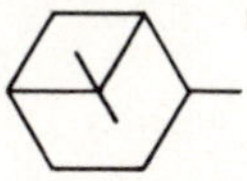	680–630 (vs)	ring deformation; six other characteristic bands between 1000 and 350 cm^{-1}
Pyridines	1620–1560 (m) 1020–980 (vs)	ring stretching ring breathing
Pyrroles	3450–3350 (s) 1420–1360 (vs)	NH stretch ring stretching
Silicon-carbon (organosilicon compounds) Si–C	1300–1200 (s)	Si–C stretch
Sulfides C–S	705–570 (s)	C–S stretch
Sulfonamides –SO$_2$NH$_2$	1155–1135 (vs)	SO$_2$ stretch
Sulfonates ROSO$_2$OR	1400–1360 (s)	SO$_2$ stretch
Sulfones	1280–1260 (m) 1150–1110 (s) 610–545 (s)	SO$_2$ antisym. stretch SO$_2$ sym. stretch SO$_2$ scissoring
Sulfonyl chlorides R–SO$_2$Cl	1230–1200 (m)	S=O stretch
Sulfoxides >S=O	1050–1010 (s)	S=O stretch
Thiocyanates –S–C≡N	650–600 (s)	S–C stretch
Thiols RSH	2590–2560 (vs) 700–550 (vs) 340–320 (vs)	S–H stretch C–S stretch S–H out-of-plane bend

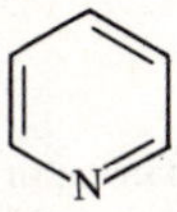

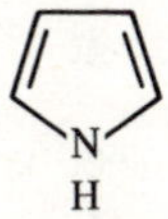

(table continues)

Table 4–3. *(cont.)*

Group or Class	Range (cm^{-1}) and Intensity	Assignment and Remarks
Thiophenes	740–680 (vs)	C–S–C stretch
	570–430 (s)	ring deformation
Tin–carbon (organotin compounds) Sn–C	600–450 (s)	C–Sn stretch
Xanthates	670–620 (vs)	C=S stretch; not seen in ir
	480–450 (vs)	C–S stretch

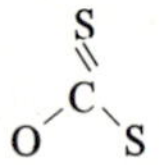

Table 4–4. *Alphabetic Listing of Functional Groups and Classes of Compounds with Their Absorption Frequencies in the Far Infrared*

Group or Class	Range (cm^{-1}) and Intensity	Assignment and Remarks
Acetylenes –C–C≡C–	680–580 (s)	–C≡CH bend
	350–300 (m–s)	–C–C≡C bend
Acid chlorides 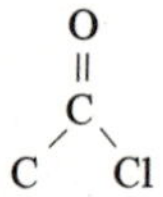	440–420 (s)	C–C=O bend
Alcohols R–OH	680–620 (s, broad)	C–OH deformation
Aldehydes	695–635 (s)	C–C–CHO bend
	565–520 (s)	C–C=O bend
Alicyclic compounds	250–50 (w)	ring puckering; several bands in vapor phase
Alkenes RCH=CH$_2$	640–630 (s) and 555–550 (s)	C=CH$_2$ twisting
Alkyl C$_n$H$_{2n+1}$–	565–440 (w–m)	chain deformations; two bands
Aldoximes 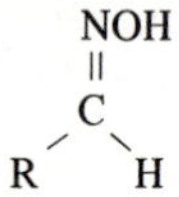	375–355 (vs)	OH torsion
Aluminum–carbon Al–C	720–660 (m) and 585–540 (m)	AlC$_2$ stretch in Al$_2$R$_6$ compounds
Amides	630–570 (s)	N–C=O bend
	615–535 (s)	C=O out-of-plane bend
	480–420 (m–s)	C–C=O bend
Amines		
primary R–NH$_2$	500–440 (w)	overtone of torsional mode
	280–200 (vs, broad)	NH$_2$ torsion
secondary R′–NHR	455–410 (w, broad)	C–N–C bend
tertiary R′–NR$_2$	510–480 (s)	C–N–C bend
Amino acids H$_3$$\overset{+}{N}$–C–COO$^-$	560–500 (s)	COO$^-$ rocking

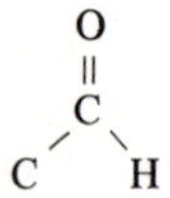

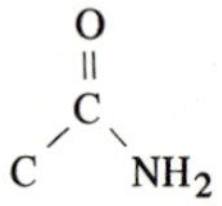

Table 4-4. *(cont.)*

Group or Class	Ranges (cm^{-1}) and Intensity	Assignment and Remarks
Antimony–carbon Sb–C	520–480 (m–s)	C–Sb stretch
Arsenic–carbon As–C	580–560 (vs) 240–220 (vs)	As–C stretch AsC_2 deformation
Benzenes (mono-, di-, and trisubst.)	625–495 (w–m) 560–420 (s)	in-plane ring deformation out-of-plane ring deformation
Bromo C–Br	650–490 (s) 310–270 (s)	C–Br stretch C–C–Br bend
Carboxylic acids 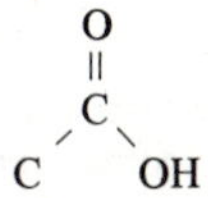	700–590 (s) 550–465 (s)	O–C=O bend C–C=O bend
Chloro C–Cl	760–510 (s) 340–290 (s)	C–Cl stretch C–C–Cl bend
Cyclohexanes C_6H_{11}–	570–430 (s)	ring deformation
Cyclopentanes C_5H_9–	580–490 (s)	ring deformation
Cyclopropanes C_3H_5–	540–500 (m–s)	ring deformation
Disulfides R–S–S–R	520–430 (w)	S–S stretch, lower end of range when R is aromatic
Esters –COOR	645–575 (s) 350–300 (s)	O–C–O bend C–O–C bend
Ethers C–O–C	520–430 (m–s)	C–O–C bend
Furans 	610–590 (s)	ring deformation
Ferrocenes	520–465 (s)	ring deformations; two bands
Gallium–carbon Ga–C	560–530 (s)	Ga–C stretch; alkyl Ga compounds
Germanium–carbon Ge–C	600–540 (m–s)	Ge–C stretch; lower frequencies for unsaturated groups
Hydrogen bonds X–H $\cdots$ X	200–100 (broad)	H $\cdots$ X stretch
Imidazoles 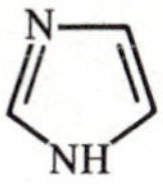	700–600 (m–s)	ring deformation; one or two bands
Iodo C–I	600–465 (s) 290–265 (m)	C–I stretch C–C–I bend
Isothiocyanates –N=C=S	650–600 (m–s) 560–510 (s) 470–440 (s)	C=N and C=S stretches N=C=S bend
Ketones 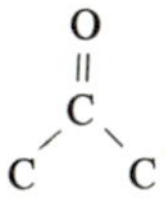	630–565 (s) 560–510 (s)	C–C–C bend C–C=O bend
Lead–carbon Pb–C	480–420 (s)	C–Pb stretch
Mercury–carbon CH_3–Hg	585–515 (s)	C–Hg stretch

(table continues)

Table 4-4. *(cont.)*

Group or Class	Ranges (cm^{-1}) and Intensity	Assignment and Remarks
Methyl		
aromatic Ar–CH$_3$	390–260 (m)	Ar–CH$_3$ deformation; mono- and disubst. benzenes
general	250–100 (variable)	CH$_3$ torsion; may not be observed
Naphthalenes	645–615 (m–s)	in-plane ring bending
	545–520 (s)	
	490–465 (variable)	out-of-plane ring bending
Nitriles R–C≡N	580–530 (m–s)	C–C–CN bend
	390–350 (m–s)	C–C≡N bend
Nitro R–NO$_2$	650–600 (s)	NO$_2$ bend, aliphatic compounds
	580–520 (m)	NO$_2$ bend, aromatic compounds
	530–470 (m–s)	NO$_2$ rocking
Oximes C=NOH	375–355 (vs)	OH torsion
Phenols Ar–OH	720–600 (s, broad)	OH out-of-plane deformation
	450–375 (w)	C–O–H deformation
Phenyl–metal		
Ar–Al	700-660 (w)	metal–carbon stretch; one of the monosubstituted benzene ring modes is sensitive to substituent
Ar–Ge	480–450 (m–s)	
Ar–Pb	450–430 (m–s)	
Ar–Sn	460–440 (m–s)	
Purines	690–540 (s)	ring deformations; two bands
Pyridines	635–605 (m–s)	in-plane ring bending
	420–385 (s)	out-of-plane ring bending
Pyrimidines	685–630 (m–s)	ring deformations; three bands in 2- and 4-substituted pyrimidines
	580–480 (m–s)	
	500–440 (m–s)	
Sulfides C–S	710–570 (m)	C–S stretch
Sulfones ⟩S⟨$_O^O$	610–545 (m–s)	SO$_2$ scissoring
Sulfonyl chlorides –SO$_2$Cl	775–650 (m)	C–S stretch
	610–565 (vs)	SO$_2$ deformation
	570–530 (vs)	SO$_2$ rocking
Tin–carbon Sn–C	600–450 (s–vs)	sym. and antisym. C–Sn stretches (two bands); aliphatic compounds
Thiocyanates –S–C≡N	650–600 (w)	S–C stretch
	405–400 (s)	S–C≡N bend
Thiophenes	570–430 (w–m)	ring modes; one or two bands

Table 4-4. *(cont.)*

Group or Class	Ranges (cm^{-1}) and Intensity	Assignment and Remarks
Trifluoromethyl Ar–CF_3	650–500 (s)	CF_3 deformations; two or three bands
–CF_3	below 65 (variable)	CF_3 torsion
Trifluoromethylmercury compounds	275–255 (s)	C–Hg stretch

Table 4-5. *Alphabetic Listing of Functional Groups and Classes of Compounds with Their Absorption Frequencies in the Near Infrared*

Group or Class	Range (cm^{-1})	(microns)	Assignment and Remarks
Acetylenes –C≡CH	9800–9550	(1.02–1.05)	second overtone of ≡CH stretch
	6580–6450	(1.52–1.55)	overtone of ≡CH stretch
Alcohols –OH (nonhydrogen-bonded)	10,800–10,400	(0.93–0.96)	second overtone of –OH stretch
	7,140–7040	(1.40–1.42)	overtone of –OH stretch
Aldehydes $\quad$ O $\quad\ \ \ $ ‖ $\quad\ \ \ $ C $\quad$ R $\quad$ H	4570–4500	(2.19–2.22)	combination of C=O stretch and CH stretch of CHO group
Alkanes –CH_2– and –CH_3 groups	11,630–10,530	(0.86–0.95)	third overtone of CH stretch
	8950–8350	(1.12–1.20)	second overtone of CH stretch, most useful bands in this region
	5850–5620	(1.71–1.78)	overtones of CH stretch
	4480–4100	(2.23–2.44)	combination of stretch and deformation of –CH_2– and –CH_3 groups
Alkenes =CH_2	11,900–11,110	(0.84–0.90)	third overtone of CH stretch
	9260–8700	(1.08–1.15)	second overtone of CH stretch
	7580–7400	(1.32–1.35)	combination of overtone of CH stretch and C=C stretch
	6170–6060	(1.62–1.65)	overtone of CH stretch
	4760–4700	(2.10–2.13)	combination of CH streteh and C=C stretch
cis-RCH=CHR′ (trans isomers have no unique bands)	9710–9350	(1.03–1.07)	combination of overtones of =CH stretch and C=C stretch
	4720–4610	(2.12–2.17)	combination of C=C stretch and CH stretch
Amides primary $\quad$ O $\quad\ \ \ $ ‖ $\quad\ \ \ $ C $\quad$ RR $\quad$ NH_2	10,000–9350	(1.00–1.07)	second overtone of NH stretch
	6945–6540	(1.44–1.53)	overtone of NH stretch
	5240–4810	(1.91–2.08)	three peaks: second overtone of C=O stretch; second overtone of NH deformation, combination of C=O stretch and NH stretch
secondary $\quad$ O $\quad\ \ \ $ ‖ $\quad\ \ \ $ C $\quad$ R $\quad$ NHR′	6850–6760	(1.46–1.48)	overtone of NH stretch
	5050–4900	(1.98–2.04)	combination of NH stretch and NH bend

(table continues)

Table 4-5. *(cont.)*

Group or Class	Range		Assignment and Remarks
	(cm^{-1})	*(microns)*	
Amines −NH$_2$ and −NHR	10,000−9260	(1.00−1.08)	second overtone of NH stretch
	6990−6370	(1.43−1.57)	overtone of NH stretch
	5155−4925	(1.94−2.03)	combination of NH stretch and NH deformation
Aromatic compounds	8740−8670	(1.14−1.15)	second overtone of CH stretch
	6130−5920	(1.63−1.69)	first overtone of CH stretch
Aromatic aldehyde ArCHO	4560−4440	(2.19−2.25)	combination of CH stretch and C=O stretch of CHO group
Carbonyl C=O	5200−5100	(1.92−1.96)	second overtone of C=O stretch
Carboxylic acids 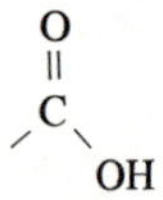	7000−6800	(1.43−1.47)	overtone of OH stretch
Glycols −C(OH)−C(OH)−	7140−7040	(1.40−1.42)	overtone of OH stretch
Imides 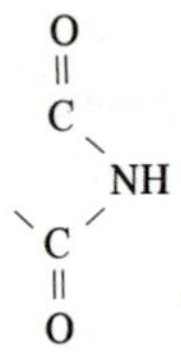	9900−9620	(1.01−1.04)	second overtone of NH stretch
	6540−6370	(1.53−1.57)	overtone of NH stretch
Methyl −CH$_3$	8380−8350	(1.19−1.20)	second overtone of CH stretch
Methylene −CH$_2$−	8260−8200	(1.21−1.22)	second overtone of CH stretch
Nitriles −C≡N	5290−5210	(1.89−1.92)	combination of C≡N stretch and CH stretch
Oximes C=NOH	7090−7000	(1.41−1.43)	overtone of OH stretch
Phosphines −PH−	5350−5260	(1.87−1.90)	overtone of P−H stretch
Phenols Ar−OH nonbonded	7140−6800	(1.40−1.43)	overtone of OH stretch
	5000−4950	(2.00−2.02)	combination of OH stretch and OH deformation
intramolecularly bonded	7000−6700	(1.43−1.49)	overtone of OH stretch
Thiols −SH	5080−5020	(1.97−1.99)	overtone of S−H stretch (weak band)

For Carboxylic acids the group structure shown is:

O
‖
C
/ \
OH

For Imides the group structure shown is:

O
‖
C
\
NH
/
C
‖
O

BIBLIOGRAPHY

A. Infrared Group Frequencies

1. M. St. C. Flett, *Characteristic Frequencies of Chemical Groups in the Infra-Red*, Elsevier, Amsterdam, 1963.
2. L. J. Bellamy, *The Infrared Spectra of Complex Molecules*, 2nd ed., Methuen, London, 1958.
3. L. J. Bellamy, *Advances in Infrared Group Frequencies*, Methuen, London, 1968.
4. R. N. Jones and C. Sandorfy, "The Application of Infrared and Raman Spectrometry to the Elucidation of Molecular Structure," in *Technique of Organic Chemistry*, Vol. IX, A. Weissberger, ed., Interscience, New York, 1956.
5. N. B. Colthup, L. H. Daly, and S. E. Wiberly, *Introduction to Infrared and Raman Spectroscopy*, Academic, New York, 1964.
6. G. C. Pimentel and A. L. McClellan, *The Hydrogen Bond*, Freeman, San Francisco, 1960.
7. N. L. Alpert, W. E. Keiser, and H. A. Szymanski, *IR—Theory and Practice of Infrared Spectroscopy*, 2nd ed., Plenum, New York, 1970.
8. K. Nakanishi, *Infrared Absorption Spectroscopy*, Holden-Day, San Francisco, 1962.
9. H. A. Szymanski, *A Systematic Approach to the Interpretation of Infrared Spectra*, Hertillon Press, Buffalo, N.Y., 1967.
10. C. E. Meloan, *Elementary Infrared Spectroscopy*, Macmillan, New York, 1963.
11. N. B. Colthup, *J. Opt. Soc. Amer.*, **40**, 397 (1950).
12. G. Varsanyi, *Vibrational Spectra of Benzene Derivatives*, Academic, New York, 1969.

B. Raman Group Frequencies

1. J. A. Hibben, *The Raman Effect and Its Chemical Applications*, American Chemical Society Monograph Series No. 80, Reinhold, New York, 1939.
2. Jones and Sandorfy (see reference A4).
3. Colthup, Daly, and Wiberley (reference A5).
4. M. C. Tobin, *Laser Raman Spectroscopy*, Wiley–Interscience, New York, 1971.
5. H. A. Szymanski, *Correlation of Infrared and Raman Spectra of Organic Compounds*, Hertillon Press, Cambridge Springs, Pa., 1969.
6. H. J. Sloane, "The Use of Group Frequencies for Structural Analysis in the Raman Compared to the Infrared," in *Polymer Characterization*, C. D. Craver, ed., Plenum, New York, 1971, pp. 15–36.
7. F. E. Dollish, W. G. Fateley, and F. F. Bentley, *Group Frequencies in the Laser Raman Spectra of Organic Compounds*, Wiley–Interscience, New York, 1973.
8. J. L. Koenig, *J. Pol. Sci., Part D, Macromolecular Reviews*, **6**, 59 (1972).

C. Far Infrared Group Frequencies

1. F. F. Bentley, L. D. Smithson, and A. L. Rozek, *Infrared Spectra and Characteristic Frequencies 700–300 cm^{-1}*, Interscience, New York, 1968.
2. J. E. Stewart, "Far Infrared Spectroscopy," in *Interpretive Spectroscopy*, S. K. Freeman, ed., Reinhold, New York, 1965, Ch. 3.
3. K. D. Möller and W. G. Rothschild, *Far Infrared Spectroscopy*, Wiley–Interscience, New York, 1971, Chs. 5–8.
4. F. A. Miller, "Far Infrared Spectroscopy," in *Molecular Spectroscopy*, The Institute of Petroleum, London, 1968, pp. 1–27.
5. J. W. Brasch, R. J. Jakobsen andY. Mikawa, *Appl. Spectrosc.*, **22**, 641 (1968).
6. D. O. Hummel, *Infrared Analysis of Polymers, Resins and Additives*, Vol. 1, Wiley-Interscience, New York, 1971.
7. S. M. Craven and F. F. Bentley, *Appl. Spectrosc.*, **26**, 449, 484 (1972).

D. Near Infrared Group Frequencies
  1. W. Kaye, *Spectrochim. Acta*, **6**, 257 (1954).
  2. R. F. Goddu and D. A. Delker, *Anal. Chem.*, **32**, 140 (1960).
  3. J. D. McCallum, "Chemical Analysis in the Near Infrared," application data sheet, UV–8085, Beckman Instruments, Inc., Fullerton, Ca. 92634.
  4. O. H. Wheeler, *Chem. Rev.*, **59**, 629 (1959).

STRUCTURE DETERMINATION

5-1 Introduction

The discussion in this chapter falls into two categories. First, we examine the deduction of the structure of an unknown compound from its infrared and Raman spectra, with the help of other information such as elemental analysis and molecular weight. In most cases, the nmr, uv, and mass spectra will also be needed to confirm or assist in the determination of the structure; of course, X-ray crystallography can give complete details of bond lengths, bond angles, and nonbonded internuclear distances. Second, we look at some examples of ways in which infrared and Raman spectra can give details of atomic arrangements in compounds of known gross structure. Information can be obtained on conformations, isomerism, tautomerism, orientation of polymer chains, hydrogen bonding, and crystal structure. The presence of impurities can be detected in the infrared and Raman spectra of a known compound. Studies of equilibria in solutions, the formation of complexes, and chemical kinetics are examples of further applications of vibrational spectroscopy.

It is assumed in what follows that the organic chemist has available a low or medium resolution infrared spectrometer and, when necessary, can obtain higher resolution infrared, far infrared, and Raman spectra. The first and most important thing to do is to ensure that the sample is reasonably pure, since even 5% of impurity could give rise to spurious peaks in the spectra. Gas chromatography will usually give an indication of the purity of a compound, but even here, caution must be exercised. As an example, a small amount of ethyl ether (molecular weight 74, bp 35°) present as an impurity in t-butylacetylene (molecular weight 70, bp 39°) cannot be readily detected by gas chromatography. However, a band due to the C—O—C stretching mode of the ether is observed at 1120 cm^{-1} in the infrared spectrum. This is normally the strongest band in the infrared spectra of ethers so that it has considerable intensity even when only a few percent are present as an impurity (Figure 5-1).

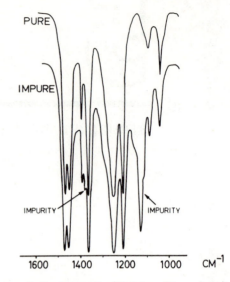

FIGURE 5-1. Part of the infrared spectra of thick films of two samples of t-butylacetylene. The presence of an impurity is clearly indicated by the strong band at 1120 cm^{-1} and the peak at 1380 cm^{-1}.

The preliminary infrared spectrum of a solid is recorded using mulls (Nujol or Fluorolube) or a KBr pellet; that of a liquid, as a film between two NaCl or KBr plates. When hydrogen bonding is evident from very broad bands, especially between 3500 and 2500 cm^{-1}, spectra are obtained from dilute solutions in CCl$_4$ or some other suitable solvent (see section 2-3e). For a detailed study, several spectra should be obtained from solutions of various concentrations, or from mulls or pellets containing various relative amounts of sample, or from liquid films of various thicknesses. The variation of concentrations, etc., should cover at least an order of magnitude, so that the weak features are brought out as well as the very strong absorptions.

When Raman spectra of pure liquids or solids are recorded, the problems of pathlength and concentration do not arise. For Raman spectra of solutions, however, the highest concentrations possible are used since the spectra are inherently weak and one does not have the option of increasing the pathlength.

The next step is to make a list of the main peaks in the infrared spectrum and to note any changes in frequency, intensity, or width of bands in going from pure compound to dilute solutions. Now we are ready to begin deduction of the structure.

Preliminary information such as analysis for C, H, N, and other elements and molecular weight, together with uv and nmr data, are gathered. A rapid examination of the infrared spectrum can reveal the type of X–H bonds present; the presence of triple bonds, conjugated double bonds, and carbonyl groups; and the presence of an aromatic ring. Various possible structures can then be suggested. Finally, the infrared, far infrared, and Raman spectra can be searched in detail for bands due to the various groups in the postulated structures. Comparison with published infrared spectra of compounds having the suggested or a similar structure will also be helpful.

5-2 Comparison of Prism and Grating Spectra

Most of the published work on infrared group frequencies is illustrated by spectra obtained using prism instruments. In earlier work, the phrase *rock salt region* was frequently used for the 2–15 μ wavelength region of the infrared spectrum. The appearance of spectra linear in wavenumber, obtained using grating instruments, is considerably different from prism spectra.

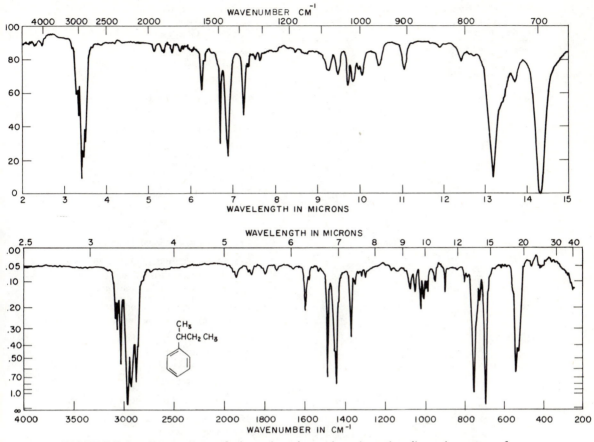

FIGURE 5-2. Comparison of the prism (upper) and grating (lower) spectra of sec-butylbenzene. [© Sadtler Research Laboratories, Inc. (1967).]

A comparative illustration is given in Figure 5-2 for *sec*-butylbenzene. The most striking features are that the center part of the spectrum is displaced to the right and the low frequency bands are much narrower in the grating spectrum. Closer examination shows that the actual appearance of the central part of the spectrum is essentially the same in both spectra. However, the high frequency region near 3000 cm^{-1} is better resolved and more spread out in the grating spectrum. These features must be borne in mind when making visual comparisons with literature spectra. As we shall see later, aromatic compounds have one or two very strong absorptions between 900 and 700 cm^{-1}. These bands are very prominent in prism spectra (Figure 5-2a) and give an immediate indication of the presence of a benzene ring. They are not quite so prominent in the grating spectrum (Figure 5-2b). This spectrum is discussed in section 5-6b.

5-3 Preliminary Analysis

The infrared spectrum is arbitrarily divided into several regions. The presence of bands in these regions gives immediate information. The absence of bands in certain regions is also important, since many groups can be excluded from further analysis, provided the factors

Table 5-1. *Regions of the Infrared Spectrum for Preliminary Analysis*

Region (cm^{-1})	Group	Possible Compounds and Remarks
3700–3100	—OH, —NH ≡C—H	alcohols, aldehydes, carboxylic acids amides, amines acetylenes
3100–3000	Ar—CH =CH$_2$ or —CH=CH—	aromatic compounds (may be weak) alkenes or unsaturated rings
3000–2800	—CH, —CH$_2$—, —CH$_3$	aliphatic groups
2800–2700	—CHO	aldehydes (Fermi doublet)
2700–2400	—POH —SH —PH	phosphorous oxyacids mercaptans phosphines
2400–2000	—C≡N —N—N≡N —C≡C—	nitriles azides acetylenes (may be weak or absent)
1870–1650	C=O	acid halides, aldehydes, amides, amino acids, anhydrides, carboxylic acids, esters, ketones, lactams, lactones, quinones
1650–1550	C=C, C=N, and NH	unsaturated aliphatics, aromatics, unsaturated heterocyclics, amides, amines, amino acids
1550–1200	NO$_2$, B—O, and B—N CH$_3$ and CH$_2$	nitro compounds, organoboron compounds alkanes, alkenes, etc.
1300–1000	C—O—C and C—OH S=O, P=O, C—F	ethers, alcohols, sugars sulfur, phosphorus and fluorine compounds
1100–800	Si—O and P—O	organosilicon and phosphorus compounds
1000–650	=C—H —NH	alkenes and aromatic compounds aliphatic amines
800–650	C—Cl and C—Br	may be below 650 cm^{-1}, especially C—Br

discussed in section 4-2 have been considered. A list of these regions, together with absorbing groups and possible types of compounds, is given in Table 5-1, which is a condensation of Table 4-2. A similar table could be drawn up for Raman spectra.

Certain types of compounds give strong, broad absorptions that are very prominent in the infrared spectrum. These very characteristic bands are noted in Table 5-2. It is worthwhile to look through one of the available collections of infrared spectra (see section A-1b) to familiarize oneself with these bands. The Aldrich Library of Infrared Spectra is particularly valuable for this purpose, since the spectra are arranged by class, eight to a page.

In the spectra of certain compounds there are weak, but characteristic bands that are known to be due to overtones or combinations. Some of these bands are listed in Table 5-3.

A preliminary exercise that is often useful is to determine the degree of unsaturation or number of rings in a molecule by noting the hydrogen deficiency.[1] The number of hydrogen atoms in a molecule is decreased by two from a completely saturated noncyclic structure for each double bond or ring present in the structure. When the molecule is a hydrocarbon, it is a simple matter to decide the degree of unsaturation. For example, C_7H_{12} has a hydrogen

[1]The term *index of hydrogen deficiency* is used in the text book *Organic Chemistry*, by J. B. Hendrickson, D. J. Cram, and G. S. Hammond, 3rd ed., McGraw-Hill Book Company, New York (1970).

Table 5-2. *Characteristic Broad Absorption Bands*

Band Center or Range (cm^{-1})	Classes of Compounds	Assignment and Remarks
3600–3200 (vs)	alcohols, phenols	OH stretch
3400–3000 (vs)	primary amides	usually a doublet
3400–2400 (s)	carboxylic acids and other compounds with OH groups	H-bonded OH stretch
3300–2200 (vs)	azoles (–NH– tautomers)	NH stretch similar to OH stretch of carboxylic acid dimers
3200–2400 (vs)	amino acids (zwitterion), amine hydrohalides, purines, and pyrimidines	very broad asymmetric band
3000–2800 (vs)	hydrocarbons, all compounds containing CH_3 and CH_2 groups	CH stretch
1700–1250 (s)	amino acids	broad region of absorption with much structure
1650–1500 (vs) 1400–1300 (s)	salts of carboxylic acids	two bands due to antisym. and sym. stretch of $-C\diagup\diagdown\raisebox{0.5ex}{O}\raisebox{-0.5ex}{O}$
ca. 1250 (vs)	perfluoro compounds	may cover the whole region 1400–1100 cm^{-1} with several bands
ca. 1200 (vs)	esters, phenols	ester linkage (not always broad) with much structure
ca. 1150 (vs)	sulfonic acids	with structure
1150–950 (vs)	sugars	with structure
ca. 1100 (vs)	ethers	C–O–C linkage
1100–900 (s)	diols	with structure
ca. 1050 (vs)	anhydrides	not always reliable
1050–950 (vs)	phosphites and phosphates	often two bands
ca. 920 (ms)	carboxylic acids	H-bonded C–OH deformation
ca. 830 (vs)	primary aliphatic amines	may cover the region 1000–700 cm^{-1}
ca. 730 (s)	secondary aliphatic amines	may cover the region 800–650 cm^{-1}

deficiency of four, which implies that two double bonds or rings are present. The compound could be methylcyclohexene, dimethylcyclopentene, etc., or a noncyclic diene. Oxygen atoms singly bonded to carbon atoms in a chain or ring cause no extra hydrogen deficiency, but a carbonyl group has the same effect as a double bond. Thus the compound $C_5H_8O_3$, with a hydrogen deficiency of four, most likely contains two carbonyl groups and one singly bonded C–O group.

When the molecule has two or more unsaturated centers, the presence of conjugation can be detected from the uv absorption spectrum. To decide whether the hydrogen deficiency is due to rings or double bonds, the compound can be hydrogenated. The number of moles of hydrogen taken up per mole of unsaturated compound indicates the number of C=C or C=N

Table 5-3. *Characteristic Overtone or Combination Bands*

Range (cm^{-1})	Classes of Compounds	Assignment or Remarks
ca. 3450	esters	overtone of C=O stretch
3100–3060	secondary amides	overtone of NH deformation
ca. 2700	cyclohexanes	combination of CH bending modes, enhanced by Fermi resonance
ca. 2700	aldehydes	part of a Fermi doublet at 2800–2700 cm^{-1}
2200–2000	amino acids and amine hydrohalides	combination of $\overset{+}{N}H_3$ torsion and $\overset{+}{N}H_3$ antisym. deformation
2000–1650	aromatic compounds	several bands (see Table 5–8)
1990–1960 and 1830–1800	vinyl compounds	high frequency band is stronger
ca. 1820	imidazoles	origin unknown
1800–1780	vinylidine compounds	overtone of CH_2 wag

double bonds.   Rings do not normally open under the reaction conditions used.   Each C≡C and C≡N bond will take up two moles of hydrogen per mole of compound, but the presence of these groups can easily be detected from the infrared or Raman spectra.

After these preliminary steps, the infrared spectrum is then examined in more detail. In the next few sections, some of the detailed structural information that can be obtained from such an examination are discussed.   In many cases, Raman and far infrared data will provide important additional information.

5-4   The CH Stretching Region (3340–2700 cm⁻¹)

5-4a   Introduction

Details of CH stretching frequencies in various compounds are summarized in Table 5-4. This is usually the first region that one examines carefully since certain structural features are immediately revealed by the position of the CH stretching bands.   A band at the high frequency end of the range indicates the presence of an acetylenic hydrogen atom, whereas a band near 2710 cm^{-1} usually means that there is an aldehyde group in the molecule.   Absorption above 3000 cm^{-1} indicates unsaturation, or small rings, while bands below 3000 cm^{-1} arise from saturated parts of a molecule.   We now examine this region in more detail.

5-4b   Acetylenes

A terminal ≡CH group gives rise to a sharp band between 3340 and 3270 cm^{-1}.   Usually this observation can be confirmed by a small sharp peak in the infrared near 2100 cm^{-1} due to the −C≡C− stretch (see section 5-9c).   An example is given in Figure 5-3.

5-4c   Aromatic Compounds

Aromatic compounds have one or more sharp peaks of weak or medium intensity between 3100 and 3000 cm^{-1}.   A word of warning here.   These bands may appear only as shoulders on a very strong CH_3/CH_2 stretching band, which may well cover the region from 3000 to 2800

Table 5–4. *CH Stretching Frequencies*

Range (cm^{-1}) and Intensity	Group or Class	Assignment and Remarks
3340–3270 (w–m)	≡CH	sharp band
3100–3000 (w–m)	aromatic compounds	several bands, often weak
3100–3070 (m)	cyclopropanes (CH_2 groups)	antisym. CH_2 stretch
3090–3070 (m–s)	=CH_2	vinylidine group
3080–3040 (m)	epoxides	sharp band
3035–2995 (m)	cyclopropanes (CH_2 groups)	sym. CH_2 stretch
3030–3000 (m–s)	=CHR	unsaturated aliphatic compounds
3000–2970 (m)	cyclobutanes (CH_2 groups)	antisym. CH_2 stretch
2970–2950 (vs)	–CH_3 group (in alkyl groups)	antisym. stretch
2960–2950 (vs)	cyclopentanes (CH_2 groups)	antisym. CH_2 stretch
2940–2915 (vs)	–CH_2– (in alkyl groups)	antisym. stretch
2925–2875 (vs)	cyclobutanes (CH_2 groups)	sym. CH_2 stretch
2885–2860 (vs)	–CH_3 group (in alkyl groups)	sym. stretch
2875–2855 (vs)	cyclobutanes (CHR groups) and cyclopentanes (CH_2 groups)	CH (R) stretch sym. CH_2 stretch
2870–2840 (vs)	–CH_2– group	sym. stretch
2850–2820 (m) and 2750–2720 (m)	aromatic aldehydes	Fermi doublet
2835–2815 (m–s)	CH_3–O–	methoxy group
2830–2810 (m) and 2725–2700 (m)	aliphatic aldehydes	Fermi doublet

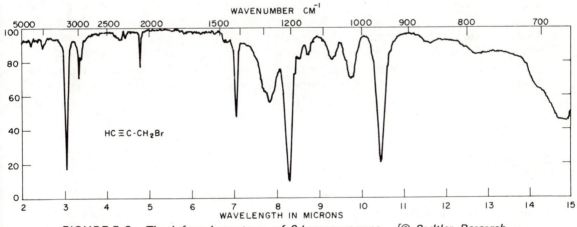

FIGURE 5–3. The infrared spectrum of 3-bromopropyne.   [© Sadtler Research Laboratories, Inc. (1965).]

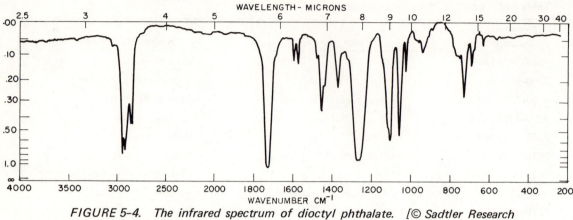

FIGURE 5-4. *The infrared spectrum of dioctyl phthalate.* [© *Sadtler Research Laboratories, Inc. (1968).*]

cm^{-1}. This problem is illustrated by the spectrum of dioctyl phthalate shown in Figure 5-4 and discussed in section 5-6a.

5-4d Unsaturated Nonaromatic Compounds

Other unsaturated compounds and three-membered ring compounds show absorption due to CH stretching in the 3100-3000 cm^{-1} region. Compounds containing the vinylidine ($=CH_2$) group absorb near 3080 cm^{-1}; di- and trisubstituted ethylenes absorb at lower frequencies nearer to 3000 cm^{-1} and the band may be overlapped by the CH_3 absorption. Cyclopropane derivatives have a band between 3100 and 3070 cm^{-1}, while epoxides absorb between 3060 and 2990 cm^{-1}.

5-4e Saturated Groups: CH_3, CH_2, and CH

Saturated compounds can have methyl, methylene, or methine groups, each of which has characteristic CH stretching frequencies. CH_3 groups absorb near 2960 and 2870 cm^{-1} and the CH_2 bands are at 2930 and 2850 cm^{-1}. In many cases, only one band in the 2870-2850 cm^{-1} region can be resolved when both CH_3 and CH_2 groups are present in the molecule. Figure 5-5 shows the CH stretching region of high resolution spectra of three normal paraffin molecules. As the carbon chain becomes longer, the CH_2 group bands increase in intensity relative to the CH_3 group absorptions. The doublet band at 2960 cm^{-1} is the antisymmetric CH stretching mode, which would be degenerate in a free CH_3 group, or in a molecule in which the C_{3v} symmetry of the group is maintained, e.g., CH_3Cl. The degeneracy is removed in the paraffin molecules, and consequently two individual antisymmetric CH stretching bands are observed.

5-4f Aldehydes

The CH stretching mode of the aldehyde group appears as a Fermi doublet near 2820 and 2710 cm^{-1} (see section 3-8). The 2710 cm^{-1} absorption is very useful and characteristic of aldehydes, but the higher frequency component often appears as a shoulder on the 2850 cm^{-1} CH_2 band. The frequencies of the doublet are about 20 cm^{-1} higher in aromatic aldehydes, with the high frequency component stronger and broader. Of course, when methyl or methylene groups are also present, it is again the low frequency band that identifies the compound as an aldehyde.

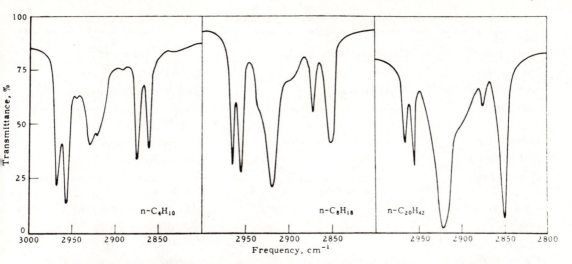

FIGURE 5-5.   The C–H stretching region of n-C_4H_{10}, n-C_8H_{18}, and n-$C_{20}H_{42}$.
[Reproduced with permission from R. G. Snyder and J. H. Schachtschneider,
Spectrochim. Acta, **19**, 115 (1963).]

5-5   The Carbonyl Stretching Region

5-5a   General

If we were to include metal carbonyls and salts of carboxylic acids, the region of the C=O
stretching mode would extend from 2200 down to 1350 cm^{-1}. Most organic compounds
containing the C=O group, however, show very strong infrared absorption in the range
1870–1650 cm^{-1}. The actual position of the peak (or peaks) within this range is characteristic
of the type compound. At the upper end of the range are found bands due to anhydrides
(two bands) and four-membered ring lactones (β-lactones), whereas amides and substituted ureas
absorb at the lower end of the range. Table 5–5 summarizes the ranges in which compounds
with a single carbonyl group absorb. Table 5–6 lists compounds that have two interacting
carbonyl groups. In these latter compounds, out-of-phase and in-phase (antisymmetric and
symmetric) vibrations can occur with the out-of-phase mode at higher frequencies.

5-5b   Compounds Containing a Single C=O Group

The type of functional group usually cannot be identified from the C=O stretching
band alone, since there can be several carbonyl-containing functional groups that absorb
within a given frequency range. Nevertheless, an initial separation into possible compounds
can be achieved using Tables 5–5 and 5–6. Other parts of the spectrum must then be consulted
together with other information to reduce the possibilities further. For example, a single
carbonyl peak in the region 1750–1700 cm^{-1} could indicate an ester, an aldehyde, a ketone
(including cyclic ketones), a large ring lactone, a urethane derivative, an α-haloketone,
or an α-halocarboxylic acid. The presence of the halogen could be checked from the
elemental analysis, an aldehyde would be identified by a peak near 2700 cm^{-1}, esters and
lactones give a strong band near 1200 cm^{-1} that is often quite broad in esters. Urethanes
would have an N–H stretching band.

Table 5-5. *Carbonyl Stretching Frequencies for Compounds Having One Carbonyl Group*

Range (cm^{-1}) and Intensity	Classes of Compounds	Remarks
1840–1820 (vs)	β-lactones	Four-membered ring
1810–1790 (vs)	acid chlorides	saturated aliphatic compounds
1800–1760 (vs)	aromatic and unsaturated esters	the C=C stretch is higher than normal (1700–1650 cm^{-1})
1800–1740 (s)	carboxylic acid monomer	only observed in dil soln
1790–1740 (vs)	γ-lactones	five-membered ring
1790–1760 (vs)	aromatic or unsaturated acid chlorides	second weaker combination band near 1740 cm^{-1}
1780–1700 (s)	lactams	position depends on ring size
1770–1745 (vs)	α-halo esters	higher frequency due to electronegative halogen
1750–1740 (vs)	cyclopentanones	unconjugated structure
1750–1730 (vs)	esters and δ-lactones	aliphatic compounds
1750–1700 (s)	urethanes	R–O–(C=O)–NHR compounds
1745–1730 (vs)	α-haloketones	noncyclic
1740–1730 (vs)	phenyl esters	esters of aromatic carboxylic acids
1740–1720 (vs)	α-halocarboxylic acids	20 cm^{-1} higher frequency if halogen is fluorine
1740–1720 (vs)	aldehydes	aliphatic compounds
1730–1700 (vs)	ketones	aliphatic and large ring alicyclic
1720–1680 (vs)	aromatic aldehydes	also α, β-unsaturated aliphatic aldehydes
1720–1680 (vs)	carboxylic acid dimer	broader band
1710–1640 (s)	thiol esters and aromatic esters	lower than normal esters
1700–1680 (vs)	aromatic ketones	position affected by substituents on ring
1700–1650 (vs)	conjugated ketones	check C=C stretch
1700–1680 (vs)	aromatic carboxylic acids	dimer band
1700–1670 (s)	primary and secondary amides	in dilute solution
1690–1660 (vs)	quinones	position affected by substituents on ring
1680–1630 (vs)	amides (solid state)	note second peak due to N–H deformation near 1625 cm^{-1}
1670–1660 (s)	diaryl ketones	position affected by substituents on ring
1670–1640 (s)	ureas	second peak due to N–H deformation near 1590 cm^{-1}
1670–1630 (vs)	ortho OH or NH$_2$ aromatic ketones	frequency has been lowered by chelation with ortho group

Table 5-6.   *Carbonyl Stretching Frequencies for Compounds Having Two Interacting Carbonyl Groups*

Ranges (cm⁻¹) and Intensity	Classes of Compounds	Remarks
1870–1840 (m–s) 1800–1770 (vs)	cyclic anhydrides	low frequency band is stronger
1825–1815 (vs) 1755–1745 (s)	acyclic anhydrides	high frequency band is stronger
1780–1760 (m) 1720–1700 (vs)	imides	low frequency band is broad, high frequency band may be obscured
1760–1740 (vs)	α-keto esters	usually only one band
1740–1730 (vs)	β-keto esters (keto form)	may be a doublet due to two C=O groups
1660–1640 (vs)	β-keto esters (enol form)	may be a doublet due to a C=O and a C=C group
1710–1690 (vs) 1640–1540 (vs)	diketones (keto form) diketones (enol form)	
1690–1660 (vs)	quinones	frequency depends on substituents
1650–1550 (vs) 1440–1350 (s)	carboxylic acid salts	broad bands

5-5c   Compounds Containing Two C=O Groups

Compounds with two coupled carbonyl groups are fewer in number, but again ambiguities can exist and other parts of the infrared spectrum must be analyzed in conjunction with the positions of the carbonyl bands in order for a particular structural grouping to be identified. An example of this type of problem is provided by the spectrum shown in Figure 5-6. An elemental analysis indicates that only C, H, and O are present and the formula is $C_5H_8O_3$. Looking at the CH stretching region (Table 5-4), we conclude that there is no unsaturated group present. This conclusion is confirmed by the absence of absorption between 1670 and 1540 cm⁻¹. A CH₃ group or groups is present (2995 cm⁻¹) and possibly a CH₂ group (2940 cm⁻¹). Now, turning to the carbonyl stretching region, we note from Tables 5-5 and 5-6 that the compound could be a cyclic ketone, an ester, a lactone, an aldehyde, or an

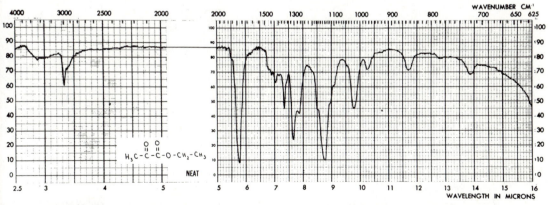

FIGURE 5-6.   *The infrared spectrum of an α-keto ester, ethyl pyruvate. [Courtesy of Aldrich Chemical Company, Inc.]*

α-keto ester. We can eliminate aldehyde (no band near 2700 cm^{-1}) and lactones or cyclic ketones (not possible with three oxygens and a methyl group). The band at 1150 cm^{-1} is due to an ester linkage. The compound is then most likely an α-keto ester. Two possible structures are **1** and **2**. Structure **2** is more likely because an O—CH$_3$ group would give a

band near 2830 cm^{-1}. This example is a case where a higher resolution spectrum than that of Figure 5–6 is desirable for the CH stretching region.

5-6 Aromatic Compounds

5-6a General

As previously stated in the discussion of the CH stretching region, absorption in the 3100–3000 cm^{-1} range gives an indication of the presence of benzene rings. However, this absorption may sometimes be weak. An example of a very weak aromatic CH stretching band is shown in Figure 5–4. A grating spectrum is shown because it gives better resolution in the CH stretching region. The weak absorption at 3050 cm^{-1} is an aromatic CH band. The strong doublet at 2940 and 2910 cm^{-1} indicates CH$_3$ and CH$_2$ groups, as does the 2850/2840 cm^{-1} doublet. Neither doublet is resolved in the prism spectrum.

Confirmation that the very small peak at 3050 cm^{-1} is due to a fundamental is found in the Raman spectrum shown in Figure 5–7. The sharp doublet at 1595/1575 cm^{-1} (Figure 5–4) and the shoulder at 1480 cm^{-1}, together with the fairly strong band at 730 cm^{-1}, provide further evidence for an aromatic compound. The 730 cm^{-1} band stands out in the prism spectrum and is indicative of ortho disubstitution. Benzene derivatives also have several sharp bands (often weak) between 1275 and 1000 cm^{-1}. The present spectrum exhibits such peaks, but there are many other groups that could also absorb between 1300 and 1000 cm^{-1}, so the evidence is not convincing in this region. Part of a spectrum of a thick film of the com-

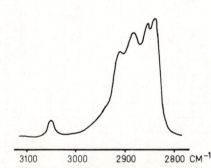

FIGURE 5–7. Part of the Raman spectrum of dioctyl phthalate.

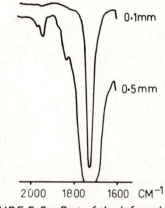

FIGURE 5–8. Part of the infrared spectrum of a thick film of dioctyl phthalate.

pound is shown in Figure 5-8. This spectrum reveals some very weak absorptions between 2000 and 1800 cm^{-1}, which are also indicative of a 1,2-disubstituted benzene derivative.

We see, then, that there are four regions in addition to the CH stretching region where bands characteristic of a substituted benzene ring occur. There are usually two sharp bands near 1600 and 1500 cm^{-1} in most benzene derivatives. These may be weak or hidden in carbonyl compounds or in amines, for which very strong C=O stretching and NH deformation bands are nearby. The 1600 cm^{-1} band is sometimes a doublet in condensed ring systems or when there is a substituent that can conjugate with the ring. This region gives no indication of the kind of substitution on the ring. However, the 2000–1700 cm^{-1} and 900–690 cm^{-1} regions contain absorption bands that are very characteristic of the substitution pattern. The 1275–1000 cm^{-1} region does not have great utility.

5-6b Substituted Aromatic Compounds

The monosubstitution pattern is easily recognized by two very strong bands near 750 and 700 cm^{-1}. These bands are most striking in prism spectra, as illustrated in Figure 5-2, which will now be recognized as a monosubstituted benzene derivative. In addition, monosubstitution is indicated by four weak but clear absorptions between 2000 and 1700 cm^{-1}. Again, Figure 5-2 illustrates this pattern. By using thicker solutions the bands in this region can be very useful in deciding the type of substitution. In the 900–690 cm^{-1} region the more highly substituted compounds have bands at the highest frequencies. The characteristic bands of substituted aromatic compounds are summarized in Table 5-7.

The substitution pattern in the 2000–1700 cm^{-1} region is summarized in Table 5-8. The spectra *must* be recorded from thick films or rather concentrated samples since the bands in this region are due to combinations or overtones. After consulting Table 5-8, one can then go back to Table 5-7 and look for confirmation in the 900–690 cm^{-1} region.

5-6c Raman Spectra of Aromatic Compounds

The Raman spectra of aromatic compounds also contain several characteristic bands that can be very useful in cases where the infrared spectrum gives ambiguous results. The most prominent band in monosubstituted benzene derivatives, found near 1000 cm^{-1}, is due to a ring-breathing vibration. All aromatic compounds show a strong Raman line near 1600 cm^{-1} that is usually seen in the infrared, but may be weak or obscured. Monosubstituted benzene rings also have a Raman line of medium intensity between 1030 and 1010 cm^{-1} due to in-plane CH bending and a weak depolarized line between 625 and 605 cm^{-1} due to an in-plane ring-bending mode.

A few monosubstituted benzene derivatives such as nitrobenzene and benzoyl chloride are hard to characterize from their infrared spectra, but the Raman spectra give a clear indication of the structure of these compounds.

5-7 Compounds Containing Methyl Groups

5-7a General

In section 3-10 the vibrations of a methyl group are discussed in terms of the C_{3v} symmetry of the group. These vibrations (stretching, bending or deformation, rocking, and torsion) give rise to infrared absorption and Raman scattering in four regions. The frequencies of CH stretching vibrations of methyl groups in various environments are discussed in section 5-4.

Table 5-7. *Absorption Bands Characteristic of Substitution in Benzene Rings*

Type of Substitution	Range (cm^{-1}) and Intensity	Remarks
monosubstitution	770–730 (vs)	very characteristic
	710–690 (s)	
	2000–1700 (w)	four prominent bands in thicker samples
orthodisubstitution	770–730 (vs)	a single strong band
	1950–1650 (w)	two prominent bands in thick samples near 1900 and 1780 cm^{-1} with two weaker bands near 1680 and 1940 cm^{-1}
metadisubstitution	810–750 (vs)	higher frequency than monosubstitution
	725–680 (m–s)	more variable in position and intensity than monosubstitution
	1930–1740 (w)	three prominent bands in thick samples; the lowest frequency band may be broader with some structure
paradisubstitution	860–800 (vs)	a single strong band
	1900–1750 (w)	two bands; the higher frequency one is usually stronger
1,3,5-trisubst.	865–810 (s)	two bands with wider separation than mono- or *m*-disubstitution
	765–730 (s)	
	1800–1700 (w)	one fairly broad band with a much weaker band near 1900 cm^{-1}
1,2,3-trisubst.	780–760 (s)	similar to *m*-disubstituted
	745–705 (s)	
	2000–1700 (w)	similar to monosubstituted, but only three bands
1,2,4-trisubst.	885–870 (s)	two bands at higher frequencies than mono-, *m*-di-, and the other trisubst. compounds
	825–805 (s)	
	1900–1700 (w)	two prominent bands near 1880 and 1740 cm^{-1} with a much weaker one between
tetrasubst. (three isomers)	870–800 (s)	a single band in this region might be confused with *p*-disubst.
	1900–1700 (w)	a single prominent band with a much weaker one near 1950 (1,2,3,5-subst.), 1790 (1,2,4,5-subst.), or 1730 cm^{-1} (1,2,3,4-subst.)
pentasubst.	900–850 (s)	variable position and intensity
	2000–1700 (w)	very similar to 1,2,3,5-tetrasubst.
hexasubst.	–	no band in the 900–700 cm^{-1} region
	2000–1700 (w)	very similar to 1,2,3,5-tetrasubst.

Antisymmetric deformation of the HCH angles of a CH_3 group gives rise to very strong infrared absorption and Raman scattering in the 1470-1440 cm^{-1} region. Unfortunately, methylene ($-CH_2-$) groups also give rise to a band in the same region. However, the symmetric CH_3 deformation gives a strong band between 1380 and 1360 cm^{-1}, which appears as a doublet or multiplet when more than one CH_3 group is present. This multiplicity provides a good indication of the presence of isopropyl or *t*-butyl groups.

The CH_3 rocking vibrations are usually coupled with skeletal modes and may be found anywhere between 1250 and 800 cm^{-1}. Medium to strong bands in both infrared and Raman

Table 5-8. *Summary of the Substitution Patterns in the 2000–1700 cm^{-1} Region of Aromatic Compounds*

Number of Bands	Type of Substitution	Remarks
4	mono-	equally spaced, similar intensities
3	1,2,3-tri-	equally spaced, similar intensities
3	meta di-	the third band near 1750 cm^{-1} is usually broader and may have shoulders
2+	ortho di-	two main bands with several other weaker peaks or shoulders
2	1,2,4-tri-	there may be a much weaker band between the two main bands
1	para di	near 1900 cm^{-1}
	1,3,5-tri-	near 1750 cm^{-1} (broader)
	1,2,3,4- and 1,2,4,5-tetra, penta, and hexa-	near 1730 cm^{-1}
	1,2,3,4-tetra-	near 1860 cm^{-1}

spectra may be observed, but these are of little use for structure determination. The methyl torsion vibration has a frequency between 250 and 100 cm^{-1}, but is often not observed in either infrared or Raman spectra. In cases where torsional frequencies can be observed or estimated, information on rotational isomerism, conformation, and barriers to internal rotation can be obtained.[2]

5-7b   Isopropyl and *t*-Butyl Groups

Since isopropyl and *t*-butyl groups in hydrocarbon chains give characteristic bands in their spectra, it might be possible to identify these structural components in some cases. Table 5–9 gives typical frequencies for these groups. The ethyl group is not usually distinguishable.

Table 5-9. *Typical Frequencies for Isopropyl and* t-*Butyl Group Vibrations*

Vibration	Isopropyl			t-Butyl		
	cm^{-1}	ir	R	cm^{-1}	ir	R
sym. CH$_3$ deformations	1385	vs	w	1395	m	w
	1370	vs	w	1370	s	w
CH$_3$ rocking coupled with skeletal modes	–	–	–	1260–1250	m–s	m
	–	–	–	1220–1200	m–s	m
	1150–1140	s	m	–	–	–
	960–950	m	m	–	–	–
	930–910	m–s	m	930–925	m	m
	835–795	m	s	750–710	w	s pol.

When a methyl group is attached to an atom other than carbon, there is a significant shift in the symmetric CH$_3$ deformation frequency. Table 5–10 lists some typical ranges.

[2]A recent issue of the *Journal of Molecular Structure* was devoted to studies of these and related topics: *J. Mol. Struct.*, **6**, 1–84 (1970).

Table 5-10. *Frequencies of the Symmetric CH_3 Deformation in Various Compounds*

Compounds	Group	Range (cm^{-1})
esters, ethers, etc.	$O-CH_3$	1460–1430
amines, amides	$N-CH_3$	1440–1410
hydrocarbons	$C-CH_3$	1400–1360
sulfoxides, thioethers, etc.	$S-CH_3$	1330–1290
phosphines	$P-CH_3$	1310–1280
silanes	$Si-CH_3$	1280–1250
arsines	$As-CH_3$	1260–1240
germanes	$Ge-CH_3$	1240–1220
organomercury compounds	$Hg-CH_3$	1210–1190
organotin compounds	$Sn-CH_3$	1200–1180
organolead compounds	$Pb-CH_3$	1170–1150

5-8 Compounds Containing Methylene Groups

5-8a Introduction

There are two kinds of methylene groups, the $-CH_2-$ group in a saturated chain and the terminal $=CH_2$ group in vinyl, allyl, or vinylidene compounds. Diagrams of stretching, bending, wagging, twisting, and rocking motions of a CH_2 group are given in Figure 3-18. The CH stretching vibrations are covered in section 5-4. Bending, wagging, and rocking modes also give rise to important group frequencies.

5-8b CH_2 Bending (Scissoring)

The bending or scissoring motion of saturated $-CH_2-$ groups gives a band of medium to strong intensity between 1480 and 1440 cm^{-1}. The band is often obscured by the even stronger CH_3 absorption in this region. When the $-CH_2-$ group is adjacent to a carbonyl or nitro group, the frequency is lowered to 1430–1420 cm^{-1}.

5-8c CH_2 Bending in Vinyl Groups

A vinyl $=CH_2$ group gives a band of medium intensity between 1420 and 1410 cm^{-1}. This band is sometimes assigned as an in-plane deformation, since the two hydrogen atoms are in the same plane as the C=C group.

5-8d CH_2 Wagging and Twisting

The $-CH_2-$ wagging and twisting frequencies in saturated groups are observed between 1350 and 1150 cm^{-1}. The infrared bands are weak unless an electronegative atom such as a halogen is attached to the same carbon atom. The CH_2 twisting modes occur at the lower end of the frequency range and give very weak infrared absorption.

5-8e CH_2 Rocking

A very useful band is observed near 725 cm^{-1} in the infrared spectrum when there are four or more $-CH_2-$ groups in a chain. The intensity increases with increasing chain length. This frequency is assigned to CH_2 rocking. One should be careful in aromatic compounds since the out-of-plane bending of the ring hydrogens gives rise to bands in this region for some substitution patterns (see Table 5-7 and Figure 5-4).

5–8f CH$_2$ Wagging in Vinyl and Vinylidene Compounds

The CH$_2$ wagging modes in vinyl and vinylidene compounds are found at much lower frequencies than in saturated groups. In vinyl compounds, a strong band is observed in the infrared between 910 and 900 cm^{-1}. For vinylidene compounds the frequency range is 10 cm^{-1} lower. The overtone of the CH$_2$ wag often can be clearly seen as a band of medium intensity near 1820 cm^{-1} for vinyl and 1780 cm^{-1} for vinylidene compounds. These frequencies are raised above the normal range by halogens or other functional groups on the α-carbon atom. More will be said about vinyl and vinylidene groups in the next section.

5–8g Relative Numbers of CH$_2$ and CH$_3$ Groups

One further useful observation can be made concerning the H$-$C$-$H deformation bands in saturated parts of a molecule. When there are more $-$CH$_2-$ groups than $-$CH$_3$ groups present, the 1480–1440 cm^{-1} band will be stronger than the 1380–1360 cm^{-1} band (sym. CH$_3$ deformation). The relative intensities of these two bands, together with the 725 cm^{-1} band and the bands in the CH stretching region (see Figure 5-5), can give information on the relative numbers of $-$CH$_2-$ and CH$_3$ groups as well as the unsaturated carbon chain length. If the 1480–1440 cm^{-1} and 1380–1360 cm^{-1} bands are of comparable intensity, it can be concluded that there are two or fewer $-$CH$_2-$ groups in the molecule.

5–9   Unsaturated Compounds

5–9a   The C=C Stretching Mode

Stretching of the C=C bond gives rise to infrared and Raman bands in the region 1690–1560 cm^{-1}. The band is often very weak in the infrared and sometimes is not observed at all. Other weak absorptions may occur in the region and could be mistaken for a C=C stretching mode. An example was given in section 4-2a (see Figure 4-1). In such cases a Raman spectrum will confirm the presence or absence of a C=C bond because this group always gives a strong Raman line in the 1690–1560 cm^{-1} region (see Figure 4-2).

The exact frequency of the C=C stretching mode gives some additional information on the environment of the double bond. Tri- or tetraalkyl-substituted groups and trans-disubstituted alkenes have frequencies in the 1690–1660 cm^{-1} range. These bands are weak or absent in the infrared, but strong in the Raman. Vinyl and vinylidene compounds as well as cis alkenes absorb between 1660 and 1630 cm^{-1}. Fluorinated alkenes have very high C=C stretching frequencies (1800–1730 cm^{-1}). Chlorine and other heavy substituents, on the other hand, usually lower the frequency.

5–9b Cyclic Compounds

Cyclic compounds have various C=C stretching frequencies depending on ring size and substitution. Cyclobutene, for example, has its C=C stretching mode at 1565 cm^{-1}. This frequency is increased by substitution. Cyclopentene and cyclohexene have C=C stretching frequencies of 1610 and 1645 cm^{-1}, respectively. These frequencies are also increased by substitution. Bridged rings with unsubstituted double bonds also have low C=C stretching frequencies.

5-9c   The C≡C Stretching Mode

A small sharp peak due to C≡C stretching is observed in the infrared spectra of terminal acetylenes near 2100 cm^{-1} (see Figure 5-3). In substituted acetylenes the band is shifted 100-150 cm^{-1} to higher frequencies. If the substitution is symmetric, no band will be observed in the infrared (see section 4-2a). Even when the substitution is unsymmetric, the band may be very weak in the infrared and could be missed. In such cases a Raman spectrum is very valuable, since the C≡C stretching mode always gives a strong line.

5-9d   =CH and =CH₂ Bending Modes

The CH and CH_2 wagging or out-of-plane bending modes are very important for structure identification in unsaturated compounds. They occur in the region between 1000 and 650 cm^{-1}. The trans CH wagging of a vinyl group gives rise to a strong infrared band between 1000 and 980 cm^{-1}. A trans-disubstituted alkene group is characterized by a very strong band in the frequency range 980-950 cm^{-1}. Electronegative substituents tend to lower this frequency. The cis-disubstituted alkenes give a medium to strong, but less reliable band between 750 and 650 cm^{-1}.

The CH_2 out-of-plane wagging vibration of vinyl and vinylidene compounds gives a strong band between 910 and 890 cm^{-1}. This band, coupled with the trans CH wagging, gives a very characteristic doublet and makes vinyl compounds readily identifiable. The overtone of the CH_2 out-of-plane wagging mode is unusually strong near 1800 cm^{-1}. This band provides a useful confirmation of the structural grouping.

A trisubstituted alkene usually shows a band of medium intensity between 830 and 780 cm^{-1}. This band, however, may be obscured by other absorptions in this region.

The cis hydrogen atoms on a double bond in a cyclic structure produce a strong band between 700 and 650 cm^{-1}. An example of this type of compound is given in Figure 5-9. Other cyclic unsaturated compounds have characteristic bands in the 750-650 cm^{-1} region. There are always several bands of medium intensity between 1200 and 800 cm^{-1} in the infrared spectra of cycloalkanes and cycloalkenes due to $-CH_2-$ rocking modes.

A summary of the out-of-plane CH bending or wagging modes in alkenic compounds is given in Table 5-11.

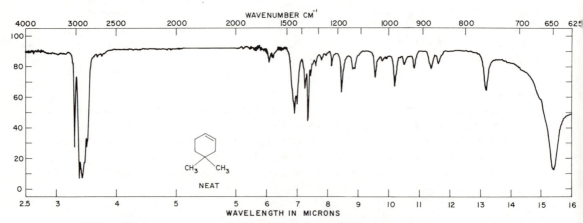

FIGURE 5-9. The infrared spectrum of a cyclic unsaturated compound, 4,4-dimethyl-1-cyclohexene, with cis hydrogen atoms on the double bonds. [Courtesy of Aldrich Chemical Company, Inc.]

Table 5-11. *CH$_2$ Wagging and CH Out-of-Plane Bending Frequencies of Alkenic Compounds in the Infrared*

Range (cm^{-1}) and Intensity	Group or Class	Assignments and Remarks
1000–980 (s)	vinyl group	trans CH wagging, lower in vinyl esters
980–950 (vs)	disubstituted	trans isomers, frequency lowered by halogen
920–900 (s)	vinyl group	CH$_2$ out-of-plane wagging, higher in vinyl ketones or acrylates, but lower in vinyl esters
900–880 (s)	vinylidene group	terminal =CH$_2$ out-of-plane wagging
830–780 (m–s)	trisubstituted	may be weak or obscured
750–650 (m–s)	disubstituted and cyclic compounds	cis CH wagging

5-10 Compounds Containing C—O—C, C—O—H, or N—O—H Groups

5-10a General

Carboxylic acids and anhydrides, alcohols, phenols, sugars, and carbohydrates all give strong, often broad, infrared absorption bands somewhere between 1400 and 900 cm^{-1}. In oximes the band is at higher frequencies between 1500 and 1400 cm^{-1}. These bands are associated with stretching of the C—O—C or C—OH bonds or bending of the C—O—H or N—O—H group. The position and multiplicity of the absorption together with evidence from other regions of the spectrum can help to distinguish the particular functional group. The usual frequency ranges for these groups in various compounds are summarized in Table 5-12.

5-10b Ethers

The simplest structure with the C—O—C link is the ether group. Aliphatic ethers absorb near 1100 cm^{-1}, whereas alkyl aryl ethers have a very strong band between 1280 and 1220 cm^{-1} and another strong band between 1050 and 1000 cm^{-1}. In vinyl ethers the C—O—C stretching mode is shifted to about 1200 cm^{-1}. Vinyl ethers can be further distinguished by a very strong C=C stretching band and the two out-of-plane CH deformation vibrations that are observed at frequencies below the usual ranges near 960 and 820 cm^{-1}. Examples of these three types of ether are compared in Figure 5-10. Cyclic saturated ethers such as tetrahydrofuran have a strong antisymmetric C—O—C stretching band in the range 1250–1150 cm^{-1}, while unsaturated cyclic ethers have their C—O—C stretching modes at lower frequencies.

5-10c Alcohols

In simple alcohols, the type of substitution on the C—OH carbon atom can often be learned from the frequency of the C—OH stretching band (Table 5-12). Sugars and carbohydrates give very broad absorption centered in the 1150–950 cm^{-1} region. Phenols absorb near 1350 cm^{-1}, a band due to the C$_6$H$_5$—OH deformation and give a second broader, stronger band centered near 1200 cm^{-1}. This second band always has fine structure because of underlying aromatic CH in-plane deformation vibrations.

5-10d Carboxylic Acids and Anhydrides

Carboxylic acids exist mostly as dimers. There are three vibrations associated with the C—OH group in these compounds: a band of medium intensity near 1430 cm^{-1}, a stronger

Table 5-12. *C—O—C and C—O—H Group Vibrations*

Range (cm^{-1}) and Intensity	Classes of Compounds	Assignments and Remarks
1440–1400 (m)	carboxylic acids	C—O—H deformation may be obscured by CH$_3$ and CH$_2$ deformation bands
1430–1280 (m)	alcohols	C—O—H deformation gives a broad band under CH$_3$ and CH$_2$ deformation bands
1390–1310 (m–s)	phenols	C—O—H deformation
1340–1160 (vs)	phenols	C—O stretch gives a broad band with structure
1310–1250 (vs)	aromatic esters	C—O—C antisym. stretch
1300–1200 (s)	aromatic carboxylic acids	C—O stretch
1300–1100 (vs)	aliphatic esters	see below for more specific frequency ranges
1280–1220 (vs)	alkyl aryl ethers	aryl C—O stretch
1270–1200 (s)	vinyl ethers	a second band near 1050 cm^{-1}
1265–1245 (vs)	acetate esters	C—O—C antisym. stretch
1250–900 (s)	cyclic ethers	position varies with compound
1230–1000 (s)	alcohols	broad band; see below for more specific frequency ranges
1200–1180 (vs)	formate and propionate esters	C—O—C stretch
1180–1150 (m)	alkyl phenols	C—O stretch
1150–1050 (vs)	aliphatic ethers	usually centered near 1100 cm^{-1}
1150–1130	tertiary alcohols	lowered by chain branching or adjacent unsaturated groups
1110–1090 (s)	secondary alcohols	lowered 10–20 cm^{-1} by chain branching
1060–1030 (s)	primary alcohols	lowered by unsaturation
1060–1020 (s)	saturated cyclic alcohols	not cyclic propanols or cyclic butanols
1050–1000 (s)	aromatic ethers	alkyl C—O stretch
960–900 (m–s)	carboxylic acids	C—O—H deformation of dimer

band near 1240 cm^{-1}, and another band of medium intensity near 930 cm^{-1}. The presence of an anhydride functionality is detected by the very characteristic absorption in the C=O stretching region (a doublet with one band at unusually high frequency). The C—O—C stretch gives rise to a band near 1190 cm^{-1} in open chain compounds and at higher frequencies in cyclic anhydrides.

5-10e Esters

Esters are distinguished from other compounds containing C—O—C linkages by the very strong and often broad band centered near 1200 cm^{-1}. The actual frequency of the maximum

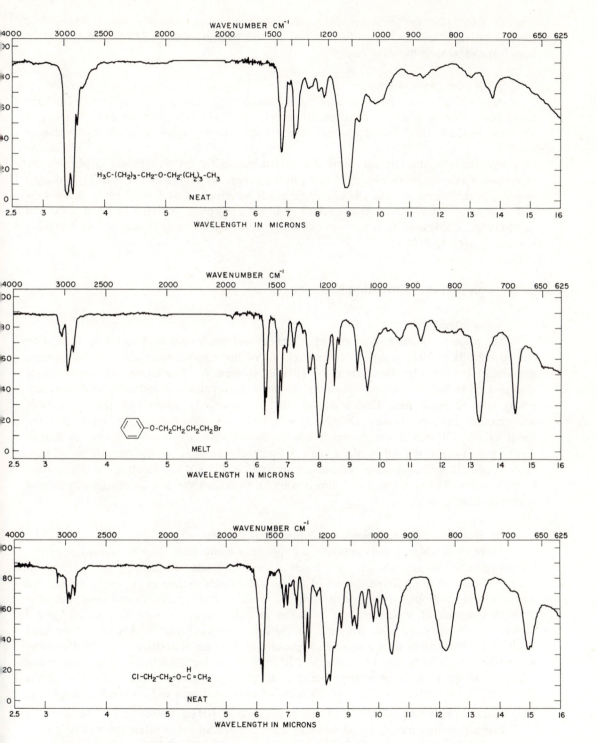

FIGURE 5-10. The infrared spectra of three types of ethers: (top) a simple aliphatic ether, (middle) an alkyl aryl ether, and (bottom) a vinyl ether. [Courtesy of Aldrich Chemical Company, Inc.]

can vary from 1290 cm^{-1} in benzoate and dialkyl carbonate esters down to 1180 cm^{-1} in formate esters. There will often be structure on this band because of CH deformation vibrations, which absorb in the same region.

5-10f OH Stretching Bands

Alcohols, phenols, carboxylic acids, and oximes in the pure liquid or solid states all have broad bands due to hydrogen bonded OH stretching. Table 5-2 lists some frequency ranges for these bands in the infrared spectra. These bands are very weak or absent in the Raman spectra.

For alcohols, the OH stretching is centered near 3300 cm^{-1}, whereas in phenols the absorption maximum is 50–100 cm^{-1} lower in frequency. In dilute solution in CCl$_4$, hydrogen bonding is eliminated and the OH stretching bands are observed near 3650 cm^{-1} in alcohols and 3600 cm^{-1} in phenols. Oximes show a broad absorption centered between 3300 and 3150 cm^{-1}. Carboxylic acids have an extremely broad band due to H-bonded OH stretching that may extend to 2400 cm^{-1}.

5-11 Compounds Containing Nitrogen

5-11a General

The presence of primary or secondary amines and amides can be detected by absorption due to the NH or NH$_2$ groups in several regions of the infrared spectrum. Tertiary amines and amides on the other hand are more difficult to identify. Nitriles and nitro compounds also give characteristic infrared absorption bands. Isocyanates and carbodiimides have very strong infrared bands near 2260 and 2140 cm^{-1}, respectively, where very few absorptions due to other groupings occur. Oximes, imines, and azo compounds give weak infrared bands in the 1700–1600 cm^{-1} region due to the –C=N– or –N=N– group. A Raman spectrum is usually needed in these cases. Some characteristic group frequencies of non-heterocyclic nitrogen-containing compounds are listed in Table 5–13. Much of the information is duplicated in Tables 4–1 and 4–2, but it is useful to collect the data on nitrogen-containing groups in a single place.

5-11b Amino Acids, Amines, and Amine Hydrohalides

Three classes of nitrogen-containing compounds (amino acids, amines, and amine hydrohalides) give rise to very characteristic broad absorption bands. Perhaps the most striking of these are found in the infrared spectra of amino acids, which contain an extremely broad band centered near 3000 cm^{-1}, often extending as low as 2200 cm^{-1}, with some structure. The infrared spectrum of an amino acid is shown in Figure 5-11. Amine hydrohalides give a similar, very broad band with structure on the low frequency side. The center of the band tends to be lower than in amino acids, especially in the case of tertiary amine hydrohalides where the band center may be as low as 2500 cm^{-1}. In fact, this band gives a very useful indication of the presence of a tertiary amine. Both types of compound also have a weak but characteristic band between 2200 and 2000 cm^{-1} that is believed to be a combination of the $\overset{+}{N}H_3$ torsional mode near 500 cm^{-1} and the antisymmetric $\overset{+}{N}H_3$ deformation near 1600 cm^{-1}.

Primary amines have a broad band in their infrared spectra centered near 830 cm^{-1}, whereas the frequency for secondary amines, is about 100 cm^{-1} lower. This band is not

Table 5-13. *Details of Infrared Frequencies of Nitrogen-Containing Groups*

Group and Class	Range (cm^{-1}) and Intensity	Assignment and Remarks
The $-NH_2$ group		
amides	3530–3520 (s)	dil soln; NH_2 antisym. stretch
primary	3400–3390 (s)	dil soln; NH_2 sym. stretch
	3360–3340 (vs)	solid state; NH_2 stretching
	3190–3170 (vs)	frequencies lowered by hydrogen bonding
aliphatic primary	1650–1620 (s)	dil soln or solid state; NH_2
	1630–1610 (s)	deformations; may not be resolved from C=O stretch
amines		
aromatic primary	3500–3390 (m)	NH_2 antisym. stretch
	3420–3300 (m)	NH_2 sym. stretch
aliphatic primary	3400–3330 (s–vs)	NH_2 antisym. stretch
	3330–3250 (s–vs)	NH_2 sym. stretch
	850–750 (s, broad)	NH_2 wagging and stretching
primary	1650–1590 (s)	NH_2 deformation
The $-NH-$ group		
secondary amides	3450–3400 (m)	dil soln; NH stretch
	3350–3300 (s)	solid state; NH stretch
	3100–3060 (w)	overtone band
	1560–1530 (s)	coupled NH deformation and C–N stretch
	750–650 (s, broad)	NH wag
secondary amines	3350–3300 (m)	NH stretch
	750–650 (s, broad)	NH wag
The C≡N group		
nitriles		
saturated aliphatic nitriles	2260–2240 (w)	C≡N stretch; a Raman spectrum may be needed
unsaturated aliphatic nitriles adjacent to a double bond	2230–2220 (m)	C≡N stretch; doublet when the adjacent double bond is disubstituted
aromatic nitriles	2240–2220 (variable)	C≡N stretch; stronger than in saturated aliphatic nitriles
isonitriles		$\overset{+}{}\overset{-}{}$
alkyl	2180–2150 (w–m)	$-N≡C$ stretch; strong in the Raman spectrum
aryl	2130–2110 (w–m)	$\overset{+}{}\overset{-}{}$ $-N≡C$ stretch; strong in the Raman spectrum
The C=N group		
isocyanates	2280–2260 (vs)	N=C=O antisym. stretch
aliphatic Schiff bases	1680–1660 (m–s)	RC=NR stretch
aromatic Schiff bases	1640–1620 (m–s)	Ar–C=NAr stretch
oximes	1690–1620 (w–m)	–C=NOH stretch
amidines	1680–1580 (s)	–N–C=N– stretch

(table continues)

Table 5-13. *(cont.)*

Group and Class	Range (cm⁻¹) and Intensity	Assignment and Remarks
The C–N group		
primary aliphatic amines	1140–1070 (m)	antisym C–C–N stretch
secondary aliphatic amines	1190–1130 (m–s)	antisym. C–N–C stretch
primary aromatic amines	1330–1260 (s)	phenyl–N stretch
secondary aromatic amines	1340–1250 (s)	phenyl–N stretch, two bands
lactams	1350–1300 (s)	C–N stretch
The NO_2 group		
aliphatic nitro compounds	1560–1530 (vs)	antisym. NO_2 stretch
	1390–1370 (m–s)	sym. NO_2 stretch
aromatic nitro compounds	1540–1500 (vs)	antisym. NO_2 stretch
	1370–1330 (s–vs)	sym. NO_2 stretch
nitrates R–O–NO_2	1660–1620 (vs)	antisym. NO_2 stretch
	1300–1270 (s)	sym. NO_2 stretch
	710–690 (s)	NO_2 deformation
The NO groups ($N=O, \overset{+}{N}-\overset{-}{O}, \overset{+}{N}-\overset{-}{O}$)		
nitrites R–ONO	1680–1650 (vs)	N=O stretch; often a weaker band is seen between 1630 and 1600 cm⁻¹
nitrates R–O–NO_2	870–840 (s)	N–O stretch
azoxy compounds $\overset{\diagdown}{N=\overset{+}{N}}\overset{\diagup \overset{\diagup}{O^-}}{}$	1340–1260 (vs)	$\overset{+}{N}-\overset{-}{O}$ stretch
monomeric nitroso compounds		
aliphatic	1620–1540 (vs)	N=O stretch
aromatic	1520–1450 (vs)	N=O stretch
dimeric nitroso compounds		
cis isomers	1420–1320 (vs)	N=O stretch, two bands
trans isomers	1300–1180 (vs)	N=O stretch, one band
oximes	965–930 (s)	N–O stretch
N-oxides		
aromatic	1300–1200 (vs)	$\overset{+}{N}-\overset{-}{O}$ stretch
aliphatic	970–950 (vs)	$\overset{+}{N}-\overset{-}{O}$ stretch
The N≡N group		
azides	2120–2160 (variable)	N≡N stretch, strong in Raman
The N=N group		
azo compounds	1450–1400 (vw)	N=N stretch, strong in Raman

present in the spectra of tertiary amines or amine hydrohalides. In Figure 5–12, the infrared spectra of diethylamine and its hydrochloride are compared. In both Figure 5–12b and Figure 5–11, it should be remembered that the presence of the mulling material (Nujol) makes the absorption maximum appear to be near 2900 cm⁻¹ in each case. For the amino acid the true maximum is probably near 3100 cm⁻¹, whereas for the amine hydrochloride band the maximum is near 2750 cm⁻¹.

In anilines the characteristic broad band shown by aliphatic amines in the 830–730 cm⁻¹ region is not present, so that the out-of-plane CH deformations of the benzene ring can be observed. The ring substitution pattern therefore can be determined. Of course, when an

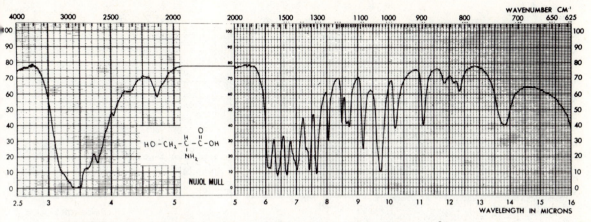

FIGURE 5-11. Infrared spectrum of the amino acid DL-serine. [Courtesy of Aldrich Chemical Company, Inc.]

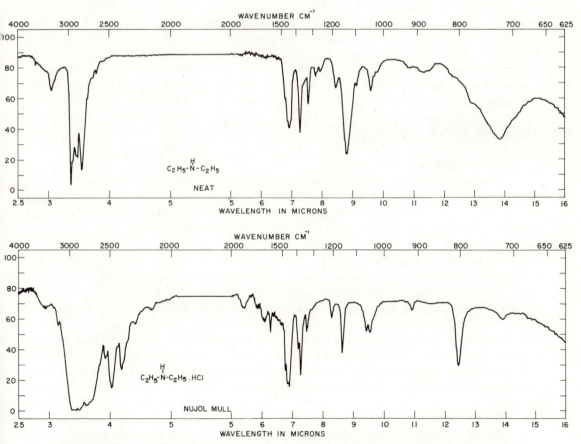

FIGURE 5-12. Infrared spectra of an amine and its hydrochloride: (upper) diethylamine and (lower) the hydrochloride. [Courtesy of Aldrich Chemical Company, Inc.]

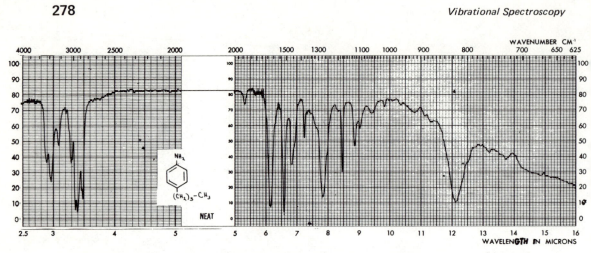

FIGURE 5-13. Infrared spectrum of an aniline derivative, p-n-butylaniline. [Courtesy of Aldrich Chemical Company, Inc.]

aliphatic amine is joined to a benzene ring through a carbon chain, both the characteristic amine band and the CH deformation pattern will be present. These two cases are illustrated by the infrared spectrum of an aniline derivative in Figure 5-13 and the spectrum of an aliphatic amine joined to a benzene ring in Figure 5-14. The presence of the benzene ring is identified in both compounds by bands between 3100 and 3000 cm^{-1} and by the substitution patterns between 2000 and 1700 cm^{-1}. In the CH out-of-plane bending region, the single band at 825 cm^{-1} in Figure 5-13 indicates the presence of a para-disubstituted benzene ring, and the doublet at 740 and 700 cm^{-1} in Figure 5-14 indicates monosubstitution (see section 5-6).

5-11c Nitriles
Saturated nitriles absorb weakly in the infrared near 2250 cm^{-1}. The band is strong in the Raman spectrum. Unsaturated or aromatic nitriles, in which the double bond or ring is adjacent to the CN group, absorb more strongly in the infrared than saturated compounds and the band occurs at somewhat lower frequencies near 2230 cm^{-1}.

5-11d Nitro Compounds
Nitro compounds have two very strong absorption bands due to symmetric and anti-symmetric NO_2 stretching. In aliphatic compounds, the frequencies are near 1550 and 1380 cm^{-1}, whereas in aromatic compounds, the bands are observed near 1520 and 1350 cm^{-1}. These frequencies are somewhat sensitive to nearby substituents. In particular, the 1350 cm^{-1} band in aromatic nitro compounds is intensified by electron-donating substituents in the ring. The out-of-plane CH bending patterns of ortho-, meta-, and para-disubstituted benzene rings are often perturbed in nitro compounds. Other compounds containing N—O bonds have strong characteristic infrared absorption bands (see Table 5-13).

5-11e Amides
Secondary amides (N-monosubstituted amides) usually have their NH and CO groups trans to each other as in 3. The carbonyl stretching frequency occurs as a very strong infrared band between 1680 and 1640 cm^{-1}. This absorption is known as the amide I band. A second,

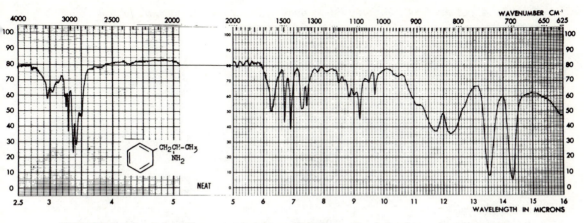

FIGURE 5-14.   Infrared spectrum of an aliphatic amine joined to a benzene ring
(amphetamine).   [Courtesy of Aldrich Chemical Company, Inc.]

3

very strong absorption, which occurs between 1560 and 1530 cm^{-1}, is known as the amide II
band. It is believed to be due to coupling of the NH bending and C–N stretching vibrations.
The trans amide linkage (3) also gives rise to absorption between 1300 and 1250 cm^{-1} and to
a broad band centered near 700 cm^{-1}. Occasionally, the amide linkage is cis in cyclic compounds such as lactams. In such cases, a strong NH stretching band is seen near 3200 cm^{-1}
and a weaker combination band (involving C=O stretching and NH bending) is observed near
3100 cm^{-1}. The amide II band is absent, but a cis NH bending mode absorbs between 1500
and 1450 cm^{-1}. This may be confused with the CH$_3$ and CH$_2$ deformation bands.

5-11f   The C–N Stretching Mode

The C–N stretching vibration is likely to be coupled with C–C stretching and C–C–H
bending modes in most molecules. One expects to find infrared absorption or Raman scattering
as a result of these vibrations in the region 1350-1050 cm^{-1}. Aromatic amines and lactams
have frequencies at the upper end and aliphatic amines at the lower end of this range. However,
the C–N group frequencies are not really very useful for structure determination because of
the large number of other groups or compounds that have vibrational frequencies in the same
region.

5-12   Compounds Containing Phosphorus

5-12a   General

The presence of phosphorus in organic compounds can be detected by the infrared
absorption bands arising from the P–H, P–OH, P–O–C, O–P–O, P=O, and P=S groups.
A phosphorus atom directly attached to an aromatic ring is also well characterized. The

Table 5-14. *Characteristic Infrared Frequencies of Groups Containing Phosphorus*

Group and Class	Range (cm^{-1}) and Intensity	Assignment and Remarks
The P–H group		
phosphines	2320–2270 (m)	P–H stretch, sharp band
	1090–1080 (m)	PH$_2$ deformation
	990–910 (m–s)	P–H wag
The P–OH group		
phosphinic, phosphoric,	2700–2550 (w, broad)	OH stretch, a broad and often very
phosphorous acids,	2300–2100 (w, broad)	weak band
esters, and salts	and	
	1040–920 (s)	P–O stretch
aromatic compounds	2800–2200 (w, broad)	might not be recognized
The P–O–C group		
aliphatic compounds	1050–950 (vs)	antisym. P–O–C stretch
	830–750 (s)	sym. P–O–C stretch (methoxy and ethoxy phosphorus compounds only
aromatic compounds	1250–1160 (vs)	aromatic C–O stretch
	1050–870 (vs)	O–P stretch
The P–C group		
aromatic compounds	1450–1430 (s, sharp)	P joined directly to a ring
quaternary aromatic	1110–1090 (s, sharp)	P$^+$ joined directly to a ring
The P=O group		
aliphatic compounds $R{-}O{-}\overset{\displaystyle O}{\overset{\|}{P}}{-}$	1260–1240 (s)	strong, sharp band
aromatic compounds $Ar{-}O{-}\overset{\displaystyle O}{\overset{\|}{P}}{-}$	1350–1300 (s)	lower frequency (1250–1180 cm^{-1}) when OH group is attached to a P atom
phosphine oxides	1200–1140 (s)	P=O stretch
The O–P–O group		
polynucleotides	1110–1080 (Raman)	O$=$P$=$O sym. stretch
	820–780 (Raman)	O–P–O sym. stretch
The P–CH$_3$ and P–CH$_2$– groups		
P–CH$_3$	1310–1280 (m)	sym. CH$_3$ deformation
	960–860 (s)	CH$_3$ rock
P–CH$_2$–	1440–1400 (m)	CH$_2$ deformation, lower than the usual CH$_2$ deformation range

usual frequencies of these groups in various compounds are listed in Table 5-14. Most of these groups absorb strongly or very strongly in the infrared with the exception of P=S. The Raman spectrum is valuable for detecting this group, which has a frequency between 700 and 600 cm^{-1}. There is no characteristic P–C group frequency in aliphatic compounds so that the spectrum of a trisubstituted phosphine resembles that of the substituents.

5-12b Phosphorus Acids and Esters

Several phosphorus acids have P–OH groups, which give one or two broad bands of medium intensity between 2700 and 2100 cm^{-1}. Esters and acid salts that have P–OH

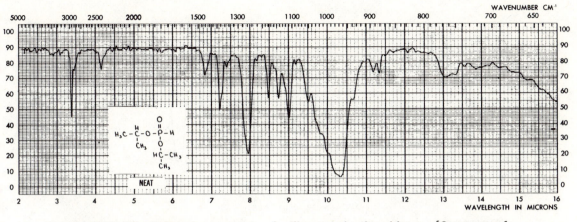

FIGURE 5-15. Infrared spectrum of diisopropyl phosphite. [Courtesy of Aldrich Chemical Company, Inc.]

groups also absorb in this region. In ethoxy and methoxy phosphorus compounds, as well as other aliphatic compounds having a P–O–C linkage, a very strong and quite broad infrared band is observed between 1050 and 950 cm^{-1}. The presence of a P=O bond is indicated by a strong band very close to 1250 cm^{-1}. An example of the spectrum of an aliphatic compound containing P–H, P=O, and P–O–C groups is given in Figure 5-15.

5-12c Aromatic Phosphorus Compounds

Aromatic phosphorus compounds have several characteristic group frequencies. A fairly strong, very sharp infrared peak is observed near 1440 cm^{-1} in compounds in which a phosphorus atom is attached directly to a benzene ring (Figure 5-16). A quaternary phosphorus atom attached to a benzene ring has a very characteristic strong, sharp band near 1100 cm^{-1}. The P–O group attached to an aromatic ring gives rise to two strong bands between 1250 and 1160 cm^{-1} and between 1050 and 870 cm^{-1}. The P=O group, when attached to a ring through an oxygen atom, absorbs at a frequency 50 to 100 cm^{-1} higher

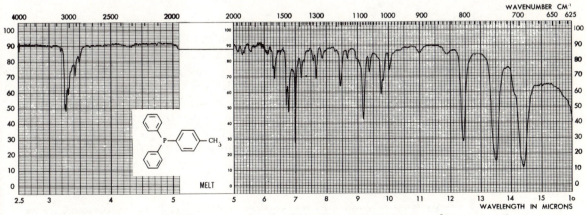

FIGURE 5-16. Infrared spectrum of diphenyl-p-tolylphosphine. [Courtesy of Aldrich Chemical Company, Inc.]

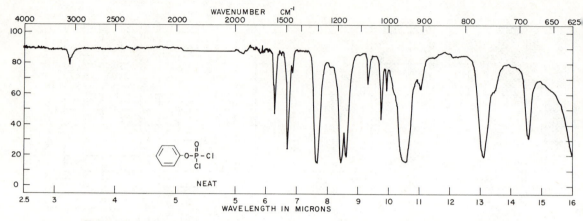

FIGURE 5-17. Infrared spectrum of phenyl dichlorophosphate. [Courtesy of Aldrich Chemical Company, Inc.]

than the usual 1250 cm^{-1} region. Some of these features are illustrated in Figure 5-17. If an OH group is also attached to the P atom, the P=O stretching frequency is observed between 1250 and 1180 cm^{-1}. The P–OH group itself might not be noticed in the infrared spectrum because it gives rise to only a weak, broad absorption.

5-12d   Nucleic Acids
DNA, RNA, and related molecules consist of heterocyclic bases linked through a sugar to phosphoric acid. Raman spectra of these compounds have characteristic bands due to symmetric stretching of O–P–O linkages between 820 and 780 cm^{-1} and bands due to O=P=O symmetric stretching between 1110 and 1080 cm^{-1}. The first of these bands is sharp and strongly polarized in helical structures and is useful in studying the conformation of polynucleotide molecules.[3]

5-13   Compounds Containing Sulfur

5-13a   General
The SO$_2$ and SO groups give rise to very strong infrared bands in various compounds, but other bonds involving sulfur, such as C–S, S–S, and S–H, give very weak infrared absorption and a Raman spectrum is needed to identify these groups. Characteristic frequencies of some sulfur-containing groups are listed in Table 5-15. The C=S group has been omitted from the table because the C=S stretching vibration is invariably coupled with vibrations of other groups in the molecule. Frequencies in the range 1400-850 cm^{-1} have been assigned to this group with thioamides at the low frequency end of the range. The infrared bands involving C=S groups are usually weak, except for xanthates, R–O–C(=S)–S–, which have strong bands between 1250 and 1000 cm^{-1}.

[3] For further published information on this important application of Raman spectroscopy, see the work of Peticolas and others in the journal *Biopolymers*.

Table 5-15. *Characteristic Infrared Frequencies of Groups Containing Sulfur*

Group and Class	Range (cm^{-1}) and Intensity	Assignment and Remarks
The S–H group		
thiols (mercaptans)	2580–2550 (w)	strong in Raman
The S–CH$_2$ group		
thiols and thioethers	1440–1410 (s)	CH$_2$ deformation, lower frequency than usual
	1270–1220 (s)	CH$_2$ wag
The S–CH$_3$ group	1330–1320 (m)	sym. CH$_3$ deformation, lower frequency than usual
The C–S group	700–600 (w)	strong in Raman
The S–S group		
disulfides	550–450 (vw or absent)	strong in Raman
$\overset{\displaystyle S}{\overset{\displaystyle \|}{}}$ *The O–C–S group*	1250–1000 (s)	several bands
xanthates	670–620 (vw or absent)	strong in Raman
	470–440 (w)	strong in Raman
The S=O group		
sulfoxides	1060–1020 (vs)	S=O stretch
dialkyl sulfites	1220–1190 (vs)	S=O stretch
sulfinic acids	1090–900 (vs)	S=O stretch
	870–810 (s)	S–O stretch
sulfinic acid esters	1140–1120 (s)	S=O stretch
The SO$_2$ group		
sulfones, sulfonamides	1390–1290 (vs)	antisym. SO$_2$ stretch
sulfonic acids, sulfonates, and sulfonyl chlorides	1190–1120 (vs)	sym. SO$_2$ stretch
dialkyl sulfates and sulfonyl fluorides	1420–1390 (vs)	antisym. SO$_2$ stretch
	1220–1190 (vs)	sym. SO$_2$ stretch
The S–O–C group		
dialkyl sulfites	1050–850 (vs)	two bands
sulfates	1050–770 (vs)	two or more bands

5-13b C–S, S–S, and S–H Groups

Raman spectra of compounds containing C–S and S–S bonds contain very strong lines due to these groups between 700 and 600 and near 500 cm^{-1}, respectively. These group frequencies, especially S–S, are either absent or appear only very weakly in the infrared. The S–H stretching band near 2500 cm^{-1} is normally quite weak in the infrared, but shows a high intensity in the Raman spectrum. As mentioned above, xanthates contain the –O–C(=S)–S– group, which also has two characteristic stretching frequencies near 650 and 460 cm^{-1}. Only the lower frequency mode is observed as a weak-to-medium intensity band in the infrared, but both frequencies appear very strongly in the Raman.

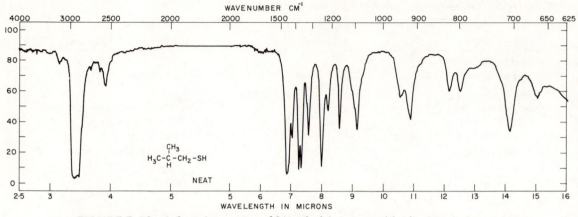

FIGURE 5-18. Infrared spectrum of 2-methyl-1-propanethiol (isobutyl mercaptan).
[Courtesy of Aldrich Chemical Company, Inc.]

5-13c S—CH$_3$ and S—CH$_2$ Groups

Methyl and methylene groups attached to a sulfur atom can be recognized by shifts from the normal frequencies exhibited by these groups. The CH$_3$ and CH$_2$ deformation modes are observed at lower frequencies than usual, and the CH$_2$ wagging vibration gives rise to a strong infrared band near 1250 cm^{-1}. The spectrum of a simple aliphatic thiol is shown in Figure 5-18.

5-14 Heterocyclic Compounds

5-14a General

Heterocyclic compounds and their substituted derivatives may exhibit three kinds of group frequencies: those involving CH or NH vibrations, those involving motion of the ring or rings, and those due to the group frequencies of the substituents. The identification of a heterocyclic compound from its infrared and Raman spectra is not an easy task. In many cases, the spectra of a heterocyclic compound resemble those of the corresponding non-heterocyclic compound. In some cases, however, characteristic features are apparent. Characteristic frequencies for some heterocyclic compounds are collected in Table 5-16, but this table covers only a small fraction of known heterocyclic systems.[4]

5-14b Aromatic Heterocycles

A classification of heterocyclic compounds into aromatic or nonaromatic can be made. Hydrogen atoms attached to carbon atoms in an aromatic ring will give rise to CH stretching modes in the usual 3100-3000 cm^{-1} region, or a little higher in furans, pyrroles, and some other compounds. Characteristic ring stretching modes, similar to that of benzene, will be observed, and the out-of-plane CH deformation vibrations give rise to strong infrared bands in the 1000-650 cm^{-1} region. In some cases these patterns are characteristic of the type of substitution in the heterocyclic ring. The in-plane CH bending modes also give several bands in the 1300-1000 cm^{-1} region in aromatic heterocyclic compounds. The CH vibrations in benzene derivatives can be correlated with the analogous modes in related heterocyclic compounds

---

[4]For a comprehensive coverage, see A. R. Katritzky and A. P. Ambler in *Physical Methods in Heterocyclic Chemistry*, A. R. Katritzky, ed., Academic Press, New York, 1963, vol. II, pp. 161–360.

Table 5-16. *Characteristic Frequencies for Heterocyclic Compounds*

Classes of Compounds	Range (cm^{-1}) and Intensity	Assignment and Remarks
Azoles (imidazoles, isoxazoles, oxazoles, pyrazoles, triazoles, tetrazoles)	3340–2500 (s, broad) 1650–1380 (m–s) 1040–980 (s)	NH stretch resembles carboxylic acids three ring-stretching bands ring breathing
Benzofurans	3175–3115 (w–m) 1640–1615 (s) 1275–1250 (s) 1110–1090 (s)	two bands due to CH stretch ring stretch in-plane CH deformation
Carbazoles	3490–3470 (vs)	NH stretch, dil soln in nonpolar solvents
Coumarans (2,3-dihydrobenzofurans)	1640–1600 (w) 1500–1480 (s) 1260–1210 (s) 900–650 (s)	ring stretch C–O stretch out-of-plane CH deformation pattern
1,4-Dioxanes	1460–1440 (vs) 1400–1150 (s) 1130–1000 (m) 850–830 (w)	CH_2 deformation CH_2 twist and wag ring mode, s in Raman vs in Raman
1,3-Dioxolanes	1170–1150 (vs) and 1100–1050 (vs) 940 (variable)	ring stretching modes ring breathing
Furans general	3140–3120 (m) 1600–1400 (m–s) 770–720 (vs)	CH stretch, higher than most aromatics ring stretch, three bands band becomes weaker as number of substituents increases
2-subst. furans	1020–1000 (s–vs) 930–910 (s)	ring breathing CH out-of-plane deformation
Imidazoles	1830–1810 (w)	combination band
Indoles	3470–3450 (vs) 1600–1500 (m–s) 900–660 (vs)	NH stretch two bands substitution patterns due to both six- and five-membered rings
Purines (and other compounds with several heterocyclic N atoms)	2850–2200 (broad) 900–850 (s)	H-bonded NH stretch CH out-of-plane deformation
Pyridines general	3080–3020 (w–m) 2080–1670 (w) 1615–1565 (s) 1030–990 (s)	CH stretch, several bands combination bands two bands, due to C=C and C=N stretch in ring ring breathing
2-subst. pyridines	780–740 (s) 630–605 (m–s) 420–400 (s)	CH out-of-plane deformation in-plane ring deformation out-of-plane ring deformation
3-subst. pyridines	820–770 (s) 730–690 (s) 635–610 (m–s) 420–380 (s)	CH out-of-plane deformation ring bend in-plane ring deformation out-of-plane ring deformation

(table continues)

Table 5–16. *(cont)*

Classes of Compounds	Range (cm^{-1}) and Intensity	Assignment and Remarks
4-subst. pyridines	850–790 (s)	CH out-of-plane deformation
disubst. pyridines	830–810 (s) and 740–720 (s)	CH out-of-plane deformations
trisubst. pyridines	730–720 (s)	CH out-of-plane deformation
Pyrimidines	1590–1370 (m–s)	ring stretching, four bands
2-subst. pyrimidines	650–630 (m–s) and 580–480 (m–s) and 520–440 (m–s)	ring deformation modes
4-subst. pyrimidines	685–660 (vs)	ring deformation
Pyrroles	3480–3430 (vs)	NH stretch, often a sharp band
	3130–3120 (w)	CH stretch, higher than normal for aromatics
	1560–1390 (variable)	ring stretch, three bands
	770–720 (s, broad)	out-of-plane CH deformation
Quinolines monosubst. quinolines	1310–1250 (m–s) and 1140–1110 (m–s) and 1040–1010 (m–s)	in-plane CH deformation
	970–950 (m)	ring mode
disubst. quinolines	1180–1130 (m–s) and 1080–1060 (m–s)	in-plane CH deformation
	1050–1030 (w–m)	ring mode
	830–790 (s–vs)	out-of-plane CH deformation
Thiazoles	1540–1470 (s) and 1400–1380 (s)	ring stretching modes
	900–650 (m–s)	several bands
Thiophenes general	1590–1400 (m) and 1420–1400 (s–vs) and 1390–1350 (m–s) and 1250–1220 (s)	several bands due to ring stretching modes
2-subst. thiophenes	860–820 (s)	usually two bands
	700–680 (vs)	CH out-of-plane deformation, lower than in pyrroles and furans
3,5-disubst. thiophenes	810–790 (s)	CH out-of-plane deformation
Triazines	1560–1520 (vs) and 1420–1400 (s)	ring stretch
	820–740 (s)	out-of-plane ring deformation

and may be useful in structure determination. Nonaromatic heterocyclic compounds will usually have one or more CH_2 groups present. The stretching and deformation (scissoring) modes give rise to bands in the usual regions (see section 5–8). The wagging, twisting, and rocking modes, however, often interact with skeletal ring modes and may be observed over a wide range of frequencies.

In aromatic heterocyclic compounds involving nitrogen, the coupled C=C and C=N stretching modes give rise to several characteristic vibrations that are similar in frequency to their counterparts in the corresponding nonheterocyclic compounds. Ring stretching modes are found in the 1600–1300 cm^{-1} region. Other skeletal ring modes may be characterized, such as ring breathing modes near 1000 cm^{-1} that give very strong Raman bands, in-plane ring deformations between 700 and 600 cm^{-1}, and out-of-plane ring deformation modes that may be observed between 700 and 300 cm^{-1}.

5-14c NH Stretching Bands

Spectra of heterocyclic nitrogen compounds contain bands due to the secondary or tertiary amine group. Pyrroles, indoles, and carbazoles in nonpolar solvents have their NH stretching vibration between 3500 and 3450 cm^{-1} and the band is very strong in the infrared. In saturated heterocyclics such as pyrrolidines and piperidines, the band is at lower frequencies.

Various azoles such as imidazoles, pyrazoles, triazoles, and tetrazoles have a very broad hydrogen-bonded NH stretching band between 3300 and 2200 cm^{-1}. This band might be confused with the broad OH stretching band of carboxylic acid dimers. See, for example, Figure 5–19, where an imidazole is compared with a simple carboxylic acid.

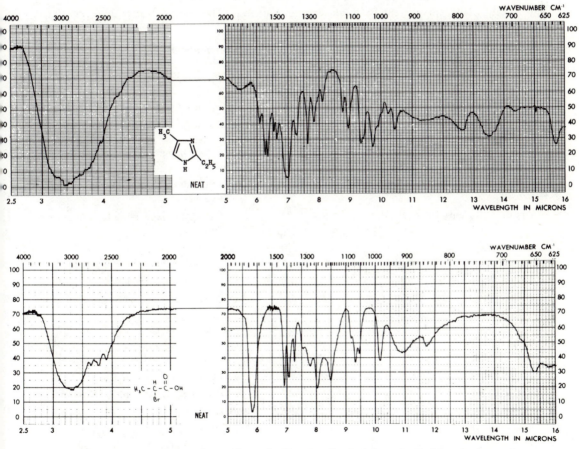

FIGURE 5-19. Comparison of an imidazole, 2-ethyl-4-methylimidazole (upper) with a carboxylic acid, 2-bromopropionic acid (lower). [Courtesy of Aldrich Chemical Company, Inc.]

5-14d Substituent Bands

Substituents retain their own characteristic group vibrations. Changes in both frequency and intensity can give important information concerning the heterocyclic ring to which the substituent is attached. An increase in C=O or C=N frequencies gives an indication of the electron-releasing power of a ring. Also, changes in intensities of substituent bands, which are related to changes in dipole moment, can be correlated with the electronic nature of the heterocyclic ring.

5-15 Compounds Containing Halogens

5-15a General

A halogen atom attached to a carbon atom adjacent to a functional group often causes a significant shift in the group frequency. Some such examples are listed in Table 5-17. Fluorine is particularly important in this regard and special care must be exercised in conclusions drawn from infrared and Raman spectra when this element is present. Carbon-fluorine stretching bands are very strong in the infrared, usually between 1350 and 1100 cm^{-1}, but they are extremely weak in the Raman. The presence of other functional

Table 5-17. *The Effect of Halogen Substituents on Group Frequencies*

Group or Class	Range (cm^{-1})	Assignment or Remarks
Fluorine		
$-FCH-$ and CF_2H	3010–2990	CH stretch, higher frequency than normal
$F_2C=C$	1760–1730	C=C stretch, much higher frequency than normal
$F_2C=C$ F	1870–1800	C=C stretch, much higher frequency than normal
$F-C=O$	1900–1820	acid fluorides, the highest carbonyl group frequency
$-CF_2CO-CH_2$ and $-CF_2CO-CF_2$	1800–1770	Normal C=O stretch range for ketones is 1730–1700 cm^{-1}
$-CF_2-COOH$	1780–1740	normal C=O stretch range for carboxylic acids (dimers) is 1720–1680 cm^{-1}
$-CF_2-C\equiv N$	2280–2260	C≡N stretch, 20 cm^{-1} higher than normal
$-CF_2-CONH_2$	1730–1700	C=O stretch, 30 cm^{-1} higher than normal
Chlorine, Bromine, and Iodine		
CH_2Cl	1300–1240	CH_2 wag, strong in ir
CH_2Br	1240–1190	CH_2 wag, strong in ir
CH_2I	1190–1150	CH_2 wag, strong in ir
$Cl-C=O$	1810–1790	acid chlorides, C=O stretch
α-halo esters	1770–1745	C=O stretch, 20 cm^{-1} higher than normal
noncyclic haloketones	1745–1730	C=O stretch, 15 cm^{-1} higher than normal
α-halocarboxylic acids	1740–1720	normal C=O stretch range is 1720–1680 cm^{-1}
chloroformates	1800–1760	C=O stretch, near 1720 cm^{-1} in formate esters
α-chloroaldehydes	1770–1730	C=O stretch, higher than normal aldehydes

Table 5-18. *Infrared Stretching Frequencies of Bonds Involving Halogens*

Group	Range (cm^{-1}) and Intensity		Assignment and Remarks
$-AsCl$	400–360	(vs)	v strong in both ir and Raman
$-AsBr$	280–260	(vs)	v strong in both ir and Raman
$-AsI$	230–200	(vs)	v strong in both ir and Raman
$-BF$	1400–1000	(vs)	v strong and broad in BF_4^- compounds
$-BCl$	1000–600	(s)	variable position
$-GeCl$	ca. 375	(m)	phenylgermanium compounds
$-GeBr$	ca. 310	(m)	phenylgermanium compounds
$-HgCl$	340–310	(s)	organomercury (R–Hg–X) compounds
$-HgBr$	240–210	(s)	organomercury (R–Hg–X) compounds
$-PF$	890–800	(s)	pentavalent phosphorus
$-PCl$	590–430	(vs)	
$-SO_2F$	820–750	(vs)	sulfonyl fluorides
$-SO_2Cl$	600–530	(s)	sulfonyl chlorides
$-SiF$	920–820	(s)	single fluorine atom
$-SiF_2$	950–910	(s)	antisym. SiF_2 stretch
	910–870	(m)	sym. SiF_2 stretch
$-SiF_3$	980–940	(s)	antisym. SiF_3 stretch
	910–860	(m)	sym. SiF_3 stretch
$-SiCl$	550–470	(s)	single Cl atom
$-SiCl_2$	600–530	(s)	antisym. $SiCl_2$ stretch
	540–460	(m)	sym. $SiCl_2$ stretch
$-SiCl_3$	620–570	(s)	antisym. $SiCl_3$ stretch
	540–450	(m)	sym. $SiCl_3$ stretch
$-SnCl$	360–300	(s)	Sn–Cl stretch in organotin compounds
$-SnBr$	260–240	(s)	Sn–Br stretch in organotin compounds
$-SnI$	240–180	(s)	Sn–I stretch in organotin compounds

groups that absorb in this region of the spectrum can often be detected in the Raman spectrum. It should be mentioned that there are many known cases of symmetrical C–F stretching modes at frequencies much lower than the "usual" 1350-1100 cm^{-1}. The usual regions for the C–X stretching and bending vibrations have been given previously in Tables 4-1 to 4-4. Some of the stretching frequencies of halogens attached to elements other than carbon in organic compounds are listed in Table 5-18.

5-15b CH₂X Groups

The CH_2 wagging mode in compounds with a CH_2X group gives rise to a strong band whose frequency depends on X. When X is Cl, the range is 1300-1250 cm^{-1}; for Br, the band is near 1230 cm^{-1}; and for I, a still lower frequency near 1170 cm^{-1} is observed. Halogen atoms attached to aromatic rings are involved in certain vibrations that are sensitive to the mass of the halogen atom. One of the benzene ring vibrations that involves motion of the substituent atom gives rise to bands between 1250 and 1100 cm^{-1} when the substituent is fluorine, between 1100 and 1040 cm^{-1} for chlorine, and between 1070 and 1020 cm^{-1} for bromine.

5-15c Haloalkyl Groups

In haloalkyl groups, the presence of more than one halogen atom on a single carbon atom shifts the C—X stretching frequency to the high wave number end of the range. The CCl_3 group antisymmetric stretching frequency is found in the 830-700 cm^{-1} range. Branching of a carbon chain at the atom adjacent to the C—X group causes a lowering of the C—X stretching frequency.

5-16 Organometallic Compounds

5-16a Metal-Carbon Stretching Frequencies

In this section we look at some of the characteristic frequencies of groups containing metals and borderline elements such as boron and silicon. The field of spectroscopy of organometallic compounds is very large, and in recent years the Chemical Society (London) has published several Specialist Periodical Reports on "Spectroscopic Properties of Inorganic and Organometallic Compounds." This continuing series provides an excellent guide to the literature. Metal–carbon stretching frequencies are found in the range 800-400 cm^{-1}. Many of these are listed in Table 5-19. Boron–carbon and silicon–carbon stretching modes are not

Table 5-19. *Metal-Carbon Group Frequencies*

Metal	*Range (cm^{-1})*	*Assignment and Remarks*
Aluminum		
Al–C	780–530	bridged structures
Arsenic		
As–C	580–560	very strong in infrared
AsC_2	240–220	AsC_2 deformation
Gallium		
Ga–C	560–530	alkyl Ga compounds
Germanium		
Ge–C	600–540	(m–s) in infrared
Gold		
Au–C	540–480	square planar complexes
Lead		
Pb–C	480–420	lead alkyls and alkyl halides
Mercury		
Hg–C	580–510	alkylmercury compounds
$Hg–CF_3$	270–250	extremely low frequency
Palladium		
Pd–C	540–430	square planar complexes
Platinum		
Pt–C	580–510	square planar complexes
Tin		
Sn–C	580–510	alkyltin halides, etc.
SnC_2	610–500	antisym. Sn–C stretch
	530–450	sym. Sn–C stretch
	150–130	SnC_2 deformation
Titanium		
Ti–C	460–400	$Ti(CH_3)_n Cl_{4-n}$ compounds

usually identifiable since they are coupled with other skeletal modes. The C—B—C antisymmetric stretching mode in phenylboron compounds gives a strong infrared band between 1280 and 1250 cm^{-1}. A silicon atom attached to an aromatic ring gives two very strong bands near 1430 and 1110 cm^{-1}. The B—O and B—N bonds in organoboron compounds give very strong infrared bands between 1430 and 1330 cm^{-1}. The Si—O—C vibration gives a very strong infrared absorption, which is often quite broad, in the 1100–1050 cm^{-1} range. Some characteristic frequencies for boron and silicon compounds are listed in Table 5-20.

Table 5-20. *Infrared Group Frequencies in Boron and Silicon Compounds*

Group	Range (cm^{-1}) and Intensity		Assignment and Remarks
Boron			
—BOH	3300–3200	(s)	broad band due to H-bonded OH stretch
—BH and —BH$_2$	2650–2350	(s)	doublet for —BH$_2$ stretch
	1200–1150	(ms)	—BH$_2$ deformation or B—H bend
	980–920	(m)	—BH$_2$ wag
—B—Ar	ca. 1430	(m–s)	benzene ring vibration
B—N	1460–1330	(vs)	borazines and aminoboranes
B—O	1380–1310	(vvs)	borates, boronates, boronic acids, and anhydrides; absent in N—B—O coordination compounds
C—B—C	1280–1250	(vs)	C—B—C antisym. stretch
Silicon			
—SiOH	3700–3200	(s)	similar to alcohols
	900–820	(s)	Si—O stretch
—SiH, —SiH$_2$, and —SiH$_3$	2150–2100	(m)	Si—H stretch
	950–800	(s)	Si—H deformation and wag
Si—Ar	ca. 1430	(m–s) and	ring modes
	1100	(vs)	
Si—O—C (aliphatic)	1100–1050	(vvs)	antisym. Si—O—C stretch
Si—O—Ar	970–920	(vs)	Si—O stretch
Si—O—Si	1100–1000	(s)	asym. Si—O—Si stretch

The B—CH$_3$ and Si—CH$_3$ symmetric CH$_3$ deformation modes occur at 1330–1280 cm^{-1} and 1280–1250 cm^{-1}, respectively. Other characteristic CH$_3$ deformation frequencies in M—CH$_3$ groups have been listed previously in Table 5-10.

5-16b Metal Carbonyls
Metal carbonyls absorb very strongly in the infrared between 2200 and 1700 cm^{-1}. When only one metal atom is bonded to C=O, the frequencies are in the upper end of this range, whereas when carbonyl groups are involved in bridging, bands are observed between 1900 and 1700 cm^{-1}.

5-16c Sandwich Compounds
Sandwich compounds such as the ferrocenes and dibenzenechromium have the characteristics of the cyclopentadiene and benzene rings, but there are in addition several new modes associated with these structures. Antisymmetric ring-tilting modes have been observed between 530 and 380 cm^{-1} in a larger number of M(C$_5$H$_5$)$_2$ compounds. The M—R$_2$ stretching mode varies with the metal in the frequency range 450–300 cm^{-1}.

5–16d Aromatic Compounds

The infrared spectra of aromatic organometallic compounds usually contain a fairly strong, sharp band near 1430 cm^{-1} due to a benzene ring vibration. This band has been observed for compounds in which As, Sb, Sn, Pb, B, Si, and P atoms are attached directly to the ring.

5–16e Organomercury Compounds

The mercury-carbon bond in aliphatic organomercury compounds is characterized by a very strong Raman band between 550 and 500 cm^{-1}. For aromatic compounds, a band between 250 and 200 cm^{-1} is assigned to the phenyl-Hg stretch. These bands are so strong that they can be seen in dilute aqueous solutions, with concentrations of organomercury compounds as low as 25 ppm.

5–17 Carbohydrates

5–17a General

The region between 1500 and 700 cm^{-1} contains a large number of bands in the spectra of carbohydrates. Localized group frequencies seldom can be identified in this region. There are, however, certain bands in the spectra of most carbohydrates that also are observed in the spectra of simpler related compounds. For example, several bands in the infrared spectrum of tetrahydropyran, between 1700 and 700 cm^{-1}, are attributed to ring vibrations. These are often observed at similar frequencies in pyranoses.

5–17b Anomers

An important use of vibrational spectra is in distinguishing different anomers. Spectra of anomers differ in the 960–730 cm^{-1} region for pyranoses because of the different disposition of bonds at the anomeric carbon atom. It is possible to distinguish between the α (D and L) or β (D and L) configurations by comparison with reference compounds of known configuration.

Table 5–21. *Group Frequencies in Carbohydrates*

Range (cm^{-1})	Assignment	Remarks
3800–3200	OH stretch	H-bonded; disappears on complete acetylation
2880–2840	CH stretch of CH$_3$O	absent in C-methyl or C$_2$H$_5$O groups
1500–1200	CH, CH$_2$, and OH bending	several bands in cellulose and xylan
1370–1200	CH$_2$ wagging and twisting	several bands in deoxy sugars
1200–1030	C–O stretch and C–O–C stretch	complex bands; appear simpler in polysaccharides
930–900 and 785–755	pyranose ring modes	α and β configurations have peaks at different frequencies
900–880	C$_1$H deformation	α-(equatorial) pyranoses
860–840	C$_1$H deformation	β-(axial) pyranoses
935–915	ring mode	
855–835	CH deformation	α-polysaccharides
800–750	ring breathing mode	
600–250	pyranose ring modes	several bands; can be correlated with α or β conformation

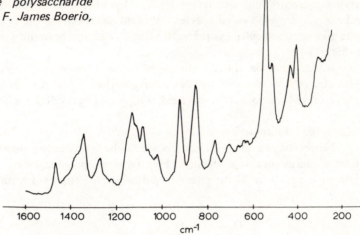

FIGURE 5-20. Raman spectrum of a 20% aqueous solution of the polysaccharide dextran. [Courtesy of Dr. F. James Boerio, University of Cincinnati.]

The differences between anomers unfortunately cannot be generalized. In aldopyranoses and acetylated pyranoses, however, certain groups of absorption bands shift or particular bands change intensity between anomers.

5-17c Characteristic Frequencies

Some characteristic frequencies for carbohydrates are listed in Table 5-21. It should be emphasized, however, that studies of the structures of carbohydrates are very empirical and rely heavily on collections of spectra. One such collection is included in the book by Zhbankov (see section A-1b, no. 12).[5]

5-17d Raman Spectra

Laser Raman spectra of carbohydrates are relatively easy to obtain from either solids or aqueous solutions. Infrared spectra of aqueous solutions are usually poor, and several regions are completely obscured. The Raman spectrum of water, on the other hand, is very weak. Also, bands involving OH stretching and bending are very broad in the infrared spectra of carbohydrates and much information is lost. An example of the Raman spectrum of a polysaccharide is shown in Figure 5-20. The compound dextran is a polysaccharide of high molecular weight containing linear chains of mainly 1-6'-linked α-D-glucopyranose residues as the dominant structural feature, with some 1-3', 1-4', and less frequently 1-2' linkages. The spectrum in Figure 5-20 was obtained using the 6471 Å line of a Kr^+ laser operating at a power of 500 milliwatts.

5-18 Detailed Structural Studies

5-18a Introduction

In this section, some applications of vibrational spectroscopy in detailed structural and conformational studies will be outlined. Geometrical isomers, rotational isomers, and tautomers

---

[5]Further discussion and references can be found in a review by H. Spedding in *Advances in Carbohydrate Chemistry*, **19**, 23 (1964).

can be observed in infrared and Raman spectra and the equilibria involved can be studied by varying concentrations and temperature. Monomer–dimer equilibria can be studied similarly, including hydrogen-bonded species. Vibrational spectroscopy is a valuable tool in studies of inter and intramolecular hydrogen bonding. Both the near and far infrared regions are useful in this area.

Information on the structures and conformations of polymers and biological macro-molecules can be obtained by means of polarized infrared studies. Raman spectroscopy has only recently been applied in this field, but promising results have been obtained.

5–18b Tautomerism

Numerous examples of tautomerism can be found in the literature. Infrared spectroscopy offers a useful means of distinguishing between possible tautomeric structures. A simple example is found in β-keto esters or β-diketones. The keto form has two C=O groups with

keto   enol

separate stretching frequencies, so that a doublet is often observed in CCl_4 solution in the usual ketone carbonyl stretching region near 1730 cm^{-1}. The enol form, on the other hand, has only one carbonyl group, the frequency of which is lowered by hydrogen bonding by 80–100 cm^{-1}. This structure also has an ethylenic double bond, which should give a band between 1650 and 1600 cm^{-1}. The C=O and C=C peaks may then appear as a doublet. In Figure 5–21, the compound ethyl stearoylacetate clearly shows keto and enol forms.

Heterocyclic compounds with a carbonyl group in the ring frequently exhibit tautomerism. Interesting examples are found in the isoxazoles. 3-Hydroxyisoxazoles (4) exist in the enol form, but 5-hydroxyisoxazoles (5) are usually unstable with respect to the keto forms (6, 7). The

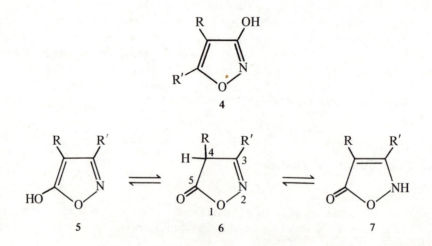

4-H form (6) is characterized by a very strong C=O stretching band between 1810 and 1790 cm^{-1} and a weak C=N stretching mode near 1620 cm^{-1}. The C=N stretching frequency

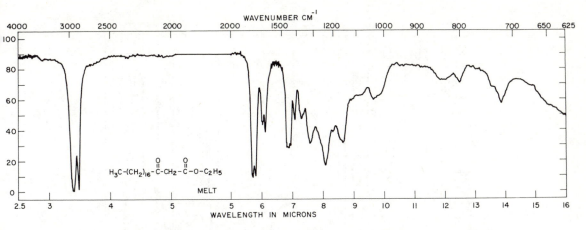

FIGURE 5-21.   Infrared spectrum of a β-keto ester, ethyl stearoylacetate.   [Courtesy of Aldrich Chemical, Inc.]

is lowered considerably by conjugation when the substituent R' is a phenyl group. The 2-H form (7) is characterized by a normal carbonyl stretching frequency, 1740-1720 cm^{-1}, and a strong C=C stretching mode between 1645 and 1615 cm^{-1}. Numerous other examples in heterocyclic molecules have been discussed in the chapter by Katritzky and Ambler (see footnote 4 on page 284).

5-18c Structural Isomerism

Structural isomers will usually differ in the functional groups present, so that their vibrational spectra will differ considerably. Some examples are the amino acid α-alanine (D and L), the ester ethyl carbamate (urethane), and the nitro compound 1-nitropropane. All three compounds have the same empirical formula, $C_3H_7O_2N$. The infrared survey spectra of these three compounds are shown in Figure 5-22.

Compounds with the formula $C_3H_7O_2N$ can also provide examples of structural isomers with similar functional groups. There are two isomers of nitropropane, $CH_3CH_2CH_2NO_2$ and $CH_3CHNO_2CH_3$. The spectra of these two compounds are shown in Figure 5-23. It is hard to decide which is the *n*-propyl and which is the isopropyl compound on the basis of these two spectra alone. The main differences lie in the CH_3 and CH_2 group vibrations. Both isomers have CH_3 groups, but only the *n*-propyl molecule has CH_2 groups. Unfortunately, the NO_2 symmetric stretch falls in the same region as the symmetric CH_3 deformation mode. The absorptions in the region between 1500 and 1400 cm^{-1}, however, make a distinction possible. In the spectrum of Figure 5-23b, a doublet is noted with the low frequency component stronger. This doublet is most likely due to both methyl and methylene groups. A nitro group adjacent to a CH_2 group is known to lower the deformation frequency and to intensify the band. The spectrum of Figure 5-23a, on the other hand, has only one strong band near 1460 cm^{-1}, so that this compound most likely has the two methyl groups.

Perhaps the best example of the importance of vibrational spectra in differentiating between structural isomers is found in ortho-, meta-, and para-disubstituted benzenes. The CH out-of-plane deformation patterns in the 850-700 cm^{-1} region (discussed in section 5-6) are different for each isomer. Substituted pyridines, pyrimidines, and other heterocyclic compounds provide further examples of structural isomers that can be distinguished by their vibrational spectra.

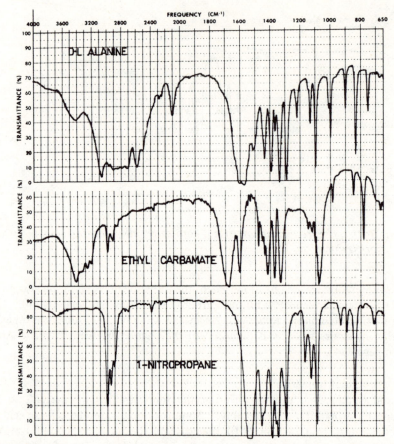

FIGURE 5–22.   *The infrared spectra of three compounds of formula* $C_3H_7O_2N$.

5-18d   Rotational Isomerism

In open chain compounds, the barriers to internal rotation about one or more carbon-carbon single bonds may be too high for rapid interconversion between different conformations. In such cases, two or more different rotational isomers can exist, and their presence may be detected in their infrared or Raman spectra. The restriction of rotation about double bonds can be thought of as an extension of the above concept. In this case, very high barriers are involved and cis and trans compounds result. (This form of isomerism is discussed separately in section 5-18e). The axial-equatorial conformations of cyclohexane and cyclopentane derivatives are examples of another kind of conformational (rotational) isomerism.

In noncyclic structures, rotation about a single bond can produce an infinite number of arrangements. Some of these are energetically favored (energy minima). The simplest examples are the substituted ethanes (CH_2XCH_2Y), where there are several preferred staggered conformations, two of which (**8, 9**) are illustrated, together with the high energy (in this case) eclipsed conformation **10**.

When there is a stabilizing interaction, the eclipsed form may be one of the stable conformations. Many such examples are known for rotation about sp^2–sp^3 bonds, as in α-haloketones, esters, acid halides, and amides. In these compounds, the halogen atom is believed to be either cis (eclipsed) or gauche (staggered) with respect to the carbonyl group.

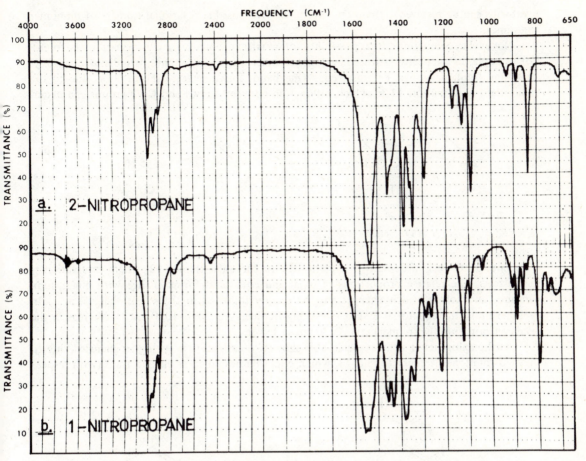

FIGURE 5-23. Infrared spectra of the two isomeric nitropropanes.

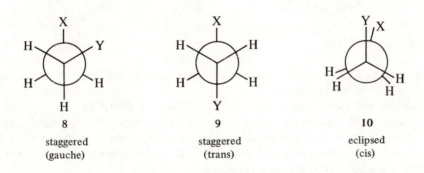

	8	9	10
	staggered (gauche)	staggered (trans)	eclipsed (cis)

Two C=O stretching bands are observed in such cases. One is at higher than normal frequencies because of the eclipsed interaction between the halogen atom and the C=O group. The other is found at the normal frequency. In α-haloketones, substitution of a second halogen on the other side of the carbonyl group leads to three preferred isomers, eclipsed/eclipsed, eclipsed/gauche, and gauche/gauche. Three C=O stretching frequencies can be observed in such cases.

For α-halocarboxylic acids, multiple carbonyl bands are not usually observed because of the complications due to hydrogen bonding. An exception is found in FCH_2COOH, for which five bands can be resolved. These arise from the five possible rotational isomers of monomer and dimer.

In cyclic compounds, the possibility of axial and equatorial conformations is found. For example, in α-chlorosubstituted cyclopentanones or cyclohexanones, two distinct carbonyl stretching frequencies can be observed. One band, found near 1745 cm^{-1}, is due to the equatorial conformation **11** in which interaction between the Cl and C=O groups can occur.

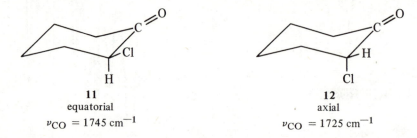

11	**12**
equatorial	axial
$\nu_{CO} = 1745$ cm^{-1}	$\nu_{CO} = 1725$ cm^{-1}

A second band near 1725 cm^{-1} is attributed to the axial isomer **12**, where interaction is minimized. The relative proportions of axial and equatorial forms change with phase, temperature, and solvent; such changes can be readily followed in the vibrational spectra. In cyclohexanols, the equatorial C–OH stretching frequency is 1050–1030 cm^{-1} whereas in the axial conformation the frequency is 10–30 cm^{-1} lower.

Ortho-halogenated benzoic acids also show two carbonyl stretching frequencies, due to the two rotational isomers **13** and **14**, which could be described as cis and trans (with respect

13	**14**
cis	trans

to the halogen and C=O groups). There is no hydrogen bonding between the halogen and hydrogen atoms in these compounds.

In alcohols and sterols the nonhydrogen-bonded O–H stretching band (in dilute CCl$_4$ solutions) is asymmetric and can be resolved by computer methods[6] into two bands.

For ethanol, the normal OH stretch is located at 3636 cm^{-1} with a weaker band at 3622 cm^{-1} due to the OH stretch of a different rotational conformation. These conformers are described as staggered (**15**) and skew (**16**).

Vinyl ethers show a doublet for the C=C stretching mode at 1640–1620 and 1620–1610 cm^{-1}. These bands correspond to rotational isomers about the C–O bond. The two

[6]R. N. Jones et al., *Computer Programs for Absorption Spectrophotometry*, N. R. C. Bulletin No. 11, National Research Council of Canada, Ottawa; J. T. Bulmer and H. F. Shurvell, *Can. Spectrosc.*, **16**, 94 (1971).

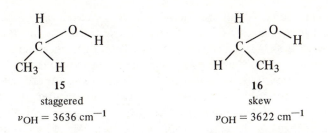

15	16
staggered	skew
$\nu_{OH} = 3636 \text{ cm}^{-1}$	$\nu_{OH} = 3622 \text{ cm}^{-1}$

bands show variations in intensity with temperature. The CH_2 deformation band is also found to be a doublet that is due to the two different rotational isomers.

5-18e Geometrical Isomers

Infrared and Raman spectroscopy is useful in distinguishing between cis and trans isomers. Absorption of infrared radiation by a molecule can only occur if there is a change in dipole moment accompanying a vibration. For cis isomers a dipole moment change occurs for most of the normal vibrations. However, trans isomers usually have higher symmetry, which leads to a zero or very small dipole moment change for some vibrations, so that they are not observed in the infrared spectrum. This phenomenon is illustrated in Figure 5-24 for the C—Cl stretching vibrations of *cis-* and *trans*-dichloroethylene.

It is seen that in case (b), of Figure 5-24 the symmetric vibration involving stretching of the C—Cl bonds produces no change in dipole moment. If the chlorine atoms were to be replaced by similar, but not identical groups, then a small dipole moment change would be produced during the vibration. This oscillating dipole, however, might be too weak to give rise to an observable infrared absorption. Thus we can conclude that trans compounds often have simpler infrared spectra than the cis isomers.

For a vibration to be active in the Raman Spectrum, there must be a change in polarizability during the vibration. The highly symmetrical modes *always* produce a change in polarizability, whereas less symmetrical vibrations sometimes give rise to no change and therefore are not seen in the Raman spectrum.

For the C—Cl stretching modes of the two dichloroethylenes of Figure 5-24, both vibrations of the cis isomer (point group C_{2v}) are seen in the Raman spectrum. But only the symmetric mode of the trans compound (point group C_{2h}) is Raman active. Once again we conclude that trans isomers have simpler spectra than the cis compounds. We also note that for

FIGURE 5-24. The C—Cl stretching modes of cis- and trans-dichloroethylenes (the plane of the molecule is assumed to be the xz plane). (a) Dipole moment changes (observed in both ir and Raman at 710 cm⁻¹). (b) No change in dipole moment (not observed in the ir but observed in the Raman at 845 cm⁻¹; see Figure 4-2). (c) Dipole moment changes (observed in both ir and Raman at 860 cm⁻¹). (d) Dipole moment changes, but no change in polarizability (observed only in the ir at 920 cm⁻¹; see Figure 4-1).

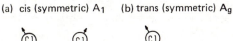

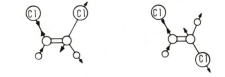

the trans compound some vibrations not seen in the infrared spectrum are Raman active (Figure 5-24b), and conversely certain frequencies not seen in the Raman spectrum are infrared active (Figure 5-24d). Trans-substituted alkenes are characterized by a very strong ir band near 970 cm^{-1} due to the wagging vibration. For the cis isomer a medium-to-strong band is observed between 730 and 650 cm^{-1} for this mode. Another point that is helpful in determining which structural isomer is present is the observation that for a trans isomer the antisymmetric C–X stretching mode in the XC=CX structure is observed at frequencies 20-40 cm^{-1} higher than for the corresponding cis isomer. Many such examples are found in the spectra of long-chain acids, alcohols, and esters. A similar observation has been made concerning the C=C stretch in unsaturated ketones and unsaturated hydrocarbons. Again the trans isomer has a slightly higher (5-10 cm^{-1}) frequency. The infrared absorption may be very weak for the trans compound for the reasons noted above. The lower frequency in the cis compound is probably due to the lower symmetry (or pseudo-symmetry), which gives rise to greater coupling of –C–C= or –C=C– vibrations with other lower frequency skeletal modes of the molecule.

5-18f Hydrogen Bonding

Hydrogen bonding manifests itself in very broad OH and NH stretching bands at frequencies considerably lower than normal. Changes in intensity of these bands can be brought about by changes in temperature and concentration. In solutions of carboxylic acids in an inert solvent such as CCl$_4$, the presence of monomer, dimer, and polymeric species can be identified in the carbonyl stretching region.

In addition to the H–bonded OH and NH stretching bands ($\bar{\nu}_{OH}$ or $\bar{\nu}_{NH}$) between 3500 and 2500 cm^{-1}, the R–OH or R–NH bending modes can also be observed between 1700 and 1000 cm^{-1}. The torsional motion of the R–OH or R–NH bonds gives rise to absorption between 900 and 300 cm^{-1}. Stretching of the hydrogen bond itself has been observed in the far infrared in many cases between 200 and 50 cm^{-1}, and bending of the hydrogen bond occurs at very low frequencies, usually below 50 cm^{-1}.

Overtones of the OH and NH stretching modes can be conveniently studied in the near infrared. These modes are quite anharmonic (see section 3-7c) in the free state. Hydrogen bonding often reduces the anharmonicity so that the overtones are observed at $2\times, 3\times$, etc., the fundamental frequencies.

In addition to the well-known cases of –OH and –NH bonding to O or N, there is evidence for hydrogen bonding by ≡CH and –SH to various atoms and by –OH to halogens and to π-electron clouds.[7] Acetylenes form H-bonded complexes with acceptor groups in solutions of inert solvents, so that a splitting of the $\bar{\nu}_{CH}$ band is observed. The relative intensities of the bands due to free and associated CH groups change with concentration. A unique example of a hydrogen bond to a carbon atom is found in the complex formed between phenylacetylene and benzylisocyanide. Acetylene itself can form 1/1 or 1/2 H-bonded complexes with acetone. The C≡C stretching mode is forbidden by symmetry in the infrared (see section 3-10), but in the unsymmetrical 1/1 complex with acetone the vibration becomes allowed. The 1/2 complex has a center of symmetry and the C≡C stretch again is not seen in the infrared.

In thiophenols self-association, possibly due to hydrogen bonding, occurs to a small

---

[7]Other C–H bonds can be involved in hydrogen bonding, and a recent book by R. D. Green, *Hydrogen Bonding by C–H Groups*, John Wiley & Sons, New York, 1974, is devoted to this subject.

extent. The SH group can form hydrogen bonds with pyridine and other compounds, but frequency shifts of $\bar{\nu}_{SH}$ are usually small. Both intra- and intermolecular hydrogen bonding can occur between OH groups in alcohols or phenols and halogen atoms. In 2-chloroethanol, for example, an intramolecular hydrogen bond stabilizes the gauche rotational isomer. The free $\bar{\nu}_{OH}$ in the anti conformation absorbs at 3623 cm^{-1}, while for the H-bonded isomer the frequency is 3597 cm^{-1}. Halophenols also show two $\bar{\nu}_{OH}$ bands separately by 50–100 cm^{-1} due to bonded and nonbonded conformations. Infrared spectra also indicate that intermolecular OH$\cdots$halogen bonding occurs between alkyl halides and phenols or alcohols.

Well-defined hydrogen bonds can be formed between OH groups and π-electron clouds. Shifts of $\bar{\nu}_{OH}$ are quite small and only rarely exceed 50 cm^{-1}. Inter- and intramolecular bonds have been studied in phenols and aromatic alcohols. An interesting case is o-allylphenol, in which only for the cis arrangement can an intramolecular hydrogen be formed between the OH group and the π-electron cloud of the double bond. In the trans isomer, the double bond is too far away from the OH group for interaction to occur. The OH stretching frequency of the associated conformation is 60 cm^{-1} lower than that of the unassociated molecule.

In this section, only a very brief mention has been made of some of the cases in which hydrogen bonding is found. The reader is referred to the books by Bellamy and by Pimental and McClellan (references A3 and A6, Chapter 4) for further discussion and references to the voluminous literature on the subject.

5-19   Interpretation of Spectra: Examples

5-19a   Introduction

To deduce the structure of an unknown from its infrared and Raman spectra alone is not easy; in fact, for large complicated molecules it is not possible. The best way to learn how to extract structural information from spectra is by practice. A list of sources of interpreted spectra and problems is given in the Bibliography (references A1 to A9). A few examples will be given here of an approach that might be taken to a problem.

It is useful to start with a checklist of basic structural questions: Which elements are present? What is the molecular formula? What other information is available from other instrumental or chemical methods? Further clues than can be gathered from an infrared survey spectrum. Again, a checklist is useful: Are there any broad absorption bands (see Table 5–2)? Is there an aromatic ring present (see section 5–6)? What kind of X—H bonds are present? Is there a carbonyl group in the molecule? Finally, a systematic analysis of the bands in the infrared spectrum can be made, first using Table 5–1, then Table 4–2. Tentative assignments can be made to each band, cross-checking where possible in other regions of the spectrum and referring to Tables 4–1 and 5–3 to 5–21. A Raman spectrum may be useful in many cases to confirm or eliminate certain assignments (Table 4–3). This method will be illustrated with some examples.

5-19b   Example 1:   An Unsaturated, Conjugated Compound

From elemental analysis and molecular weight determination the antibiotic mycomycin was found to have the formula $C_{13}H_{10}O_2$. It is obviously unsaturated and hydrogenation showed that there are eight multiple bonds. The infrared and Raman spectra indicated the presence of a terminal acetylenic CH group (3280 cm^{-1}) (see section 5–4b) and the presence of two triple bonds (2040 and 2200 cm^{-1}) (see section 5–9b). The triple bonds are probably conjugated with each other because otherwise the $-C{\equiv}C-$ frequencies would be closer together. A strong band near 1920 cm^{-1} indicates the presence of an allene $\diagup\!C{=}C{=}C\!\diagdown$ linkage (see section 4-2b). A very strong band near 1700 cm^{-1} indicates the presence of a carboxylic acid (see section 5-5b). The methyl ester was prepared and gave a carbonyl absorption at 1735 cm^{-1}, a frequency that implies a normal carboxyl group (not attached to an unsaturated C atom). A possible structure can now be written.

$$H-C{\equiv}C-C{\equiv}C-CH{=}C{=}CH-CH{=}CH-CH{=}CH-CH_2-COOH$$

The presence of the conjugated diene is characterized by two bands at 950 and 980 cm^{-1} due to CH vibrations (see section 5-9c). The positions of these bands also indicate a cis, trans arrangement for the double bonds.

5-19c Example 2: An Aromatic Unsaturated Ketone

A compound of formula $C_{10}H_{10}O$ was shown from the nmr spectrum to contain a CH_3 group. The infrared spectrum shown in Figure 5-25 indicates the presence of a monosubstituted benzene ring (very strong bands at 750 and 690 cm^{-1} and sharp bands at 1575 and 1495 cm^{-1}). The strong band at 1610 cm^{-1} and the very strong absorption centered at 985 cm^{-1} are characteristic of a trans alkene. The very strong band at 1670 cm^{-1} is then most likely due to a conjugated ketone or an aromatic ketone. Two possible structures are **17** and **18**.

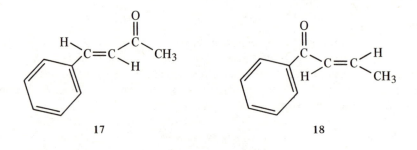

17   18

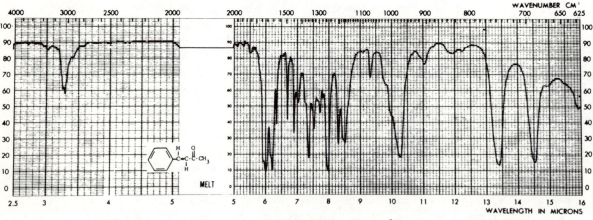

FIGURE 5-25. Infrared spectrum of benzalacetone. [Courtesy of Aldrich Chemical Company, Inc.]

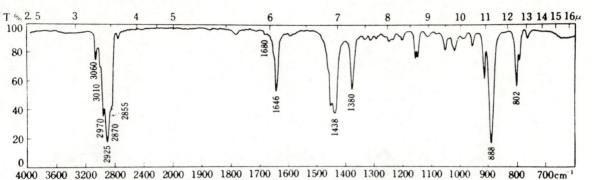

FIGURE 5-26. *Infrared spectrum of a liquid film of an optically active essential oil $C_{10}H_{16}$. [Reproduced with permission from K. Nakanishi, Infrared Absorption Spectroscopy, Nankodo Publishers Co., Ltd., Tokyo, and Holden-Day, Inc., San Francisco, 1962.]*

5-19d   Example 3: An Optically Active Essential Oil[8]

The problem is to deduce the structure of an optically active essential oil of formula $C_{10}H_{16}$ from the infrared spectrum shown in Figure 5-26. The uv spectrum shows no conspicuous maximum, indicating that the compound does not contain a conjugated system, although the degree of unsaturation (see section 5-3) is three. The bands at 888 and 802 cm^{-1} are probably due to a terminal methylene group and a trisubstituted double bond respectively (see Table 5-11). The third degree of unsaturation could be accounted for by a ring. There are no vinyl or disubstituted alkenic groups (absence of strong bands between 1000 and 900 cm^{-1} and of a medium-to-strong band between 730 and 650 cm^{-1}). There are no isopropyl or *t*-butyl groups (the single band at 1380 cm^{-1} indicates isolated CH_3 groups). A possible structure is **19**.

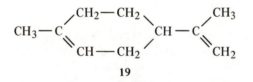

19

It should be noted that there is no direct evidence from the infrared spectrum that the compound is a cyclohexene derivative. However, the absence of a vinyl group or disubstituted olefin groups makes linear nonconjugated structures impossible. The reader should verify this claim by writing out noncyclic structures for $C_{10}H_{16}$.

5-19e   Example 4: An Amine Hydrochloride[9]

Figure 5-27 shows the infrared spectrum of a Nujol mull of an amine hydrochloride of formula $C_{10}H_{16}N_2 \cdot HCl$. The broad band centered near 2500 cm^{-1} indicates that it is the hydrochloride of a tertiary amine (see section 5-11b). The bands at 3450 and 3230 cm^{-1} suggest that a primary amine is also present. Absorption at 1615 and 1520 cm^{-1} indicates a benzene ring, and the strong band at 835 cm^{-1} is characteristic of para disubstitution.

[8]Taken with permission from K. Nakanishi, *Infrared Absorption Spectroscopy*, Nankodo Publishers Co., Ltd., Tokyo, and Holden-Day, Inc., San Francisco, 1962, problem 3, p. 121.
[9]*Ibid.*, problem 34, p. 158, with permission.

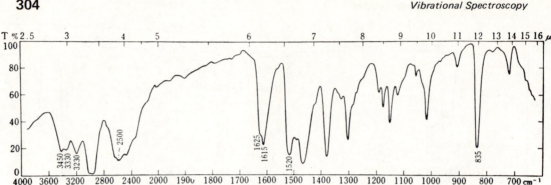

FIGURE 5–27. Infrared spectrum of a Nujol mull of $C_{10}H_{16}N_2 \cdot HCl$. [Reproduced with permission from K. Nakanishi, Infrared Absorption Spectroscopy, Nankodo Publishers Co., Ltd., Tokyo, and Holden-Day, Inc., San Francisco, 1962.]

The uv spectra of the mono- and dihydrochlorides of the compound are different. This observation indicates that both amine groups are attached to the benzene ring. A possible structure for the amine is **20**.

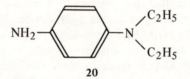

20

It should be noted that the spectrum in Figure 5–27 is of a Nujol mull and the bands at 2900, 1460, and 1380 cm^{-1} are due to the Nujol. The absorptions from the CH$_2$ and CH$_3$ groups are completely obscured.

5–19f Example 5: Using Both Infrared and Raman Spectra

Figure 5–28 shows the infrared and Raman spectra of a compound of formula $C_9H_{10}O_3$. The formula tells us that the compound is unsaturated. Several conclusions can be drawn from a preliminary examination of the infrared spectrum. The broad band centered at 3180 cm^{-1} indicates an OH group of a phenol (see section 5–10f). Absorption between 3000 and 2850 cm^{-1} is due to aliphatic CH stretching. The very strong band at 1670 is characteristic of a carbonyl group. The sharp doublet at 1610 and 1580 cm^{-1} is the first indication of the presence of a benzene ring, which we suspect from the formula.

Further indications of the benzene ring are both the sharp band at 3070 cm^{-1} and the intense, very strongly polarized band at 1030 cm^{-1} in the Raman spectrum. The peak at 755 cm^{-1} in the infrared spectrum indicates ortho disubstitution, although the presence of a second band at 700 cm^{-1} could mean monosubstitution (see section 5–6b). The position of the carbonyl stretching band indicates an aromatic ester group. This is supported by the strong band at 1210 cm^{-1}. Since we have already decided that the compound is probably a phenol, we can now conclude that it has an ortho-disubstituted benzene ring.

The broad background absorption between 1400 and 1100 cm^{-1} is probably due to hydrogen-bonded OH deformation of the phenol group. This broad background is absent in the Raman spectrum. We note also that the OH stretching band is not seen in the Raman spectrum.

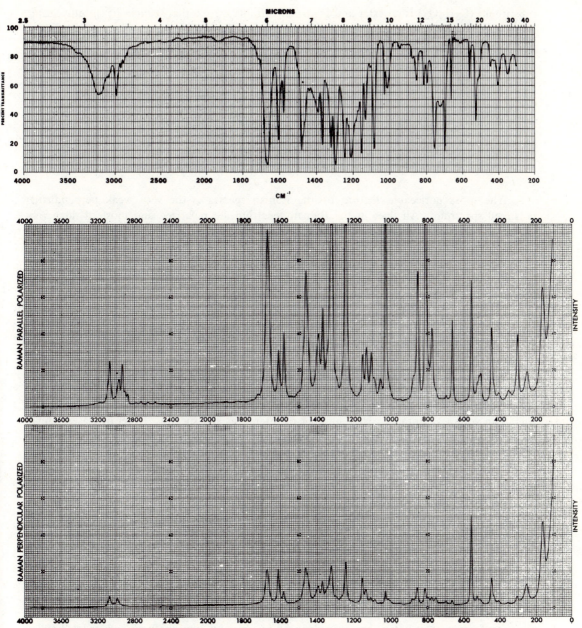

FIGURE 5–28. *Infrared and Raman spectra of a compound of formula* $C_9H_{10}O_3$.
[© *Sadtler Research Laboratories, Inc. (1973).*]

The strong band at 1480 cm^{-1} and the strong sharp peak at 1370 cm^{-1} in the infrared spectrum are characteristic of a CH$_3$ group. From the Raman lines at 1460 and 2930 cm^{-1} we conclude that there is probably also a CH$_2$ group in the molecule.

Summarizing the evidence, we conclude that the molecule is a substituted phenol (C$_6$H$_4$OH), with the second group attached to the benzene ring in the ortho position. The substituent ($C_3H_5O_2$) is an ester and from the formula we see that it must be an ethyl ester. From the infrared and Raman spectra alone we conclude that the compound is the ethyl ester of salicylic acid (**21**).

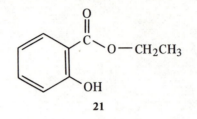

21

The most noticeable feature of the Raman spectra is the very weak perpendicular polarized spectrum. All lines in the Raman spectrum are polarized and most of them very strongly polarized.[10] Hence, there is only one symmetry type and the point group must be C_1. In other words, the molecule has no symmetry.

A second indication of low symmetry is the fact that every line in the Raman spectrum (above 300 cm^{-1}) is seen on careful examination to have a corresponding band in the infrared. It is also seen that in many cases strong bands in one spectrum are weak in the other.

5-20 Polymers

Infrared spectra have been used extensively as an aid in the determination of the structure of polymers. Little more can be done here than give references to books and review articles on this work (references B1 to B9). The comprehensive treatment by Hummel and Scholl in their two volumes contains a large number of references and spectra. Recently, laser Raman spectroscopy has been used to study polymers. This technique is particularly useful for hard, insoluble polymers, since in many cases no sample preparation is required. Conformational studies in polynucleotides have been made using Raman spectroscopy. Infrared studies on some biopolymers have been carried out.

Generally, group frequencies can be found in spectra of polymers. This is useful for identification purposes. Polarized infrared measurements (see section 2–5f) can provide information on chain conformation and structure in polymers. Folded (α) and extended (β) forms of proteins and polypeptides can be distinguished. For example, the NH stretching bands at 3310 and 3060 cm^{-1} of α- and β-keratins show marked infrared dichroism. In the α form both bands show parallel dichroism, whereas for the β form the dichroism is perpendicular. The NH deformation modes behave similarly.

The fine structure of cellulose has been studied by the polarized infrared technique. The chain conformation in xylans has been deduced by this and other methods to be a left-handed helix. Other examples of studies by polarized infrared spectroscopy include the crystal structure of the aminopolysaccharide chitin and the detection of crystallinity in synthetic polymers.

[10]The bands at 165 and 560 cm^{-1} appear at first sight to be depolarized, but each has a depolarization ratio less than 0.75 (see section 3–11b), so they, too, are polarized.

5-21 Applications of Near Infrared Spectroscopy

The region of the infrared spectrum between 12,500 and 4000 cm^{-1} can be useful for the detection and analysis of functional groups containing hydrogen. Some specific cases are terminal methylene and epoxide groups, cis double bonds, and phenolic OH groups. The characteristic frequencies listed in Table 4-5 can be used to supplement information obtained from the mid-infrared and Raman spectra to identify the groups present in compounds of unknown structure.

Near infrared measurements have been used to study natural polymers. For example, a combination of the NH stretching and amide II fundamentals gives rise to a band near 4800 cm^{-1} in the spectra of proteins and polypeptides that can be used to identify α or β forms by the use of polarized radiation.

The intensity of an absorption band can be used to determine quantitatively the number of groups in a molecule. For example, the second overtones of the CH stretching vibrations in CH_3 and CH_2 groups are observed at 8400 cm^{-1} and 8260 cm^{-1}, respectively. The intensities of these bands are proportional to the number of CH_3 or CH_2 groups present. Similarly, the overtone of the NH stretching mode near 6670 cm^{-1} has an intensity of $\epsilon =$ ca. 0.4 per NH group in aliphatic amines and $\epsilon =$ ca. 0.6 in simple primary aromatic amines.

Many other applications of near infrared spectroscopy to structural studies are given in references C1 to C5.

5-22 Applications of Far Infrared Spectroscopy

5-22a Introduction

The far infrared, as an extension of the normal infrared region, is useful for detection of fundamentals involving heavy atoms, skeletal deformations, torsional modes, and other low frequency vibrations. A list of group frequencies in the far infrared was given in Table 4-4. An important use of the far infrared is the study of hydrogen bonding. The H$\cdots$X stretching mode can be observed between 200 and 50 cm^{-1} and in some cases the bending vibration may be observed below 50 cm^{-1}.

5-22b Torsions

Torsional vibrations involving CH_3 groups, although very weak in the infrared, have been observed between 250 and 100 cm^{-1} (see also section 5-7). The CF_3 torsion occurs at still lower frequencies (ca. 50 cm^{-1}). In aliphatic halogen compounds the C–X bending modes occur in the far infrared between 350 and 250 cm^{-1} so that trans and gauche rotational isomers in these compounds can be studied. Ring puckering modes in compounds such as trimethylene oxide (53 cm^{-1}) and 3-methyleneoxetane (122 cm^{-1}) have been reported, and a pseudorotation process in tetrahydrofuran has been observed.

5-22c Lattice Modes in Crystals

Lattice vibrations in crystals and polymers can be recorded in the far infrared. In polyethylene, for example, lattice modes occur at 95 and 65 cm^{-1}. Infrared active lattice modes in orthorhombic polyoxymethylene have been observed at 130, 89, and 83 cm^{-1}.

5-22d Metal-Carbon Frequencies

Metal-carbon stretching frequencies are found in the region 800–400 cm^{-1} and many of these were listed in Table 5-19. Deformation modes of groups involving metal-carbon bonds occur at lower frequencies. Some further references to applications of far infrared spectroscopy are given in references D1 to D7.

BIBLIOGRAPHY

A. Sources of Interpreted Spectra and Problems

1. T. Cairns, *Spectroscopic Problems in Organic Chemistry*, Heyden, London, 1964.
2. A. J. Baker, T. Cairns, G. Eglinton, and F. J. Preston, *More Spectroscopic Problems in Organic Chemistry*, Heyden, London, 1967.
3. *Unknown Spectra*, Sadtler Research Laboratories, Inc., Philadelphia, Pa. 19104. A collection of 99 spectra taken from the Sadtler collection. An answer sheet is supplied.
4. K. Nakanishi, *Infrared Absorption Spectroscopy*, Holden-Day, San Francisco, 1962. Contains 85 problems with detailed solutions.
5. D. Steele, *The Interpretation of Vibrational Spectra*, Chapman and Hall, London, 1971. Contains 26 infrared (and other) spectra of organic molecules with interpretation.
6. R. K. Smalley and B. J. Wakefield, "Infrared Spectroscopic Problems and Answers," in *An Introduction to Spectroscopic Methods for the Identification of Organic Compounds*, Vol. 1, *Nuclear Magnetic Resonance and Infrared Spectroscopy*, F. Scheinmann, ed., Pergamon, Oxford, 1970. The chapter contains 14 problems, followed by detailed answers.
7. H. A. Szymanski, *Interpreted Infrared Spectra* 3 vols., Plenum, New York, 1964, 1966, and 1967.
8. N. B. Colthup, L. H. Daly, and S. E. Wiberley, *Introduction to Infrared and Raman Spectroscopy*, Academic, New York, 1964. Contains 624 partially interpreted infrared spectra.
9. R. T. Conley, *Infrared Spectroscopy*, 2nd ed., Allyn and Bacon, Boston, 1972. Chapter 9 contains some interpreted spectra.

B. Vibrational Spectra of Polymers

1. D. O. Hummel and F. Scholl, *Infrared Analysis of Polymers, Resins, and Additives: An Atlas*, 2 vols., Wiley–Interscience, New York, 1971.
2. A. Elliott, *Infrared Spectra and Structure of Organic Long-Chain Polymers*, Edward Arnold, London, 1969.
3. D. O. Hummel, *Infrared Spectra of Polymers*, Interscience, New York, 1966.
4. A. Davis, *Progr. Infrared Spectrosc.*, **1**, 17 (1962).
5. S. Hanlon, in *Spectroscopic Approaches to Biomolecular Conformation*, D. W. Urry, ed., American Medical Association, Chicago, 1970.
6. R. F. Schaufele, *Trans. N. Y. Acad. Sci.*, **30**, 69 (1967).
7. M. J. Gall and P. J. Hendra, "Laser-Raman Spectroscopy of Synthetic Polymers," *The Spex Speaker*, vol. XVI, No. 1, March (1971). (Available from Spex Industries Inc., Metuchen, N. J. 08840.)
8. J. L. Koenig, "Polymer Characterization by Raman Spectroscopy," *Spectra-Physics Raman Technical Bulletin*, No. 2. (Available from Spectra-Physics, Mountain View, Ca. 94040).
9. E. W. Small and W. L. Peticolas, *Biopolymers*, **10**, 1377 (1971).

C. Near Infrared Spectroscopy

1. O. H. Wheeler, *Chem. Rev.*, **59**, 629 (1959).
2. W. Kaye, *Spectrochim. Acta*, **6**, 257 (1954).

3. R. F. Goddu and D. A. Delker, *Anal. Chem.*, **32**, 140 (1960).
4. J. D. McCallum, *Chemical Analysis in the Near Infrared*, Application Data Sheet UV-8085, Beckman Instruments Inc., Fullerton, Ca. 92634.
5. J. D. McCallum, *Progr. Infrared Spectrosc.*, **2**, 227 (1964).

D. Far Infrared Spectroscopy

1. F. F. Bentley, L. D. Smithson, and A. L. Rozek, *Infrared Spectra and Characteristic Frequencies 700-300 cm^{-1}*, Interscience, New York, 1968.
2. J. E. Stewart in *Interpretive Spectroscopy*, S. K. Freeman, ed., Reinhold, New York 1965.
3. A. Finch, P. N. Gates, K. Radcliffe, F. N. Dickson, and F. F. Bentley, *Chemical Applications of Far Infrared Spectroscopy*, Academic, London, 1970.
4. F. A. Miller in *Molecular Spectroscopy*, P. Hepple, ed., The Institute of Petroleum, London, 1968 p. 5.
5. J. W. Brasch, Y. Mikawa, and R. J. Jakobsen, *Appl. Spectrosc. Rev.*, **1**, 187 (1968).
6. R. K. Harris and R. E. Witkowski, *Spectrochim. Acta*, **20**, 1651 (1964).
7. K. Fukushima, *Bull. Chem. Soc. Japan*, **38**, 1694 (1965).

Appendix

USING THE LITERATURE ON VIBRATIONAL SPECTRA

Vibrational spectra are extremely useful for checking for the presence of particular functional groups in a compound. This capability is, of course, very valuable to the synthetic chemist, who should routinely run an infrared spectrum at each step of a synthesis. Unequivocal identification of a compound can be obtained by direct comparison of its infrared spectrum with the spectrum of the known compound. To do this, rapid access to spectra collections and references in the literature is needed. At the time of writing, published spectra, either in the literature or in collections of spectra, are available for about 120,000 compounds.

This Appendix presents sources of collections of spectra, sources of literature, references to spectra, and some methods of searching for spectra.

A-1 Collections of Spectra

A-1a Introduction

In addition to the very extensive Sadtler Standard Infrared Spectra Collection and other large collections of infrared spectra, there are several small but useful collections that can be obtained from the literature or by writing to the companies indicated. Perhaps the most useful of these is the Aldrich Library of Infrared Spectra.

A-1b Infrared Spectra

1. *Sadtler Standard Infrared Spectra*, Sadtler Research Laboratories, Inc., 3316 Spring Garden Street, Philadelphia, PA 19104. The main collection consists of over 40,000 prism or small grating spectra in loose leaf volumes, containing 1000 spectra per volume. Approximately 2000 new spectra are added each year. The format of these spectra is linear in wavelength (2–15 microns). Other Sadtler publications include infrared grating spectra, linear in wavenumber (4000–200 cm^{-1}), and infrared spectra of commercial products, biochemical, and pharmaceutical compounds.

2. *Documentation of Molecular Spectra* (DMS), Butterworth and Co. Ltd., London. A collection of data cards with spectra, frequencies, and structural information. The spectra are presented in a linear wavenumber format and the cards can be rim punched for sorting. There are over 20,000 cards in the collection and 1600 new cards are added annually.

3. *Selected Infrared Spectral Data*, American Petroleum Institute (API), Research Project 44, Department of Chemistry, Texas A&M University, College Station, TX 77843. A large collection of spectra that is continually updated. The presentation is usually linear in wavelength on the older entries in the collection, but linear in wavenumber on more recent spectra.

4. Infrared Data Committee of Japan (IRDC cards), Nankodo Co., Bunkyo-ku, Tokyo, Japan. A collection of spectra on 5″ × 8″ cards with structural and other data; similar to the DMS Collection.

5. R. Mecke and F. Langenbucher, *Infrared Spectra of Selected Chemical Compounds*, Heyden and Sons Ltd., London, 1965. A collection of 1880 spectra, with tables of frequencies; uses a format linear in wavelength.

6. C. J. Pouchert, *The Aldrich Library of Infrared Spectra*, Aldrich Chemical Company, 940 West St. Paul Avenue, Milwaukee, WI 53233. A collection of 8000 spectra in one volume. The format is linear in wavelength (2.5–16 microns), and there are 8 spectra to a page (12″ × 9″). Despite the small size of the spectra, the book is very useful.

7. K. Dobriner, E. R. Katzenellenbogen, and R. N. Jones, *Infrared Absorption Spectra of Steroids: An Atlas*, Interscience Publishers, New York, 1953. Also Vol. II by G. Roberts, B. S. Gallagher, and R. N. Jones, 1958.

8. J. E. Stewart, *Infrared Spectra of Primary and Secondary Amines*, Application Data Sheet IR 8059–C, Beckman Instruments, Inc., Fullerton, CA 92634. A collection of 25 infrared spectra of amines.

9. N. B. Colthup, L. H. Daly, and S. E. Wiberly, *Introduction to Infrared and Raman Spectroscopy*, Academic Press, New York, 1964. This book includes 624 spectra that have been redrawn with a reduced ordinate scale; the abscissa is linear in wavenumber.

10. R. H. Pierson, A. N. Fletcher, and E. St. Clair Gantz, *Anal. Chem.*, **28**, 1218 (1956). A collection of 66 infrared spectra of gases and vapors.

11. D. O. Hummel, *Infrared Spectra of Polymers*, Interscience Publishers, New York, 1966. A collection of 192 spectra of monomers, polymers, and solvents; many spectra obtained using internal reflectance techniques.

12. R. G. Zhbankov, *Infrared Spectra of Cellulose and Its Derivatives*, Consultants Bureau, New York, 1966. A collection of 210 infrared spectra of mono-, di-, and polysaccharides and derivatives; polyhydric alcohols; celluloses and their hydrolysis, ethanolysis, and oxidation products; and cellulose ethers and esters.

13. H. A. Szymanski, *Interpreted Infrared Spectra*, 3 vols., Plenum Press, New York, 1964, 1966, and 1967. These volumes contain 635 spectra interpreted in terms of group frequencies. The presentation of most spectra is linear in wavelength.

14. *Eastman Spectrophotometric Solvents*, Distillation Products Industries, Rochester, NY 14603. A collection of 36 infrared and ultraviolet spectra of solvents with linear wavelength presentation.

15. *Spectroquality Solvents*, MC/B Manufacturing Chemists, 2909 Highland Avenue, Norwood, OH 45212. A collection of 52 infrared, near infrared, and ultraviolet spectra of solvents. The infrared spectra are in linear wavenumber format (4000–250 cm^{-1}).

16. F. S. Parker, *Progr. Infrared Spectrosc.*, **3**, 89 (1967). A collection of 23 infrared spectra of dipeptides in acidic, basic, and isoionic aqueous solutions.

17. R. W. Hannah and S. C. Pattacini, *Drug Identification from Their Infrared Spectra*, Infrared Bulletin No. 16, The Perkin–Elmer Corporation, Norwalk, CT 06852, 1972. This article contains a catalogue of 55 spectra of drugs and common diluents. The spectra were

recorded in the 4000–250 cm^{-1} range, linear in wavenumber, on a high resolution instrument.

18. D. O. Hummel and F. Scholl, *Infrared Analysis of Polymers, Resins, and Additives: An Atlas*, Wiley–Interscience, New York, 1971. This two-volume book contains textual material and a large collection of spectra, linear in wavelength.

19. J. Holubek and O. Strouf, eds., *Spectral Data and Physical Constants of Alkaloids*, Heyden and Sons Ltd., London, Vol. 1, 1965, and subsequent volumes. By 1971 six volumes, containing the infrared spectra of 800 compounds, had been published.

A-1c Near Infrared Spectra

Spectroquality Solvents, MC/B Manufacturing Chemists, 2909 Highland Avenue, Norwood, OH 45212. The near infrared spectra of 52 solvents are presented, linear in wavelength from 750 to 2500 nanometers (13,333–4000 cm^{-1}).

A-1d Far Infrared Spectra

1. *Sadtler Standard Infrared Grating Spectra*, Sadtler Research Laboratories, Inc., 3316 Spring Garden Street, Philadelphia, PA 19104. These spectra cover the far infrared region out to 200 cm^{-1}.

2. F. F. Bentley, L. D. Smithson, and A. L. Rozek, *Infrared Spectra and Characteristic Frequencies 700–300 cm^{-1}*, Interscience Publishers, New York, 1968. This book includes a collection of spectra of 1500 organic compounds.

3. D. O. Hummel, *Infrared Spectra of Polymers*, Interscience Publishers, New York, 1966. A collection of 192 far infrared spectra of polymers, resins, and related substances in the 700–250 cm^{-1} region.

A-1e Raman Spectra

1. *Selected Raman Spectra Data*, American Petroleum Institute (API), Research Project 44. This compilation is produced in the same format as the API infrared spectra (section A-1b, no. 3). (Mercury arc excitation.)

2. M. R. Fenske, W. G. Braun, R. V. Wiegand, D. Quiggle, and D. H. Rank, *Anal. Chem.*, **19**, 700 (1947). Raman spectra of 172 pure hydrocarbons. (Mercury arc excitation.)

3. W. G. Braun, D. F. Spooner, and M. R. Fenske, *Anal. Chem.*, **22**, 1074 (1950). Raman spectra of 119 hydrocarbons and oxygenated compounds. (Mercury arc excitation.)

4. F. E. Dollish, W. G. Fateley, and F. F. Bentley, *Group Frequencies in the Laser-Raman Spectra of Organic Compounds*, Wiley–Interscience, New York, 1973. This work includes some representative spectra.

5. *Sadtler Standard Raman Spectra*, Sadtler Research Laboratories, Inc., 3316 Spring Garden Street, Philadelphia, PA 19104. A new collection, started in 1973, consisting initially of 2000 compounds in five volumes, grouped into approximately 50 chemical classes. For each compound three spectra are given: (a) the infrared (grating) spectrum, linear in wavenumber, (b) the Raman parallel polarized spectrum, and (c) the Raman perpendicular polarized spectrum.

6. K. F. W. Kohlrausch, *Ramanspektren*, Heyden and Sons Ltd., London, 1972, reprinted from the original German edition. This work contains data on virtually all Raman spectra recorded up to 1940. (Mercury arc excitation.)

A-2 Sources of References to Published Spectra

A-2a Infrared Spectra

1. *Molecular Formulae List of Compounds, Names, and References to Published IR and Far IR Spectra*, American Society for Testing and Materials (ASTM), 1916 Race Street, Philadelphia, PA 19103, 1969. An index to 92,000 published ir spectra, listed in increas-

ing number of C atoms, H atoms, etc. Also available is an alphabetical list of compounds, formulas, and references to published spectra.

2. H. M. Hershenson, *Infrared Absorption Spectra*, Indexes for 1945–1957 and 1958–1962, Academic Press, New York, 1959 and 1964. A total of 36,000 references to ir absorption spectra. The indexes are alphabetic and references are made to 66 journals and one collection of spectra.

3. *An Index to Published Infrared Spectra*, H. M. Stationary Office, London. An index compiled according to chemical structure.

4. *Current Literature Lists of IR, Raman, and Microwave Spectra*, Butterworth and Co. Ltd., London, 1967 onward. A list of references issued periodically by the publishers of the DMS collection. A "peephole" index for the literature lists is supplied in the form of computer-size cards. The index cards are consolidated every two years on larger cards. Each biannual set of cards indexes a set of 5000 DMS spectral cards as well as the literature lists.

5. W. F. Ulrich, *Bibliography of Infrared Applications,* Beckman Instruments, Inc., Fullerton, CA 92634. A bibliography covering the period from the late 1940's to early 1963. The references are alphabetic by author names, but there is also an extensive alphabetic cross-reference index.

A–2b Far Infrared Spectra

1. The ASTM index (section A–2a, no. 1) includes far infrared spectra. The entries carry a suffix when they refer to far infrared.

2. F. F. Bentley, L. D. Smithson, and A. L. Rozek, *Infrared Spectra and Characteristic Frequencies 700–300 cm^{-1}*, Interscience Publishers, New York, 1968. This book contains a bibliography of about 2000 references to the literature through December, 1966. It is compiled according to chemical classes and all references have spectral data below 700 cm^{-1}.

A–2c Raman Spectra

1. *Bibliography of Raman Publications, 1968–69,* Spectra-Physics, Inc., 1250 West Middlefield Road, Mountain View, CA 94040.

2. G. Herzberg, *Infrared and Raman Spectra of Polyatomic Molecules*, Van Nostrand, New York, 1945. In addition to the treatments of symmetry, theory of infrared and Raman spectroscopy, and molecular rotation and vibrations, this book contains a large section on infrared and Raman spectra of individual molecules and numerous references to published Raman spectra.

A–3 Spectra Searching by Manual and Computer Methods

There are several methods available for retrieval of infrared data by searching the compilations of spectra described in the previous two sections. The DMS cards can be searched by inserting a needle through various holes in the perimeter of the deck of DMS cards. Those compounds containing certain combinations of functional groups can be quickly removed from the collection and studied.

Another method is the Sadtler Spec-Finder, which must be used in conjunction with the Sadtler Standard Spectra Collection. All spectra in the Sadtler Collection are included in the Spec-Finder. The spectra are arranged according to a systematic classification of major absorption bands. The spectrum of an unknown is coded by selecting the strongest band in each 1 micron region. Frequencies in cm^{-1} have to be converted to microns before this method can be used. The strongest band in the spectrum is noted and compared with entries in the Spec-Finder list. With a special coding slip, the entries can be rapidly scanned and serial numbers of Sadtler spectra obtained. These are then looked up and compared with the spectrum of the unknown..

A third method makes use of a mechanical sorter. A collection of computer cards have been prepared by the American Society for Testing and Materials. Basically, this compilation is an index to collections of spectra, mainly the Sadtler. The cards are sorted by machine to yield the serial number of a spectrum in a published collection. They contain information on chemical classification, melting point and boiling point, number of C, N. O, S, and Si atoms, as well as the positions of the absorption bands in microns. Further details of this method are available from Sadtler Research Laboratories.

Recently, a fully computerized method of searching has been described by D. S. Erley.[1] This approach is essentially a "fingerprinting" device for identifying unknowns by comparing their spectra with the spectra in the ASTM infrared data file. The unknown spectrum is "described" to the computer by entering the positions of its strong absorption bands. Regions where no absorption is observed are entered together with other data such as elements present and functional groups if known. The computer then compares this information to the 102,000 ASTM standards that have been coded and stored in the computer. The serial numbers of those compounds that most closely match the unknown are printed.

At the time of writing, the above system is available from DNA Systems, Inc., 2415 West Stewart Avenue, Flint, MI 48504. The computer program is called FIRST-2 and is available on a perpetual lease basis for IBM 1130, 1800, and 360 computers. The searching can also be done by teletype for a fee.

---

[1] Fast Searching System for the ASTM Infrared Data File, D. S. Erley, *Anal. Chem.*, **40**, 894 (1968); *Appl. Spectrosc.*, **25**, 200 (1971).

Part Three

ELECTRONIC SPECTROSCOPY:
ULTRAVIOLET-VISIBLE AND
CHIROPTICAL

1

INTRODUCTION

1-1 Light Absorption

All organic compounds absorb light in the ultraviolet or visible region of the electromagnetic spectrum. Absorption of uv or visible light[1] by molecules will occur only when the energy of the incident radiation is the same as that of a possible electronic transition in the molecules involved. Such absorption of energy is termed *electronic excitation* and corresponds to promotion of an electron from the ground state to a higher electronic excited state. The excitations can be considered *approximately* in terms of the transition of a single electron from a filled to a vacant molecular orbital. The electronic excitations of greatest importance in organic chemistry usually involve the transition of an electron from the highest occupied to the lowest vacant MO.

Molecules continually vibrate and rotate about the axes between atoms. Since these vibrational and rotational motions are quantized, absorption of light also causes transitions to higher vibrational and rotational energy levels. Thus all electronic transitions are necessarily accompanied by corresponding transitions in the vibrational and rotational energy levels of the molecule. The total excitation energy ΔE is a sum of three terms, electronic, vibrational, and rotational (eq. 1-1). The energy changes involved in the three terms of eq. 1-1 decrease

$$\Delta E = \Delta E_{\text{elec}} + \Delta E_{\text{vib}} + \Delta E_{\text{rot}} \qquad \text{1-1}$$

on going from left to right as written. Rotational energy changes in a molecule may be observed

---

[1]The terms *light* and *electromagnetic radiation* will be used interchangeably. Note that there is no fundamental difference between the visible and ultraviolet regions. The distinction is a physiological one, based on the fact that the human eye is only sensitive between about 400 and 750 nm (25,000 and 13,000 cm^{-1}).

as pure transitions only in the microwave spectral region. Changes in vibrational levels require more energy and are observed in the infrared region (see Part Two). In this part we focus on the chemical information obtainable by the measurement of the electronic transitions in the ultraviolet and visible regions.

1-2 Some Concepts and Delineations

Electromagnetic radiation may be described in terms of the frequency ν (sec^{-1}) of its waves or the wavelength λ between them or by the wavenumber $\bar{\nu}$ (see Part Two, Chapter 1). These quantities are related to each other by eq. 1-2, in which c is the velocity of light (3×10^8 m sec^{-1}),

$$c = \lambda \nu \qquad\qquad 1\text{-}2$$

and eq. 1-3. Commonly used wavelength units in the ultraviolet and visible regions are either

$$\bar{\nu} = \frac{1}{\lambda} \qquad\qquad 1\text{-}3$$

nanometers (nm; 1 nm = 10^{-9} m) or Ångstrom units (Å; 1 Å = 10^{-10} m). The older unit millimicron (mμ) has been replaced in the SI system.

According to the Planck equation (1-4), frequency is directly proportional to energy.

$$E = h\nu \qquad\qquad 1\text{-}4$$

However, it is common practice to use units of either wavelength or wavenumber. Table 1-1 lists the units commonly used for λ, ν, and $\bar{\nu}$, and Table 1-2 gives some useful conversion factors. Equation 1-5 is convenient for calculation of energies in the familiar units of

$$E = \frac{28,635}{\lambda} \text{ kcal mole}^{-1} \quad \text{(for } \lambda \text{ in nm)} \qquad\qquad 1\text{-}5$$

kcal mole^{-1}. Hence, light of 300 nm wavelength corresponds to an energy of 95.4 kcal mole^{-1}.

Table 1-1. *Definitions of Terms and Equations*

Quantity	Equation	Unit	Dimensions
Wavelength, λ	–	nanometer, nm Ångstrom, Å	1 nm = 10^{-9} m 1 Å = 10^{-10} m
Wavenumber, $\bar{\nu}$	$\nu = \dfrac{1}{\lambda}$	reciprocal centimeter, cm^{-1}	vibrations per second; the wavenumber is the reciprocal of the wavelength in centimeters.
Frequency, ν	$\nu = \dfrac{c}{\lambda}$	hertz, Hz, or sec^{-1}	cycles per second
Energy	$E = h\nu, \dfrac{hc}{\lambda}, hc\bar{\nu}$	depends on the units of of h	–

Table 1-2. *Useful Conversion Factors*

cm^{-1}	Hz	$kcal\ mole^{-1}$
1	3.00×10^{-10}	2.86×10^{-3}
3.33×10^{-11}	1	9.53×10^{-14}
3.50×10^2	1.05×10^{13}	1

Combination of eqs. 1-2, 1-3, and 1-4 leads to eqs. 1-6 and 1-7.

$$E = hc\bar{\nu} \qquad\qquad\qquad 1\text{-}6$$

$$E = 28.635 \times 10^{-4} \times \bar{\nu}\ kcal\ mole^{-1} \quad (\text{for } \bar{\nu} \text{ in } cm^{-1}) \qquad 1\text{-}7$$

Wavenumbers are thus directly proportional to energy so that the same number of reciprocal centimeters (cm^{-1}) represents the same energy anywhere in the electromagnetic spectrum. For example, a shift of λ_{max} of 700 cm^{-1} anywhere in the spectrum corresponds to 1.95 kcal mole^{-1}. On the other hand, wavelength is inversely proportional to energy and thus the relationship is not linear. As an example, an energy change of 1.95 kcal mole^{-1} at 200 nm corresponds to a shift of 2.7 nm, but the same energy change at 800 nm corresponds to a shift of approximately 4.4 nm.

At the lower end of the visible spectrum, below 400 nm, is the uv region. It is convenient to divide the uv into two parts, the near uv, 190–400 nm (53,000–25,000 cm^{-1}), and the far or vacuum uv, below 190 nm (53,000 cm^{-1}). The reason for this seemingly arbitrary division is due mainly to the fact that atmospheric oxygen begins to absorb around 190 nm. Oxygen must be removed from the spectrophotometer, either by using a vacuum instrument or by vigorous purging with, e.g., nitrogen. The uv-vis spectroscopic regions and some associated energies are given in Figure 1-1.

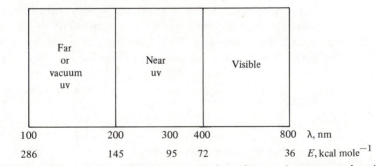

FIGURE 1-1. *Ultraviolet-visible spectroscopic regions and some associated energies.*

1-3 The Shape of Absorption Curves; Franck–Condon Principle

For simplicity, let us look at the ground and excited electronic states of a diatomic molecule. The case for a polyatomic molecule is similar but more difficult to visualize since it requires the superposition of many two-dimensional potential energy surfaces.

In the more common case, the bond strength in the excited electronic state will be

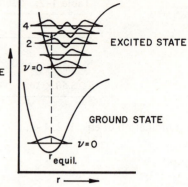

FIGURE 1-2. *Potential energy diagram for a diatomic molecule illustrating Franck–Condon excitation. The equilibrium separation is longer in the excited than in the ground state.*

less than that in the ground state and the equilibrium internuclear distance will be longer than in the ground state, giving the potential curves shown in Figure 1-2.

If the vibrational frequency is fairly high, essentially all the molecules will exist in their ground vibrational state. Excitation can occur to any of the excited state vibrational levels[2] so that the absorption due to the electronic transition consists, in theory, of a large number of lines. In practice, for most organic molecules the lines overlap so that a continuous band is observed. Hence the shape of an absorption band may be considered to be determined by the spacing of the vibrational levels and by the distribution of the total band intensity over the vibrational subbands. The intensity distribution is determined by the *Franck–Condon principle*, which states that *nuclear motion may be considered negligible during the time required for an electronic excitation*. For example, the time required for an electron to circle a hydrogen nucleus can be calculated from Bohr's model to be about 10^{-16} sec, whereas a typical molecular vibration is about 10^{-13} sec, about a thousand times longer. Another statement of the Franck–Condon principle based on classical mechanics is that the most probable vibrational component of an electronic transition is one that involves no change in the position of the nuclei, a so-called *vertical transition* represented by the vertical arrow in Figure 1-2. The most probable transition is to the excited $\nu = 3$ state. This state has a maximum at the same internuclear distance as that corresponding to the starting point of the transition. Figure 1-3 shows the vibrational-electronic (vibronic) spectrum corresponding to Figure 1-2, with the 0–3 band the most intense one. Note that the other transitions, including the 0–0 band, have significant probabilities. This result is not necessarily due to *nonvertical* transitions but

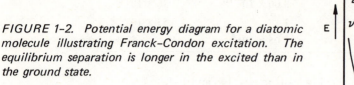

FIGURE 1-3. *Intensity distribution among vibronic bands as determined by the Franck–Condon principle.*

---

[2]Excitation also occurs to various excited state rotational levels, but the rotational fine structure is almost never resolved. It contributes only a bandwidth to each vibrational subband.

may be due to the fact that even in the ground electronic state (zeroth vibrational level), the internuclear distance is described by a probability distribution (Figure 1-2). Therefore, transitions may originate from $v = 0$ over a range of r values so that more than one band originating from $v = 0$ may be observed.

According to classical mechanics, a vibrating system spends most of its time at the ends of the vibration, where the system moves most slowly. Quantum mechanically, this situation corresponds to having greater amplitudes at the ends of the vibration than at the middle. Note that the functions are oscillatory and extend appreciably beyond the classical turning points. As the number of nodes increases in the higher vibrational states, the amplitudes of the wave functions in the middle of the vibration are less and at the ends of the vibration are greater. If the reasonable assumption is made that the population ratio of the ground vibronic state is essentially unity, then quantum mechanics leads to the following statement of the Franck-Condon principle: the intensity of a vibrational band in an electronically allowed transition is proportional to the absolute square of the overlap integrals of the vibrational wavefunctions of the initial and final states. Note that this statement does not require that transitions be *vertical*, but only that transitions will be favored when the overlap between the two vibrational wavefunctions is large. In addition, the statement that electronic transitions are fast compared to nuclear motion would suggest a measurement that violates the Heisenberg uncertainty principle.[3]

Sometimes on raising the temperature, the vibrational structure of a band will be lost. This band broadening is due to the population of several ground vibrational states at higher temperature so that a larger number of possible vibrational transitions can occur upon electronic excitation. Featureless or broad bands are also observed at ambient temperatures, usually in solution spectra where solute–solvent vibrational interactions become important.

1-4 Singlet and Triplet States

Most molecules have ground and first excited electronic states in which all electron spins are paired, even though in the excited state there may be two orbitals each possessing only one electron. Such states are known as *singlet states* since, with no net spin angular momentum, imposition of a reference direction by an applied magnetic or electric field can produce only the single component of zero angular momentum in the field direction.

For molecules having an even number of electrons, regardless of whether or not the ground state is a singlet, there will be excited states in which a pair of electrons have their spins parallel, giving the molecule a net spin angular momentum. Angular components along a given direction can now have values of $+1, 0,$ and -1 times the angular momentum. Such an electronic configuration is known as a *triplet state*.

1-5 Measurement of Light Absorption

The laws of Lambert, Bouger, and Beer (also see Part Two) state that the proportion of light absorbed by a transparent medium is independent of the intensity of the incident light and proportional to the number of absorbing molecules through which the light passes, according

---

[3]Further details may be found in an interesting article by S. E. Schwartz, *J. Chem. Educ.*, **50**, 608 (1973).

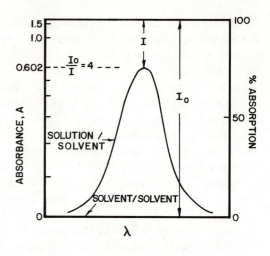

FIGURE 1-4. Measurement of solute absorbance A by a double-beam spectrophotometer.

to eq. 1-8, where I_0 is the intensity of incident light, I the intensity of transmitted light,

$$I = I_0 \, 10^{-alc} \quad \text{or} \quad \log \frac{I_0}{I} = alc \qquad\qquad 1\text{-}8$$

a the absorptivity, l the pathlength (cm), and c the concentration. Since the absorbance A is the quantity actually measured, eq. 1-8 is rewritten as eq. 1-9. When concentration is in units

$$\log \frac{I_0}{I} = A = alc \qquad\qquad 1\text{-}9$$

of moles liter^{-1}, the molar absorption coefficient (molar a) is denoted by ϵ (eq. 1-10).

$$A = \epsilon cl \qquad\qquad 1\text{-}10$$

The units of ϵ are cm^2 mole^{-1} (liters mole^{-1} cm^{-1}) but are usually omitted.

    In practice, the quantities actually measured are the relative intensities of the light beams transmitted by a reference cell containing pure solvent and by an identical cell containing the solution. When the intensities are taken as I_0 and I, respectively, the resulting absorption is that of the dissolved solute only (Figure 1-4).

1-6 Absorption Intensity; Dipole and Oscillator Strengths

As we have seen in section 1-3, the shape of an electronic absorption band is due primarily to the vibrational sublevels of the electronic states. The intensity of the interaction causing the electronic transition should therefore be proportional to the *area* under the particular band rather than to any particular value of the absorbance. A convenient unit characterizing the intensity of this interaction for the Kth electronic transition has been given by Mulliken and is termed the dipole strength D_K (eq. 1-11) in which h is Planck's constant, c is the speed

$$D_K = \frac{3hc}{8\pi^3 N_i} \int k_K(\lambda) \frac{d\lambda}{\lambda} = \frac{3000 \, hc}{8\pi^3 N} \int \epsilon_K(\lambda) \frac{d\lambda}{\lambda} \qquad\qquad 1\text{-}11$$

of light in vacuum, N_i is the number of absorbing molecules per milliliter (cm^3) of solution, N is Avogadro's number, k_K is the partial absorption coefficient for the Kth band in cm^{-1}, ϵ_K is the molar absorption coefficient in units of cm^2 $mole^{-1}$ for the Kth band, and λ is the wavelength of the light. Application of the definition of D_K to typical experimental curves reveals that the dipole strength varies from 10^{-34} to 10^{-38} erg $\cdot$ cm^3. The dipole strength is also a convenient unit for the expression of theoretical calculations on the intensities of electronic transitions.

Another unit that is used to characterize the intensity of the Kth transition is the oscillator strength f, which is related to ϵ_K by eq. 1-12, where m and e are the mass and charge of the elec-

$$f = \frac{2303mc^2}{N\pi e^2} \int \epsilon_K d\bar{\nu} = 4.315 \times 10^{-9} \int \epsilon_K d\bar{\nu} \qquad \text{1-12}$$

tron, $\bar{\nu}$ is the frequency of light in cm^{-1}, and the other units are defined above. The oscillator strength is unitless; its value generally falls between zero and one.

Having said all this, one must now confess that it is a widespread practice among many chemists to describe the intensities of absorption bands in terms of the molar absorption coefficients ϵ_{max}. The quantities are readily obtainable from the determination of the wavelength of maximum absorption. Unfortunately, the value of ϵ_{max} *is not directly related to any quantity obtainable from theory.*

1-7 Allowed and Forbidden Transitions

Electronic transitions may be classed as intense or weak. These correspond to "allowed" or "forbidden" transitions. Allowed transitions are those for which

(a) there is no change in the orientation of electron spin,
(b) the change in angular momentum is 0 or ±1, and
(c) the product of the electric dipole vector and the group theory representations of the two states is totally symmetric.

The first rule is the spin selection rule and may be stated as follows: transitions between states of different spin multiplicities are invariably forbidden since electrons cannot undergo spin inversion except for spin-orbit and spin-spin interactions. The second rule usually presents no problem since most states are within one unit of angular momentum of each other. The last rule is the symmetry selection rule. If the direct product of the representations to which the initial and final state functions belong is different from all the representations to which the coordinate axes belong, the transition moment of that transition is zero. Such a transition is said to be symmetry-forbidden. For most organic molecules, such forbidden transitions are usually observable but of weak intensity. They arise because the intensity of the electronic absorption band really depends on the average of the electronic transition moment over all the nuclear orientations of the vibrating molecule and this average is not necessarily zero. When the symmetry of a molecule is periodically changed by some vibration that is not totally symmetric, the symmetry of the electronic wavefunctions is also periodically changed since the electrons adapt instantaneously to the motion of the nuclei. Hence a symmetry-forbidden transition may become allowed. The intensity of a transition that is symmetry-forbidden but has become vibrationally allowed will be much less than that of an ordinarily allowed transition. Such vibrational contributions will be temperature dependent.

1–8 Classification of Electronic Transitions

The wavelength of an electronic transition depends on the energy difference between the ground state and the excited state. It is a useful *approximation* to consider the wavelength of an electronic transition to be determined by the energy difference between the molecular orbital originally occupied by the electron and the higher orbital to which it is excited. Saturated hydrocarbons contain only strongly bound σ electrons. Their excitation to antibonding σ^* orbitals (σ-σ^*) requires relatively large energies, corresponding to absorption in the far uv region. One exception is cyclopropane, which has λ_{max} at 190 nm. Contrast this cycloalkane to propane, which has λ_{max} about 135 nm.

Electronic transitions commonly observed in the readily accessible uv (above $\sim$190 nm) and visible regions have been grouped into several main classes.[4]

n-π^* *transitions.* These transitions can be considered to involve the excitation of an electron in a nonbonding atomic orbital—i.e., unshared electrons on O, N, S, or halogen atom—to an antibonding π^* orbital associated with an unsaturated center in the molecule. The transitions occur with compounds possessing double bonds involving heteroatoms, e.g., C=O, C=S, N=O. A familiar example is the low intensity absorption in the 285–300 nm region of saturated aldehydes and ketones.

π-π^* *transitions.* Molecules that contain double or triple bonds or aromatic rings can undergo transitions in which a π electron is excited to an antibonding π^* orbital. Although ethylene itself does not absorb strongly above about 185 nm, conjugated π-electron systems are generally of lower energy and absorb in the accessible spectral region. An important application of uv-vis spectroscopy is to define the presence, nature, and extent of conjugation. Increasing conjugation generally moves the absorption to longer wavelengths and finally into the visible region; this principle is illustrated in Table 1–3.

Table 1–3. *Effect of Extended Conjugation in Alkenes on Position of Maximum Absorption*

n *in* $H(CH=CH)_n H$	λ_{max} *(nm)*	ϵ_{max} *(liter mole^{-1} cm^{-1})*	*Color*
1	162	10,000	colorless
2	217	21,000	colorless
3	258	35,000	colorless
4	296	52,000	colorless
5	335	118,000	pale yellow
8	415	210,000	orange
11	470	185,000	red
15	547[a]	150,000	violet

[a] Not a maximum.

n-σ^* *transitions.* These transitions, which are of less importance than the first two classes, involve excitation of an electron from a nonbonding orbital to an antibonding σ^* orbital. Since n electrons do not form bonds, there are no antibonding orbitals associated

---

[4]In addition to that given here, several other systems of classification exist. For example, the designation *N–V* is used to describe transitions from a bonding to an antibonding orbital (σ-σ^*, π-π^*). The term *N–Q* designates transitions from a nonbonding atomic orbital to a higher energy molecular orbital (n-σ^*, n-π^*). Burawoy has termed π-π^* transitions *K*-bonds (from the German *Konjugation*) and n-π^* transitions *R*-bands (from an early theory that the excited state was a radical). Numerous terms also exist in the literature for the classification of ground and excited states on the basis of symmetry.

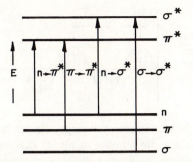

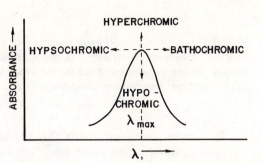

FIGURE 1-5. Relative electronic orbital energies and selected transitions in order of increasing energy.

FIGURE 1-6. Terminology of shifts in the position of an absorption band.

with them. Some examples of n–σ^* transitions are CH_3OH (vapor), λ_{max} 183 nm, ϵ 150; trimethylamine (vapor), λ_{max} 227 nm, ϵ 900; and CH_3I (hexane), λ_{max} 258 nm, ϵ 380.

Rydberg transitions are mainly to highly excited states. For most organic molecules, they occur at wavelengths below about 200 nm. A Rydberg transition is often part of a series that terminates at a limit representing the ionization potential of the molecule. Usually the positions of the bands in the series, in wavenumbers, can be expressed by eq. 1–13,

$$\nu = \nu_{IP} - \frac{R}{(n-a)^2} \qquad\qquad 1\text{-}13$$

where ν_{IP} is the ionization energy in cm^{-1}, R is a screening constant, n is an integral quantum number, and a is an empirical constant. A Rydberg series may also be recognized by rather sharp bands that are in contrast to the diffuse bands of other transitions in the same region.

The relative orbital energies of some selected electronic transitions are depicted in Figure 1–5.

Groups that give rise to electronic absorption are known as *chromophores* (color bearer, from an early theory of color). The term *auxochrome* (color increaser) is used for substituents containing unshared electrons (OH, NH, SH, halogens, etc.). When attached to π-electron chromophores, auxochromes generally move the absorption maximum to longer wavelengths (lower energies). Such a movement is described as a *bathochromic* or *red shift*. The term *hypsochromic* denotes a shift to shorter wavelength (*blue shift*). Increased conjugation usually results in increased intensity termed *hyperchromism*. A decrease in intensity of an absorption band is termed *hypochromism*. These terms are summarized in Figure 1–6.

1–9 Chiroptical Methods: Optical Rotatory Dispersion and Circular Dichroism

The light beam used in uv–vis spectroscopy is essentially unpolarized. Use of linearly polarized light (sometimes less rigorously referred to as plane-polarized light) to investigate optically active (chiral) molecules is a powerful technique for obtaining structural and stereochemical information.

Figure 1–7 considers light in the context of a wave phenomenon caused by transverse vibrations of the electric field vector (vertical arrows). There is an associated magnetic field vector perpendicular to the oscillating electric field vector, but we can ignore it for purposes

FIGURE 1-7. Wave motion propagated in the x direction by transverse vibration; λ is the wavelength. The arrows denote the electric field vector at a given instant as the light wave progresses along the x axis.

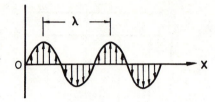

of the present discussion. Note that the electric field vector vibrates perpendicular to the direction of travel of the light wave. Now there are an infinite number of planes that we can pass through the line *OX* in Figure 1-7. Ordinary light consists of different wavelengths vibrating in many different planes.

If we could place ourselves at the point *X* and look toward *O*, we would see the cross section of the light wave depicted in Figure 1-8, a schematic representation of unpolarized and linearly polarized light. The radial electric field vectors (the arrows) are meant to indicate that no single direction predominates in completely unpolarized light.

FIGURE 1-8. Schematic representation of unpolarized and linearly polarized light. (a) Cross section of a narrow beam of ordinary light traveling directly toward the observer. Vibration of the light may be in any direction that is perpendicular to the direction of travel, as indicated by the numerous arrows. (b) A beam of polarized light has vibration in only one direction. This direction is the plane of polarization.

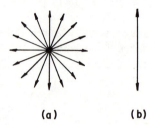

(a)   **(b)**

Even if we were to use unpolarized light of a single wavelength, it would still consist of waves vibrating in many planes at right angles to the direction of propagation. Since several directions of propagation are possible within a plane, it is correct to refer to light traveling in a specified direction as linearly polarized rather than plane polarized.

In addition to linearly polarized light, other kinds of polarized light exist. Figure 1-9 shows light beams that are polarized (a) linearly and horizontally, (b) linearly and vertically, (c) right circularly, and (d) left circularly.

In the case of circularly polarized light, the transverse vibrations trace out a helix as a function of time. The helix may be either right-handed (Figure 1-9c) or left-handed (Figure 1-9d). Viewed in cross section, i.e., as if an observer were situated on the *X* axis looking toward the light source, the transverse vibrations trace out a circle. Light whose electric field vector traces out a right-handed helical pattern is termed *right circularly polarized light.* The cross sectional appearance of clockwise rotation of the electric field vector is obtained by pushing the helix forward through a perpendicular plane without rotating it. In other words, the helix is moved forward, but it is not turned like a mechanical screw.

Another type of polarized light, which we have not pictured, resembles a flattened helix and has a cross section that is an ellipse. Elliptically polarized light may also be right-handed or left-handed.

The French physicist Biot discovered early in the nineteenth century that certain naturally occurring organic compounds possessed the unusual property of rotating the plane of polarization of a linearly polarized incident light beam. A few years later, in 1817, Biot and his countryman Fresnel independently found that the extent of optical rotation of a compound increases as one uses light of increasingly shorter wavelength for the measurement. The change in optical rotation with wavelength is termed *optical rotatory dispersion* (ord).

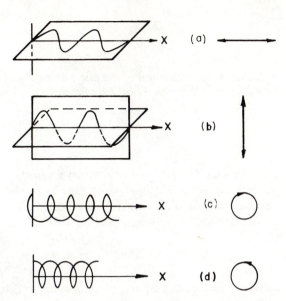

FIGURE 1-9. *Linearly and circularly po-larized radiation. (Left) The light wave as a function of time. (Right) Cross section of the light wave. Light polarized (a) linearly and horizontally, (b) linearly and vertically, (c) right circularly, (d) left circularly. Light having a right-handed helical pattern is termed right circularly polarized. The cross sectional clockwise rotation of the electric field vector is obtained as the helix is moved forward without rotation through a perpendicular plane.*

Thirty years later, Haidenger reported his observations on the differences in the absorption of the left- and right-handed components of circularly polarized light by crystals of amethyst quartz. Such differential absorption of left and right-handed circularly polarized light is termed *circular dichroism* (cd).

Since both ord and cd involve optical measurements on chiral molecules, they have been termed chiroptical methods.

As a useful model for conceptualizing the rotation of linearly polarized light, consider the light as composed of two oppositely rotating coherent beams of circularly polarized light. The linearly polarized light is then the vector sum of the left and right circularly rotating components as shown in Figure 1-10. The vector sums are indicated at points *A* to *E* with the resultant vectors having the properties of a linearly polarized light wave.

FIGURE 1-10. *A representation of linearly polarized radiation as the vector sum of two oppositely rotat-ing beams of circularly polarized radiation. [From J. D. Roberts and M. C. Caserio, Basic Princi-ples of Organic Chemistry, copy-right © 1964 by W. A. Benjamin, Inc., Menlo Park, California.]*

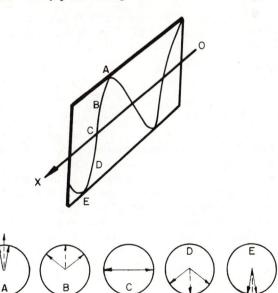

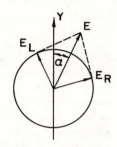

FIGURE 1–11. *Rotation of the plane of polarized light as the result of a change in the velocity of E_R relative to E_L.*

Fresnel, in 1825, postulated that when the circularly polarized light beams pass through an optically active medium, which may be a solid, liquid, or gas, the refractive index for one circularly polarized component will be different from that for the other. The medium is said to be *circularly birefringent* and to have the property given by eq. 1-14. Differences in

$$n_L - n_R \neq 0 \qquad\qquad 1\text{-}14$$

refractive indices correspond to differences in light velocities. Consequently, one of the two circularly polarized components of the linearly polarized light becomes retarded with respect to the other. Upon emerging from the optically active medium, the two components are no longer in phase and the resultant vector has been rotated by the angle α to the original plane of polarization (Figure 1-11).

In the region of an absorption band, the two circularly polarized components, in addition to suffering a differential retardation due to the circular birefringence of the medium, also are absorbed to different extents. In other words, the optically active medium has an unequal molar absorption coefficient ϵ for left and for right circularly polarized light. This difference in molar absorptivity (eq. 1-15) is termed *circular dichroism*.

$$\Delta\epsilon = \epsilon_L - \epsilon_R \neq 0 \qquad\qquad 1\text{-}15$$

Upon emerging from the optically active medium, the two circularly polarized components not only are out of phase but are of unequal amplitude. The resultant vector no longer oscillates along a single line but now traces out an ellipse, as shown in Figure 1-12. The linearly

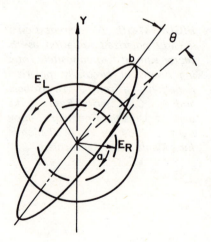

FIGURE 1–12. *Elliptically polarized light caused by the unequal speed and unequal absorption of left and right circularly polarized light by a chiral medium. The tangent of the ratio of the minor axis a to the major axis b is θ, the angle of ellipticity. The major axis of the ellipse forms the angle of rotation α to the original plane of polarization, the Y plane.*

polarized light beam has been converted to elliptically polarized light by the unequal absorption of its two circularly polarized components.

Note that ord involves measurement of a rotation, whereas cd involves an absorption measurement, namely the differential absorption of left- and right-handed circularly polarized radiation. Hence cd occurs only in the vicinity of an absorption band, whereas ord is theoretically finite everywhere.

1-10 ORD and CD Quantities

The angle of rotation α, in degrees per decimeter, is given by eq. 1-16. We may note that for

$$\alpha = \frac{1800}{\lambda \, (\text{cm})} \, (n_L - n_R) \qquad \text{1-16}$$

an observed optical rotation of $1°$ at 360 nm in a 1 decimeter (10 cm) cell, $n_L - n_R$ is 2×10^{-8}. Typical indices of refraction are of the order of unity. Hence the difference in refractive indices is extremely small, of the order of one millionth of one per cent.

Spectropolarimeters, as polarimeters able to make measurements at a variety of wavelengths are termed, record the angle of rotation α as a function of wavelength. Equation 1-17

$$[\alpha]_{\lambda}^{t} = \frac{\alpha}{cl} = \frac{\text{observed rotation (degrees)}}{\text{concn. (g ml}^{-1}) \times \text{length of sample tube (dm; } = 0.1\text{m)}} \qquad \text{1-17}$$

is used to calculate the specific rotation $[\alpha]$. For ord work, it is more common to use the molar rotation $[\Phi]$, which is simply the specific rotation multiplied by the molecular weight M over 100 (eq. 1-18). The units of $[\Phi]$ are degrees $\cdot$ cm^2 dmole^{-1}.

$$[\Phi] = \frac{[\alpha] \, M}{100} \qquad \text{1-18}$$

Just as $(n_L - n_R)$ is small in magnitude compared to the mean index of refraction, so the differences $(\epsilon_L - \epsilon_R)$ between the absorption coefficients for left and right circularly polarized light is of the order of 10^{-2} to 10^{-3}. Most cd instruments measure the differential absorbance, $\Delta A = A_L - A_R$. This quantity is related to the difference in molar absorption coefficients, $\Delta \epsilon = \epsilon_L - \epsilon_R$, by eq. 1-19, in which c is in moles liter^{-1} and l is the pathlength in cm.

$$\Delta A = \Delta \epsilon \, cl \qquad \text{1-19}$$

By analogy with the molar rotation, the molar ellipticity $[\theta]$ is defined by eq. 1-20 in which

$$[\theta] = 3300 \Delta \epsilon \qquad \text{1-20}$$

$[\theta]$ has the units degrees $\cdot$ cm^2 dmole^{-1}.

1-11   ORD and CD Spectra

Measurement of the angle of rotation α of a chiral compound as a function of wavelength gives an ord curve (literally, the dispersion of the optical rotation). A typical ord curve for a dextrorotatory ketone having an absorption maximum at 295 nm is shown in Figure 1-13. The part of the ord curve above about 325 nm, labeled plain ord region, is characteristic of compounds that have no optically active absorption bands in the spectral region being measured. The smoothly rising part of the curve is referred to as a plain dispersion curve; in the example shown it is a plain, *positive* dispersion curve. A plain *negative* curve would be one whose rotational values fall or become increasingly more negative on going toward shorter wavelength. Such plain dispersion curves can be described by a single-term Drude equation in which λ_0, the wavelength of the chromophore controlling the dispersion, lies far from the measured wavelengths, i.e., eq. 1-21, in which the A_i are constants characteristic of the chromophores

$$[\Phi] = \sum_i \frac{A_i}{\lambda^2 - \lambda_i^2} \quad (i = 0, 1, 2, \ldots, n) \qquad \text{1-21}$$

responsible for the observed rotations, λ is the wavelength of the measurement, and λ_i are the wavelengths of absorption maxima of the compound.

In terms of chemical information, the most useful curves are anomalous rotatory dispersion curves. The term *anomalous* is used because the curves do not obey a one-term Drude equation.

The ord curve in Figure 1-13 arises from a dextrorotatory compound containing a ketone function with an absorption maximum near 295 nm. As measurements of optical rotation are made to shorter wavelengths, the rotation increases. It is found to increase rapidly as the absorption maximum is approached. Somewhat before this maximum, rotational values reach a maximum (termed a *peak*), then drop drastically, *going through zero rotation*, until another inflection point (termed a *trough*) is reached. The rotation will then tend to increase again.

FIGURE 1-13. Optical rotatory dispersion curve of a typical saturated ketone. The amplitude of the Cotton effect a is defined as shown. The crossover point from positive to negative rotational values, $\lambda_0 = 295$ nm, corresponds closely to the absorption maximum of the ketone.

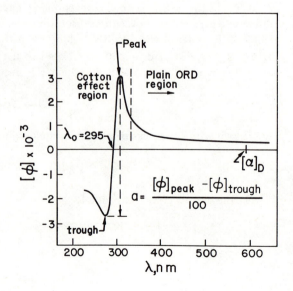

FIGURE 1-14. Positive (a) and negative (b) circular dichroism Cotton effects of an isolated absorption band with their corresponding optical rotatory dispersion curves (dashed lines). The ord and cd are related to each other by the Kronig–Kramers transform.

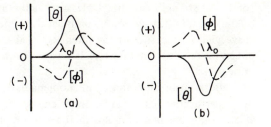

In the ideal case where the molecule possesses no other absorption bands near the one measured, λ_0 will closely correspond to λ_{max}, the maximum of the absorption band.

The vertical distance between the peak and trough divided by 100 is termed the amplitude (eq. 1-22).

$$a = \frac{[\Phi]_1 - [\Phi]_2}{100} \qquad\qquad 1\text{-}22$$

The S-shaped curve is known as a Cotton effect, in honor of the French physicist Aimé Cotton, who observed both ord and cd phenomena beginning in 1896. If the peak precedes the trough on measuring from longer to shorter wavelength, the Cotton effect is termed positive. Conversely, if a trough precedes a peak, it is a negative Cotton effect.

In another type of dispersion curve, termed a multiple Cotton effect curve, one can observe several small Cotton effects due to vibrational transitions within the main electronic transition.

As discussed above, in the region of an absorption band, left-handed and right-handed circularly polarized light beams are not only propagated with different velocities, but they are also absorbed to different extents. A cd curve is a plot of $\Delta\epsilon$ or $[\theta]$ versus wavelength. Positive and negative cd curves are shown in Figure 1-14 together with their corresponding ord Cotton effects.

The shape and appearance of a cd curve closely resembles that of the isotropic absorption curve of the electronic transition to which it corresponds. Unlike absorption curves, however, cd curves may be positive or negative. Both positive and negative cd peaks are referred to as maxima. Whereas only the anomalous part of an ord curve is termed the Cotton effect, all cd curves, by definition, are Cotton effects (cd Cotton effects are also termed ellipticity bands or Cotton bands).

For every Cotton effect in the ord curve of a compound, there exists a corresponding Cotton effect of the same sign in the cd spectrum (Figure 1-14). Note that the maximum of the cd peak corresponds closely to λ_0 in the ord Cotton effect. The ord and cd curves are related by integral transforms of the Kronig–Kramers type (section 1-13).

1-12 The Rotational Strength

In the absence of effects from overlapping electronic transitions, the integrated area under a cd curve is proportional to the rotational strength R of the transition (eq. 1-23). The

$$R = \frac{3hc\,(10)^3\,(\ln 10)}{32\pi^3 N} \int \frac{\Delta\epsilon}{\nu}\,d\nu \qquad\qquad 1\text{-}23$$

rotational strength is a direct measure of the intensity of the dichroism (i.e., of the magnitude of the optical rotatory power) in much the same way as the integrated area under an absorption curve yields a measure of the intensity of absorption (dipole strength, section 1-6). However, unlike the dipole strength, which is always positive, the rotational strength may be either positive or negative.

Both absorption and cd phenomena have their origin in the charge displacements induced by a perturbing light wave. Such charge displacements lead to the occurrence of induced electric and magnetic dipoles in the chromophore. The rotational strength R_K of the Kth transition of a chromophore is the scalar product of these induced electric and magnetic dipole moments, according to eq. 1-24 in which θ is the angle between the two transition moments

$$R_K = \mu_e{}^K \cdot \mu_m{}^K \cos \theta \qquad\qquad 1\text{-}24$$

and $\mu_e{}^K$ and $\mu_m{}^K$ have the dimensions of electric and magnetic dipole moments, respectively.

If a molecule possesses certain symmetry elements, such as a center of inversion or a reflection plane of symmetry, the rotational strength will be equal to zero and the molecule will be optically inactive. The rotational strength also will be equal to zero for the following three cases:

(a) $\mu_e{}^K \neq 0$ but $\mu_m{}^K = 0$.

(b) $\mu_e{}^K = 0$ but $\mu_m{}^K \neq 0$.

(c) Directions of $\mu_e{}^K$ and $\mu_m{}^K$ are at right angles so that their scalar product is zero ($\mu_e{}^K$ orthogonal to $\mu_m{}^K$).

Case (a) refers to an *electric-dipole-allowed, magnetic-dipole-forbidden* transition. Case (b) refers to a transition designated as *electric dipole forbidden, magnetic dipole allowed*.

For electronic transitions that are both electric dipole and magnetic dipole allowed, if we take μ_e to be of the order of 1 Debye (about 10^{-18} cgs units) and μ_m to be of the order of 1 Bohr magneton (about 10^{-20} cgs unit), we may expect rotational strength values to be of the order of 10^{-38} cgs unit. Such values are observed in chromophores possessing large rotatory powers, e.g., the skewed biphenyls and hexahelicene. However, for many other groups of interest such as the benzene chromophore, rotational strength values are much weaker, of the order of 10^{-42} cgs unit. As we shall see shortly, a weak cd Cotton effect corresponds to a weak Cotton effect in the ord curve of the same molecule.

1-13 Relationship Between ORD and CD; The Kronig–Kramers Transform

The relationship between absorption and dispersion may be expressed in terms of integral transforms known as the Kronig–Kramers relations. Thus ord and cd are related through a Kronig–Kramers type of relationship, and the corresponding ord curve may be calculated from the experimentally determined cd spectrum and vice versa.

The Kronig–Kramers relations lead to an expression, derived from the n–π* transition of saturated ketones, which relates the amplitude a of an ord Cotton effect (eq. 1-22) to $\Delta\epsilon$ of the corresponding cd peak (eq. 1-25).

$$a = 40.28\Delta\epsilon = (1.22 \times 10^{-2}) \, [\theta] \qquad\qquad 1\text{-}25$$

2

EXPERIMENTAL METHODS

2-1   Ultraviolet–Visible Instrumentation

For organic structure elucidation, the instrument of choice is an automatic recording, photo-electric spectrophotometer. Such instruments, capable of covering both the uv and visible regions, are characterized by high sensitivity, speed, and ease of operation. Some of the wide variety of currently available photoelectric spectrophotometers are listed in Table 2-1.

A simplified diagram of the optical system of a typical high quality double-beam, double monochromator uv-vis recording spectrophotometer is shown in Figure 2-1. The instrument may be divided into five components: light source, monochromator, photometer (the electronics), sample compartment, and detector.

Energy from either the deuterium light source (uv region) or the tungsten–halogen lamp (visible region) enters the monochromator through entrance slit S_1 and is dispersed by double monochromators. A narrow band of dispersed light passes through the exit slit S_3 and is alternated between the reference and sample cells at a rate of 60 Hz. These reference and sample beams alternately strike a single photoelectric detector where their optical energy is converted to an electric signal. The two signals are amplified, and finally the ratio of the sample signal to the reference signal is presented on a recorder output, usually as a plot of absorbance versus wavelength.

2-2   ORD and CD Instruments

2-2a   Spectropolarimeters

The number of firms offering instruments for the measurement of ord or cd is small, but the available instruments are reliable and of good quality. Most ord and cd instruments are sold

Table 2–1. *Some Representative UV–Visible Spectrophotometers*

Model	Manufacturer	Wavelength Range (nm)	Monochromator (Single = 1; Double = 2)
PE 202	Perkin–Elmer	190–750	1
PE 402		190–850	1
PE 323		185–2600	1
PE 356		185–850	2
Hitachi 139		195–800	1
Coleman Model 124		190–800	1
Cary 14	Varian Instruments	186–2650	2
Cary 15		185–800	2
Cary 17		186–2650	2
Cary 118		185–800	2
Jasco UV–175	Japan Spectroscopic Co. Ltd.	175–700	2
MPS–50L	Shimadzu	195–2500	1
PMQ–II	Carl Zeiss	200–1100	1
Spectronic 200	Bausch & Lomb	190–800	1
Spectronic 505		200–800	2
Spectronic 600		200–800	2
Acta CII	Beckman Instruments	190–800	1
Acta CIII		190–800	1
Acta CV		190–800	2
DK–2A		185–3500	1
Pye–Unicam SP500	Phillips Electronic	186–1000	1
Pye–Unicam SP1800	Instruments	190–700	1
DW–2	American Instrument Co.	200–850	1

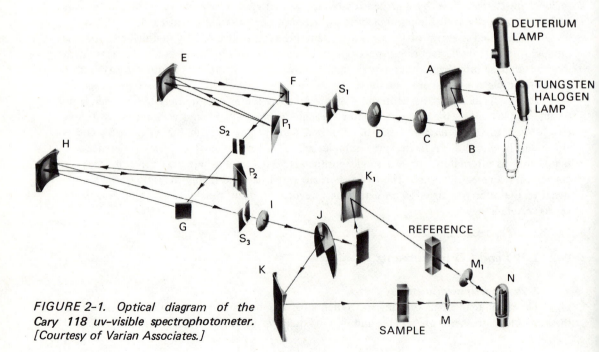

FIGURE 2–1. Optical diagram of the Cary 118 uv-visible spectrophotometer. [Courtesy of Varian Associates.]

FIGURE 2-2. *Schematic diagram of a photo-electric spectropolarimeter.*

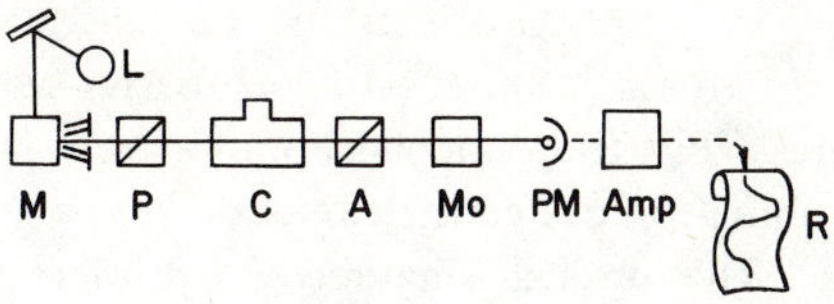

as separate units; but at least one company manufactures an instrument that contains both capabilities.

For reasons discussed in section 2-4, cd instruments have gradually replaced spectropolarimeters. If one can purchase only one instrument for chiroptical measurements, it should be a cd apparatus because of its wide variety of use.

A few instruments are capable of rotational measurements at several predetermined wavelengths (usually 5), such as the Perkin–Elmer Model 241. However, of greatest interest to chemists working on structure elucidation are the continuously recording, photoelectric spectropolarimeters. A schematic diagram is shown in Figure 2-2. The light source L is typically a 450 watt high pressure xenon arc lamp, which has a relatively high intensity continuum in the visible and uv spectral regions. The light is dispersed by the monochromator M. Most instruments use double monochromators of fused silica that reduce stray light levels (section 2-3e) to negligible amounts. The monochromatic light is passed through an exit slit and is linearly polarized by the polarizing prism P and then passes through the sample cell C. The plane of polarization is detected by the analyzer prism A, followed by conversion of the signal from dc to ac by the modulator Mo. The modulation may be achieved electrically by means of the Faraday effect, as in the Cary 60 instrument. Another method for achieving signal modulation, not depicted in Figure 2-2, is by a mechanical oscillation of the polarizing prism (Perkin–Elmer, Jasco instruments). The modulated signal is detected by the photomultiplier PM, and then fed to a phase-sensitive amplifier Amp, and from there to the recorder R. Some spectropolarimeters in current use are listed in Table 2-2.

Table 2-2. *Representative ORD and CD Instruments*

Type of Instrument	Model No.	Manufacturer	Wavelength Range (nm)	Modulation Type
ord	Cary 60[a]	Varian Instruments	185–600	Faraday cell
cd	Cary 61		185–800	electro-optic
ord-cd	J–20	Jasco	185–800	oscillating polarizer (ord) electro-optic (cd)
cd	J–40		185–1000	electro-optic
ord	P–22[a]	Perkin–Elmer	210–600	oscillating polarizer
cd	CD–185[a]	Jouan–Quentin	185–600	electro-optic
cd	Mark III	CNRS–Roussel–Jouan	180–800	elasto-optic
ord	Spectropol 1	FICA (France)	200–600	Faraday cell

[a] These and some other models such as the Jasco ORD/UV-5 have been discontinued recently, but many of them continue to be in service.

2-2b CD Spectrometers

Instruments for measuring circular dichroism are not graced by a single name but are variously termed dichrographs, dichrometers, or cd spectrometers. Some manufacturers call their cd instruments by the misnomer *recording spectropolarimeters.*

In 1960, Grosjean and Legrand (reference 1) introduced the use of the Pockels cell as an electro-optic modulator or quarter-wave retarder for the production of circularly polarized light. Thus they opened up a new era in cd instrumentation. Figure 2–3 shows a picture of a combination ord-cd instrument.

The optical diagram of a typical cd recorder is given in Figure 2–4. Monochromatic radiation from the 450 watt xenon lamp passes through an achromat lens into a linear polarizer. The ordinary and extraordinary rays of the linearly polarized light then are passed through the electro-optic modulator, after which the extraordinary ray is masked. The electro-optic modulator consists of a thin slice of a uniaxial crystal, such as ammonium dihydrogen phosphate or potassium dideuterium phosphate, which has been cut perpendicular to the z axis and coated on both sides with a conductive material. Upon application of an electric field the crystal becomes biaxial. This is known as the Pockel's effect and results in a quarter-wave retardation of the light beam and the production of circularly polarized light. The polarity of the voltage applied to the electro-optic modulator is varied periodically in order to convert the linearly polarized light to alternately left and right circularly polarized light. The magnitude of the voltage is a function of the wavelength of the light. The alternating left and right circularly polarized light then traverses the sample cell and is focused on the photomultiplier tube. If the sample exhibits circular dichroism, the output voltage of the photomultiplier tube will contain an alternating voltage superimposed on a steady component. The phase and magnitude of the alternating voltage correspond to the sign and magnitude, respectively, of the cd. The ratio of the alternating and constant voltages (ac and dc outputs) is amplified and transmitted to the recorder, which gives a direct reading of differential absorption ΔA by the sample of left and right circularly polarized light. Many cd instruments are calibrated in degrees of

FIGURE 2–3.   A combination ord–cd instrument. [Courtesy of Jasco, Inc.]

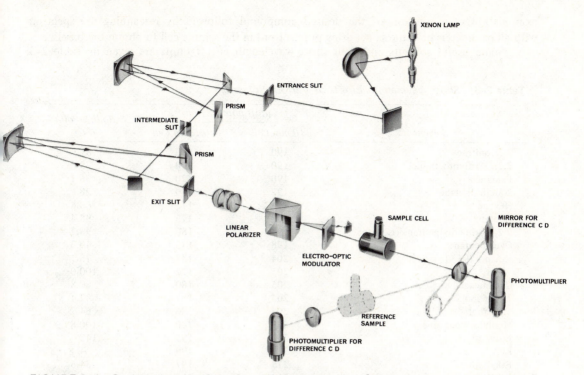

FIGURE 2-4. Optical diagram of the Cary 61 cd instrument. [Courtesy of Varian Associates.]

ellipticity θ. The ellipticity is related to the molar ellipticity $[\theta]$ (eq. 1-20) by eqs. 2-1 and 2-2, in which c has units of moles liter^{-1} and l of cm.

$$\theta = 33\Delta A \qquad\qquad\qquad 2\text{-}1$$

$$[\theta] = \frac{10^2 \cdot \theta}{cl} \qquad\qquad\qquad 2\text{-}2$$

2-3 Experimental Aspects

2-3a Solvents

The methods and procedures for uv–visible (isotropic absorption) and ord-cd (chiroptical) measurements are virtually the same. Most measurements are carried out on fairly dilute solutions (10^{-2} to 10^{-6} M) of the sample in an appropriate solvent. Such a solvent should not interact with the solute, should not absorb in the spectral region of interest, and, in the case of chiroptical measurements, should not be optically active. It is well to point out that in both isotropic absorption and chiroptical techniques, measurements may be made on pure liquids, gases, and solids.

An important difference between uv–vis spectrophotometers and ord-cd instruments is that the former are almost always double-beam instruments, whereas the latter, at the present writing, are invariably single-beam instruments. Hence chiroptical measurements involve the examination of a solution of the desired compound followed by rescanning the spectrum with all parameters held the same, using pure solvent in the sample cell to obtain the baseline.

Some useful solvents and their short wavelength cutoff limits are given in Table 2–3.

Table 2–3. *Short Wavelength Limits of Various Solvents*

Solvent	Cutoff point, λ (nm)[a]		Boiling Point (°C)
	10 mm Cell	0.1 mm Cell	
Acetonitrile	190	180	81.6
2,2,2-Trifluoroethanol	190	170	79
Pentane	190	170	36.1
2-Methylbutane	192	170	28
Hexane	195	173	68.8
Heptane	197	173	98.4
2,2,4-Trimethylpentane (isooctane)	197	180	99.2
Cyclopentane	198	173	49.3
Ethanol (95%)	204	187	78.1
Water	205	172	100.0
Cyclohexane	205	180	80.8
2-Propanol	205	187	82.4
Methanol	205	186	64.7
Methylcyclohexane	209	180	100.8
Butyl ether	210	195	142
EPA[b]	212	190	–
Ethyl ether	215	197	34.6
1,4-Dioxane	215	205	101.4
Bis(2-methoxyethyl) ether (glyme)	220	199	162
1,1,2-Trichlorotrifluoroethane	231	220	47.6
Dichloromethane	232	220	41.6
Chloroform	245	235	62
Carbon tetrachloride	265	255	76.9
N,N-Dimethylformamide	270	258	153
Benzene	280	265	80.1
Toluene	285	268	110.8
Tetrachloroethylene	290	278	121.2
Pyridine	305	292	116
Acetone	330	325	56
Nitromethane	380	360	101.2
Carbon disulfide	380	360	46.5

[a]The cutoff point is taken as the wavelength at which the absorbance in the indicated cell is about one.
[b]5/5/2 by volume mixture of ethyl ether, isopentane, and ethanol.

Note the significant advantage in uv penetration to be gained by the use of short pathlength cells (1 mm or less).

It is of the utmost importance to use solvents or other reagents with known purity. Examples of two types of solvent contamination are shown in Figures 2–5 and 2–6. Commercially available solvents with such designations as "spectral grade" and "for spectroscopy" are not necessarily pure but have been specially prepared to ensure the absence of impurities *absorbing* in the uv–visible region. Nonabsorbing contaminants may well be present.

Commonly used polar solvents are 95% ethanol, water, and methanol. Aliphatic hydro-

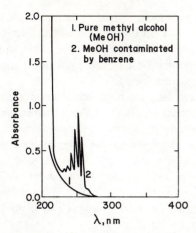

FIGURE 2-5. Ultraviolet spectra of pure methanol (10 mm cell) and methanol contaminated by benzene. [Reproduced with permission from F. Grum in Techniques of Chemistry, A. Weissberger and B. W. Rossiter, eds., Wiley-Interscience, New York, 1972, Vol. 1, Pt. 3B, ch. 3.]

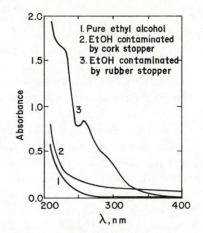

FIGURE 2-6. Ultraviolet spectra of pure ethanol (10 mm cell) and contaminated by a cork stopper and a rubber stopper. [Reproduced with permission from F. Grum in Techniques of Chemistry, A. Weissberger and B. W. Rossiter, eds., Wiley-Interscience, New York, 1972, Vol. 1, Pt. 3B, ch. 3.]

carbons (hexane, heptane, cyclohexane, etc.) are examples of nonpolar solvents that allow good uv penetration (Table 2-3) and have boiling points high enough so that solvent evaporation does not become a problem. However, they must be rigorously purified since these hydrocarbons may contain alkenic impurities or traces of aromatic compounds. It has been observed that fluoroalkanes have enhanced transparency relative to the alkanes, and a similar finding has been made for the fluorinated alcohols such as 2,2,2-trifluoroethanol.

Organic cyanides such as acetonitrile and propionitrile are aprotic polar solvents having excellent spectral transparency. A widely used polar aprotic solvent is 1,4-dioxane, transparent to about 205 nm. A highly transparent liquid for use in special cases is anhydrous sulfuric acid.

Several mixed solvents have found use in spectroscopic studies at very low temperatures, usually down to liquid nitrogen temperatures, about −190°. These solvent systems do not crystallize when cooled but instead become viscous and glassy. Low temperature solvents include (1) EPA, a 5/5/2 by volume mixture of diethyl ether, isopentane, and ethanol, (2) methanol-glycerol, 9/1 v/v, (3) tetrahydrofuran-diglyme, 4/1 v/v, and (4) methylcyclohexane-isopentane, 1/3 v/v. A table of the percent degree of contraction of these solvents over the range 25 to −190° has been published (reference 2).

2-3b Cells

Cells for use in the visible region are usually made of Pyrex or similar glass which is transparent to about 380 nm. In the uv region, quartz cells are necessary, and those made of high purity fused silica (Ultrasil, Spectrosil, Supersil) are recommended. Rectangular cuvettes are routinely used in isotropic absorption work. The preferred pathlength is 10 mm since l in eq. 1-10 will then be equal to unity.

Short pathlength cells are essential when solvent absorption must be minimized. Cells in the range 0.01–2 mm are used for work in the far uv region. Fused cylindrical cells with fixed windows are commonly used for chiroptical work. If cells are to be used in ord work, they should be checked for stress birefringence by observing if the empty cell exhibits optical rotation. Cells are calibrated by the manufacturer but may be checked by determining the absorbance or optical activity of known substances (see section 2–3d).

For most purposes cells may be cleaned by rinsing *at least ten times* with the solvent used in the measurement. A final rinse with methanol or acetone is suitable if the cells are to be air dried. In general, oven drying is recommended, particularly for short pathlength cells. In this case the final two rinses should be with distilled water so that no traces of a flammable organic liquid remain. Although a drying oven set at a high temperature will not damage the cells (remember that they have been fused during manufacture at temperatures above 1000°), removal of a fragile quartz cell from a hot oven can be troublesome. A fast and convenient drying method is to place the cell in a vacuum desiccator, heated to about 50° if possible. In this manner, cells may be dried in less than five minutes.

Above all, cell windows should never be touched with the fingers. It is good practice when cleaning a cell to rinse the outside a few times. Any material that still remains on the outside optical faces should be wiped off with a lintless wiper, e.g., Kimwipes, *soaked in solvent*. Never use a dry cloth or tissue.

2–3c Sample Preparation

Quantitative analytical techniques are applied to sample preparation. Volumetric glassware and cells must be clean and dry. Solid samples should be dried to constant weight in a desiccator in order to remove adhering water or solvent. A typical 1–10 mm cell holds anywhere from 0.2 to 3 ml of solution so that an appropriate amount of stock sample solution should be prepared. Since the measurement is nondestructive, the sample may be recovered by evaporation of the solvent. If the amount of sample available permits, 10–25 ml of solution is a convenient size to prepare. The amount of sample required for this volume is sufficient to minimize errors associated with weighing small quantities. Naturally, one must use an analytical balance capable of weighing directly to at least 0.1 mg. As most compounds being measured probably will not have been run previously by the operator, an initial sample concentration to begin with is approximately 0.05% (about 0.5 mg ml^{-1}) for small molecules and about 0.005% for polypeptides and large macromolecules. A peak absorbance in the range 0.7–1.2 absorbance units is desirable for most instruments since it gives a good pen deflection and the electronics are usually most sensitive in this range.

2–3d Calibration of ORD and CD Instruments

Spectropolarimeters may be calibrated with aqueous sucrose solutions or with a standardized quartz plate. The data for sucrose solutions are readily available in the reprint of Lowry's classic book (reference 3).

Recently, Cassim and Yang (reference 3) have calibrated a cd instrument against a standardized spectropolarimeter by application of the Kronig–Kramers transform (section 1–13). The compounds adopted as standards were (+)-camphor (**1**) and (+)-camphor-10-sulfonic acid (**2**). For the latter compound the interrelation between the ord and cd values was determined to be given by eq. 2–3.

$$f = 1.055 = \frac{2.20}{\Delta\epsilon_{290.5}} \times \frac{[\Phi]_{306}}{4120} \qquad\qquad 2\text{–}3$$

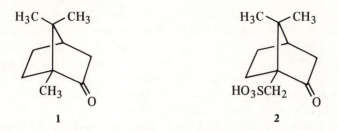

1   2

DeTar (reference 3) gives eq. 2-4, which allows one to calculate the value of $\Delta\epsilon$ by

$$\Delta\epsilon = 2.38 \times 10^{-4} \, ([\Phi]_{305} - [\Phi]_{270}) \qquad\qquad 2\text{-}4$$

measuring the ord peak and trough rotations for (+)-camphor-10-sulfonic acid.

The cd and ord calibration data for **1** and **2** are given in Table 2-4.

Table 2-4. *Calibration Data for ORD and CD Instruments[a]*

| | | CD | | | ORD | | | |
| | | | | | Peak | | Trough | |
Solute	*Solvent*	$\Delta\epsilon$	*[θ]*	*λ_{max} (nm)*	*[φ]*	*λ (nm)*	*[φ]*	*λ (nm)*
(+)-Camphor-10-sulfonic acid	water	+2.20	+7260	290.5	+4120	306	−5300	270
(+)-Camphor	methanol	+1.54	+5080	295.5	+3150	313	−3370	275
(+)-Camphor	dioxane	+1.69	+5570	299.5	+3980	316	−3500	277

[a] J. Y. Cassim and J. T. Yang, *Biochemistry*, 8, 1947 (1969).

2-3e Possible Sources of Error

Errors may arise from the nature of the sample being examined, from the instrument, and last but not least, from the operator. The simplest error may be one in which the Lambert–Bouger–Beer law (eq. 1-9) is not obeyed. In practice, Beer's law, which states that the absorbance is proportional to the number of absorbing molecules, has been found to hold for a large number of compounds over a considerable range of concentrations. However, since the molar absorption coefficient ϵ depends upon wavelength, Beer's law can be strictly true only for pure monochromatic light. True deviations may be expected when the concentration of absorbing molecules is so high that they interact with each other. Effects such as association and dissociation are common causes of deviations from the Lambert–Bouger–Beer law. For example, compounds that tend to form dimers, such as aqueous dye solutions, seldom follow Beer's law over any extended concentration range. A test of Beer's law can be made by dilution of the sample solution to a different volume, which should then show the correct absorbance corresponding to the dilution.

The problem of stray light occurs in most double monochromator instruments at high sample absorbances. Stray light can be defined as the ratio of spurious light to the desired wavelength. It can also arise from scattering of the light beam from any of the surfaces that it encounters: lenses, prisms, slit edges, etc., as well as dirty cell windows. Turbid solutions scatter light, and the scattering becomes more important at shorter wavelengths. Sometimes dust particles are responsible for the scattering. Such solutions should be passed through a fine filter, such as Millipore, or centrifuged.

In ord and cd instruments, stray light may be produced by nonpolarized radiation reaching the detector. A loose sample compartment cover or other light leak may allow room light to enter the instrument and cause stray light effects. It is a common occurrence for stray light effects to occur when solvent absorption becomes severe toward the short wavelength end of a scan. The slits are then wide open to pass the maximum amount of energy, and the absorbance appears to drop sharply. It is important to recognize this effect in order to know the absorbance limits to which valid measurements may be made. Rescanning the spectrum is of no help since stray light effects are usually highly reproducible!

Oxygen has a series of absorption bands that begin at about 195 nm and extend to shorter wavelength. Hence, to work in this spectral region air must be excluded from the instrument. Otherwise much of the light will be lost in both the sample and the reference beam because of oxygen absorption. Oxygen is most easily removed by flushing the entire optical path of the instrument with pure, dry nitrogen. A liquid nitrogen cylinder having a gaseous take-off valve provides a convenient source of high purity nitrogen. Optimum flow rates can be determined by observing the disappearance of the oxygen absorption spectrum.

2-4 Relative Advantages of ORD and CD

The most significant difference between ord and cd is that cd will give a signal only in the vicinity of an optically active absorption band. In contrast, the change in optical rotation with wavelength is everywhere finite. Thus, even chromophores in the far uv will contribute to the sign and magnitude of the ord signal. In addition, the rather broad S-shaped form of the ord Cotton effect often causes difficulties in separating the contributions of neighboring electronic transitions and in detecting vibrational fine structure. The dispersive nature of ord can constitute an advantage in cases such as obtaining information on Cotton effects just beyond the wavelength limits of the instrument by determining the sign and magnitude of the ord curve at these limits.

The capability to measure rotations at various wavelengths can be useful for comparison purposes with data in the literature. Compounds devoid of chromophores in the accessible spectral region will be transparent by cd. A related problem occurs with compounds of low optical activity, e.g., chiral hydrogen–deuterium compounds or substances of low optical purity. In these cases, ord measurements are necessary.

Generally speaking, however, it seems fair to state that if a choice must be made, then cd is preferred over ord. Both phenomena are related by the Kronig-Kramers transform, but the inherent simplicity of cd spectra recommends them over ord for the majority of applications.

BIBLIOGRAPHY

1. Circularly Polarized Light: M. Grosjean and M. Legrand, *C. R. Acad. Sci.*, Paris, **251**, 2150 (1960); M. Legrand and L. Velluz, *Angew. Chem., Int. Ed. Engl.*, **4**, 838 (1965).
2. Solvent Contraction: O. Korver and J. Bosma, *Anal. Chem.*, **43**, 1119 (1971).
3. CD Instrument Calibration: T. M. Lowry, *Optical Rotatory Power*, Dover, New York, 1964; J. Y. Cassim and J. T. Yang, *Biochemistry*, **8**, 1947 (1969); D. F. DeTar, *Anal. Chem.*, **41**, 1406 (1969).

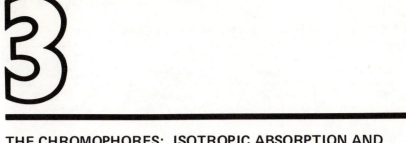

THE CHROMOPHORES: ISOTROPIC ABSORPTION AND CHIROPTICAL SPECTRA

A *chromophore* is an atomic grouping that can undergo an electronic transition. For organic molecules, typical chromophores are functional groups: carbonyl, nitro, carboxyl, double and triple bonds, aromatic rings, etc. Electronic absorption data for some common chromophores are given in Table 3-1. *Auxochromes* are nonchromophoric groups, usually containing unshared electrons (−Cl, −NR$_2$, −OH, etc.), that when attached to a chromophore, cause a shift in the position and usually an increase in the intensity of the absorption maximum of the chromophore.

An *optically active chromophore* is one that gives rise to a Cotton effect in the ord or cd spectrum of the compound. All chromophores in a chiral molecule should exhibit Cotton effects in the region of their electronic absorption bands. However, rotatory strength falls off with distance from a chiral center so that a distant chromophore may have a Cotton effect too weak to measure.

It is useful to divide optically active chromophores into two broad classes: (a) the inherently chiral chromophore and (b) the inherently achiral but chirally perturbed chromophore. The original classification of Moscowitz (reference 1) used the terms dissymmetric and symmetric for chiral and achiral. The optical activity of compounds belonging to the first class is inherent in the geometry of the chromophore. Some examples of inherently chiral chromophores are given in Figure 3-1. Inherently achiral but chirally perturbed chromophores include most of the common functional groups; some examples of this class are given in Figure 3-2. Inherently chiral chromophores are characterized by large rotational strengths, on the order of 10^{-38} cgs unit, whereas the achiral but chirally perturbed chromophores are characterized by rather weak Cotton effects. Their rotatory strength is generally less than 10^{-41} cgs unit.

Table 3-1. *Electronic Absorption Data for Isolated Chromophores*

Chromophore	Example	Solvent	λ_{max} (nm)	ϵ_{max} (liters mole^{-1} cm^{-1})
C=C	1-hexene	heptane	180	12,500
$-C\equiv C-$	1-butyne	vapor	172	4,500
C=O	acetaldehyde	vapor	289	12.5
			182	10,000
	acetone	cyclohexane	275	22
			190	1,000
	camphor	hexane	295	14
−COOH	acetic acid	ethanol	204	41
−COCl	acetyl chloride	heptane	240	34
−COOR	ethyl acetate	water	204	60
$-CONH_2$	acetamide	methanol	205	160
$-NO_2$	nitromethane	hexane	279	15.8
			202	4,400
$\overset{+}{=}N\overset{-}{=}N$	diazomethane	ethyl ether	417	7
−N=N−	*trans*-azomethane	water	343	25
$>C=N-$	$C_2H_5CH=NC_4H_9$	isooctane	238	200
benzene ring	benzene	water	254	205
			203.5	7,400
	toluene	water	261	225
			206.5	7,000

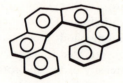

HEXAHELICENE

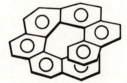

HEPTAHELICENE

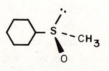

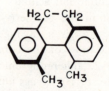

FIGURE 3-1. Some inherently chiral chromophores.

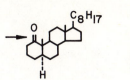

FIGURE 3-2. Some inherently achiral but chirally perturbed chromophores.

3-1 Alkenes

Ethylene itself absorbs well outside of the generally accessible uv region with a broad absorption maximum at about 162 nm. The rather intense absorption ($\epsilon \sim 10,000$) is attributed to a $\pi-\pi^*$ transition in which the ground state is planar and the excited state nonplanar. Increasing alkyl substitution at the double bond results in a bathochromic shift of the $\pi-\pi^*$ absorption maximum. The electronic spectra of ethylene and its alkyl derivatives has been reviewed in detail by Merer and Mulliken (reference 2).

Geometrical isomers of disubstituted alkenes often can be differentiated by the longer wavelength absorption of the trans compounds. Solution spectra are displaced to longer wavelength than the corresponding vapor spectra. When two alkenic chromophores in the same molecule are insulated from each other by saturated carbons, their spectrum approximates the sum of the two chromophores. Conjugation of π-electron systems of double bonds results in dramatic bathochromic shifts and increased intensities. Spectral data illustrating some of the above effects are given in Table 3-2.

Table 3-2. *Absorption Data for Alkenes*

Compound	Solvent	λ_{max} (nm)	ϵ_{max} (liters mole^{-1} cm^{-1})
Ethylene	vapor	162	10,000
cis-2-Butene	vapor	174	–
trans-2-Butene	vapor	178	13,000
1-Hexene	vapor	177	12,000
	hexane	179	–
Allyl alcohol	hexane	189	7,600
Cyclohexene	vapor	176	8,000
	cyclohexane	183.5	6,800
Cholest-4-ene	cyclohexane	193	10,000
1,5-Hexadiene	vapor	178	26,000
1,3-Butadiene	vapor	210	–
	hexane	217	21,000
1,3,5-Hexatriene	isooctane	268	43,000
1,3,5,7-Octatetraene	cyclohexane	304	–
1,3,5,7,9-Decapentaene	isooctane	334	121,000
1,3,5,7,9,11-Dodecahexaene	isooctane	364	138,000

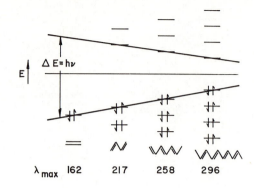

FIGURE 3-3. Schematic Hückel MO diagram illustrating the effect of conjugation on the π–π^* absorption maximum.

The bathochromic shifts resulting from increased double bond conjugation may be illustrated in terms of Hückel molecular orbital theory (more sophisticated theories lead to the same result). The molecular orbitals and their relative energies for ethylene, 1,3-butadiene, and two higher conjugated homologues are depicted in Figure 3-3. Note that the energies of the highest occupied molecular orbitals (HOMO) increase while those of the lowest unfilled molecular orbitals (LUMO) decrease with increasing conjugation. The observed transition involves promotion of an electron from the HOMO to the LUMO. One notes from Figure 3-3 that as the conjugated π system increases in length, the energy required for the transition becomes less, i.e., a bathochromic shift results.

Extensive studies of the uv spectra of alkenes, in particular terpenes and steroids, led Woodward and then the Fiesers (references in Table 3-3) to formulate empirical rules for the prediction of the absorption maxima of various dienes. These rules have proved quite useful in the solution of structural problems, particularly in the natural products field. The procedure involves beginning with a base absorption maximum for the parent chromophore and then incrementing it by values corresponding to each·substituent attached to the parent π-electron system. The values used in the Woodward–Fieser rules for diene absorption are given in Table 3-3.

Let us use the rules to calculate λ_{max} for ergosta-4,6,22-trien-3-one enol acetate (**1**):

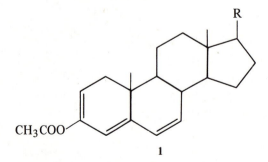

1

Parent diene (homoannular)	253 nm
Exocyclic double bond	5
Alkyl substituents, 3 × 5	15
Conjugated double bond	30
λ_{max} Calculated	303 nm
Observed (EtOH)	304 nm

The predicted values are generally within a few nanometers of the observed ones.

Information on steric effects may sometimes be gained from absorption intensities. For example, in the butadiene series s-trans forms (s = single bond) usually have ϵ values in the range of 14,000–30,000, whereas s-cis conformers have values of about 3,000–12,000.

Table 3-3.   *Woodward–Fieser Rules for the Calculation of Absorption Maxima of Dienes and Polyenes*[a]

	λ *(nm)*
Parent heteroannular diene or acyclic polyene of the type:	214

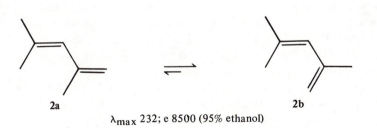

Parent homoannular diene or polyene of the type:	253

Note: In cases where both types of diene systems are present, the longer wavelength one is chosen as the parent system.

Each additional conjugated double bond	30
Exocyclic double bond	5
Each alkyl substituent	5
Each polar group	
—O-acyl	0
—OR	6
—SR	30
—Cl, —Br	5
—NR$_2$	60
Solvent correction	0

[a] R. B. Woodward, *J. Amer. Chem. Soc.*, **63**, 1123 (1941); **64**, 72, 76 (1942); L. F. Fieser and M. Fieser, *Natural Products Related to Phenanthrene*, Reinhold, New York, 1949.

1,3-Butadiene has an ϵ of 21,000 and is known to exist primarily in the s-trans form with a small amount (ca. 2.5%) in the somewhat more strained s-cis form.   The molar absorption coefficient for the butadiene derivative for **2** indicates that a substantial fraction exists in ethanol solution in the s-cis conformation (**2b**).

2a   ⇌   **2b**

λ_{max} 232; e 8500 (95% ethanol)

   The double bond, when rendered optically active by a chiral environment, gives rise to Cotton effects.   While the uv spectra of the isomeric steroidal alkenes cholest-4-ene (**3**) and cholest-5-ene (**4**) are almost identical, the ord curves are quite different.   The data are presented in Figure 3-4.   The shape of the ord curve in the Cotton effect region is characteristic of the

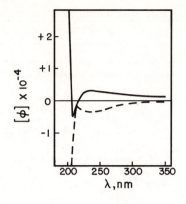

FIGURE 3-4. The ord's in cyclohexane of cholest-4-ene (3) —— and cholest-5-ene (4) – – – –. [Reproduced with permission from A. Yogev, D. Amar, and Y. Mazur, Chem. Commun., 339 (1967).]

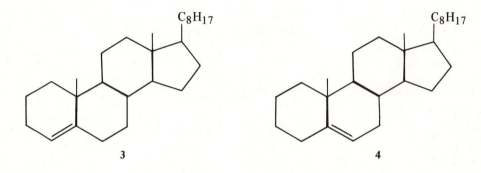

3   4

position of the double bond for various steroidal alkenes. Hence, once the ord curves are known for these model compounds, they can be used to locate the position of double bonds in unknown compounds so long as the latter do not contain other chromophores that absorb in the same spectral region as the alkenic chromophore or interact with it.

The uv spectra of unconjugated alkenes have been shown to exhibit more than just the high intensity $\pi-\pi^*$ transition in the accessible spectral region. Within the past decade other bands have been observed in the vapor phase spectra of alkyl-substituted alkenes: higher energy Rydberg transitions and a weaker band somewhat above 200 nm, which has been assigned to a $\pi-\sigma^*$ transition (reference 2). Recent studies by Yogev, Sagiv, and Mazur (reference 2) based on linear dichroism (the measurement of $\epsilon_{\parallel}$ and $\epsilon_{\perp}$ of an oriented molecule) have shown that the broad uv band of **3** and **4** near 190 nm is actually composed of two transitions, a weaker one at about 203 nm (ϵ 4,000) and a more intense band at approximately 185 nm (ϵ 8,000). The first band is polarized at an angle of about 16.5° to the double bond axis, whereas the second transition is polarized along the direction of the double bond. The observed and resolved uv spectra, together with the cd curve of cholest-4-ene (**3**), are shown in Figure 3-5. This example illustrates the usefulness of cd in detecting transitions that are not resolved in the isotropic absorption spectrum. The cd spectrum of **3** reveals a negative Cotton effect associated with the longer wavelength transition and, overlapping it on the shorter wavelength side, a more intense positive Cotton effect whose maximum was not reached but which presumably is associated with the second transition. A caveat should be given here. Oppositely signed, overlapping Cotton effects will appear essentially as the algebraic sum of the component Cotton effects. Hence, depending upon their relative magnitudes and proximity to each other, the observed bands will be shifted in position and diminished in intensity relative to the isolated Cotton effects.

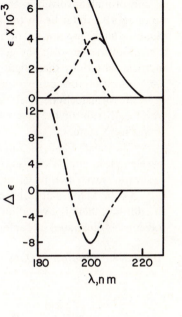

FIGURE 3-5. Isotropic absorption ———, resolved iso-tropic absorption - - - -, and cd — · — spectra of cholest-4-ene (3) in cyclohexane. [Reproduced with permission from A. Yogev, J. Sagiv, and Y. Mazur, Chem. Commun., 411 (1972).]

3-2   Carbonyl Compounds

3-2a   Aldehydes and Ketones

The longest wavelength transition in aliphatic aldehydes and ketones, the n–π* band, is probably the best studied of any transition. It is a low intensity ($\epsilon \sim$10–20) and rather broad band, occurring in the neighborhood of 270–300 nm. As we shall see, its position is quite solvent sensitive.

A second carbonyl band, attributed to an n–σ* transition, occurs near 180 nm and is considerably more intense than the n–π* transition. This wavelength region is just beyond the range of most uv instruments so that only the beginning of the band will be observed as so-called end absorption.

Transitions at wavelengths shorter than about 180 nm are most likely due to π–π* and Rydberg transitions. A schematic energy level diagram for the carbonyl group is shown in Figure 3-6. The lowest energy transition involves the promotion of an electron from the

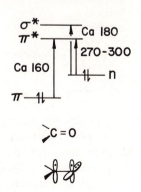

FIGURE 3-6. Schematic energy levels for the n–π*, n–σ*, and π–π* transitions of the carbonyl group. Note that the p orbital on oxygen containing the unshared electron pair is orthogonal to the plane of the carbonyl π and π* orbitals.

nonbonding p orbital on oxygen to the antibonding π^* orbital associated with the entire carbonyl group. The transition is symmetry-forbidden (see section 1–7; formaldehyde and symmetrically disubstituted ketones belong to the C_{2v} point group) and thus is of low intensity, whereas the n–σ^* transition is allowed and is relatively intense. The symmetries of the orbitals involved in the transitions are important. The bonding π orbital and the antibonding π^* orbital lie in the same plane, whereas the nonbonding orbital is in an orthogonal plane. Hence promotion of an electron from the nonbonding orbital is not possible without a significant change in the geometry of the molecule. The weak intensity is due to nonsymmetrical vibrations that slightly deform the molecule and lower its symmetry, allowing the n–π^* transition to acquire a finite probability.

In formaldehyde itself, a very weak absorption system has been observed on the long wavelength side of the n–π^* transition and is assigned to a transition of an n electron to a *triplet* n–π^* state.

The n–π^* band of formaldehyde occurs at 304 nm in the vapor phase and at longer wavelength in aliphatic solvents (Table 3–4). Progressive alkyl substitution displaces the band to shorter wavelength for aldehydes and to still shorter values for ketones.

Cyclic ketones absorb at longer wavelength than the corresponding open chain analogues. In addition, there is a variation in the position of the absorption band with ring size in nonpolar solvents, as illustrated in Table 3–4.

Table 3–4. *Absorption Data for Aliphatic Aldehydes and Ketones*

		n–π^* Transition		n–σ^* Transition	
Compound	*Solvent*	λ_{max} *(nm)*	ϵ_{max} *(liter mole^{-1} cm^{-1})*	λ_{max} *(nm)*	ϵ_{max} *(liter mole^{-1} cm^{-1})*
Formaldehyde	vapor	304	18	175	18,000
	isopentane	310	5		
Acetaldehyde	vapor	289	12.5	182	10,000
Acetone	vapor	274	13.6	195	9,000
	cyclohexane	275	22	190	1,000
Butanone	isooctane	278	17		
2-Pentanone	hexane	278	15		
4-Methyl-2-pentanone	isooctane	283	20		
Cyclobutanone	isooctane	281	20		
Cyclopentanone	isooctane	300	18		
Cyclohexanone	isooctane	291	15		
Cycloheptanone	isooctane	292	17		
Cyclooctanone	isooctane	291	14		
Cyclononanone	isooctane	293	15		
Cyclodecanone	isooctane	288	15		
2-Chloro-4-*t*-butylcyclohexanone					
equatorial Cl	isooctane	286	17		
axial Cl	isooctane	306	49		

In addition to alkyl groups, other substituents α to the carbonyl group affect the position of the n–π^* transition. The presence of an α bromine in the cyclohexanone series causes a bathochromic shift of λ_{max} of about 23 nm when the bromine is axial but a 5 nm shift when it is equatorial. Equatorially substituted 2-chloro-4-*t*-butylcyclohexanone has its n–π^* maximum at a slightly shorter wavelength than that of the parent ketone, whereas the axial chlorine isomer has a more intense absorption band at considerably longer wavelength (Table 3-4).   The strong bathochromic and hyperchromic effect of an α-halo substituent also is observed in steroidal ketones, but a satisfactory explanation for the phenomenon has not yet been advanced.

In general, n–π* transitions are easily recognizable by their low intensities, by the spectral shifts caused by substitution, and by the sensitivity of the position of the band to solvent effects.

Shifts in the position of absorption bands on going from the vapor phase to solution or from one solvent to another are mainly due to differences in the solvation energies of the solute in the ground and excited electronic states.

The effect of solvent on the position of the n–π* absorption has served as an important diagnostic tool.   Indeed, the fundamental role of the unshared electron pair in this transition can be demonstrated by the disappearance of the n–π* band in acid solution in which the unshared pair is protonated.

Changing from a nonpolar solvent to a polar one results in a significant hypsochromic shift in the position of the n–π* transition.   It has been shown that hydroxylic solvents of comparable dielectric constant cause a larger blue shift than do aprotic polar solvents.   The larger shifts occasioned by hydroxylic solvents are attributable in part to greater hydrogen bonding to the carbonyl oxygen lone pairs than to the π* electrons, thus lowering the energy of the ground state relative to the excited state.   An example of solvent effects on the n–π* band of acetone is shown in Figure 3-7.

There are, in addition, two so-called Franck–Condon effects that sometimes result in solvent shifts.   One effect, *orientation strain*, occurs when the vertical electronic transition produces an excited-state molecule that has a geometry and solvent orientation corresponding to equilibrium in the ground state but actually is a highly strained configuration of the excited state.   In this situation, the excited-state solvation energy is raised and the transition occurs at higher energy than if the Franck–Condon factor were absent.   The second effect, *packing strain*, occurs when the molecule in the excited state is substantially larger than in the ground state, resulting in compression by the ground state solvent arrangement and an increased excited-state energy with its associated blue shift.   The effect of packing strain is usually negligible for most organic compounds, since they normally are large enough that changes in molecular dimensions upon excitation are very small.

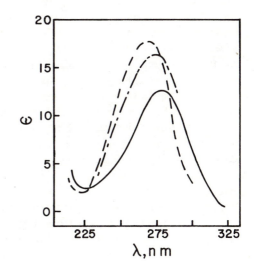

FIGURE 3-7.   Solvent effects on the n–π* transition of acetone in hexane ———, 95% ethanol — · —, and water - - - -.

The renaissance of the chiroptical techniques in the mid-1950's began with Djerassi's studies of the ketone n–π* transition. Optical rotatory dispersion was first applied to stereochemical and configurational investigations of compounds such as the 5α-cholestanones (**5, 6, 7**). The n–π* carbonyl band is an example of an electric-dipole-forbidden, magnetic-

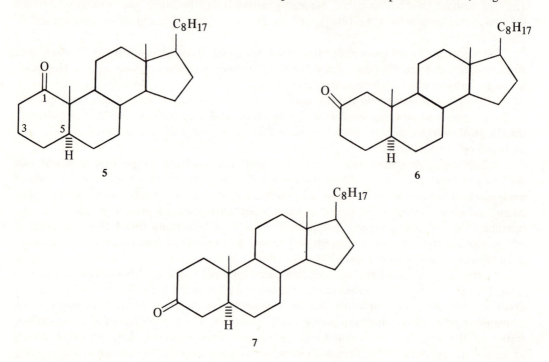

5

6

7

dipole-allowed transition (see section 1-12) that is rendered optically active by its chiral environment. It has a low intensity of absorption as measured by ε but a relatively strong rotational strength as measured by Δε, so that the *anisotropy ratio Δε/ε* of the carbonyl group is fairly large. In practice, this means that optically active ketones are generally favorable compounds to measure, giving strong Cotton effects and good signal (Δε)/noise (ε) ratios.

The ord curves of the isomeric 1-, 2-, and 3-keto steroids (**5, 6,** and **7**) are shown in Figure 3–8. While **6** and **7** exhibit positive Cotton effects, the 1-keto isomer (**5**) shows only a weak negative n–π* Cotton effect ($a = -25$) superimposed on a positive background rotation due to more intense Cotton bands at shorter wavelength.

The sign and magnitude of the Cotton effects are due to the chiral environment in the vicinity of the carbonyl chromophore. In principle, then, optically active chromophores in a molecule can be used as stereochemical probes.

Distinguishing between a 2- and a 3-keto steroid by means of their uv spectra is practically impossible; even the ir spectra of such compounds show only slight differences. However, ord allows a clear distinction. As seen in Figure 3–8, the magnitudes of the two positive Cotton effects are quite different; $a = +121$ for the 2-keto isomer (**6**) and $+55$ for the 3-isomer (**7**). Since these compounds are derived from natural products, they are optically pure and the amplitudes of their 290 nm Cotton effect can serve to differentiate them.

The corresponding cd spectra of **5** and **7** are shown in Figure 3–9. The negative Cotton band of the 1-keto steroid (**5**) is seen much more clearly than by ord, owing to the absence of background effects.

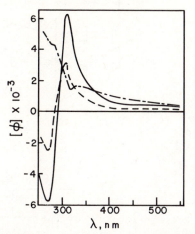

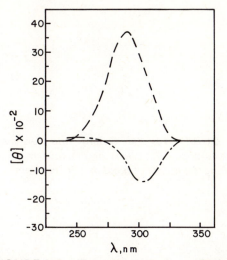

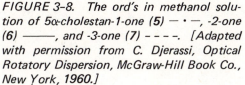

FIGURE 3-8.   The ord's in methanol solution of 5α-cholestan-1-one **(5)** — · —, -2-one
(6) ———, and -3-one **(7)** - - - -.   [Adapted
with permission from C. Djerassi, Optical
Rotatory Dispersion, McGraw-Hill Book Co.,
New York, 1960.]

FIGURE 3-9.   The cd spectra in methanol
solution of 5-cholestan-1-one **(5)** — · — and
-3-one **(7)** - - - -.   [Reproduced with the
permission of the American Chemical Society
from C. Djerassi, H. Wolf, and E. Bunnenberg, J. Amer. Chem. Soc., **84**, 4552 (1962).]

By 1961 a considerable amount of chiroptical data had accumulated on the carbonyl
chromophore, and a semiempirical generalization, the *octant rule*, was proposed (reference 3).
The rule allows the sign of the n–π* Cotton effect to be deduced from the contributions of
perturbing groups in each of eight sectors (octants) about the carbonyl group. Hence the octant
rule allows one to decide not only upon the absolute configuration but also on a likely conformation of the molecule. Conversely, configuration and conformation may be deduced from
the sign of the Cotton effect.

In order to understand the octant rule, let us consider the symmetry properties of the
orbitals that are involved in the n–π* transition near 290 nm. As depicted in Figure 3-10(a),
the n orbital on oxygen has a nodal plane that bisects the R—C—R′ bond angle and is perpendicular to the plane of the carbonyl group. For the π* orbital, the plane containing the

FIGURE 3-10.   Nodal surfaces for a saturated ketone, R—CO—R′.   (a) The nodal
plane of the n orbital.   It bisects the
R—C—R′ angle and is perpendicular to the
plane of the ketone.   (b) The nodal
surfaces of the π* orbital.   The plane of
the carbonyl group is a nodal plane and
there is another nodal surface, not necessarily a plane, perpendicular to the C—O
axis and intersecting it between the carbon
and oxygen atoms.

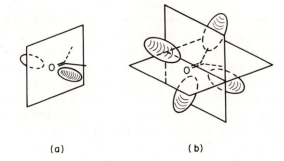

(a)   (b)

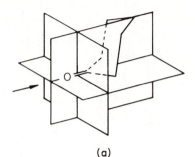

FIGURE 3-11. (a) Octant rule for saturated cyclo-hexanones. (b) Signs of the four rear octants viewed along the carbonyl bond axis from O to C.

(a)

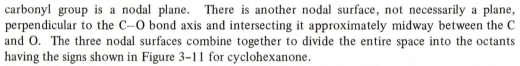

(b)

carbonyl group is a nodal plane.   There is another nodal surface, not necessarily a plane, perpendicular to the C—O bond axis and intersecting it approximately midway between the C and O.   The three nodal surfaces combine together to divide the entire space into the octants having the signs shown in Figure 3–11 for cyclohexanone.

The octant rule states that substituents lying in the nodal planes make no contributions to the n–π^* Cotton effect.   Substituents within an octant contribute the sign of that octant to the overall sign of the Cotton effect.   Since most substituents are usually on the same side of the nodal surface as the carbonyl carbon, the octant diagram is simplified by considering only the four rear octants.   Relative intensities are determined qualitatively.   For example, when both negative rear octants are occupied, the magnitude of the Cotton effect is enhanced.

A few substituents such as $-$F, $-\overset{+}{\text{N}}\text{Me}_3$, and cyclopropyl exhibit *antioctant* behavior, and several breakdowns of the octant rule for certain aliphatic ketones have recently been reported (reference 3).   Ketones having an axial α-chloro or α-bromo substituent display Cotton effects whose signs are determined only by the octant location of the halogen, even if most of the molecule occupies an oppositely signed octant; this is the *axial haloketone rule* of Djerassi and Klyne (reference 3).

As an illustration of the use of the octant rule, consider (+)-3-methylcyclohexanone, which exhibits a positive n–π^* Cotton effect.   We would predict that the chair form with the methyl group equatorial (Figure 3–12a) would be energetically favored over the conformer with the axial methyl group.   In the octant projection (+)-3-methylcyclohexanone (Figure 3–12d), carbon atoms 2, 4, and 6 lie in nodal planes and thus make no contribution to the sign of the Cotton effect.   The contribution of C-3 is canceled by that of C-5, which is equal but opposite.   However, the methyl group at C-3 lies in a positive octant and is responsible for the sign of the positive Cotton effect.   The absolute configuration of (+)-3-methylcyclohexanone is deduced to be *R*, since the enantiomeric molecule would lead to an octant prediction of a negative n–π^* Cotton effect.

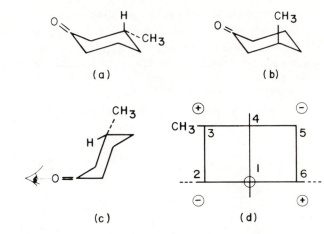

FIGURE 3-12. Equatorial and axial conformations of (+)-3-methylcyclohexanone and the octant rule projection of the equatorial conformer.

        The octant rule is also of great value in the field of natural products.   We saw in Figure 3-8 that the Cotton effect of cholestan-2-one (**6**) was about twice as large as that of the 3-isomer (**7**).   The octant diagrams for the two steroids are shown in Figure 3-13.   In the top octant diagram for **7**, the contributions of rings A and C are seen to cancel.   The two angular methyl groups lie in the vertical nodal plane, so that only the methylene groups C-6 and C-7 (and C-15 and C-16, these two being very remote from the chromophore) contribute to the Cotton effect.   According to the octant diagram, these contributions should be positive. In the case of cholestan-2-one (Figure 3-13b), the positive contributions are very strong and give rise to the relatively intense Cotton effect.

        It is generally found that the contribution of a substituent to the total Cotton effect will depend upon its octant for the sign and upon the distance from the chromophore for its magnitude.

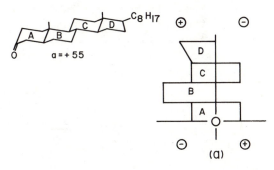

FIGURE 3-13. Octant diagrams for (a) cholestan-3-one and (b) cholestan-2-one.

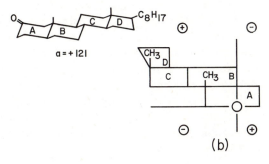

3–2b Carboxylic Acids, Esters, Acid Halides, and Amides

Introduction of a heteroatom at the carbon atom of the carbonyl chromophore causes a large blue shift of the n–π* band, although the intensity remains about the same as for aldehydes and ketones. The effect of such substituents on the n–π* transition is given in Table 3-5.

Table 3-5. *Effect of Heteroatom Substituents on the Carbonyl n–π* Transition*

X in $CH_3\overset{\displaystyle O}{\overset{\|}{C}}{-}X$	Solvent	λ_{max} (nm)	ϵ_{max} (liters mole^{-1} cm^{-1})
−H	vapor	290	10
−CH$_3$	hexane	279	15
	95% ethanol	272.5	19
−OH	95% ethanol	204	41
−SH	cyclohexane	219	2200
−OCH$_3$	isooctane	210	57
−OC$_2$H$_5$	95% ethanol	208	58
	isooctane	211	58
−O−$\overset{\displaystyle O}{\overset{\|}{C}}$−CH$_3$	isooctane	225	47
−Cl	heptane	240	40
−Br	heptane	250	90
−NH$_2$	methanol	205	160

The heteroatom at the carbonyl group in carboxylic acids, esters, acid chlorides, amides, etc., can donate electron density by conjugation to the carbonyl function. The energy of the antibonding π* orbital will be raised by interaction with the unshared pair of the substituent, whereas the p orbital occupied by the n electrons in the ground state is not affected. The n–π* transition energy is thus raised, and the absorption shifts to shorter wavelength.

Two bands have been identified in the uv spectra of aliphatic carboxylic acids and esters. The first is an n–π* transition in the vicinity of 210 nm (ε = 40–60). The second band corresponds to a π–π* transition at about 165 nm (ε = 2500–4000).

Carboxylic acids are extensively dimerized in the liquid state and in nonpolar solvents and even to some extent in the vapor phase (eq. 3-1). Hence, depending upon solvent and

$$2 \text{ RCOOH} \rightleftharpoons R{-}C\underset{O{-}H\bullet\bullet\bullet\bullet O}{\overset{O\bullet\bullet\bullet\bullet H{-}O}{\Big<\quad\Big>}}C{-}R \qquad 3\text{-}1$$

concentration, one may be dealing with pure dimer, pure monomer, or a mixture of the two. In the case of carboxylic acids, it is recommended that spectra, either isotropic absorption or chiroptical, be determined at more than one concentration. Protic solvents such as alcohols shift the above equilibrium toward monomer. Conversion of the acid to an ester obviates the possibility of hydrogen bonding according to eq. 3-1. Essentially identical spectra from a carboxylic acid and, for example, its methyl or ethyl ester are indicative that dimerization is absent or is not spectroscopically important.

The uv and chiroptical spectra of amides have received considerable attention because of the importance of the amide chromophore in polypeptides and proteins. At least five transitions of the amide group have been identified (reference 4). The two lowest energy ones

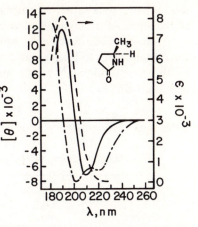

FIGURE 3-14. Isotropic absorption - - - - and cd ———— spectra in water and cd spectrum in cyclohexane — · — of γ-valerolactam. [Adapted with the permission of the American Chemical Society from D. W. Urry, J. Phys. Chem., **72**, 3035 (1968).]

are a weak n–π* band ($\epsilon \sim 100$) near 220 nm in nonpolar solvents and a relatively strong π–π* transition in the 173-200 nm region ($\epsilon \sim 8{,}000$). Both of these transitions are strongly perturbed by the nitrogen 2p orbital of the conjugated π system, which extends over the N, C, and O atoms. The n–π* amide transition exhibits the usual solvent effect, being blue-shifted on going from nonpolar to hydroxylic solvent.

The question of whether aliphatic amides protonate on oxygen or nitrogen was examined recently by uv spectroscopy. Benderly and Rosenheck (reference 4) reasoned that if protonation occurs on nitrogen, the nitrogen atom would be removed from the amide system and the spectral properties of the simple carbonyl group should result, i.e., a large red shift of the n–π* band and a large blue shift of the π–π* band. Studies of the acidity dependence of the π–π* absorption of N,N-dimethylacetamide near 195 nm showed no significant shifts, indicating that the oxygen-protonated amide is the dominant species in dilute acid solutions.

Urry (reference 4) has studied the uv and cd spectra of the cyclic amide γ-valerolactam in cyclohexane and in aqueous solution (Figure 3-14). In water solution the amide shows a negative n–π* Cotton effect at 210 nm and a positive π–π* band at 190 nm. In cyclohexane solution, the 210 nm Cotton band is split into two, at 218 and 202 nm, because of dimer formation. It is instructive to note that the n–π* band is absent in the isotropic absorption but is clearly observed in the CD spectrum.

3-3 Unsaturated Carbonyl Compounds

Many organic molecules of interest contain both carbonyl and alkene chromophores. If the groups are separated by two or more sigma bonds, there is generally (but with important exceptions, see below) little electronic interaction and the effect of the two chromophores on the observed spectrum is essentially additive. Compounds in which the double bond is conjugated to a carbonyl group exhibit spectra in which both the alkenic π–π* and the carbonyl n–π* absorption maxima of the isolated chromophores have undergone bathochromic shifts of 15-45 nm, although each band is not necessarily displaced by an equal amount. Photoionization data indicate that the n orbital energy is relatively constant, so that the red shift is most probably caused by a lowering of the energy of the π* orbital.

Crotonaldehyde ($CH_3CH{=}CH{-}CH{=}O$) in ethanol solution has an intense band at 220 nm ($\epsilon = 15{,}000$) and a weak band at 322 nm ($\epsilon = 28$). The low intensity and hypsochromic shift in hydroxylic solvents suggest that the 322 band is the n–π* transition, and the bathochromic shift relative to a saturated carbonyl indicates that the excited π^* orbital now extends over all the atoms of the conjugated carbonyl group.

The solvent effect on the π–π^* transition is opposite to that on the n–π^* peak; the π–π^* absorption shifts to longer wavelength with increasing solvent polarity. The effect of solvent on the n–π^* and π–π^* transitions of mesityl oxide, $(CH_3)_2C{=}CH{-}CO{-}CH_3$, is illustrated in Table 3-6.

Table 3-6. *Solvent Dependence of the π–π^* and n–π^* Absorption Maxima of*

$$\underset{\text{Mesityl Oxide, } (CH_3)_2C{=}CH{-}\overset{\displaystyle O}{\overset{\displaystyle \|}{C}}{-}CH_3}{}$$

	π–π^* Transition		n–π^* Transition	
Solvent	λ_{max} (nm)	ϵ (liters mole^{-1} cm^{-1})	λ_{max} (nm)	ϵ (liters mole^{-1} cm^{-1})
Hexane	229.5	12,600	327	40
Ethyl ether	230	12,600	326	40
Ethanol	237	12,600	325	90
Methanol	238	10,700	312	55
Water	244.5	10,000	305	60

As the number of double bonds conjugated with the carbonyl increases, the π–π^* transition shifts to longer wavelength and its intensity increases, with the result that the much weaker n–π^* absorption appears as a shoulder or becomes completely obscured by the more intense overlapping π–π^* band.

Alkyl substitution shifts the π–π^* and n–π^* maxima in opposite direction, the π–π^* being displaced to longer wavelength. Such effects of substitution on the position of the π–π^* transition can be predicted through the use of empirical rules similar to those already discussed for dienes and also first formulated by R. B. Woodward, then modified by the Fiesers. These rules, which have played an important role in assigning the structures of steroids and other natural products, are given in Table 3-7. Similar rules have been formulated for α,β-unsaturated acids (reference 5), which generally show intense absorption in the 205–225 nm region (ϵ 10,000–20,000).

The following calculation of λ_{max} for the π–π^* transition of cholest-4,6-diene-3-one (**8**) illustrates the rules for α,β-unsaturated ketones.

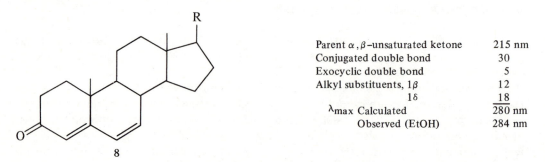

Parent α,β–unsaturated ketone	215 nm
Conjugated double bond	30
Exocyclic double bond	5
Alkyl substituents, 1β	12
1δ	18
λ_{max} Calculated	280 nm
Observed (EtOH)	284 nm

Both the π–π^* and n–π^* transitions of α,β-unsaturated aldehydes and ketones give rise to Cotton effects. In rigid systems such as the steroids, the electronic interactions of two *achiral* chromophores, the alkene and the carbonyl, can result in an inherently *chiral*

Table 3-7. *Rules for the Calculation of the Position of π–π^* Absorption of Unsaturated Carbonyl Compounds[a]*

$$\begin{array}{cccc} \delta & \gamma & \beta & \alpha \end{array}$$
$$-C=C-C=C-C=O$$
$$\qquad\qquad\quad |$$
$$\qquad\qquad\quad R$$

		λ *(nm)*
Parent α,β-unsaturated carbonyl compound (acyclic, six-membered or larger ring ketone)		215
α,β bond in five-membered ring		−13
Aldehyde		−6
Each additional conjugated double bond		30
Homodiene compound		39
Exocyclic double bond		5
Each alkyl substituent	α	10
	β	12
	γ and higher	18
Each polar group		
—OH	α	35
	β	30
	δ	50
—O-Ac	$\alpha, \beta,$ or δ	6
—OR	α	35
	β	30
	γ	17
	δ	31
—SR	β	85
—Cl	α	15
	β	12
—Br	α	25
	β	30
—NR$_2$	β	95
Solvent correction		
ethanol, methanol		0
chloroform		1
dioxane		5
ethyl ether		7
hexane, cyclohexane		11
water		−8

[a] R. B. Woodward, *J. Amer. Chem. Soc.*, **63**, 1123 (1941); **64**, 72, 76 (1942); L. F. Fieser and M. Fieser, *Natural Products Related to Phenanthrene*, Reinhold, New York, 1949; A. I. Scott, *Interpretation of the UV Spectra of Natural Products*, Pergamon, New York, 1964.

α,β-unsaturated ketone chromophore. As a consequence, the octant rule, based on the concept of the inherently achiral carbonyl group, cannot be generally applied.

The π–π^* transition near 240 nm usually exhibits a Cotton effect of opposite sign to that of the n–π^* band. Recently several authors have reported a new Cotton effect of relatively high intensity near 215 nm, which overlaps the π–π^* transition and sometimes obscures its sign. Thus considerable caution should be used in the analysis of chiroptical spectra for α,β-unsaturated ketones.

Unsaturated carbonyl compounds such as β,γ-unsaturated ketones can also form an inherently chiral chromophoric system. Spectroscopically, in these compounds there is a chromophoric interaction, *coupling*, in which the forbidden n–π^* transition borrows intensity

FIGURE 3-15. Correlation between chirality and sign of Cotton effect for inherently chiral β,γ-unsaturated ketones. The (+) and (−) signs refer to the n–π Cotton effect near 300 nm. [Adapted with the permission of the American Chemical Society from A. Moscowitz, K. Mislow, M. A. W. Glass, and C. Djerassi, J. Amer. Chem. Soc., **84**, 1945 (1962).]*

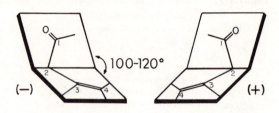

from the π–π* band resulting in a substantial increase in the n–π* Cotton effect near 310 nm.

A chirality rule, depicted in Figure 3-15, has been proposed (reference 6) for this chromophore to correlate the sign of the n–π* Cotton band. The plus and minus signs refer to the sign of the n–π* Cotton effect. An application of this chirality rule for β,γ-unsaturated ketones is given in Figure 3-16, which shows the uv and cd spectra of (+)-2-benzonorbornenone (reference 6). The high rotational strength expected of such inherently chiral chromophores is reflected in the large molar ellipticity, $[\theta]_{307.5}$ +62,000. In order to have the orientation corresponding to that shown in the chirality rule (Figure 3-15), the structure shown on the uv-cd spectra would need to be turned over. The positive n–π* Cotton effect centered at 307.5 nm then corresponds to the positive geometry in Figure 3-16. Note the three fingers and a shoulder of this Cotton effect, due to vibrational transitions within the n–π* electronic transition. On the basis of the chiroptical correlation, which was in agreement with chemical evidence, Mislow and co-workers (reference 6) were able to assign the 1R-configuration to (+)-2-benzonorbornenone.

*FIGURE 3-16. The uv - - - - and cd ——— spectra in isooctane of (1R)-(+)-2-benzonorbornenone, a homoconjugated system. [Adapted with the permission of the American Chemical Society from D. J. Sandman, K. Mislow, W. P. Giddings, J. Dirlam, and G. C. Hanson, J. Amer. Chem. Soc., **90**, 4877 (1968).]*

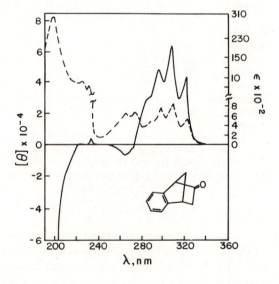

3-4 Aromatic Compounds

3-4a Benzene and Its Derivatives

The benzene ring ranks with the carbonyl group as one of the most widely studied chromophores. The spectrum of benzene above 180 nm consists of three well-defined absorption bands due to π–π* transitions (Figure 3-17). An intense structureless band occurs

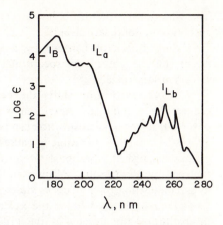

FIGURE 3-17. The uv spectrum of benzene in hexane.

at about 185 nm, with a somewhat weaker band (λ_{max} ∼200) of poorly resolved vibrational structure overlapping the 185 nm absorption. The longest wavelength transition is a low intensity system centered near 255 nm and exhibiting characteristic vibrational structure. Data on the benzene absorption bands and some of the various nomenclature systems used to describe them are given in Table 3-8.

Table 3-8. *Notation Systems for Benzene Absorption Bands*

λ_{max} (ϵ_{max}) in Hexane			
184 nm (68,000)	*204 nm (8,800)*	*254 nm (250)*	*Origin of the Spectral Notation*
^1B	^1L_a	^1L_b	a
β	p (para)	α	b
$^1E_{2u}$	$^1B_{1u}$	$^1B_{2u}$	c
Second primary	primary	secondary	d
	K	B	e
	(conjugation)	(benzenoid)	

[a] Platt free-electron method notation: J. R. Platt, *J. Chem. Phys.*, **17**, 484 (1949).

[b] Empirical notation of Clar based on behavior of bands with temperature: E. P. Clar, *Aromatische Kohlenwasserstoffe*, Springer Verlag, Berlin, 1952.

[c] Molecular orbital approach based on the group theoretical notation of the transitions.

[d] Empirical notation: L. Doub and J. M. Vandenbelt, *J. Amer. Chem. Soc.*, **69**, 2714 (1947); **71**, 2414 (1949).

[e] Early empirical notation: A. Burawoy, *J. Chem. Soc.*, 1177 (1939); 20 (1941).

The benzene absorptions at 254 and 204, termed ^1L_b and ^1L_a in the Platt notation (reference 7), are both forbidden, but the ^1L_a is able to "borrow" intensity from the allowed ^1B transition, which overlaps it at shorter wavelength. The different transition probabilities relate to configuration interaction because of the degeneracy of the highest occupied and lowest vacant orbitals in benzene. The superscript one indicates that the transition is to a singlet excited state. Benzene belongs to the D_{6h} point group, and the intensity of the symmetry-forbidden 254 nm ^1L_b transition should be zero. However, vibrational distortions from hexagonal symmetry result in a small net transition dipole moment and the observed low intensity.

The ^1L_b absorption, sometimes called the benzenoid band, is usually easily identifiable; it has about the same intensity in benzene and its simple derivatives, ϵ 250–300. It also has similar well-defined vibrational structure with up to six vibrational bands, as in benzene itself.

The vibrational structure is less evident in polar solvents and more sharply defined in vapor spectra or in nonpolar solvents.

Substitution of benzene by auxochromes, chromophores, or fused rings has different effects on the absorption spectrum. Because of their importance, we shall consider these effects in some detail.

Alkyl substitution shifts the benzene absorption to longer wavelengths and tends to reduce the amount of vibrational structure. Increases in band intensities are commonly observed. In general, substitution can perturb the benzene ring by both inductive and resonance effects. A methyl substituent causes the largest wavelength shift, hyperchromism, and the greatest change in vibrational intensities. The effect decreases as the methyl hydrogens are replaced by alkyl groups. This result is often cited as evidence for the importance of C–H hyperconjugation (σ-π electron interaction).

Data in Table 3-9 for the xylenes illustrate that bathochromic shifts caused by alkyl disubstitution are usually in the order para > meta > ortho. Alkylbenzenes, like alkyl-substituted alkenes, normally do not undergo any significant spectral changes on varying the solvent.

Introduction of polar substituents such as $-NH_2$, $-OH$, $-OCH_3$, $-CHO$, $-COOH$, and $-NO_2$ cause marked spectral changes. With these groups, the intensity of the ^1L_b band is enhanced. Much of the fine structure is lost in polar solvents, although it may be observed to some extent in nonpolar solvents. In addition, the ^1L_a band is shifted bathochromically; for example, in aniline, thiophenol, and benzoic acid it occurs in the 230 nm region (Table 3-9).

Substituents possessing nonbonding electrons can conjugate with the π system of the ring. Since the energy of the π^* state is lowered by delocalization over the entire conjugated system, the n-π^* absorption occurs at longer wavelength than in the corresponding unconjugated chromophoric substituent. For example, acetophenone, $C_6H_5-CO-CH_3$, exhibits an n-π^* absorption at 320 nm and the ^1L_b aromatic transition at 276 nm. Both bands are bathochromically shifted and considerably increased in intensity due in part to conjugation of the benzene π electrons with the π electrons of the carbonyl group.

The spectral changes found on conversion from phenol to phenoxide anion and from aniline to anilinium cation are of considerable interest and practical importance. In the case of aniline, the 280 nm band is most probably due to the benzenoid ^1L_b transition, red-shifted and enhanced by electron donation from the amino group to the ring. Resonance structures involving intramolecular charge transfer (e.g., **9**) make a substantial contribution to the ground

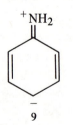

9

electronic state, but the predominant contribution is to the excited state. In general, substituted benzenes for which this type of resonance form can be written will have bathochromically shifted spectra relative to benzene and will exhibit hyperchromism. Charge transfer complexes are discussed in section 4-3.

Conversion of aniline to the anilinium cation involves attachment of a proton to the nonbonding electron pair, removing it from conjugation with the π electrons of the ring

Table 3-9. *Absorption Data for Benzene and Derivatives[a]*

Compound	Solvent	λ_{max} (nm)	ϵ_{max}	λ_{max} (nm)	ϵ_{max}	λ_{max} (nm)	ϵ_{max}	λ_{max} (nm)	ϵ_{max}
Benzene	hexane	184	68,000	204	8,800	254	250		
	water	180	55,000	203.5	7,000	254	205		
Toluene	hexane	189	55,000	208	7,900	262	260		
	water			206	7,000	261	225		
Ethylbenzene	ethanol[b]			208	7,800	260	220		
t-Butylbenzene	ethanol			207.5	7,800	257	170		
o-Xylene	25% methanol			210	8,300	262	300		
m-Xylene	25% methanol			212	7,300	264	300		
p-Xylene	ethanol			216	7,600	274	620		
1,3,5-Trimethyl-benzene	ethanol			215	7,500	265	220		
Fluorobenzene	ethanol			204	6,200	254	900		
Chlorobenzene	ethanol			210	7,500	257	170		
Bromobenzene	ethanol			210	7,500	257	170		
Iodobenzene	ethanol			226	13,000	256	800		
	hexane			207	7,000	258	610	285 (sh)	180
Phenol	water			211	6,200	270	1,450		
Phenolate ion	aq NaOH			236	9,400	287	2,600		
Aniline	water			230	8,600	280	1,400		
	methanol			230	7,000	280	1,300		
Anilinium ion	aq acid			203	7,500	254	160		
N,N-Dimethyl-aniline	ethanol			251	14,000	299	2,100		
Thiophenol	hexane			236	10,000	269	700		
Anisole	water			217	6,400	269	1,500		
Benzonitrile	water			224	13,000	271	1,000		
Benzoic acid	water			230	10,000	270	800		
	ethanol			226	9,800	272	850		
Nitrobenzene	hexane			252	10,000	280 (sh)	1,000	330 (sh)	140
Benzaldehyde	hexane			242	14,000	280	1,400	328	55
	ethanol			240	16,000	280	1,700	328	20
Acetophenone	hexane			238	13,000	276	800	320	40
	ethanol			243	13,000	279	1,200	315	55
Styrene	hexane			248	15,000	282	740		
	ethanol			248	14,000	282	760		
Cinnamic acid									
cis-	hexane	200	31,000	215	17,000	280	25,000		
trans-	hexane	204	36,000	215	35,000	283	56,000		
	ethanol			215	19,000	268	20,000		
Stilbene									
cis-	ethanol			225	24,000	274	10,000		
trans-	heptane	202	24,000	228	16,000	294	28,000		
Phenylacetylene	hexane	202	44,000	248	17,000	hidden			
2,2'-Dimethyl-biphenyl	hexane	198	43,000	228 (sh)	6,000	264	800		
Diphenylmethane	ethanol			220	10,000	262	500		

[a] If vibrational structure is present, λ_{max} refers to the subband of highest intensity.
[b] "Ethanol" should be taken to mean 95% ethanol.

(eq. 3-2). The absorption characteristics of this ion closely resemble those of benzene. The

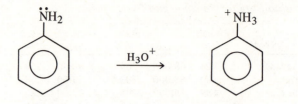

$$3-2$$

blue shift observed in the conversion of aniline to anilinium ion is typical of the spectral changes due to protonation of basic groups and can serve as a useful tool in structure elucidation.

Conversion of phenol to the phenolate anion makes an additional pair of nonbonding electrons available to the conjugated system, and both the wavelengths and the intensities of the absorption bands are increased (Table 3-9). Analogous to the information obtainable in the aniline-anilinium ion conversion, a suspected phenolic group may be determined by comparison of the uv spectrum of the compound in neutral and in alkaline (pH 13) solution.

The aniline-anilinium or phenol-phenolate conversion as a function of pH can demonstrate the presence of the two species in equilibrium by the appearance of an *isosbestic point* in the uv spectrum. If two substances, each of which obeys Beer's law, are in equilibrium, the spectra of all equilibrium mixtures at a constant total concentration intersect at a fixed wavelength. This point, termed the isosbestic point, is the wavelength at which the absorbances of the two species are equal. An example is shown in Figure 3-18 for 4-methoxy-2-nitrophenol.

In the uv spectra of the *cis*- and *trans*-cinnamic acids ($C_6H_5CH=CHCOOH$), the band at about 280 nm represents the ^1L_b benzenoid absorption displaced to longer wavelength and intensified by conjugation with the double bond and the carbonyl group of the carboxylic acid. The molar absorption coefficient of the trans isomer is more than twice that for the cis and is thought to be related to the longer chromophoric length in the trans compound. A similar relation is found in the *cis*- and *trans*-stilbenes. A generally applicable rule for many

FIGURE 3-18. The spectra of 4-methoxy-2-nitrophenol as a function of pH. [Adapted with permission from H. H. Jaffé and M. Orchin, Theory and Applications of Ultraviolet Spectroscopy John Wiley & Sons, Inc., New York, 1962, p. 562.]

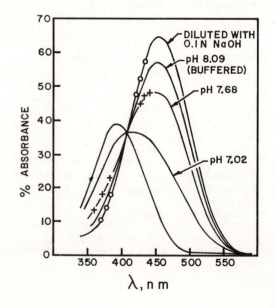

cis-trans isomer pairs is that the lowest energy π–π* transition occurs at longer wavelength and is more intense for the trans isomer.

The spectral data for biphenyl illustrate the effects of conjugation of adjacent benzene chromophores. The spectrum above 185 nm consists of two broad and intense bands at 202 (ε 44,000) and 248 nm (ε 17,000). The strong intensity of the 248 nm band indicates that it corresponds to the 205 nm ^1L_a band of benzene, shifted by conjugation between the two rings. The ^1L_b absorption is concealed beneath the broad envelope of the intense ^1L_a band.

The deviation from coplanarity in biphenyl is thought to be about 23°. Ortho substituents increase the deviation of the rings from coplanarity with concomitant loss of conjugation. This effect can be seen in the data for 2,2'-dimethylbiphenyl, which more resembles the sum of two independent alkylbenzene systems.

The presence of a saturated methylene group between two chromophores in a molecule results in almost complete loss of conjugation. This is well illustrated in the data for diphenylmethane (C_6H_5–CH_2–C_6H_5), where the ^1L_b band at 262 nm has an ε of 500, almost exactly the sum of two isolated benzene rings.

Like the carbonyl group, a monosubstituted benzene ring is an example of an inherently achiral but chirally perturbed chromophore. Chiroptical studies of benzene chromophores are surprisingly recent. Prior to 1965 only a few measurements had been reported, and disagreement existed in the literature as to whether a benzene ring in a chiral molecule could exhibit optically active transitions. The confusion was in part caused by conflicting reports on compounds such as α-phenylethanol (10). Some workers had reported that this alcohol in its

$$
\begin{array}{c}
CH_3 \\
| \\
H\!-\!\!\!-C\!-\!\!\!-OH \\
| \\
C_6H_5
\end{array}
$$

10

chiral form exhibits an ord Cotton effect in the 260 nm region, while others measured the same compound and reported no Cotton effect, only a plain ord curve. The cd of (S)-(−)-α-phenylethanol (Figure 3–19) clearly shows the presence of a positive Cotton effect containing vibrational fine structure in the 260 nm region as well as a more intense ^1L_a Cotton band near 210 nm. Both Cotton effects are due to aromatic transitions since only the benzene ring absorbs in the region above 200 nm. The 260 nm band is due to the symmetry-forbidden

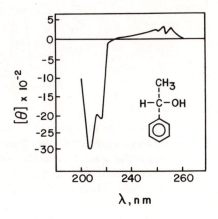

FIGURE 3-19. The cd in heptane of (S)-(−)-α-phenylethanol. [Adapted with the permission of the American Chemical Society from L. Verbit, J. Amer. Chem. Soc., 87, 1617 (1965).]

^1L_b transition. Its relative weakness, $[\theta]_{max}$ +200, explains why previous workers sometimes had difficulty in discerning this Cotton effect superposed on a steeply falling background ord curve. The background rotation is due to the more intense negative Cotton effect near 210 nm, which also determines the sign of this compound in the visible region. Hence difficulties in observing benzenoid Cotton effects were a question of instrument sensitivity rather than any intrinsic difference in the nature of the benzene ring as an optically active chromophore.

Salvadori and co-workers (reference 7) have recently reported the uv and cd spectra of a chiral aromatic hydrocarbon, (S)-(+)-2-phenyl-3,3-dimethylbutane (**11**), for which they were

$$\begin{array}{c} H \\ | \\ C_6H_5 \!\!-\!\! C \!\!-\!\! CH_3 \\ | \\ C(CH_3)_3 \end{array}$$

11

able to measure three aromatic Cotton effects in the region above 185 nm. The spectra are shown in Figure 3-20. The Cotton bands, all of which are positive, correspond to the ^1B, ^1L_a, and ^1L_b transitions of the aromatic ring. The ^1L_b band in the 260 nm region,

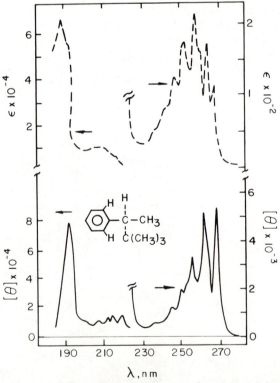

FIGURE 3-20. The - - - - (heptane) and cd ——— (methylcyclohexane–isopentane, 1/3) spectra of (S)-(+)-2-phenyl-3,3-dimethylbutane. The arrows indicate the respective scales. [Adapted with the permission of the American Chemical Society from P. Salvadori, L. Lardicci, R. Menicagli, and C. Bertucci, J. Amer. Chem. Soc., 94, 8598 (1972).]

$[\theta]_{max}$ + 5,000, is more than an order of magnitude more intense than those displayed by typical open chain phenyl compounds (compare α-phenylethanol) and suggests the possibility of restricted conformational mobility. Similar large ellipticity values are found in rigid aromatic systems such as alkaloids. Evidence that chiral 2-phenyl-3,3-dimethylbutane is constrained to a single conformation because of steric effects imposed by the bulky *t*-butyl group was obtained by measurement of the 260-nm Cotton effect at −100° (reference 7). The almost complete absence of cd changes over this temperature range indicates that the compound is already in its lowest energy conformation at room temperature, at which the spectra in Figure 3-20 were determined.

3–4b Polycyclic Aromatic Hydrocarbons

Much of our understanding of the uv spectra of this class of compounds is due to Clar, who has published a comprehensive monograph (reference 8). Clar divides the polynuclear aromatic hydrocarbons into two classes, the linear *acenes* (naphthalene, anthracene, etc.) and the angular *phenes* (phenanthrene, triphenylene, etc.). Cata–condensed hydrocarbons are those in which no carbon atom is common to more than two rings; peri–condensed compounds have carbons belonging to more than two rings. 1,2-Benzanthracene (**12**) is an

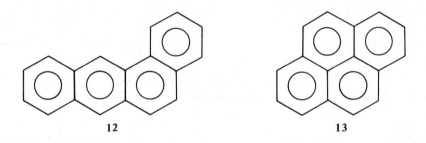

12   13

example of a cata–condensed hydrocarbon, whereas pyrene (**13**) is a peri–condensed compound. Considerable interest exists in the electronic structure of these hydrocarbons, many of which are carcinogenic.

The empirical nomenclature introduced by Clar uses the symbols α, *p* (para), and β, corresponding to the ^1L_b, ^1L_a, and ^1B benzene transitions, respectively (Table 3-8). The uv spectra of the acenes generally show the three typical bands of benzene, except that all are shifted to longer wavelength and are intensified. For example, the spectrum of naphthalene strongly resembles a displaced and intensified benzene spectrum. As the conjugated ring system increases, the ^1L_a (*p*) band shifts more than does the ^1L_b (α) band so that in anthracene and naphthacene, the ^1L_b transition is hidden by the more intense ^1L_a absorption. The ^1L_b band reemerges in the spectrum of pentacene but is now located between ^1L_a and ^1B transition. Table 3-10 lists absorption data for some polycyclic aromatic hydrocarbons. Note that naphthacene and pentacene absorb in the visible region and hence are colored compounds.

The angular hydrocarbons, the phenes, also show the three typical benzene-like bands but at shorter wavelength than in linear acenes with the same number of rings. The ^1L_b band, which is hidden under the ^1L_a absorption in anthracene (Figure 3-21), is clearly visible in the phenanthrene spectrum (Figure 3-22). The phenes exhibit an additional band in the 210–245 nm region termed β′ by Clar or ^1C_b in the Platt notation. This band is due to a transition that is forbidden in compounds with a center of symmetry but occurs near 219 nm for compounds lacking this symmetry element, e.g., phenanthrene.

Table 3–10. *Absorption Data for Polycyclic Aromatic Hydrocarbons[a]*

Compound	Solvent	^1C_b (β') λ (nm)	ε	^1B (β) λ (nm)	ε	^1L_a (p) λ (nm)	ε	^1L_b (α) λ (nm)	ε
Benzene	heptane			184	60,000	204	8,000	255	200
Acenes									
Naphthalene	isooctane			221	110,000	275	5,600	311	250
Anthracene	isooctane			251	200,000	376	5,000	hidden	
Naphthacene	heptane			272	180,000	473	12,500	hidden	
Pentacene	heptane			310	300,000	585	12,000	417	600
Phenes									
Phenanthrene	methanol	219	18,000	251	90,000	292	20,000	330	350
Triphenylene	95% ethanol			257	150,000	273	25,000	333	760
Chrysene	95% ethanol	220	37,000	267	160,000	306	15,500	360	1,000
						319	15,500		
1,2,5,6-Dibenz-anthracene	1,4-dioxane	223	40,000	299	160,000	335	17,000	394	1,000

[a] For bands containing vibrational structure, λ_{max} refers to the subband of highest intensity.

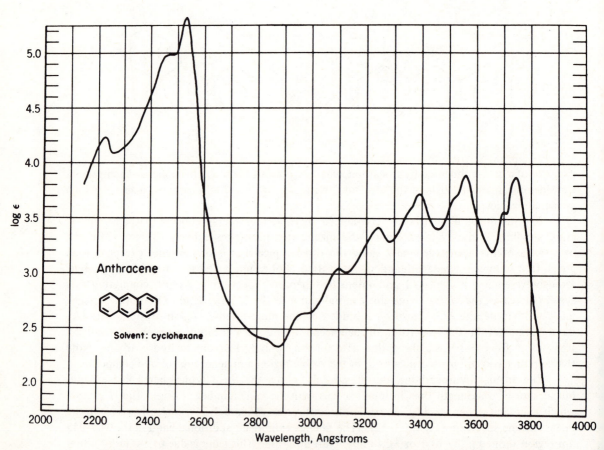

FIGURE 3–21. *Ultraviolet spectrum of anthracene. [Reproduced with permission from R. A. Friedel and M. Orchin, Ultraviolet Spectra of Aromatic Compounds, John Wiley & Sons, New York, 1951.]*

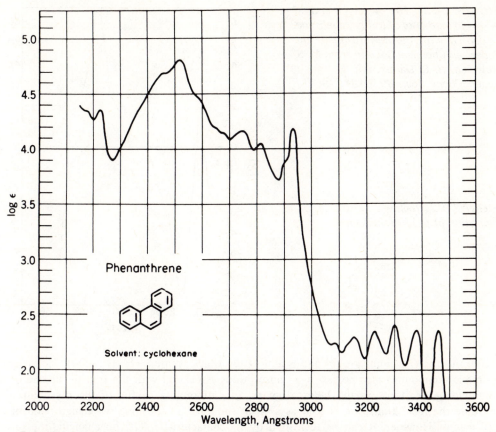

FIGURE 3-22. Ultraviolet spectrum of phenanthrene. [Reproduced with permission from R. A. Friedel and M. Orchin, Ultraviolet Spectra of Aromatic Compounds, John Wiley & Sons, New York, 1951.]

Some polycyclic hydrocarbons such as the benzpyrenes have been identified as air and water pollutants. Their characteristic absorption spectra allow them to be monitored at quite low levels (microgram quantities) by uv spectroscopy.

The polycyclic hydrocarbon hexahelicene is the classic example of an inherently chiral chromophore. Its chirality extends in a helical manner throughout the entire molecule, and its enormous D line rotation, $[\alpha]_D$ 3750°, is characteristic of this class of compounds. The uv and cd spectra of (+)-hexahelicene are shown in Figure 3–23. In 1972, Lightner and co-workers (reference 8) reported an X-ray study of optically active 2-bromohexahelicene. Their results show that (+)-hexahelicene has right-handed helicity or the P configuration in the RS nomenclature system (Figure 3–24).

3–4c   Heterocyclic Aromatic Compounds

Replacement of a CH in an aromatic hydrocarbon by N gives the corresponding isoelectronic azaaromatic compound. The π system is essentially unchanged so that strong similarities exist between the uv spectrum of a nitrogen heterocycle and its carbocyclic analogue.

The band corresponding to the symmetry-forbidden 255 nm transition of benzene is allowed in pyridine because of the lower symmetry introduced by the nitrogen atom (pyridine

*FIGURE 3-23. The uv – – – – and cd
——— in methanol of (+)-hexahelicene.
[Adapted with the permission of the
American Chemical Society from M. S.
Newman, R. S. Darlak, and L. Tsai, J.
Amer. Chem. Soc.,* **89**, *6191 (1967).]*

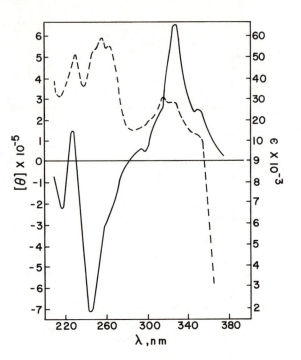

*FIGURE 3-24 (below).   Absolute con-
figuration of (+)-hexahelicene.*

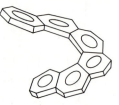

belongs to the C_{2v} point group) and is almost ten times as intense (ϵ_{251} 2,000). Heterocyclic aromatic compounds also possess n–π* absorptions associated with the transfer of one of the nonbonding electrons on the heteroatom to an antibonding orbital of the π system.   For pyridine the n–π* transition is seen as a weak shoulder on the long wavelength side of the ^1L_b band.   However, it can be identified by its characteristic shift on changing the solvent from a nonpolar to a hydroxylic one.   In a hydrocarbon solvent the n–π* transition of pyridine has been determined to occur at 270 nm (ϵ 450) with the origin (0–0 vibrational transition) at 288 nm in the vapor.   The three isomeric picolines (methylpyridines) also exhibit an n–π* absorption maximum and band origin at about the same wavelength.

  Protonation of pyridine occurs on the nitrogen atom with loss of the nonbonding electron pair.   Consistent with the above interpretation, the shoulder in the uv spectrum of pyridine assigned to the n–π* transition is absent in the spectra of pyridinium salts.

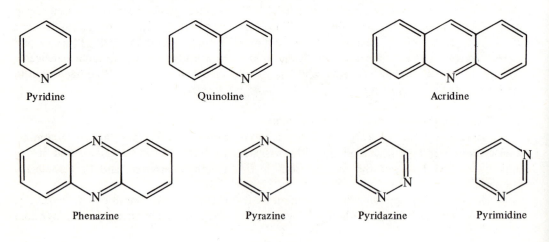

Pyridine   Quinoline   Acridine

Phenazine   Pyrazine   Pyridazine   Pyrimidine

The spectrum of quinoline has three well-defined absorption bands that strongly resemble the ^1L_b, ^1L_a, and ^1B bands of naphthalene. In addition, the spectra of acridine and phenazine are similar to that of anthracene, indicating that the nitrogen atom introduces only a small perturbation on the aromatic energy levels. Absorption data for some azaaromatic compounds are given in Table 3-11. A critical review of the electronic states of azabenzenes has been published by Innes, Byrne, and Ross (reference 9).

Table 3-11. *Absorption Data for Azaaromatic Compounds*

Compound	Solvent	λ (nm)	ε	λ (nm)	ε	λ (nm)	ε	λ (nm)	ε
Pyridine	hexane, cyclohexane	198	6,000	251	2,000	270	450^a (sh)		
Quinoline	ethanol	225	35,000	278	3,500	300	2,600	314	3,000
Acridine	ethanol					249	165,000	351	10,000
Phenazine	methanol					248	12,000	362	13,000
Pyrazine	hexane			194	6,000	260	6,000	328	$1,000^a$
Pyridazine	hexane			192	5,400	251	1,400	340	315^a
Pyrimidine	cyclohexane			189	10,000	244	2,000	298	325^a

^{a} n–π* band.

Smith and co-workers (reference 9) recently reported a uv and cd study of pyridines substituted by a chiral α-aminoethyl side chain at the 2, 3, and 4 positions. The uv and cd spectra of (S)-(−)-α-(2-pyridyl)ethylamine in methanolic potassium hydroxide is shown in Figure 3-25. On the basis of a first-order perturbation treatment of the cd data, the authors conclude that the 260 nm Cotton effect is due to the ^1L_b transition and the 240 nm negative maximum is due to the electric-dipole-forbidden n–π* transition. Furthermore, the allowed n–π* transition near 288 nm is deduced to be too weak to give a measurable cd signal.

The spectra of unsaturated five-membered heterocycles are often similar to their carbocyclic or acyclic analogs. For example, the absorption curves of pyrrole and furan

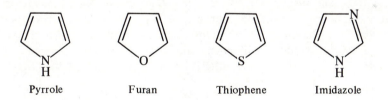

Pyrrole	Furan	Thiophene	Imidazole

resemble those of cyclopentadiene and divinyl ether, respectively (Table 3-12). The absorptions are attributed to π–π* transitions between molecular orbitals composed of the heteroatom and the diene orbitals. Although one expects n–π* transitions in these compounds, none has

Table 3-12. *Comparison of Absorption Data for Some Unsaturated Heterocyclic Compounds with Carbocyclic and Open Chain Analogues*

Compound	Solvent	λ (nm)	ε	λ (nm)	ε
Pyrrole	hexane	209	9,000		
Cyclopentadiene	hexane	191	3,000	239	3,000
Furan	vapor	204	6,500		
Divinyl ether	vapor	203.5	15,000		
Thiophene	ethanol	228	4,900	231	7,400
Divinyl sulfide	ethanol			255	5,600
Imidazole	ethanol	208	5,000		

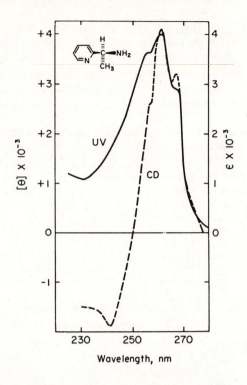

FIGURE 3-25. The uv and cd spectra of (S)-(−)-α-(2-pyridyl)ethylamine in 0.1 M methanolic KOH. [Adapted with the permission of the American Chemical Society from H. E. Smith, L. J. Schaad, R. B. Banks, C. J. Wiant, and C. F. Jordan, J. Amer. Chem. Soc., 95, 811 (1973).]

been observed as yet. The resemblance of the spectra to that of benzene decreases in the order thiophene, pyrrole, and furan, which corresponds to the order of decreasing aromaticity. Thiophene absorbs at the longest wavelength of the three, in part because sulfur has approximately the same electronegativity as carbon so that conjugation of the sulfur electrons with the π electrons of the diene system is more complete than in the oxygen or nitrogen analogue. To an extent that is not yet known the sulfur d orbitals probably play a role in the conjugation.

In a search for the elusive n–π* transition, Verbit, Pfeil, and Becker (reference 9) carried out a chiroptical study on thiophene and furan derivatives rendered optically active by a chiral side chain. The ord curve of (S)-(−)-β-(2-furyl)ethanolamine (**14**) in acetonitrile

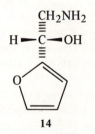

14

is shown in Figure 3-26. Only a single Cotton effect due to a π–π* transition was observed. The thiophene analogue exhibited a similar ord curve with a positive Cotton effect near 235 nm. The shapes of the ord curves of both the furan and the thiophene ethanolamine derivatives in 0.1 N HCl were not significantly different from those in acetonitrile, indicating that the nonbonding electrons of the amino group do not make an important contribution to the observed Cotton effects.

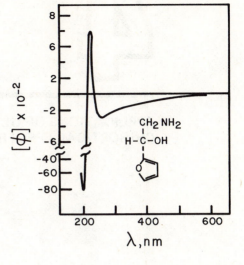

FIGURE 3-26.   *The ord of (S)-(−)-β-(2-furyl)-ethanolamine in acetonitrile.   [Adapted with the permission of Pergamon Press from L. Verbit, E. Pfeil, and W. Becker, Tetrahedron Lett., 2169 (1967).]*

BIBLIOGRAPHY

1. Classification of Chromophores: A. Moscowitz, *Tetrahedron*, **13**, 48 (1961).

2. Electronic Spectra of Alkenes: A. J. Merer and R. S. Mulliken, *Chem. Rev.*, **63**, 639 (1969); A. Yogev, J. Sagiv, and Y. Mazur, *Chem. Commun.*, 411 (1972); K. M. Wellman, P. H. A. Laur, W. S. Briggs, A. Moscowitz, and C. Djerassi, *J. Amer. Chem. Soc.*, **87**, 66 (1965).

3. The Octant Rule: W. Moffitt, R. B. Woodward, A. Moscowitz, W. Klyne, and C. Djerassi, *J. Amer. Chem. Soc.*, **83**, 4013 (1961); C. Djerassi and W. Klyne, *J. Amer. Chem. Soc.*, **79**, 1506 (1957); D. A. Lightner and D. E. Jackman, *J. Amer. Chem. Soc.*, **96**, 1938 (1974), and references cited therein.

4. Electronic Spectra of Amides: M. B. Robin, F. A. Bovey, and H. Basch in *The Chemistry of Amides*, J. Zabicky, ed., Interscience, New York, 1970, ch. 1; H. Benderly and K. Rosenheck, *Chem. Commun.*, 179 (1972); D. W. Urry, *J. Phys. Chem.*, **72**, 3035 (1968).

5. Woodward–Fieser Rules for α,β-Unsaturated Acids: A. T. Nielsen, *J. Org. Chem.*, **22**, 1539 (1957).

6. Chiroptical Spectra of β,γ-Unsaturated Ketones: A. Moscowitz, K. Mislow, M. A. W. Glass, and C. Djerassi, *J. Amer. Chem. Soc.*, **84**, 1945 (1962); D. J. Sandman, K. Mislow, W. P. Giddings, J. Dirlam, and G. C. Hanson, *J. Amer. Chem. Soc.*, **90**, 4877 (1968).

7. Electronic Spectra of Benzene Derivatives: J. R. Platt, *J. Chem. Phys.*, **17**, 484 (1949); *Ann. Rev. Phys. Chem.*, **10**, 354 (1959); P. Salvadori, L. Lardicci, R. Menicagli, and C. Bertucci, *J. Amer. Chem. Soc.*, **94**, 8598 (1972).

8. Electronic Spectra of Polycyclic Aromatic Hydrocarbons: E. P. Clar, *Polycyclic Hydrocarbons*, Academic, London, 1964; D. A. Lightner, D. T. Hefelfinger, T. W. Powers, G. W. Frank, and K. N. Trueblood, *J. Amer. Chem. Soc.*, **94**, 3492 (1972).

9. Electronic Spectra of Heterocyclic Aromatic Compounds: K. K. Innes, J. P. Byrne, and I. G. Ross, *J. Mol. Spectrosc.*, **22**, 125 (1967); H. E. Smith, L. J. Schaad, R. B. Banks, C. J. Wiant, and C. F. Jordan, *J. Amer. Chem. Soc.*, **95**, 811 (1973); G. Gottarelli and B. Samori, *Tetrahedron Lett.*, 2055 (1970); G. Fodor, E. Bauerschmidt, and J. C. Craig, *Can. J. Chem.*, **47**, 4393 (1969); L. Verbit, E. Pfeil, and W. Becker, *Tetrahedron Lett.*, 2169 (1967).

4

APPLICATIONS OF ULTRAVIOLET-VISIBLE AND CHIROPTICAL SPECTROSCOPY

4-1 Structure Investigations

One of the ways in which cd can be utilized in structural studies is illustrated by a study of quinone intermediates produced during the oxidation of dehydroabietic acid (**1**; R = H) and some related steroids. Oxidation of the methyl ester of **1** (R = CH$_3$) using 85% hydrogen peroxide in trifluoroacetic acid (*NIH-shift* conditions) gave the quinone **2**, whose uv absorption maxima at 259.5, 338, and 420 nm (95% ethanol) are characteristic of a trisubstituted *p*-benzoquinone.

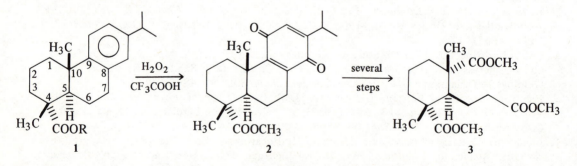

In order to demonstrate that the chiral carbons of **1** had not undergone isomerization during the course of the oxidation, **2** was converted by oxidative ozonization, and subsequent esterification with diazomethane, to the triester **3** of known absolute configuration. Thus it was determined that the acidic conditions employed in the oxidation did not affect the integrity of the chiral centers, a result that could not be assumed a priori.

The benzoquinone **2** was then utilized as a model compound for a cd investigation of the configurations of the quinones **5** and **6**, isolated from the oxidation of the estradiol derivative **4**. The cd spectra of compounds, **2**, **5**, and **6**, in 95% ethanol, are given in Figure 4-1. Note

374

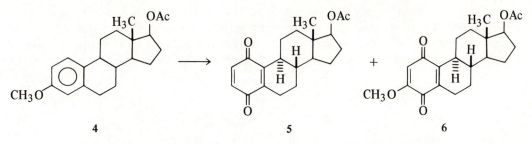

that the data are for the spectral region of the optically active *p*-benzoquinone chromophore between ca. 300 and 500 nm. The electronic transitions of this inherently achiral chromophore are perturbed by chiral centers in the molecule. The optically active ester chromophore in the molecule is transparent in this region, absorbing near 220 nm. Hence the benzoquinone ring acts as a probe of the stereochemistry of the molecule in its vicinity.

All three compounds exhibit two Cotton effects of opposite sign. The negative, long wavelength band is assigned to the n–π* transition and the higher energy, positive band is assigned to a π–π* absorption of the benzoquinone ring. The fact that the Cotton effects in the cd spectra of **5** and **6** are essentially identical to the corresponding Cotton effects of **2**, whose absolute configuration is known, is taken as evidence that all three compounds have the same stereochemistry in the vicinity of the benzoquinone ring. The structural similarities of **5** and **6** to **2** are readily seen when **2** is reoriented as **2a**.

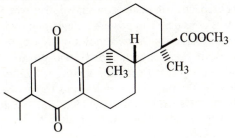

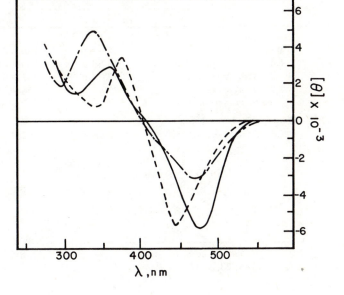

FIGURE 4-1. The cd spectra in 95% ethanol of the benzo-quinone derivatives 2 ——, 5 —·—, and 6 - - - -. [Repro-duced with permission from J.-F. Biellmann and G. Branlant, Bull. Soc. Chim. France, 2086 (1973).]

Ultraviolet spectroscopy was an important link in the chain of reasoning used in the structure elucidation of mangostin, the optically inactive, yellow coloring matter of the East Indian mangosteen tree. The uv spectrum of mangostin was found to be similar to those of known polyhydroxyl derivatives of xanthone (7) (Table 4-1). The ir spectrum confirmed the

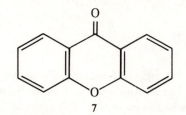

7

presence of a highly conjugated carbonyl group. The molecular formula of mangostin indicated six oxygen atoms. The xanthone nucleus accounts for two of them; chemical tests and degradation work showed the remaining oxygens to consist of three hydroxyls and one methoxyl group. Information on the location of the oxygen functions was provided by uv spectroscopy. Comparison of the spectra of xanthones having only one peri hydroxyl group (i.e., ortho to the carbonyl group) with those of xanthones having two peri hydroxyl groups (Table 4-1) indicates that introduction of a second peri hydroxyl results in bathochromic shifts and decreased intensities of the two longest wavelength bands while the short wavelength band undergoes a hypsochromic shift. The spectra of mangostin and derivatives do not show these spectral changes. Thus mangostin was considered to possess a single peri hydroxyl group.

The side chains of mangostin absorb two moles of hydrogen upon catalytic hydrogenation, indicating the presence of two double bonds (it was possible to rule out a triple bond). The double bonds cannot be conjugated with the xanthone ring or with each other since the uv spectrum of tetrahydrodimethylmangostin was found to be essentially identical to that of the unhydrogenated analogue. Hence it was concluded that the double bonds were isolated. Determination of the number of side chains showed there to be one double bond in each of

Table 4-1. *Absorption Data of Mangostin (8) and Some Polyhydroxyxanthones in 95% Ethanol[a]*

Compound	λ_{max} (nm) (log ϵ)			
One peri hydroxyl group				
Mangostin	243(4.5)	259(4.4)	318(4.4)	351(3.9)
1,6-Dihydroxyxanthone	247(4.3)	263(4.0)	305(4.1)	355(3.8)
1,3,6-Trihydroxyxanthone	237(4.6)	251(4.4)	313(4.4)	337(4.1)
1,3,7-Trihydroxy-6-methoxyxanthone	239(4.3)	256(4.5)	310(4.2)	362(4.0)
Two peri hydroxyl groups				
1,8-Dihydroxyxanthone	229(3.8)	252(3.9)	334(3.3)	380(2.9)
	231(4.0)	263(4.3)	343(3.8)	400(3.2)

[a] P. Yates and G. H. Stout, *J. Amer. Chem. Soc.*, **80**, 1691 (1958); P. Yates and A. Ault, *Tetrahedron*, **23**, 3307 (1967).

the two side chains. The overall results of the chemical and spectroscopic work led to structure 8 for mangostin.

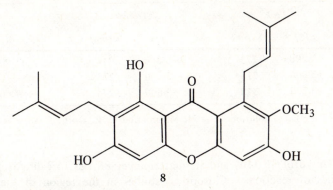

8

The recent widespread research efforts directed toward the total synthesis of prostaglandins has made important use of chiroptical techniques. The ord and cd curves of natural prostaglandin E_1 (PGE) of established configuration have been reported by Korver (reference 1).

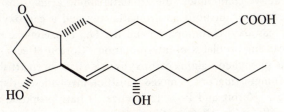

Prostaglandin E_1 (PGE)

A recent investigation by Miyano and Dorn used chiroptical methods to correlate the absolute configuration of PGE with several intermediates involved in prostaglandin synthesis. They resolved 7-(2-*trans*-styryl-3-hydroxy-5-oxocyclopentenyl)-*n*-heptanoic acid (9), the key intermediate in their total synthesis of racemic prostaglandins, and converted the resolved acids

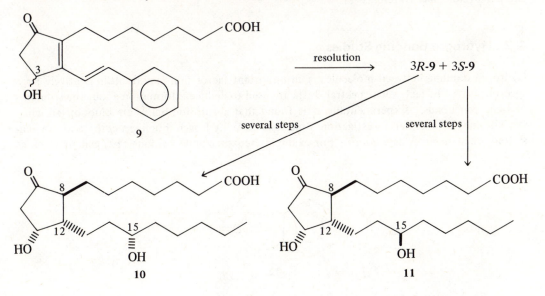

Table 4-2. *Chiroptical Data for the n–π* Absorption Region of PGE, 10, and 11*[a]

| | ORD | | | | CD | |
| | Peak | | Trough | | | |
	[φ]	λ (nm)	[φ]	λ (nm)	[θ]	λ (nm)
PGE	+7,200	272	−6,200	314	−11,000	296
10	+3,800	315	−5,100	273	––	––
11	+3,700	315	−5,600	274	+ 7,600	295

[a]Data for methanol solution. See M. Miyano and C. R. Dorn, *J. Amer. Chem. Soc.*, **95**, 2664 (1973).

into 8*S*,12*S*,15*S*-dihydroPGE (**10**) and its diastereomer 8*S*,12*S*,15*R*-dihydroPGE (**11**) by an unambiguous series of reactions. Chiroptical studies in the region of the n–π* transition (Table 4-2) showed that the ord and cd curves of **10** and **11** are both mirror images of those of natural PGE. In agreement with octant rule projections, the signs of the Cotton effects indicate that the stereochemistry about the cyclopentanone ring must be the same for both diastereomers and must be enantiomeric to that of natural PGE.

Many optically active compounds—the aromatic amino acids, substituted phenoxyacetic and mandelic acids, etc.—contain both an aromatic ring and a carboxyl function. Both the n–π* transition of the carboxyl chromophore and the ^1L_a benzene-like transition occur in the 210–220 nm region, causing problems in interpretation. Cotton effects observed in this wavelength range in various phenyl-containing carboxylic acids have been attributed to the carboxyl chromophore, to the aromatic ring, or to a mixed transition involving overlapping orbitals of both of these chromophores. Verbit and Price (reference 2) have proposed a method to identify the phenyl ^1L_a and carboxyl n–π* transitions, based on their cd band shapes. *For monosubstituted benzene derivatives attached to a chiral center, the Cotton effects due to the ^1L_a aromatic transition may be characterized by a width at half-height of 10–20 nm. The corresponding value for a n–π* transition is typically 30–35 nm.* Applying this criterion, Cotton effects due to the ^1L_a transition have been reported recently for a series of metacyclophanes and paracyclophanes (reference 2).

4-2 Hydrogen Bonding Studies

Hydrogen bonding between molecules is an important factor that may result in relatively large spectral shifts. In fact, such spectral shifts are used to deduce information about the strengths of hydrogen bonds. Experimentally it is found that absorption bands are blue-shifted when the chromophore under investigation functions as a hydrogen bond *acceptor* and are red-shifted when it serves as a *donor*. For example, benzthiazoline-2-thione (**12**) can function as

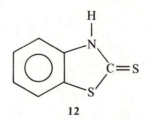

12

a hydrogen bond acceptor at the thione group or as a hydrogen bond donor at the NH group. The compound is found to exhibit blue shifts in hydroxylic solvents ($-O-H\cdots S=C$), but to undergo bathochromic shifts in the presence of acceptor molecules such as acetone, with which it functions as a hydrogen bond donor ($>N-H\cdots O=C$). In indifferent solvents such as CCl_4, where there is extensive self-association and the thione acts as both donor and acceptor, no shifts in the absorption spectra are observed. The *N*-methyl derivative of benzthiazoline-2-thione, which cannot function as a hydrogen bond donor, exhibits only a blue shift in donor solvents.

The absorption spectra of compounds engaged in *intramolecular* hydrogen bonding are generally solvent insensitive when studied in donor or acceptor solvents. Compare, for example, the behavior of *p*-nitrophenol with that of 2,4- or 2,6-dinitrophenol relative to cyclohexane solution. A substantial bathochromic shift is observed when *p*-nitrophenol is placed in the presence of a proton acceptor (from 286 to 297 nm with dioxane and from 286 to 307 nm with triethylamine). On the other hand, no spectral shifts are found for 2,4- and 2,6-dinitrophenols upon addition of these proton acceptors. A vast literature exists on spectral aspects of hydrogen bonding, and several monographs on the topic have been published (reference 3).

4–3   Charge Transfer Bands

Dissolving a sample in a solvent or mixing two compounds in an indifferent solvent for uv or chiroptical measurements may lead to the formation of a new band due to a charge transfer (ct) complex (reference 4). An excited-state ct band can be formed by the complexing of a donor molecule having a filled orbital with an acceptor molecule possessing an unoccupied orbital of appropriate symmetry at a slightly higher energy. The ct band is not present in either the isolated donor or acceptor molecule but is found, usually as a new absorption band or Cotton effect in the ct complex. Charge transfer bands are usually broad and quite strong, an important asset since often the equilibrium constants for formation of the complexes are small.

Almost any type of orbital, e.g., σ, π, d, can function as a donor or an acceptor, but the most common examples involve π orbitals. Some π donors, in increasing order of basicity are benzene $<$ mesitylene $<$ naphthalene $<$ anthracene $< N,N,N',N'$-tetramethyl-*p*-phenylenediamine. Some examples of π acceptors or π acids, in increasing order of acidity, are *p*-benzoquinone $<$ 1,3,5-trinitrobenzene $<$ chloranil $<$ tetracyanoethylene. The donor-acceptor complex is analogous to the combination of a Lewis acid with a Lewis base. In contrast to the latter complex, however, ct complexes are characterized by very weak interactions.

Matsumura and O'Brien have utilized uv spectroscopy in an investigation of the detrimental effects of chlorinated hydrocarbons on living organisms. It is known that these substances attack the central nervous system and block the transport of ions across nerve membranes. The absorption spectrum of the insecticide DDT (**13**) has a band at 240 nm,

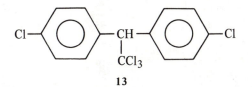

13

which in the presence of cockroach nerve axons shifts to 245 nm, with a shoulder appearing at 270 nm. DDT may function as a charge transfer acceptor with the nerve as the donor. The resultant ct complex is responsible for the deactivation of the nerve function.

The uv spectrum of (*S*)-(+)-*N*-[1-(*p*-anisyl)-2-propyl]-4-cyanopyridinium chloride **(14)**

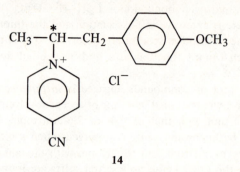

14

shows an absorption band at ca. 350 nm, which is attributed to an intramolecular ct transition. The band was shown to be optically active and to exhibit a positive Cotton effect.

Table 4–3. *Important Chromophores of Proteins*[a]

Residues	Chromophore	Location (nm)	log ϵ_{max}	Assignment
Peptide bond	CONH	162	3.8	$\pi^{+}-\pi^{-}$
		188	3.9	$\pi^{0}-\pi^{-}$
		225	2.6	$n-\pi^{-}$
Aspartic, glutamic	COOH	175	3.4	$n-\pi^{*}$
		205	1.6	$n-\pi^{*}(?)$
Aspartate, glutamate	COO^{-}	200	2	$n-\pi^{*}$
Lysine, arginine	N–H	173	3.4	$\sigma-\sigma^{*}$
		213	2.8	$n-\sigma^{*}$
Phenylalanine	phenyl	188	4.8	
		206	3.9	$\pi-\pi^{*}$
		261	2.35	
Tyrosine	phenolic	193	4.7	
		222	3.9	$\pi-\pi^{*}$
		270	3.16	
Tyrosine (ionized)	phenolate ion	200?	5	
		235	3.97	$\pi-\pi^{*}$
		287	3.41	
Tryptophan	indole	195	4.3	
		220	4.53	$\pi-\pi^{*}$
		280	3.7	
		286	3.3	
Histidine	imidazole	211	3.78	$\pi-\pi^{*}?$
CysSH	S–H	195	3.3	$n-\sigma^{*}$
CysS–	S–	235	3.5	$n-\sigma^{*}$
Cystine	–S–S–	210	3	$n-\sigma^{*}$
		250	2.5	

[a] Modified from J. Donovan, *Physical Principles and Techniques of Protein Chemistry*, Part A, S. Leach, ed., Academic Press, New York, 1969.

The resolution of hexahelicene with (+)-2-(2,4,5,7-tetranitro-9-fluorenylideneaminooxy)-propionic acid (TAPA) has been interpreted by Newman as involving diastereomeric ct complexes. A recent cd study of these complexes by Wynberg and Lammertsma clearly showed broad Cotton effects due to the ct transitions at 510 nm, with the (+)-hexahelicene-TAPA complex exhibiting a positive band and the diastereomeric complex a negative band. The observation that these complexes have ct Cotton effects that are nearly mirror images of each other was suggested to indicate that the donor hexahelicene component dominates the sign of the ellipticity.

4-4 Peptide and Protein Studies

Peptides and proteins may be considered as polymers of amino acids with the repeating group being the amide or peptide bond, $-CO-NH-$. Each amino acid residue contains at least one chiral carbon atom, with the exception of glycine. The chromophores of proteins that have significant absorption in the uv are given in Table 4-3. Aromatic amino acid side chains make important contributions to the optical activity of proteins. A summary of the cd properties of the aromatic amino acids and of cystine is given in Table 4-4. Small contributions to the optical activity also arise from the presence of carboxylate and ammonium groups. In addition, metals associated with certain proteins can modify the contributions of the chromophores with which the metals are complexed.

In 1951, Pauling, Corey, and Branson proposed the existence of a helical conformation in polypeptides and proteins that is stabilized by intramolecular hydrogen bonding. Cohen, in 1955, pointed out that changes in optical rotation accompanying the denaturation of proteins

Table 4–4. *Circular Dichroism Properties of Aromatic Amino Acids and of Cystine*

Amino Acid (as the Hydrochloride)	Chromophore	λ (nm)	[θ] (deg·cm^2 dmole^{-1})	References[a]
α-Phenylglycine	benzenoid	267, 260, 254	− 1,000	b
		218	34,000	
Phenylalanine	benzenoid	266, 263, 257	50–75	a, b
		217	14,000	
Tyrosine	phenolic	274	1,200	a, c
		225	8,000	
Tyrosine	phenolate ion	293	1,000	a, d
		230	− 2,000	
		210	7,000	
Tryptophan	indole	286, 276, 269	1,500–2,500	c
		225	20,000	
		209	−15,000	
Histidine	imidazole	216	8,000	a, c
Cystine	disulfide	250	− 2,000	a, e, f
		220	20,000	
		196	−44,000	

[a] (a) M. Legrand and R. Viennet, *Bull. Soc. Chim. France*, 479 (1965). (b) L. Verbit and P. J. Heffron, *Tetrahedron*, 24, 1231 (1968). (c) L. Verbit and P. J. Heffron, *Tetrahedron*, 23, 3865 (1967). (d) T. M. Hooker, Jr., and J. A. Schellman, *Biopolymers*, 9, 1319 (1970). (e) P. C. Kahn and S. Beychok, *J. Amer. Chem. Soc.*, 90, 4168 (1968). (f) J. P. Casey and R. B. Martin, *J. Amer. Chem. Soc.*, 94, 6141 (1972).

*FIGURE 4-2. The ord - - - - and cd
—— curves of a polypeptide having no
chromophoric side chains in (a) random
coil, (b) α helical, and (c) β conformations
[Adapted with permission from W. B.
Gratzer and D. A. Cowburn, Nature, 222,
426 (1969).]*

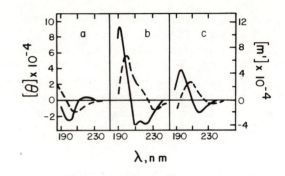

may be due to changes from a helical structure to more random conformations. Since then, several distinct conformations have been noted and characterized by their ord or cd curves.

As seen in Figure 4–2, the ord and cd curves for polypeptides in the random coil, α helix, and β conformations are distinctly different from each other. The curves indicate the general shapes, but magnitudes may vary somewhat depending upon the particular polypeptide.

Saxena and Wetlaufer have recently calculated cd spectra for the α-helical, β, and random coil forms by combining experimental cd data for three proteins with estimates of the periodic structural modes obtained from X-ray data of the same proteins. The results indicate good agreement with experiment for the α-helical spectra, fair agreement for the β conformation, and only qualitative similarities for the random coil form.

Poly-L-glutamic acid is an example of a polypeptide that is almost entirely in the α-helical form at pH 4. Three transitions are observed in the cd spectrum above 185 nm: a positive band near 192 nm and two strongly overlapped negative Cotton effects at 208 and 222 nm. The 192 and 208 nm bands arise from the split π–π^* transition of the amide chromophore and are polarized, respectively, perpendicular and parallel to the helix axis. The negative 222 nm Cotton effect is due to the n–π^* transition.

Above pH 6, poly-L-glutamic acid is in the random-coil conformation with a weak n–π^* Cotton effect near 216 nm, a still weaker band at about 235 nm, and a stronger, negative band at 198 nm, probably from the π–π^* transition, although there is a significant wavelength difference with the corresponding absorption maximum (191 nm).

In the isolation of proteins from cell membranes and cell walls, detergents such as sodium dodecylsulfate (SDS) are used. A recent cd study investigated structural changes in several proteins when exposed to SDS (reference 5). Small amounts of the detergent were found to cause marked changes in the cd spectrum of native elastase. The changes were interpreted as indicating an increased α-helical content over that present in the native protein. Dialysis of the SDS resulted in further changes of the cd curve to one indicating a substantial amount of elastase in the β conformation.

4-5 Solvent and Temperature Effects

When a ketone is dissolved in methanol and a drop of hydrochloric acid added, an equilibrium is established in which the ketone is converted to a dimethyl ketal, as shown in eq. 4–1. (An analogous reaction can be written for conversion of an aldehyde to an acetal.) Although the ketal can be isolated only after removal of the acid catalyst, its formation in solution is readily monitored by uv or chiroptical spectroscopy, since ketal formation causes the carbonyl

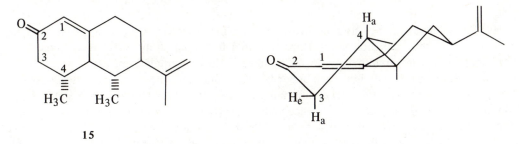

$$RR'C{=}O \xrightleftharpoons{CH_3OH/H^+} R\text{—}\underset{\underset{OCH_3}{|}}{\overset{\overset{OCH_3}{|}}{C}}\text{—}R' \quad + \quad H_2O \qquad\qquad 4\text{-}1$$

group and its associated n–π* transition to disappear.   The reaction of cholestan-3-one with acidified methanol was investigated by Zalkow and co-workers, who found that the ketone–ketal equilibrium is strongly dependent upon the amount of water present.   This result is illustrated in Figure 4–3, which also shows the use of cd as a kinetic tool.   The Cotton effect at 289 nm is due to the concentration of free ketone present; the ketal absorbs at much shorter wavelength.   Successive additions of small amounts of water shift the ketone–ketal equilibrium toward the free ketone.   The same workers also found that ketal formation depends upon the structure of the alcohol, as well as on stereochemical factors, such as the size of groups in the molecule near the carbonyl function.   Thus cholestan-3-one gives 96% of the dimethyl ketal, 84% of the diethyl ketal, and only 25% of the diisopropyl ketal.

Changes in cd curves on lowering the temperature, in the case of a conformationally mobile molecule, are assumed to be due to an increase in the population of the lowest energy conformer.   Ishida and co-workers studied the sesquiterpene nootkatone (**15**), for which three

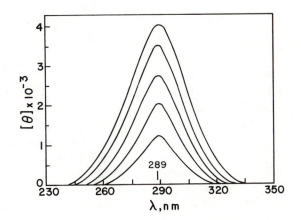

15

(+)-Nootkatone and its preferred conformation

possible conformations may exist in solution.   A variable temperature cd study from 133° in decalin solution to −144° in EPA solvent indicated the preferred conformation of nootkatone to be that shown above.

FIGURE 4–3. A cd investigation of the ketone–ketal equilibrium for cholestan-3-one in acidified methanol. The Cotton effect due to free ketone increases with successive additions of water to the ketal. [Adapted with permission from L. H. Zalkow, R. Hale, K. French, and P. Crabbé, Tetrahedron, **26**, 4947 (1970).]

Low temperature cd studies provide rather subtle information on *l*-stercobolin (**16**), one of the bile pigments. The compound had been shown to exist in a partial helical or

CH$_2$CH$_2$COOH   CH$_2$CH$_2$COOH

CH$_3$ CH$_3$

CH$_3$ C$_2$H$_5$

C$_2$H$_5$ CH$_3$

16

l-Stercobilin—internal hydrogen-bonded form

coiled form, stabilized by internal hydrogen bonding. However, in a hydrogen-bonding solvent such as the methanol–glycerol mixture used in the cd study, there is a competition between internal and external hydrogen bonding. Thus an equilibrium exists between the partially helical, internally hydrogen-bonded *l*-stercobilin and the uncoiled, solvent-bonded form. At low temperature, the equilibrium will be shifted toward the more stable form. Key information comes from the knowledge that in solvents that cannot form strong hydrogen bonds, e.g., chloroform, the coiled form is favored. A negative Cotton effect near 490 nm was found to be indicative of this conformation. In methanol–glycerol solution, this Cotton effect diminishes upon lowering the temperature from 25° to −110°. A new positive band at shorter wavelength emerges, indicating that the equilibrium is shifted to favor the uncoiled form at the expense of the partially helical conformation.

In 1967, Anand and Hargreaves (reference 6) reported that *S*-(+)-lactic acid (**17**) exhibits

CO$_2$H

HO—C—H

CH$_3$

17

a very weak negative Cotton effect in the 240–250 nm region in addition to a relatively strong positive Cotton effect near 210 nm. The absence of the negative, long wavelength band in alkaline solution caused these authors to attribute it to the n–π* transition of the carboxyl chromophore, and the stronger positive Cotton effect near 210 nm to the π–π* transition. This assignment was in direct conflict with the generally accepted assignment that the Cotton effect associated with the 210 nm absorption of carboxylic acids and esters is due to an n–π* transition. The problem was reexamined by Djerassi and co-workers (reference 6), who found that the two cd bands were not unique to lactic acid but also occurred in several esters and in *O*-ethyl ether derivatives of lactic acid. The ester and ether derivatives also ruled out the importance of hydrogen-bonding effects, both intra- and intermolecular, in the cd spectrum of lactic acid. Low temperature studies of *S*-(−)-ethyl lactate and *S*-(−)-ethyl *O*-ethyllactate showed the intensities of both Cotton effects to be temperature dependent, with the 210 nm Cotton effect increasing and the long wavelength band essentially disappearing at −192°. The data are interpreted by assuming that both cd bands have a common spectroscopic origin, the n–π* transition of the carboxyl group, and that they are due to the presence of different conformations. At −192° only the most stable conformation is expected to be appreciably populated.

4-6 Assignment of Configuration

An example of the use of model compounds of known configuration to deduce stereochemical information is provided by cafestol, a diterpene isolated from the coffee bean. Degradation of cafestol (18) gave the ketone 19 whose ord curve was found to be almost the mirror image

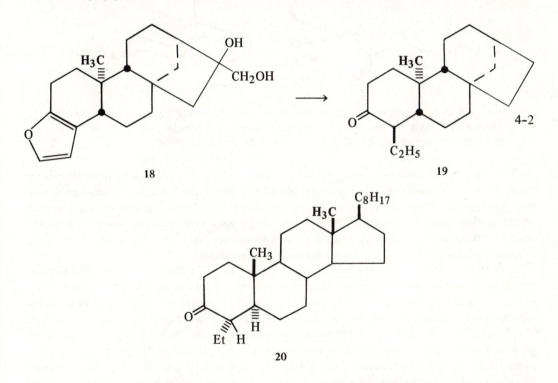

of that of 4α-ethylcholestan-3-one (20). The ord curves are shown in Figure 4-4. Since the observed Cotton effect curves are essentially mirror images and the two ketones 19 and 20 are structually the same in the vicinity of the carbonyl group, the conclusion was reached that 19, and hence cafestol, from which 19 was obtained, possessed enantiomeric sterochemistry at the A/B ring junction, as depicted in Figure 4-5.

We have seen that ord and cd methods may be used for the assignment of *relative configuration* to a molecule, that is, relative to some model compound whose *absolute configuration* has been determined by other means. A few theoretical treatments exist that are useful in certain cases for nonempirical assignments of absolute configuration. One of

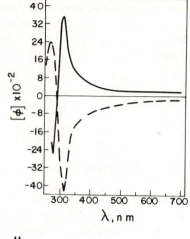

FIGURE 4-4. The ord curves of the ketone 19 from cafestol – – – – and of 4α-ethylcholestan-3-one ———. [Adapted with the permission of the American Chemical Society from C. Djerassi, M. Cais, and L. A. Mitscher, J. Amer. Chem. Soc., 81, 2386 (1959).]

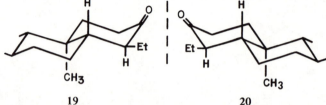

FIGURE 4-5. Conformations of the A–B ring portion of the cafestol-derived ketone (19) and of the model compound 4α-ethylcholestan-3-one (20) showing the mirror image relationship. See Figure 4-4 for the ord curves.

these is the coupled oscillator theory of Werner Kuhn, which is applicable to rigid, inherently chiral systems, in particular those that contain two identical but noncoplanar chromophores. With knowledge of the direction of the electric dipole transition moment for the chromophore one can predict the sign of the Cotton effect arising from that transition for a chosen absolute configuration. There is no need for a comparison of chiroptical data with compounds of known absolute configuration, so the method is nonempirical.

A number of natural products are essentially dimeric. An example is the indole alkaloid calycanthine, whose absolute configuration was determined by Mason and Vane (Figure 4–6). From the point of view of coupled oscillator theory, calycanthine consists of two aniline chromophores. Planes through the two aromatic rings make an angle of 61° with each other. The C_2 axis through each aniline residue makes an angle of 28° with the intersection of these planes. The rotational strengths of the two longest wavelength Cotton effects were determined theoretically from consideration of the absorption spectrum of aniline and of the calculated transition dipole interactions between the aniline chromophores in each of the two optical isomers of calycanthine. The theory predicted that for the absolute configuration of (+)-calycanthine shown in Figure 4–6, the cd spectrum should show a long wavelength positive Cotton

FIGURE 4-6. Absolute configuration of the indole alkaloid (+)-calycanthine. [After S. F. Mason and G. W. Vane, J. Chem. Soc. (B), 370 (1966).]

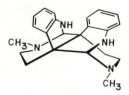

effect followed by a negative one. The experimental cd spectrum was in agreement with this prediction: the longest wavelength transition occurred near 320 nm with a molar ellipticity of about 100,000.

Several theoretical treatments for the prediction of absolute configuration have been applied to the helicenes. For example, (+)-hexahelicene is predicted to have right-handed chirality by the Fitts-Kirkwood model of pairwise interactions of benzene rings, by the free-electron-on-a-spiral method of Tinoco and Woody, by Brewster's classical treatment, and by the SCF-CI calculations of Mason and co-workers. X-ray determination of the absolute configuration of hexahelicene (see section 3-4b, reference 8 of Chapter 3, and Figure 3-24) is in agreement with these predictions.

4-7 Sector and Helicity Rules

A large number of semiempirical and empirical rules exist for the correlation of absolute configuration with the sign of an appropriate Cotton effect. Many of these are sector or symmetry rules, i.e., geometrical constructs based on the nature of the transition and the symmetry of the chromophore. The octant rule for ketones (see section 3-2) is perhaps the most successful example of a sector rule.

In recent years sector rules have been proposed for almost every appropriate chromophore. Some have been successively modified to account for new results, whereas others have been shown to be highly limited or even invalid. Recent reviews of sector and other rules may be found in reference 7. We will discuss only a few of the many useful ones.

Examination of the cd behavior of a series of lycorine alkaloids led Kuriyama and co-workers to propose an octant rule for tetrasubstituted aromatics. Their rule, keyed to the ^1L_b transition of these compounds near 290 nm, is shown in Figure 4-7. If the aromatic ring is viewed along the $-x$ to $+x$ axis, the contribution of substituents in each of the four quadrants along the positive x axis is as shown on the right side of the figure. The rule, which is similar to one deduced almost simultaneously by Brewster and Buta, is illustrated in Figure 4-8 for β-dihydrocaranine. A large negative contribution to the ^1L_b Cotton effect due to the hydroxyl-group-containing ring in a negative quadrant is predicted and is found in the cd spectrum, $[\theta]_{296} - 4100$.

Sector rules keyed to the ^1L_a aromatic transition have been put forth by De Angelis and Wildman based on a study of Amaryllidaceae alkaloids, and by Verbit and Price based on results for monosubstituted benzenes. A general octant rule for the aromatic nucleus, keyed

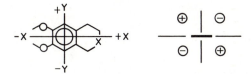

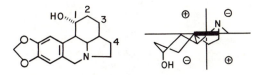

FIGURE 4-7. Octant rule for the ^1L_b transition in tetrasubstituted benzenes. Signs are for the +x quadrants. X = CH$_2$, NH, or N-alkyl. [Adapted with permission from K. Kuriyama, T. Iwata, M. Moriyama, K. Kotera, Y. Hamada, R. Mitsui, and K. Takeda, J. Chem. Soc., B, 46 (1967).]

FIGURE 4-8. β-Dihydrocaranine and its ^1L_b octant rule projection. [Adapted with permission from K. Kuriyama, T. Iwata, M. Moriyama, K. Kotera, Y. Hamada, R. Mitsui, and K. Takeda, J. Chem. Soc., B, 46 (1967).]

FIGURE 4-9. Sign of the second sphere contribution of a tetralin to the ^1K_b band Cotton effect. The arrow indicates the direction of projection; P and M refer to the chirality of the nonaromatic ring. [After G. Snatzke, M. Kajtár, and F. Werner-Zamojska, Tetrahedron, 28, 281 (1972).]

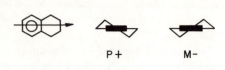

P +   M –

only to the direction of the appropriate transition moment, has been proposed recently by Brewster and Prudence (reference 7).

Snatzke has used another approach for defining sector rules for the benzene chromophore, one which is, in principle, generally applicable: the concept of chiral spheres. The molecule is divided into spheres, the aromatic chromophore being the first and an adjacent ring, if any, the second. Rings or groups attached to this second chiral sphere make up the third sphere, and so on. The contribution of the nearest chiral sphere to the chromophore is postulated to determine the sign and, to a large extent, the magnitude of the Cotton effect.

Tetralins and tetrahydroisoquinolines are examples of molecules with a chiral second sphere. A correlation has been found between the sign of the ^1L_b Cotton effect and the chirality of the nonaromatic ring; a minus or M chirality in the R, S nomenclature system results in a negative Cotton effect. The relationship is shown in Figure 4-9.

Third and higher spheres become important when the first and second spheres are achiral. In all cases, though, they contribute to the magnitude of the Cotton effects.

The absolute configurations of chiral allenes, e.g., ABC=C=CXY, are difficult to relate to other compounds by chemical means. Based on Brewster's helical model of optical activity, Lowe (reference 7) has proposed the rule illustrated in Figure 4-10 for relating the absolute configuration of an allene to its optical rotation at the sodium D line.

Recently Mason and co-workers (reference 7) have proposed a bifurcated quadrant rule relating the configuration of the allene to the sign of the Cotton effect of the long wavelength allene transition in the 220–250 nm region. This rule is shown in Figure 4-11. The signs refer to the long wavelength Cotton effect associated with groups substituted into the $+y$ hemisphere. For groups substituted into the $-y$ hemisphere, the signs are reversed.

Extensive chiroptical studies of benzoic acid esters of 1,2-dihydroxy compounds by Nakanishi and co-workers have led to the exciton chirality rule. The interacting benzoate chromophores give rise to an intramolecular charge transfer band which undergoes exciton splitting into two intense Cotton effects of equal magnitude and opposite sign. The longer wavelength band is near 233 nm, and the other is near 219 nm. By the application of molecular exciton theory, the sign of the Cotton effect couplet and hence the absolute configuration of the dibenzoate can be calculated. The exciton chirality rule, shown in Figure 4-12, states

Polarizability	Sign of D Line Rotation
A > B, X > Y	(–)
A > B, Y > X	(+)

FIGURE 4-10. Lowe–Brewster rule.

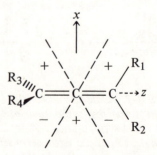

FIGURE 4-11. Bifurcated quadrant rule for allenes.

FIGURE 4-12.   Exciton chirality rule.

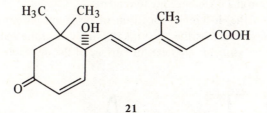

that if the chirality of the 1,2-dibenzoates is defined as being positive or negative, respectively, for clockwise or counterclockwise rotation, then the sign of the longer wavelength Cotton effect of the exciton couplet reflects the chirality.   An example of the use of the rule in correlating the absolute configuration of 2α,3β-dibenzoyloxy-5α-cholestane is shown in Figure 4-13.

The coupling between two benzoate chromophores to give split Cotton effects and the subsequent use of the exciton chirality method is not limited to 1,2-glycols, but has been successfully applied to cholestane-3β,6β-diol and extended to various triols and other aromatic systems in terpenoids, sugars, antibiotics, and alkaloids.

An important test of the exciton chirality rule occurred in the determination of the absolute configuration of the plant growth regulator (+)-abscisic acid (**21**).   The ord and cd

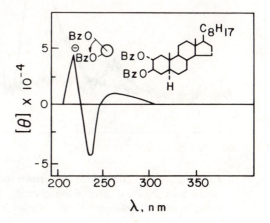

21

spectra of this compound showed typical Davydov-type split Cotton bands. This splitting was interpreted as due to the transition dipole-dipole coupling between the enone and diene-carboxylic acid systems.   Quantitative application of the exciton chirality method indicated that the S configuration should be assigned to (+)-abscisic acid (**21**).   A chemical correlation of (+)-abscisic acid with S-malic acid had been carried out simultaneously and independently of this work; happily, the results led to the same absolute configuration.

The exciton chirality method has been shown recently to have broad applications to esters, conjugated enones, and lactones.   The chromophores can be present in the molecule or may be introduced by derivatization.

FIGURE 4-13.   The cd spectra in ethanol–dioxane (9/1) of 2α,3β-dibenzoyloxy-5α-cholestane. The left-hand structure gives the exciton chirality rule illustrating negative chirality. [Adapted with the permission of the American Chemical Society from N. Harada and K. Nakanishi, Accts. Chem. Res., 5, 257 (1972).]

4-8 Special Techniques

Difference spectroscopy is used to a great extent for the detection of very small spectral differences that result from changing the temperature, pH, concentration, solvent, etc. Since small differences in absorption are involved, the technique requires a precision spectrophotometer. The difference spectrum may be obtained by measuring the sample and appropriate reference solution and then determining the baseline with the sample solution in both beams, although tandem cell techniques are usually used.

An example of difference spectroscopy is solvent perturbation, a technique widely used in studies of proteins and enzymes. In this method, the solvent composition of the sample solution is changed relative to the reference solution while the protein concentration is maintained constant in both cells. The usual technique is to add a hydroxylic organic solvent such as methanol, ethylene glycol, or glycerol to an aqueous solution of the protein; variations of the method include studies of the effects of added salts. The spectra of aromatic residues in proteins or enzymes are noticeably affected when exposed to changes in solvent composition. An important assumption of the solvent perturbation approach is that chromophores buried in the interior of the protein are not affected by the solvent change and their absorption bands cancel in the difference measurement.

Figure 4-14 shows the difference spectrum of lysozyme in aqueous phosphate buffer. Propylene glycol has been added as the perturbant. Analysis of the spectrum indicates that all six tryptophan residues in lysozyme are near the surface of the molecule.

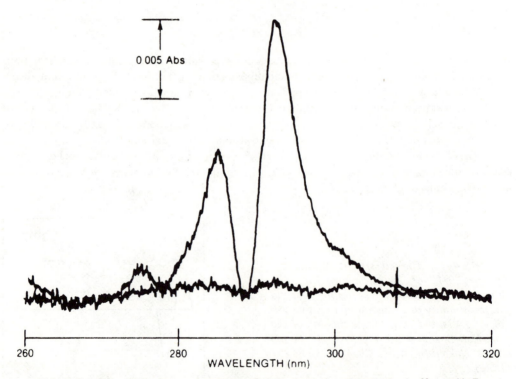

FIGURE 4-14. *Difference spectrum of lysozyme in phosphate buffer, pH 7; perturbant 20% v/v propylene glycol. [Reproduced with permission from J. O. Erikson and R. Bramston-Cook, Varian Instru. News, 7, 10 (1973).]*

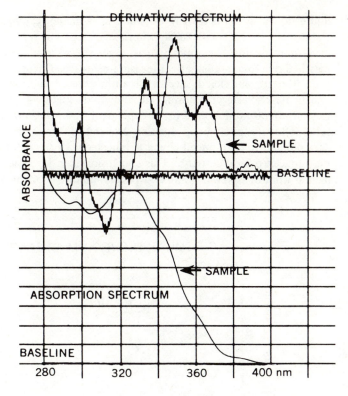

FIGURE 4-15. Absorption and first derivative spectra of testosterone in dioxane. [Courtesy of Varian Associates.]

Derivative absorption spectroscopy is becoming more widely used as a sensitive means of detecting small effects, such as overlapping bands and wavelength shifts due to temperature or solvent effects, and in the analysis of trace impurities. A derivative mode is offered in many of the newest precision spectrophotometers.

An absorption band exhibits two maxima in its first derivative spectrum: a positive one corresponding to increasing absorbance and a negative one for decreasing absorbance. In the absence of overlapping bands, the derivative will cross the baseline at the absorption maximum. The spectrum of testosterone (Figure 4-15) illustrates the sensitivity of the derivative mode in detecting the number and position of very weak shoulders on the side of the main absorption band.

4-9   Recent Developments

4-9a   Liquid-Crystal-Induced Circular Dichroism (LCICD)

It has been known for more than 65 years that cholesteric liquid crystals exhibit enormous optical rotations, of the order of several thousand degrees per millimeter (reference 8). The currently accepted model for this ordered anisotropic phase assumes it to be composed of molecular layers having a nematic arrangement, with successive layers twisted at slight angles to the preceding ones because of the chirality of the molecules. Hence, the cholesteric liquid crystal (which might better be called "twisted nematic" or "chiral nematic" since the molecules need not be steroid derivatives) is composed of a helical array of ordered molecules that can be characterized by a screw direction and a pitch.

A significant development has been the discovery that achiral molecules in cholesteric liquid crystal solvents display Cotton effects in the regions of their electronic and vibrational transitions. Hence, with an appropriate transparent liquid crystal solvent, circular dichroism measurements may be extended as a general method to *achiral* molecules, an achievement matched only by Faraday effect measurements (see Chapter 5).

The uv and lcicd spectra of anthracene in a mixture of 40:60 (wt. %) cholesteryl nonanoate–cholesteryl chloride are shown in Figure 4-16. The intensities of the induced Cotton effects were found to be dependent upon the pitch of the cholesteric helix, temperature, reflective pitch band position λ_0, and texture. In connection with other results, the data pointed to an induced helical arrangement of the solute molecules as the mechanism for the induced optical activity. This conclusion is in agreement with the theoretical and experimental work of Holzwarth and co-workers and of Sackmann and Voss.

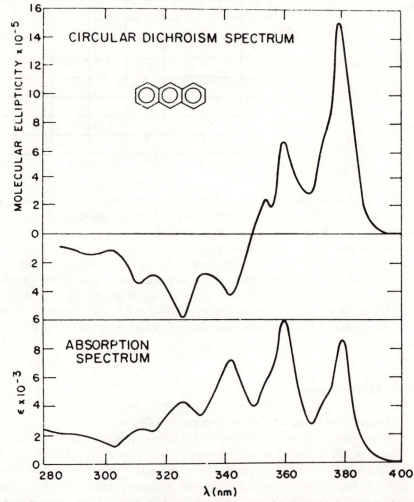

FIGURE 4-16. The uv and cholesteric liquid-crystal-induced cd spectra of anthracene in 40/60 (wt. %) cholesteryl nonanoate–cholesteryl chloride at 22°. (Grandjean texture, $\lambda_0 = 10$ μm.) [Reproduced with the permission of the American Chemical Society from F. D. Saeva, P. E. Sharpe, and G. R. Olin, J. Amer. Chem. Soc., 95, 7656 (1973).]

The cd that is induced in achiral solutes by helical solvents is evidently quite general and has been observed in the case of anthracene dissolved in the lyotropic cholesteric solvent polybenzyl-L-glutamate–dioxane. That the induced Cotton effects are associated with the cholesteric structure of the solvent is evidenced by their disappearance when the solvent is made isotropic by changing the concentration.

The solutes initially studied by this technique have been structurally anisotropic. Mason and Peacock have reported that molecules of octahedral symmetry such as molybdenum hexacarbonyl exhibit large induced Cotton effects in a cholesteric solvent. They conclude that these Cotton effects are due mainly to the differential interaction of the effective Lorentz radiation field for left- and right-handed circularly polarized light, which arises from the large circular birefringence of the solvent.

4–9b Vacuum Ultraviolet CD Studies

Considerable structural information is undoubtedly available in the so-called vacuum uv region, below ca. 185 nm. Commercial cd instrumentation for this spectral region is not available at present, but a number of vacuum cd instruments capable of measurements to about 130 nm have recently been developed (reference 9). Pioneering cd investigations in the vacuum uv have been carried out on sugars, polypeptides, ketones, alkenes, *l*-methylindane, and 2-butanol.

4–9c Infrared CD Studies

Although this Part has been concerned with spectroscopy in the region of electronic absorption, the subject matter of circular dichroism requires that mention be made of some potentially significant developments in the infrared region. Improvements in instrumentation have permitted Holzwarth, Moscowitz, and collaborators to carry out cd measurements on neat liquids and to observe Cotton effects due to *vibrational* transitions (reference 10). The compounds investigated were *S*-(+)- and *R*-(−)-2,2,2-trifluoro-1-phenylethanol (**22**, the *S* form) and *R*-(−)-neopentyl-1-*d* chloride (**23**). The *S*-(+)-trifluorophenylethanol exhibited a very

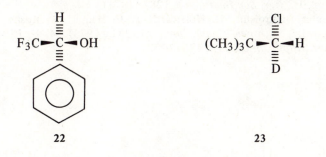

weak positive Cotton band associated with the C*H stretching vibration at 2920 cm^{-1}, about 20 cm^{-1} higher in energy than the absorption maximum. Calculations indicated that this shift most likely reflects a statistical weighting of the vibrational-rotational strengths of several conformations. No Cotton effects were found at the phenyl CH frequencies, a not unexpected result since these groups are well removed from the chiral center.

The vibrational Cotton effect at 2204 cm^{-1} associated with the C*D stretch in the deuterated chloride **23** was found to be negative but extremely weak. It was not possible to measure the C*H Cotton effect in this molecule because of overlapping neopentyl stretching modes.

BIBLIOGRAPHY

1. Chiroptical Spectra of Prostaglandin E_1:   O. Korver, *Rec. Trav. Chim. Pays-Bas*, **88**, 1070 (1969).
2. Differentiation of Phenyl and Carboxyl Transitions: L. Verbit and H. C. Price, *J. Amer. Chem. Soc.*, **94**, 5143 (1972); E. Langer and H. Lehner, *Monatsh.*, **104**, 644 (1973).
3. Hydrogen Bonding: S. N. Vinogradov and R. H. Linnell, *Hydrogen Bonding*, Van Nostrand Reinhold, New York, 1971.
4. Charge Transfer Complexes: *Molecular Complexes*, R. Foster, ed., Elek Science, London, 1973; N. Mataga and T. Kubota, *Molecular Interactions and Electronic Spectra*, Marcel Dekker, New York, 1970; R. S. Mulliken and W. Person, *Molecular Complexes*, Wiley, New York, 1969; R. Foster, *Organic Charge Transfer Complexes*, Academic, New York, 1969.
5. Effect of Sodium Dodecylsulfate on Proteins: L. Visser and E. R. Blout, *Biochemistry*, **10**, 743 (1971); B. Jirgensons and S. Capetillo, *Biochim. Biophys. Acta,* **214**, 1 (1970).
6. Temperature Effects on CD Spectra: R. D. Anand and M. K. Hargreaves, *Chem. Commun.*, 421 (1967); G. Barth, W. Voelter, E. Bunnenberg, and C. Djerassi, *Chem. Commun.*, 355 (1969).
7. Sector and Helicity Rules: *Fundamental Aspects and Recent Developments in ORD and CD*, F. Ciardelli and P. Salvadori, eds., Heyden and Son, London, 1973; P. Crabbé, *ORD and CD in Chemistry and Biochemistry*, Academic, New York, 1972; G. G. DeAngelis and W. C. Wildman, *Tetrahedron*, **25**, 5099 (1969); J. H. Brewster and R. T. Prudence, *J. Amer. Chem. Soc.*, **95**, 1217 (1973); J. H. Brewster, *Top. Stereochem.*, **2**, 33 (1967); G. Lowe, *Chem. Commun.*, 411 (1965); P. Crabbé, E. Velarde, H. W. Anderson, S. D. Clark, W. R. Moore, A. F. Drake, and S. F. Mason, *Chem. Commun.*, 1261 (1971).
8. Liquid-Crystal-Induced Circular Dichroism: A. Saupe, *Ann. Rev. Phys. Chem.*, **24**, 411 (1973); S. Chandrasekhar and N. V. Madhusudana, *Appl. Spectrosc. Rev.*, **6**, 189 (1973); R. Steinsträsser and L. Pohl, *Angew. Chem. Int. Ed. Engl.*, **12**, 617 (1973); L. Verbit, *J. Chem. Educ.*, **49**, 36 (1972); G. H. Brown, *Amer. Scientist*, **60**, 64 (1972).
9. Vacuum Ultraviolet Circular Dichroism: S. Feinleib and F. A. Bovey, *Chem. Commun.*, 978 (1968); O. Schnepp, S. Allen and E. Pearson, *Rev. Sci. Instru.*, **41**, 1136 (1970); W. C. Johnson, Jr., *Rev. Sci. Instru.*, **42**, 1283 (1971).
10. Infrared Circular Dichroism: G. Holzwarth, E. C. Hsu, H. S. Mosher, T. R. Faulkner, and A. Moscowitz, *J. Amer. Chem. Soc.*, **96**, 251 (1974); M. F. Russel, M. Billardon, and J. P. Badoz, *Appl. Optics*, **11**, 2375 (1972).

5

MAGNETIC OPTICAL ACTIVITY—THE FARADAY EFFECT

5-1 The Faraday Effect

In 1845 Faraday discovered that optical activity can be induced in substances by placing them in a magnetic field. The phenomenon, known as the Faraday effect, has proved to be a universal property of matter whether or not the sample itself possesses natural optical activity.

When linearly polarized light is passed through a substance situated in a magnetic field parallel to the direction of the light beam, the plane of polarization is rotated, and there is a difference in the absorption of left and right circularly polarized light in the region of electronic absorption by the substance. These effects, measured as a function of wavelength, are termed *magnetic optical rotatory dispersion* (mord) and *magnetic circular dichroism* (mcd), respectively.

Approximately ten years after Faraday's discovery, Verdet showed that the magnetic rotation is directly proportional to the magnetic field strength B, the pathlength l of the sample, and a frequency-dependent constant characteristic of the substance, since termed the Verdet constant, V (eq. 5-1). Early Faraday effect work centered on measurements of Verdet

$$\Phi_M = l\,V B \qquad\qquad 5\text{-}1$$

constants and attempts at their correlation with molecular structure. These attempts were considerably less fruitful than the contemporaneous studies of natural optical activity at the sodium D line.

The commercial availability during the middle 1960's of ord and cd instruments and of relatively compact superconducting magnets has led to a resurgence of interest in the Faraday effect and its applications to structural chemistry (reference 1).

Since *all* compounds exhibit optical rotation in a magnetic field, mord poses the problem of a large unwanted rotation arising from the solvent, together with the desired rotation of the solute. The background rotation may be canceled by running a blank consisting of pure solvent or by incorporating two identical magnets of opposing fields in tandem with one containing solvent only. These difficulties and others are probably best avoided by use of the alternative technique of mcd. Since mcd is becoming the clear choice for Faraday effect measurements, we will confine our discussion to this method and its applications.

5-2 Magnetic Circular Dichroism

As in cd measurements of natural optical activity, a solvent that is optically transparent in the absorption region of the solute makes no contribution to the mcd. Another similarity with natural optical activity is that mord and mcd are related by the Kronig–Kramers transform. In addition, many of the terms and relations used in natural optical activity, e.g., the Drude equation, apply also to magnetically induced optical activity. The sign convention is the same for both, when the light beam propagates in the positive direction of the magnetic field. However, the analogies end here since a fundamental difference exists. In natural optical activity, the chiral field perturbing the chromophore is contained within the molecule and thus has a fixed geometrical relation to the chromophore under investigation no matter how the molecule may translate or rotate. In magnetic optical activity, on the other hand, the perturbing force is a magnetic field external to the entire molecule, and thus mcd may not directly reflect the molecular geometry of randomly oriented molecules. Nevertheless, in many cases mcd offers spectroscopic information not available by measurement of natural optical activity.

The analogy of the normal longitudinal Zeeman effect, i.e., the splitting of a spectral line into left and right circularly polarized components, with the Faraday effect has served as a basis for the development of theoretical treatments of magnetic optical activity. Buckingham and Stephens (reference 1) have shown that the contribution of an isolated transition to the experimental mcd band can be written as a sum of three terms, A, B, and C, each with its own molecular properties, i.e., eq. 5-2, in which $[\theta]_M$ is the molar ellipticity per gauss and

$$[\theta]_M = -21.3458 \left[f_1 A + f_2 \left(B + \frac{C}{kT} \right) \right] \qquad 5\text{-}2$$

f_1 and f_2 are lineshape functions. These terms lead to three possible types of mcd curves, shown in Figure 5-1. The magnetic field causes a Zeeman splitting of degenerate ground and excited states. In the case of a degenerate ground state, one has a temperature-dependent C term of the form C/kT, since the population of the degenerate ground state and hence the magnitude of C are inversely proportional to temperature. The Faraday C term leads to a mcd curve that peaks at the maximum of the corresponding absorption band (Figure 5-1b). It should not come as a surprise that since the vast majority of organic molecules are diamagnetic and hence have nondegenerate ground states, C terms will vanish for most compounds of interest to the organic chemist.

An isolated transition will show an A term only when the excited state is degenerate. The A terms are characterized by S-shaped mcd curves (Figure 5-1a). When both the ground and excited states are degenerate, the C term is large compared to the A term. A mcd A term

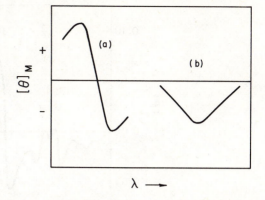

FIGURE 5-1. Wavelength dependence of mcd terms: (a) positive A term, (b) positive B or C/kT term. The signs follow from the definition of molar magnetic ellipticity (eq. 5-2).

has no counterpart in natural optical activity.   Thus, in principle, mcd can provide new information.

Group theoretical predictions indicate that for excited-state degeneracy a molecule must possess at least a threefold symmetry axis. This requirement leads to the conclusion that many organic molecules of interest will exhibit only B terms, since their molecular symmetry is usually fairly low.

The B terms arise from the mixing of electronic states by the magnetic field. They have the same mcd shape as C terms (Figure 5-1b), but are not temperature dependent. The B term is usually present whatever the degeneracies of the various states. The magnitude is quite variable from one class of compounds to another and depends on an inverse relationship between the relative energies of ground and excited states.

Any resemblance between natural and magnetically induced optical activity arises from the similar shapes of a natural Cotton effect and a mcd B term. Both result from the differential absorption of left and right circularly polarized light, but the cause of this absorption is different for the two phenomena. In the case of natural cd the perturbing force is inherent in the molecule; in the case of mcd, it is external.

5-3   Some Applications of MCD

Aromatic compounds have proved a fruitful field for mcd studies (reference 2).   Foss and McCarville investigated the mcd spectra of toluene, naphthalene, anthracene, and tetracene and found a correlation between the direction of the electric dipole transition moment and the sign of the mcd band for the lowest energy electronic transition.   In this work, it should be noted, the sign convention is opposite to that discussed above. Stephens and co-workers studied the mcd spectra of benzene, triphenylene, and coranene and treated the observed spectra theoretically.

Recent studies (reference 2) have focused on the ^1L_b transition of benzene and hexadeuteriobenzene in the vapor phase and on the longest wavelength transition of some polycyclic aromatics.   In the latter study by Larkindale and Simkin a band was found at 400 nm in the mcd spectrum of anthracene in CH_2Cl_2 that does not correspond to any band in the absorption spectrum.   It was assigned tentatively to the previously unobserved ^1L_b transition (Figure 5-2).   Note from the figure that the authors use the α and p nomenclature of Clar (Table 3-8).

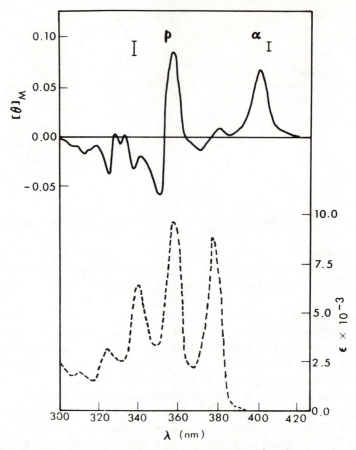

FIGURE 5-2.   The uv - - - - and mcd ―――― spectra of anthracene in methylene chloride solution.   The vertical bars represent the signal/noise ratios.   [Reproduced with permission from J. P. Larkindale and D. J. Simkin, J. Chem. Phys., 55, 5668 (1971).]

A mcd study has been reported by Foss and McCarville on a series of mono- and disubstituted benzenes. The authors found a correlation between Hammett σ_p values (related to the electron-attracting and withdrawing effects of substituents) and the sign and intensity of the mcd ^1L_b bands.

Information from mcd allowed a decision to be made regarding the structure of [18]annulene (Figure 5-3). The question to be answered was whether the hydrocarbon has aromatic character (Hückel's rule of $(4n + 2)\pi$ electrons is obeyed) or possesses alternating single and double bonds. Molecular orbital calculations based on the hypothesis of equal carbon–carbon bond lengths were unsuccessful in accounting for the observed absorption spectrum. On the other hand, X-ray diffraction results were not in agreement with the assumption of alternating single and double bonds. An explanation of this discrepancy invoked a distortion of [18]annulene due to a slight overcrowding of the internal hydrogens. This distortion would have the effect of lengthening the outer C–C bonds relative to the inner ones. If this explanation were correct, $C_{18}H_{18}$ would still have a center of symmetry and possess D_{6h} symmetry. This symmetry would lead to a degeneracy of the ground and excited

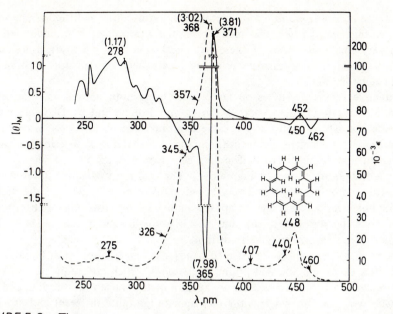

FIGURE 5-3. *The uv* – – – – *and mcd* ——— *spectra of [18]annulene in isooctane.* $[\theta]_M$ *is normalized to one gauss. [Reproduced with permission from C. Djerassi, E. Bunnenberg, and D. L. Elder, Pure Appl. Chem., 25, 57 (1971). Also see B. Briat, D. A. Schooley, R. Records, E. Bunnenberg, and C. Djerassi, J. Amer. Chem. Soc., 89, 6170 (1970).]*

electronic states and to an A term in the mcd spectrum. The uv and mcd spectra of [18]annulene in isooctane are shown in Figure 5–3. The observation of an A term corresponding to the 368 nm absorption band is in agreement with the hypothesis of a small molecular distortion and the preservation of a sixfold symmetry axis.

The Stanford school has carried out extensive mcd investigations on compounds containing the prophyrin ring system. Metalloporphyrins (Figure 5–4a) have an approximate D_{4h} symmetry. Their uv spectra characteristically show the intense Soret band near 400 nm and a weaker absorption near 580 nm, termed the Q band. Protoporphyrins (Figure 5–4b), in which the metal has been removed and the opposite nitrogen atoms protonated, possess a lower symmetry, approximately D_{2h}. The symmetry designations have been corroborated by mcd. Substantial A terms are found for metalloporphyrins with the ratio of A terms for the Q and Soret bands being about 9/1, in good agreement with the free electron model. For the protoporphyrins large B terms are observed.

FIGURE 5-4. *Basic ring structure for (a) a metallated porphyrin and (b) a protoporphyrin (metal removed, opposite nitrogens protonated).*

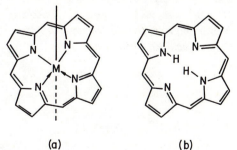

(a)　　　　　　(b)

The intense A bands exhibited by metalloporphyrins allow as little as 7 ng ml^{-1} to be detected. Such high sensitivity makes mcd a potentially powerful and unique method for the detection of very small quantities of these biological molecules. Analytical uses such as the examination of extraterrestrial samples using mcd have been carried out in connection with the Apollo 11 lunar landing.

An extensive series of studies on nucleosides using both natural and magnetic cd has been carried out by the Utah school of Eyring and Robins and by Djerassi's group. The purine base adenine, which occurs as a prosthetic group in many enzymes, has been found to have a high degree of electron delocalization, thereby enabling it to complex with various substrates in enzymatic reactions. The uv, cd, and mcd spectra of the corresponding nucleoside, adenosine, are shown in Figure 5-5. The 260 nm uv band is unresolved, and the cd spectrum exhibits only a weak and very broad negative peak. However, the mcd bands at 272 and 253 nm clearly reveal the composite nature of the 260 nm uv absorption, which contains within it the unresolved ^1L_b and ^1L_a benzene-like transitions. The 215 nm mcd band is assigned to the ^1B benzene-like transition.

The intense and oppositely signed mcd bands also can be used to differentiate between minute quantities of pyrimidine and purine nucleosides, which are frequently isolated during biochemical research. For example, the uv spectra of the nucleosides uridine and adenosine are qualitatively very similar, whereas the mcd spectra are quite different and provide a clear means of differentiation.

A recent study by Delabar and Guschlbauer of solvent effects on the cd and mcd spectra of guanosine and its derivatives has led to the formulation of an octant rule for the ^1L_b transition of guanine.

Many mcd studies have been carried out on carbonyl compounds in the hope that the

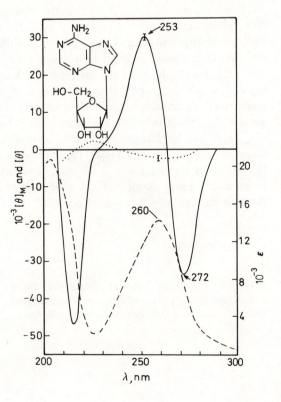

FIGURE 5-5. The uv ----, cd ······, and mcd ——— spectra of adenosine in water at pH 7. $[\theta]_M$ is based on a 49.5 kilogauss field. [Reproduced with permission from C. Djerassi, E. Bunnenberg, and D. L. Elder, Pure Appl. Chem. 25, 57 (1971). Also see W. Voelter, R. Records, E. Bunnenberg, and C. Djerassi, J. Amer. Chem. Soc., 90, 6163 (1968).]

technique would provide stereochemical information on achiral compounds analogous to that obtained earlier from ord and cd measurements.   After some initial erroneous results using mord, it was ascertained that the effects exhibited by saturated carbonyl compounds were very weak.   For example, $[\theta]_M$ for cyclohexanone is only 10^{-5} to 10^{-7} that of the strongest bands of porphyrins or of [18]annulene.   However, the use of high field superconducting magnets and more sensitive cd instruments has led to reliable mcd results for carbonyl compounds.

The mcd of over 50 aldehydes and ketones have been reported by Djerassi and co-workers (reference 3).   They found that the weak B terms associated with the 290 nm n–π^* transition exhibited a variety of bandshapes and intensities, all highly dependent on the structure of the compounds.   Indeed, these complexities indicate that mcd is highly sensitive to the molecular geometry of carbonyl compounds.   The basic reason for the observed ranges of intensities and bandshapes arises from the fact that the n–π^* transition is electric dipole forbidden and many coupling mechanisms (vibrational, solvational, etc.) may make important contributions to the observed B terms.   Recently Moscowitz and co-workers have proposed a theoretical framework for the development of symmetry rules that allows the molecular geometry of ketones to be related to their magnetic rotational strengths.

Peptides and proteins have been studied with mcd, but, in contrast to natural cd, the mcd spectra were found to be insensitive to conformational changes (reference 4).   Djerassi et al. discovered a new method for the qualitative determination of the tryptophan content of proteins.   Tryptophan residues show a positive mcd B term for the ^1L_b transition near 290 nm, where there is minimal overlap from other bands.   The intensity of this band was found to be independent of the protein conformation and of the location of the tryptophan residues and thus can be used for the determination of tryptophan in intact proteins.

Subsequently a mcd method was reported for the direct determination of the tyrosine/tryptophan ratio in proteins by ionization of the tyrosine residues to the corresponding phenolate ions.

BIBLIOGRAPHY

1. The Faraday Effect: A. J. McCaffery, *Nature*, **232**, 137 (1971); C. Djerassi, E. Bunnenberg, and D. L. Elder, *Pure Appl. Chem.*, **25**, 57 (1971); B. Briat and C. Djerassi, *Nature*, **217**, 918 (1968); D. J. Caldwell, J. M. Thorne, and H. Eyring, *Ann. Rev. Phys. Chem.*, **22**, 259 (1971); P. N. Shatz and A. J. McCaffery, *Quart. Rev. (London)*, **23**, 552 (1969); A. D. Buckingham and P. J. Stephens, *Ann. Rev. Phys. Chem.*, **17**, 399 (1966).
2. MCD of Aromatic Compounds: J. G. Foss and M. E. McCarville, *J. Chem. Phys.*, **44**, 4350 (1966); P. J. Stephens, P. N. Shatz, A. B. Ritchie, and A. J. McCaffery, *J. Chem. Phys.*, **48**, 132 (1968); I. N. Douglas, R. Grinter, and A. J. Thompson, *Mol. Phys.*, **26**, 1257 (1973); J. P. Larkindale and D. J. Simkin, *J. Chem. Phys.*, **55**, 5668 (1971); J. G. Foss and M. E. McCarville, *J. Amer. Chem. Soc.*, **89**, 30 (1967).
3. MCD of Aldehydes and Ketones: G. Barth, E. Bunnenberg, and C. Djerassi, *Chem. Commun.*, 1276 (1969); G. Barth, E. Bunnenberg, C. Djerassi, D. Elder, and R. Records, *Symp. Faraday Soc.*, **3**, 49 (1969); L. Seamans, A. Moscowitz, G. Barth, E. Bunnenberg, and C. Djerassi, *J. Amer. Chem. Soc.*, **94**, 6464 (1972); L. Seamans and A. Moscowitz, *J. Chem. Phys.*, **56**, 1099 (1972).
4. MCD of Peptides and Proteins: G. Barth, W. Voelter, E. Bunnenberg, and C. Djerassi, *J. Amer. Chem. Soc.*, **94**, 1293 (1972); G. Barth, E. Bunnenberg, and C. Djerassi, *Anal. Biochem.*, **48**, 471 (1972).

Part Four

MASS SPECTROMETRY

1

BASIC CHEMISTRY OF GASEOUS IONS

1-1 Mass Spectrometry and Ion Chemistry

The chemistry of ions in condensed media can be studied by a variety of techniques; by contrast, studies on gaseous ions invariably involve some form of mass spectrometry. Mass spectrometry not only embraces much of gaseous ion chemistry but also is a valuable analytical method for qualitative and quantitative determinations on gases, liquids, and solids and on the properties of surfaces.

The roots of mass spectrometry lie in the gifted experimental work of W. Wien and J. J. Thompson done around the turn of the century. Development of the instrument has both depended on and contributed to improvements in vacuum technology, in ion optics, and recently in data acquisition and reduction. Today's mass spectrometer accomplishes far more than the mere separation of ions according to mass, although this remains the heart of the operation.

The emphasis in this discussion of mass spectrometry will be on the underlying chemistry. Both in this chapter and in Chapter 5, questions of structure, reactivity, kinetics, and energetics are paramount. Instrumentation and experimental techniques are discussed in Chapter 3, and specific applications of mass spectrometry to the problems of the organic chemist in Chapter 4.

1-2 Ion Formation

As is the case for free radicals, the chemistry of gaseous ions involves three stages—initiation, propagation, and termination, although these terms are not usually applied (Figure 1-1). In mass spectrometry, ion neutralization is usually incidental to the analytical process of ion

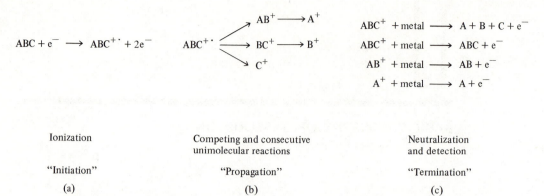

$$ABC + e^- \longrightarrow ABC^{+\cdot} + 2e^-$$

Ionization

"Initiation"

(a)

Competing and consecutive
unimolecular reactions

"Propagation"

(b)

Neutralization
and detection

"Termination"

(c)

FIGURE 1-1. *Schematic summary of the reactions occurring in the mass spectrometer. Only one type of ion formation reaction (a) is shown. The high energy ion-surface reactions (c) are not well understood.*

detection and is not of chemical interest. Ion formation, however, does merit separate discussion before ionic fragmentations are considered.

The most common method of ionization is by electron impact. Gaseous molecules that have equilibrated by collision with the walls of the ion source interact with a beam of energetic electrons that have been emitted by a hot metal filament and are accelerated toward the sample molecules (eq. 1-1). The excitation process obeys the Franck–Condon principle, the electronic

$$AB + e^- \longrightarrow AB^{+\cdot} + 2e^- \qquad\qquad 1\text{-}1$$

transition being too rapid to be accompanied by changes in molecular geometry. Even if monoenergetic reactant electrons are employed, the total energy acquired by AB can vary over a wide range since the nature of the energy transfer step will depend on the degree of interaction (impact parameter or minimum separation) between electron and molecule. The translational energies of the two electrons released (eq. 1-1) provide the necessary energy balance. The internal energy distribution $P(\epsilon)$ of the molecular ions $AB^{+\cdot}$, that is the total energy increase in AB less the ionization potential of AB, is a fundamental quantity although experimentally rather inaccessible (Figure 1-2). Because ion-ion and ion-molecule interactions are avoided in

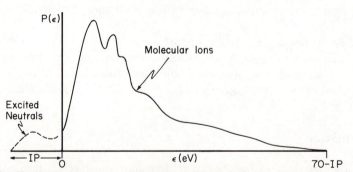

FIGURE 1-2. *Internal energy distribution $P(\epsilon)$ resulting from impact by electrons of 70 eV kinetic energy upon an organic molecule. The maximum internal energy which an ion can acquire is 70 eV, less the ionization potential (IP).*

most experimental arrangements by operation at suitably low pressures, an ion must live with the internal energy it acquires at time zero. (This conclusion ignores the possibility that energy may be depleted by relaxation processes accompanied by emission of radiation. In polyatomic ions almost nothing is known of the incidence or importance of these processes except that reasonable agreement with theory is obtained by ignoring them.) All subsequent steps are concerned with the redistribution of the internal energy ϵ; the conversion of electronic into vibrational energy and the accumulation of this energy in particular bonds are of special interest because these processes allow subsequent slow unimolecular fragmentation. "Slow" is here to be contrasted with the ionization process itself, which takes times of the order of 10^{-16} sec.

The general shape of the internal energy distribution for organic molecules given in Figure 1-2 is rationalized by the fact that an electron has a low probability of undergoing a resonance process in which it transfers all its energy to the target species. As the electron energy E_e is increased, the probability of forming the molecular ion will also increase. As it effects ionization, a high energy electron will also transfer some excess energy to the molecule, but the probability of transferring most of its energy is low. Large energy transfers would correspond to interaction with a region of high electron density in a particular orbital. Longer range interactions are much more common.

The *molecular ion* is the ion formed by loss of an electron from the molecule. A *parent ion* is any ion that fragments to produce a product or *daughter ion*. The molecular ion is just one example of a parent ion. Fragment ions formed directly from the molecular ion are primary fragments. The whole network of competing and consecutive reactions leading to the set of ions that is ultimately sampled to provide a mass spectrum is termed the *fragmentation pattern*.

Ions may be *singly* or *multiply*, *positively* or *negatively* charged. The properties of a charged species will depend strongly on whether or not it is a free radical, hence the important distinction between *odd-* and *even-electron ions*. (The free radical is always indicated as $H_2O^{+\cdot}$, $AB^{+\cdot}$, etc.) Ions containing a sequence of atoms not present in the molecule are *rearrangement ions*.

Equation 1-1 does not represent the only primary process whereby ionization is effected by electron bombardment. Multiply charged or negatively charged ions may be formed as shown in eqs. 1-2 and 1-3, while contributions from ion pair production (eq. 1-4) and predissociation (eq. 1-5) may be important in particular cases. All these processes are discussed in more detail in section 5-1.

$$AB + e^- \longrightarrow AB^{2+} + 3e^- \qquad\qquad 1\text{-}2$$

$$AB + e^- \longrightarrow AB^{-\cdot} \qquad\qquad 1\text{-}3$$

$$AB + e^- \longrightarrow A^+ + B^- + e^- \qquad\qquad 1\text{-}4$$

$$AB + e^- \longrightarrow A^+ + B^\cdot + 2e^- \qquad\qquad 1\text{-}5$$

The chemistry underlying the several major alternatives to electron impact as methods of ion formation can now be discussed (see also sections 4-1 and 5-1). Photon impact (eq. 1-6)

$$AB + h\nu \longrightarrow AB^{+\cdot} + e^- \qquad\qquad 1\text{-}6$$

is of particular interest because the technique of photoelectron spectroscopy allows the direct measurement of the kinetic energy of the liberated electron and hence of the internal energy distribution $P(\epsilon)$ of $AB^{+\cdot}$.   Charge exchange at low kinetic energies—as accomplished, for example, in the double (tandem) mass spectrometer—is the only technique that gives ions of sharply defined internal energies (eq. 1–7).   The difference between the recombination energy

$$X^{+\cdot} + AB \longrightarrow X + AB^{+\cdot} \qquad\qquad 1–7$$

of the reactant atomic ion $X^{+\cdot}$ and the ionization potential of AB appears as internal energy of $AB^{+\cdot}$.

Chemical ionization also pivots on an ion-molecule reaction, normally proton transfer from a reactant ion to the substrate (eq. 1–8).   The degree of excitation of ABH^+ depends on the relative proton affinities of the two neutral molecules AB and X.

$$XH^+ + AB \longrightarrow ABH^+ + X \qquad\qquad 1–8$$

Field ionization, a technique developed from the field ion microscope, depends on the quantum mechanical tunneling of an electron from a gaseous molecule into the conduction band of a metal.   Under the influence of very high fields, this tunneling occurs over a distance of a fraction of a nanometer.   A valence electron is usually removed, resulting in an essentially unexcited molecular ion, although several other processes also contribute to ionization in the field ion source.

1–3   Unimolecular Fragmentation

1–3a   General Considerations

The fragmentation (decomposition) reactions occurring in most analytical mass spectrometers are unimolecular and may be treated by the usual methods of unimolecular reaction kinetics, provided it is recognized that thermal equilibrium between the reactant ions is not set up.   Figure 1–3 illustrates the time scale of the mass spectrometer, an essential consideration in ion kinetics.

Even in cases where eq. 1–1 represents the only ion-forming mechanism, a complex mixture of ions normally will be present within a microsecond of ionization (this is a typical ion source residence time).   These ions will differ in mass and in the important property of whether or not they are free radicals, and they will have been formed from a network of competing and consecutive individual reactions.   The number of such reactions contributing to the mass spectrum, which approximates an instantaneous picture of this ion collection, can

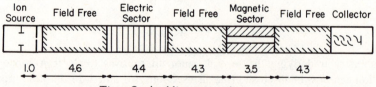

FIGURE 1–3.   *The mass spectrometer time scale.   Data refer to ions of mass 100 in the Hitachi RMH–2 mass spectrometer operating at an ion-accelerating voltage of 10 kV.   Source residence time is taken as 1 μsec.   Acceleration (0.6 μsec) occurs between the source and the first field-free region.*

reach into the hundreds even for commonplace organic compounds. It is fortunate, therefore, that only a few reactions are usually dominant and that these processes can often be predicted from a knowledge of ionic and free-radical solution chemistry. A point that has proved strangely difficult for many (especially organic chemists) to accept is that the energy available in a reacting ion is quite small. Consequently, weak bonds in the ion (usually the same bonds that are weak in the corresponding neutral molecule) are more likely to cleave than are strong bonds. Reactions that lead to electron unpairing are relatively unlikely, reactions that result in conjugation of charge or radical or those that otherwise lead to stabilized products are favored. If other requirements are satisfied (they are not for 70 eV mass spectra), bond cleavage accompanied by bond formation is more likely than simple bond cleavage.

1-3b Kinetics

Unimolecular decomposition follows formation of the activated complex, which in turn is the result of the accumulation of sufficient energy in the appropriate modes and the attainment of the appropriate geometry. The first of these twin conditions requires that the ion possess at least the activation energy for the process in question. Moreover, the greater the internal energy, the greater the probability that it will be correctly allocated for decomposition and the faster the unimolecular reaction. The time scale appropriate to most reactions occurring in mass spectrometer ion sources is of the order of 1 μsec (Figure 1–3). In this range there is an approximately linear relationship between the log of the unimolecular rate constant k and the internal energy of the ion ϵ. Some idealized k vs. ϵ curves appear in Figure 1–4. Ions may be characterized as stable, unstable, and metastable based on their rate constants, i.e., average lifetimes. The second condition for activated complex formation—appropriate geometry— merely requires time. (More complex situations in which isomerization of the reactant ion is a precondition for reaction can be temporarily ignored.) Hence the more similar the reactant and transition state geometries, the faster the reaction. Therefore simple bond cleavages, whether

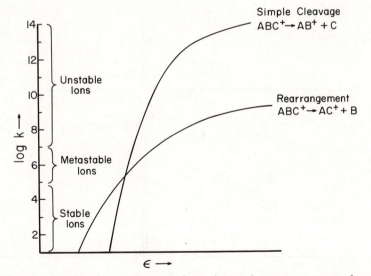

FIGURE 1–4. Idealized rate constant k vs. internal energy ϵ curves for a simple bond cleavage and a rearrangement. If both reactions are in competition, the rearrangement will occur preferentially in low energy ions but the simple cleavage will dominate in high energy ions.

endoergic or exoergic, will proceed faster than elimination reactions or other processes of comparable activation energies in which bond formation accompanies bond cleavage and adds geometrical (entropy) requirements.

None of the above is unique to mass spectrometry, but it has important effects in this field. The following generalization has proved helpful: simple bond cleavages will tend to have higher activation energies, and higher entropies of activation, than rearrangement reactions. As a consequence, simple cleavages dominate in the reactions of high energy ions when differences in activation energies are inconsequential, but rearrangements dominate in the reactions of low energy ions. The resulting crossover of rate vs. energy curves is shown in Figure 1-4. Normal analytical mass spectra result from sampling quite fast reactions and therefore tend to be due more to simple cleavages than to rearrangements. It is not difficult, however, to move in either direction from this situation: field ionization involves very fast reactions so that rearrangements are rarely observed, while the slow reactions that give rise to the so-called *metastable* peaks in normal mass spectra are, more often than not, due to rearrangement processes. This change in the dominant reaction with observation time or with its equivalent— available energy of the ion—can also be observed by the simple expedient of reducing the energy of the ionizing electron beam and so eliminating all reactant ions of high internal energy. That the remaining slow-reacting ions tend to undergo low activation energy rearrangement reactions rather than simple bond cleavages is illustrated by comparison of the spectra in Figure 1-5. All fragment ions decrease in abundance relative to the molecular ion at low energy, but the abundance of the rearrangement ion m/e 92 (due to loss of propene from the

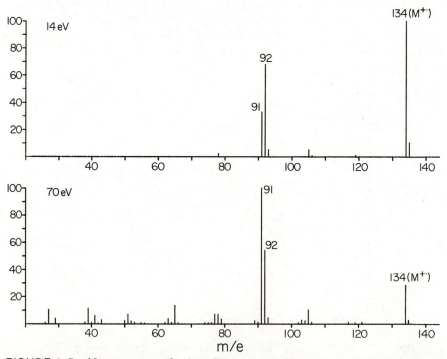

FIGURE 1-5. Mass spectra of n-butylbenzene at 70 and 14 eV, illustrating the increase at low electron energy in the abundance of m/e 92 (due to a reaction involving bond formation) relative to that of m/e 91 (due to a simple cleavage reaction).

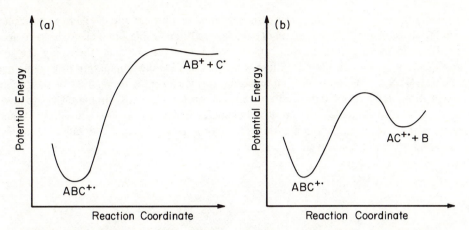

FIGURE 1-6. Potential energy diagrams typical of (a) a simple cleavage and (b) a rearrangement process. Note the lower activation energy of (b), the fragmentation that involves rearrangement.

molecular ion) exceeds that of the simple cleavage product m/e 91, reversing the situation that obtains at 70 eV.

The important factors involved in mass spectral kinetics are the $k(\epsilon)$ relationship and the $P(\epsilon)$ distribution. Ion abundances, and hence mass spectra themselves, merely represent the aggregate of individual ionic fragmentations occurring within particular time intervals after ionization. The probability that an ion will fragment in a given time period is controlled by its internal energy and by the form of the $k(\epsilon)$ relationship. The distribution of ion abundances that this relationship predicts must be convoluted over the internal energy distribution $P(\epsilon)$ in order to arrive at the mass spectrum.

1–3c Energetics

Although the potential energy profiles appropriate to ionic decompositions have been established only for relatively few reactions, the general trends seem clear. Since many ion-radical reactions have very small activation energies, it is assumed that most simple cleavages possess negligible reverse activation energies. Therefore, these reactions normally are endoergic, as illustrated in Figure 1-6a. Where a reactant ion has several simple cleavage pathways open to it, that of lowest activation energy will invariably dominate. Reactions that result in the production of stable neutral species such as CO_2, H_2O, and N_2 involve bond formation as well as bond cleavage and usually have substantial reverse activation energies. The corresponding potential energy surface for this type of ionic fragmentation will typically have the form shown in Figure 1-6b. The activation energy is deliberately shown as smaller than that for the simple cleavage. Whenever a particularly unexpected and complex ionic rearrangement is observed, it can invariably be traced to the formation of particularly stable products formed by a process having a very low activation energy.

1–3d Prediction of Fragmentation Pathways

For a gaseous ion of even moderate complexity such as any organic ion, the number of competing unimolecular reactions, together with further reactions of the daughter ions, soon becomes large. Fortunately, under given conditions only a few will be important; otherwise mass spectra would be impossibly complex. From an analytical standpoint it is obviously important to be able to establish guidelines as to which fragmentations will dominate.

These guidelines evolve from the considerations of kinetics and energetics just discussed. The conditions used for the measurement will favor either simple bond cleavages or rearrangements. If, as in most analytical applications, the former are favored, then only the simpler rearrangements need be considered. These will often involve the elimination of a stable neutral molecule (H_2O, N_2, CO_2, an olefin, an alcohol, etc.) via a simple four-, five-, or six-membered cyclic transition state. For example, diarylcarbonate molecular ions eliminate CO_2 to give ions that show all the properties of the diaryl ether molecular ions (eq. 1-9). Similarly, 1-hexanol eliminates water via a six-centered intermediate (eq. 1-10). A more detailed treatment of skeletal rearrangements is reserved for section 5-3.

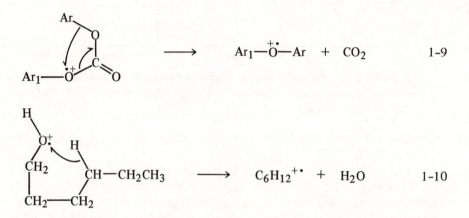

$$Ar_1 - \overset{+\cdot}{O} - Ar \ + \ CO_2 \qquad\qquad 1\text{-}9$$

$$C_6H_{12}{}^{+\cdot} \ + \ H_2O \qquad\qquad 1\text{-}10$$

The previous discussion of energetics has an important corollary: those simple cleavages that give the most stable products are favored. Since the ionic rather than the neutral product is the high enthalpy species, this corollary is equivalent to the usual generalization that mass spectral fragmentation is governed by product ion stability.

Assessment of the relative stabilities of different ions is usually easy, based as it is on chemical principles that are also applicable in solution. (Cases where gaseous ion stabilities differ from those established for solution are discussed in section 5-7). Maintenance (whenever possible) of an octet of electrons, localization of charge on the most favorable site available, resonance delocalization, and absence of electron unpairing are the fundamentals too often overlooked by the student biased in favor of the "anything goes" theory of mass spectra. These principles are worth illustrating by several examples that, taken together, form a basis from which much of ion functional group chemistry can be extrapolated.

(i)   Dialkyl ethers tend not to give RO^+ or $R-O-CH_2-CH_2{}^+$ ions. They do give $R-O-\overset{+}{C}H_2 \ \leftrightarrow \ R-\overset{+}{O}=CH_2$.

(ii) Dialkyl thioethers, however, do give RS^+ ions.

(iii) Aryl ethers give ArO^+.

(iv) Arenes give $ArCH_2{}^+$ in preference to Ar^+ or $Ar(CH_2)_2{}^+$.

(v)   The bifunctional compound $R_1-O-CH_2CH_2-N{\overset{R_2}{\underset{R_3}{\big\langle}}}$ gives $CH_2=\overset{+}{N}{\overset{R_2}{\underset{R_3}{\big\langle}}}$ rather than $R_1-\overset{+}{O}=CH_2$.

A unique generalization, which covers all the functional types under normal conditions, concerns radical loss. Of course, tertiary carbonium ion formation is favored over that of secondary or primary ions. Which, however, will be lost more readily—a large or a small radical?

The answer is illustrated in eq. 1-11 for a case in which the difference in radical size is minimal.

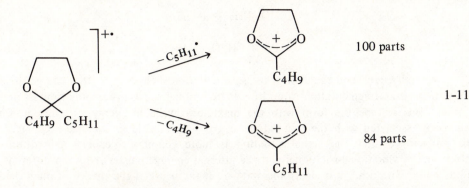

100 parts

1-11

84 parts

It is important to realize that in this case the arrows do not necessarily signify one-step reactions, since examination of normal fragment ions does not permit this distinction. (The study of metastable ions is of importance partly because they allow one to examine specific one-step reactions.)

A particularly important fact in accounting for fragmentation patterns is that molecular and other odd-electron ions may eliminate either a radical or an even-electron neutral species (eqs. 1-12), but an even-electron ion will not usually eliminate a radical to form an odd-electron

$$\text{odd} \longrightarrow \text{even} + R^{\bullet} \qquad\qquad 1\text{-}12a$$

$$\text{odd} \longrightarrow \text{odd} + N \qquad\qquad 1\text{-}12b$$

species (eq. 1-13b).   Exceptions to this last rule occur when the bond cleaved in the even-

$$\text{even} \longrightarrow \text{even} + N \qquad\qquad 1\text{-}13a$$

$$\text{even} \xrightarrow{\hspace{0.3em}\times\hspace{0.3em}} \text{odd} + R^{\bullet} \qquad\qquad 1\text{-}13b$$

electron ion is particularly weak (polybrominated compounds, for example, can undergo multiple successive $Br^{\bullet}$ losses), or when a uniquely stable odd-electron ion may result (eq. 1-14).   As a consequence of the odd-even electron generalization, the important even-

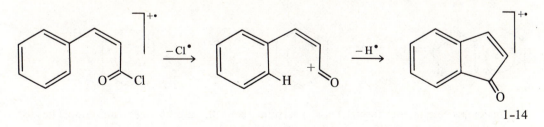

1-14

electron fragment ions $RC\overset{+}{\equiv}O$, $ArCH_2{}^{+}$, and $R\!-\!\overset{+}{O}\!=\!CH_2$ undergo further fragmentations (eqs. 1-15, 1-16, and 1-17).

$$RC\overset{+}{\equiv}O \longrightarrow R^{+} + CO \qquad\qquad 1\text{-}15$$

$$ArCH_2^+ \longrightarrow C_5H_5^+ + C_2H_2 \quad (Ar = C_6H_5) \tag{1-16}$$

$$R-\overset{+}{O}=CH_2 \quad \diagdown \quad \begin{array}{l} \overset{+}{HO}=CH_2 + alkene \\[1em] R^+ + CH_2O \end{array} \tag{1-17}$$

A well-established guideline to fragmentation behavior is the charge localization hypothesis: the fragmentation of an ion can be rationalized as proceeding from its most stable form.   This will be the form with the maximum available energy and that in which the charge is associated with the unit (atom or group) best able to stabilize a charge.   It must be emphasized that the actual situation is more complex: energy redistribution—both electronic $\rightarrow$ vibrational-rotational and the internal energy transfers between different degrees of freedom within a given electronic state—is at the heart of the theory of mass spectra (see quasiequilibrium theory, section 5-2).   Moreover, it has become apparent that in ions with sufficient energy, the charge can migrate from one site to another.   (That fragmentations may then be triggered by several charge sites in one sense is no more than an extension of the charge localization hypothesis.)

One should bear in mind, then, that localization is not unqualified and that, especially in higher energy ions, there may be no need for the extra energetic advantage that charge localization affords.   Another reason for the success of the hypothesis probably lies in the principle of the control of product ion stability.   The most stable form of the product ion will have the charge localized in the most favorable position, which will invariably be the most favorable position in the reactant ion.   Hence it is charge localization in the product ion that *impels* localization in the reactant ion.

As illustrations of the hypothesis, consider reactions 1-18, 1-19, and 1-20, each of which

$$\underset{\overset{|}{CH_3}}{C_2H_5 \overset{\frown}{\underset{}{-}}CH} \overset{+\bullet}{-}O-CH_3 \longrightarrow CH_3CH=\overset{+}{O}-CH_3 \quad + \quad C_2H_5^\bullet \tag{1-18}$$

$$C_6H_5-\overset{+\bullet}{S}\overset{\frown}{\underset{}{}}CH_3 \longrightarrow C_6H_5S^+ \quad + \quad CH_3^\bullet \tag{1-19}$$

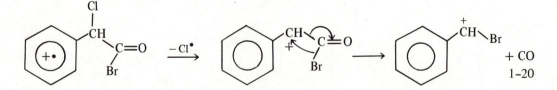

is a major process in the normal mass spectrum.   Note the use of single- and double-headed arrows to show the movements of single electrons and electron pairs, respectively.

1-3e Reactions of Metastable Ions

Metastable ions are those that—on the time scale of the instrument involved—are neither "unstable" (as are ions that react soon after formation, i.e., within the ion souce) nor "stable" (as are ions that fail to react at all prior to detection).   Metastable ions, therefore, are slowly reacting ions that possess relatively little energy in excess of the minimum required for the reaction in question. Their low energies make them more interesting for comparison with ions in solution than are normally reacting ions. Their different position on the observation time scale makes them important in kinetic studies. Their peak widths can be related to the kinetic energy released in the unimolecular reaction. This information is vital to a complete thermochemical definition of the reacting system.

Of most immediate interest, however, is the fact that the mean kinetic energies of metastable ions can be measured (either indirectly from their positions in the normal mass spectrum, or directly) and can provide evidence for a parent-daughter relationship between two ions in the mass spectrum. Proof of such a parental relationship is necessary to establish fragmentation or breakdown patterns. Fragmentation schemes in turn have both analytical and theoretical importance.

1-4   Bimolecular (Ion-Molecule) Reactions

Ion-molecule reactions in the gas phase constitute an enormous subject, and only some of the basic chemistry appropriate to this discussion of organic mass spectrometry can be given.

Reactions occurring both at kinetic energies of several thousand electron volts and low energies (thermal and above) are of interest. Reactions at the higher energies are important because they transfer energy to the incident ion and may cause it to fragment. Momentum transfer, however, is negligible, so the products of such collision-induced fragmentations are detected just like those of unimolecular metastable reactions. In addition to enhancing considerably the signals due to metastable ion decompositions, these processes reveal much about the energy states of the reactant ions and the basic physics of the collision process. This type of high energy interaction may also result in charge transfer, again with only minor changes in the relative translational energy of the incident ion. For example, doubly charged ions can be converted to singly charged ions and a doubly charged ion spectrum obtained (see section 5-4).

Low energy ion-molecule reactions are observed in chemical ionization mass spectrometry, in ion cyclotron resonance spectrometry, in ion formation by charge exchange, and in the important experimental work being done with tandem mass spectrometers. Much of the work with tandem mass spectrometers is concerned with the cross sections for collision processes and their variation with kinetic energy. Traditionally the ion-molecule reactions of interest have been charge exchange and atom transfer reactions. With the advent of the ion cyclotron resonance instrument and the growing interest of organic chemists in the technique, more complex reactants have been investigated, often with a view toward mimicking reactions that occur in solution. These efforts have been successful. Nucleophilic and electrophilic substitution at carbon, for example, has been reported in the gas phase (see section 5-7). Most significantly this type of study has begun to help elucidate solution reaction mechanisms. Similar progress through the use of ion cyclotron resonance is also seen in the determination of gas phase acidity, basicity, and nucleophilicity scales. Amine basicity in the gas phase, for example, shows simple trends with larger alkyl substituents leading to greater basic strength than smaller substituents; the order due to substitution pattern is tertiary $>$ secondary $>$ primary. This observation has prompted a reconsideration of the traditional rationalizations employed to explain the more complex solution data.

BIBLIOGRAPHY

1. General Treatments of Organic Mass Spectrometry: J. H. Beynon, *Mass Spectrometry and Its Applications to Organic Chemistry*, Elsevier, Amsterdam, 1960; K. Biemann, *Mass Spectrometry: Organic Chemical Applications*, McGraw-Hill, New York, 1962; H. Budzikiewicz, C. Djerassi, and D. H. Williams, *Mass Spectrometry of Organic Compounds*, Holden-Day, San Francisco, 1967; M. C. Hamming and N. G. Foster, *Interpretation of Mass Spectra of Organic Compounds*, Academic, New York, 1972; D. H. Williams and I. Howe, *Principles of Organic Mass Spectrometry*, McGraw-Hill, London, 1972.
2. Shorter Books on Organic Mass Spectrometry: F. W. McLafferty, *Interpretation of Mass Spectra*, 2nd ed., W. A. Benjamin, New York, 1973; H. C. Hill, *Introduction to Mass Spectrometry*, 2nd ed., Heyden and Son, London, 1972; S. R. Shrader, *Introductory Mass Spectrometry*, Allyn and Bacon, Boston, 1971.
3. Kinetics and Energetics of Organic Ions: R. G. Cooks, I. Howe, and D. H. Williams, *Org. Mass Spectrom.*, **2**, 137 (1969); F. W. McLafferty in A. L. Burlingame, ed., *Topics in Organic Mass Spectrometry*, Wiley–Interscience, New York, 1970; I. Howe *Mass Spectrom. (Spec. Period. Report)*, **1**, ch. 2 (1971).
4. Ion-Molecule Reactions: D. P. Stevenson in C. A. McDowell, ed., Mass Spectrometry, McGraw-Hill, New York, 1963, ch. 13; J. L. Franklin ed., *Ion Molecule Reactions*, Plenum, New York, 1972; E. W. McDaniel, V. Cermak, A. Dalgarno, E. E. Furguson, and L. Friedman, *Ion-Molecule Reactions*, Wiley–Interscience, New York, 1970.

THE SCOPE AND APPLICATIONS OF MASS SPECTROMETRY

Since this treatment of mass spectrometry emphasizes heavily the deduction of molecular structures and the behavior of gaseous organic ions, it seems desirable to establish some perspective by broadly sketching other uses and accomplishments of the technique.

In the years 1919-1939, after Thompson's discovery of the stable neon isotopes ^{20}Ne and ^{22}Ne, mass spectroscopists became engrossed in the search for new isotopes and the determination of their exact masses and natural abundances. Natural isotopes of almost all the lighter elements were discovered by this technique during this period. Since the instrumental criteria for accurate mass and accurate abundance measurements tend to be mutually exclusive, instruments developed largely for one or the other purpose began to appear. The use of photographic detection for mass measurement and electrical detection for abundance measurement was a part of this specialization.

Although new stable nuclides are no longer being reported, improved values for abundance ratios and masses continue to be obtained. They form the basis for the periodic revisions that are necessary in chemical atomic weights. Reliable data on isotopic masses and abundances are, of course, basic to much of physical science. A more technical application of mass spectrometry in the area of isotope science is the use of electromagnetic isotope separators, the large mass spectrometers in which the separated ions are neutralized and deposited in special collectors. Moving from isotopic enrichment to the analysis of the results of isotopic labeling experiments, one notes that mass spectrometry is the standard method of analysis for the determination of stable isotope incorporation and is particularly valuable because of the small sample size required. Recently an important extension of this technique employing high resolution has allowed the determination of both the amount and also the position of label incorporation in complex molecules. The total isotopic incorporation can be found by examining the molecular ion, and particular fragment ions reveal the extent of incorporation in different portions of the molecule. This procedure should find wide application in bio-synthetic studies, since enormous savings in time are effected by avoiding the tedious chemical degradations normally used for incorporation determination by radioisotopic methods.

The accurate determination of isotopic abundances now constitutes the basis for two important branches of geoscience. The important $^{40}K/^{40}Ar$ dating method, among others, depends on mass spectral data. In a similar vein, differences in natural abundances of stable isotopes with the nature of the source can be measured accurately, especially by the use of dual collector instruments. From the measurement of this "historical" isotope effect, the physical, chemical, or biological conditions that effected the original enrichment can some-times be inferred. For instance, the average temperature and even the seasonal temperature variation during the lifetimes of marine crustaceans can be found from the $^{16}O/^{18}O$ ratio in the calcium carbonate of their shells, since this ratio depends on the water temperature in which the organisms lived.

The challenges that still remain in the area of accurate ionic mass measurement can be illustrated by a single problem. With the increase in the application of both isotopic labeling and mass spectrometry to the study of large molecules, particularly those of biological origin, separation of the H_2/D mass doublet has become important. The mass difference, 1.7 milli-mass units, requires a resolving power of 500,000 (10% valley; see section 3-5) for an ion of mass 850. Resolution in excess of 500,000 has only been achieved in a few highly specialized research instruments. Recently introduced commercial instruments are capable of resolutions in the 100,000–200,000 range.

Following the intensive effort on ion mass and abundance measurements, rapid development occurred in the utilization of a third basic datum type obtainable from the mass spectrometer—the ionization efficiency curve, which is a plot of the abundance of an ion against the energy of the ionizing source (electron energy, photon wavelength, etc.). From this information the fundamental properties of electron affinity (EA), ionization potential (IP), and appearance potential (AP) can be derived. These quantities can be briefly defined as shown in Figure 2-1. When these values are combined with known thermochemical data on neutral species, ionic heats of formation and bond dissociation energies can be determined. This type of measurement can be done on most analytical mass spectrometers in which the ionizing electron energy can be varied. The application of IP, AP, EA, and heat of formation data in testing molecular orbital theories and in calculating molecular parameters is just one aspect of their importance. A valuable feature of these mass spectrometric measurements lies in the fact that unstable species, such as free radicals, can be studied and bonds that cannot otherwise be cleaved can be investigated. Free radicals and other transient neutrals have been studied by coupling a Knudsen cell or flash photolysis reactor to the mass spectrometer.

Recent years have seen the application of the mass spectrometer in many new ways as well as the development of new types of instrumentation with unique problem-solving capacities. Much of this growth has been directly stimulated by the role mass spectrometers played in one area of industrial production—quality control in petroleum refining. This application, which provided a powerful stimulus for structural analysis and the analysis of mixtures by mass spectrometry, led to the introduction of commercial instruments in the 1940's. Mass spectrometry still has important industrial uses, and a decline in its use in quantitative analysis has been offset by the development of such new roles as leak detection, trace element determination, silicon doping of semiconductors, and even precision machining using ion beams. Trace element determination in metallurgy requires the use of the spark ion source, in which the sample is subjected to a high voltage discharge. Photoplate detection is invariably used. Leak detectors are mass spectrometers that are usually sensitive to only one ion (*m/e* 4,

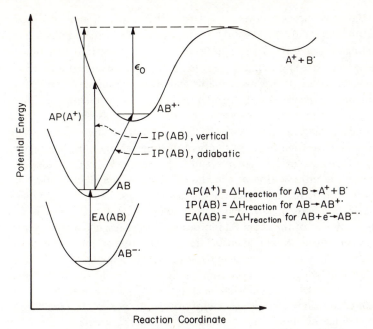

FIGURE 2-1. Some fundamental thermochemical quantities that can be determined by mass spectrometry. Note the assumption that the reverse activation energy can be ignored in the AP relationship.

$He^{+\cdot}$). They are attached to the evacuated system and respond when a jet of helium is directed over the faulty area. The applications of mass spectrometry in the semiconductor field arise because of the stringent quality control requirements. Chemical and even isotopic purity can be maintained using ion neutralization and deposition techniques.

The use of the mass spectrometer as an analytical instrument in the study of reaction mechanisms is well established, the particular advantage of the instrument being its ability to distinguish isotopically labeled species. Applications to reaction kinetics are less well known; such studies utilize the time-of-flight mass spectrometer, which can record a complete mass spectrum in 10^{-4} to 10^{-5} sec. This short recording time makes it uniquely useful for investigating intermediates in flash photolysis, shock tube experiments, and other flow systems.

Most analytical studies with the mass spectrometer are made on neutral species, the ionization of which is merely incidental to the analysis. The *direct* sampling and study of ions occur in the following situations, among others. First, ionic species formed in flames have been studied using rapid time-of-flight analysis. Second, rocket-borne mass spectrometers have made direct measurements on ions in planetary environments. The absence of vacuum systems in these instruments may be noted. Extraterrestrial mass spectrometry is actually a large and thriving branch of the science, the observation of periodic outbursts of water vapor from the moon being among its achievements.

The huge amount of data obtained when a mass spectrometer is run under high resolution conditions has resulted in a great deal of effort being spent on data acquisition and processing. In addition, the complexity of mass spectra and the highly individual approach used in solving structural problems have led to several major studies on machine learning and artificial intelligence.

Perhaps the most notable development since the application of mass spectrometry to organic chemistry in the early 1960's has been the enormous increase in its use in solving problems in the biological sciences. These applications have followed directly upon technological advances in the mass spectrometer that have made possible the examination of higher molecular weight compounds as well as relatively involatile and unstable samples. Particularly noteworthy has been the coupling of the gas chromatograph to the mass spectrometer, leading to the routine study of complex mixtures by mass spectrometry. Since gas chromatography by itself is frequently of limited value in qualitative analysis, the new technique has opened several new areas. Applications have been made in the field of drug metabolites, in forensic science, and in the analysis of archaeological artifacts and objects of art. Flavor chemistry has moved ahead sharply with the introduction of a method that allows the identification of the hundreds of volatile constituents of the freshly crushed strawberry, for example. Studies on insect and plant hormones, including the pheromones, have also been simplified. Another recent technical advance—the development of the chemical ionization source—has had almost immediate applications in medical practice. Rapid identification of the responsible agent in comatose drug overdose patients is achieved by direct analysis of the stomach contents.

Another area in which mass spectrometry is proving valuable is in the detection and identification of pollutants. Air pollution analytical equipment is usually required to monitor known compounds for long periods as inexpensively as possible. Gas chromatography is generally the technique of choice, but in water quality work a greater variety of pollutants can be expected and gc–ms is serving an important function.

The study of chemistry at the molecular level, in molecular beam work for example, frequently involves use of mass spectrometry to prepare reagents or to analyze products even if the reaction studied is not an ion-molecule reaction. The large and vigorous area of collision physics makes similar use of mass spectrometric techniques.

In concluding this survey it seems desirable to give the reader some feel for the almost vibrant technological and scientific ingenuity being displayed in the broad field of mass spectrometry. To do this, a brief description is given of two recently developed ancillary techniques—plasma chromatography and ion-scattering spectrometry. Plasma chromatography is a technique for the identification and analysis of trace constituents in gases. It operates at atmospheric pressure with a sensitivity of approximately 1 molecule in 10^{11}, which is high but still several orders of magnitude less sensitive than the canine nose. The sample undergoes low energy ion-molecule reactions with a reactant ion, and the resulting set of ions is separated by the differences in their mobilities as they drift through a voltage gradient. The ions arrive at a collector giving essentially a millisecond-scale gas chromatogram. Alternatively, the ions may be mass-analyzed prior to detection and identified in this way as well as by their characteristic drift rates.

Ion-scattering spectrometry is a technique for the analysis of surface compositions. It differs from earlier mass spectrometric methods, such as the ion microprobe (in which a primary ion beam sputters surface ions that are sampled and mass-analyzed), in that it is sensitive only to the first layer of surface atoms and therefore can be used to study gas adsorption and surface composition. Depth sampling can, however, be achieved by sputtering the surface during analysis. The method of analysis involves determination of the energy lost by incident ions that strike the surface and suffer elastic collisions. The energy lost by these ions depends on their mass and the masses of the surface atoms. The energy analysis is accurate enough to distinguish between isotopic species.

BIBLIOGRAPHY

1. Mass Spectrometry and Extraterrestrial Chemistry: L. F. Herzog, *Int. J. Mass Spectrom. Ion Phys.*, **4**, 253 (1970).
2. Analysis of Solids: A. J. Ahearn, ed., *Trace Analysis by Mass Spectrometry*, Academic, New York, 1972.
3. Nuclear Mass Determinations: F. W. Aston, *Mass Spectra and Isotopes*, 2nd ed., Edward Arnold, London, 1942; H. Hinterberger, *Nuclear Masses and Their Determination*, Pergamon, London, 1957.
4. Isotopic Abundance Measurements: A. O. Nier in H. T. Clark, ed., *Isotopes in Biology and Medicine*, University of Wisconsin Press, Madison, 1948; C. C. McMullen and H. G. Thode in C. A. McDowell, ed., *Mass Spectrometry*, McGraw-Hill, New York, 1963, ch. 10.
5. Biological Applications of Mass Spectrometry: G. R. Waller, ed., *Biochemical Applications of Mass Spectrometry*, Wiley–Interscience, New York, 1972; G. W. A. Milne in G. W. A. Milne, ed., *Mass Spectrometry: Techniques and Applications*, Wiley–Interscience, New York, 1971.
6. General discussions of the Applications of Mass Spectrometry: F. A. White, *Mass Spectrometry in Science and Technology*, Wiley, New York, 1968; R. W. Kiser, *Introduction to Mass Spectrometry and Its Applications*, Prentice-Hall, Englewood Cliffs, N.J., 1965.
7. Historical Interest, Thompson's Work on Mass Spectrometry: G. Thompson, *J. J. Thompson, Discoverer of the Electron*, Anchor Books, Doubleday, New York, 1966, ch. 9.

3

THE INSTRUMENT

3-1 Introduction

This chapter covers those aspects of instrumentation that are relevant to the analysis of organic compounds. It includes a little theory, considerable descriptive material, and some practical considerations. We have already noted that the basic operations carried out in the mass spectrometer are ionization, mass analysis, and ion detection. A more complete description requires that we also consider (a) the introduction of sample into the instrument and (b) the elements of ion optics, that is, the control and direction of ion beams.

3-2 Ion Formation: The Ion Source

The heart of the mass spectrometer is the ion source. Not only is ionization effected in this region but most of the ionic reactions occur here too. The actual reaction chamber within the source in which these processes proceed is termed the *ionization chamber* (or ion chamber). We describe here only the most widely used source, the electron impact source. Some of the other ion sources that are finding increasing application in structural organic chemistry are briefly described in section 5-1, together with the ionization phenomena that they permit.

In order to achieve maximum sensitivity with a limited amount of sample, the ionization chamber is a small (several cm^3), relatively gastight unit to which sample is supplied and from which ions are extracted. Although the ion chamber pressure cannot be measured directly in the dynamic operating mode and a distant pressure gauge will read low by a factor as high as 10, actual pressures of 5×10^{-5} torr are probably standard. Much lower pressures will often give sufficient signal while, depending on the compound, ion-molecule reactions (which change the character of the mass spectrum) can become important at higher pressures. While the efficiency of this type of ion source is relatively high, less than one molecule in 10^4 that enter the ion chamber is ionized and most are simply pumped away.

The total ion current produced by an electron impact source under normal conditions is of the order of 10^{-8} amp or less. After acceleration and passage through the source slit system, but before mass analysis, beams of 10^{-10} amp (6×10^8 ions sec^{-1}) represent a common upper limit. Further attenuation of the beam occurs in the analyzer. Nevertheless, very weak signals ($<10^{-17}$ amp) can be accurately monitored using an electron multiplier as detector. The second major parameter for judging an ion source, after efficiency, is the energy spread of the accelerated ion beam. A large energy spread is undesirable in high resolution work, and even the spread of about a volt in several thousand commonly obtained from an electron impact source is undesirable in some applications, including measurement of the kinetic energy release in the fragmentation of metastable ions. The third important characteristic of the ion source is the electron energy distribution. This property is particularly important for all thermochemical measurements. Modified sources in which the effective electron energy spread is decreased give far more reliable ionization and appearance potentials than do the standard types.

We turn now from the operating parameters to the methods used to achieve them. Electrons are generated by thermionic emission from an incandescent rhenium or tungsten filament. The temperature of the metal ($\sim 1800^{\circ}$C) is responsible for the distribution in electron energy, the average energy being determined by the potential difference between the filament and the ion chamber. (In the usual arrangement the filament is external to the chamber, and the electrons enter through a slit.) The ionizing electron beam, which usually takes the form of a wide but shallow band, traverses the chamber under the influence of a small external magnetic field. This field constrains the electrons to a shallow equipotential region in the chamber (thereby reducing the energy spread in the accelerated ion beam) and causes them to move in helical paths (thereby increasing the possibilities of ionizing collisions). The electron beam passes through the ion chamber and is collected at a trap or anode that is positive with respect to the chamber. In order to achieve even moderate precision in ion abundance measurements, the ionizing electron current must be kept constant. A typical value is 100 μamp. The fraction of the current reaching the trap is controlled by a feedback loop to the filament that controls the total emission.

A schematic diagram of an ion source is given in Figure 3-1. It is worth taking the time to grasp this arrangement in three dimensions; the use of x, y, and z in magnetic mass spectrometry is standardized. The x direction is always the direction of travel of the ion beam, and the y direction is that in which deflection of the ion beam by the magnetic field occurs. The source slit width therefore controls the ion beam width in the y direction, while the height of the slit controls the beam width in the z direction.

Removal of the ions from the ion chamber is achieved by gently repelling the ions by applying 0-10 V to a repeller plate (Figure 3-1). The ion source is held at a high potential (typically 3-10 kV) with respect to the analyzer so that the ions acquire kinetic energy as they leave it. The ion accelerating voltage penetrates to a small extent into the ion chamber and assists in ion removal.

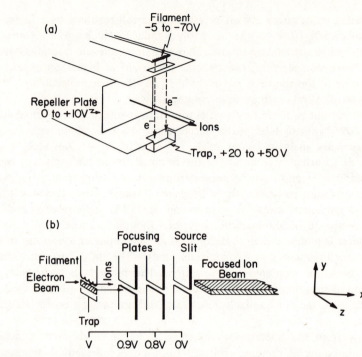

FIGURE 3-1. The ion source. All voltages in (a) are relative to that of the ion chamber, which is floated off ground at the ion-accelerating potential V.

3-3 Directing Ions with Electric and Magnetic Fields

Analogies can be found between ion optics and other more familiar forms of optics, and they are helpful in approaching the subject of this section. Most mass spectrometers have little or no focusing action in the z direction. The lenses are cylindrical rather than spherical. Hence for maximum signal it is as well to employ a band of ions rather than a point source by using long ($\sim$1 cm) narrow slits. The first such slit, the source slit, is positioned soon after the ion beam exits from the ion chamber. In theory, infinite resolution should be possible by use of sufficiently narrow slits if the beam were homogeneous in energy. Adequate signal is of paramount importance, however, so that focusing becomes a prime consideration. In order to lose as little of the ion beam as possible, ions whose paths diverge from the main ion beam axis should, as far as possible, be collected. As in optics such dispersion occurs after passage through a slit. If a component of the system can refocus these ions to an image, then it is said to have the property of *direction focusing*. This property is possessed, for example, by sector magnetic and electrostatic fields. The direction-focusing properties that result when electric and magnetic sectors are used in tandem are illustrated in Figure 3-2, which shows a monoenergetic beam of ions all of one mass. Note that each field is individually capable of bringing a diverging ion beam to a point of focus.

Because of the kinetic energy distribution of the original neutral molecules in the ion source, because of the occurrence of ionization at different points in the ion chamber, and because of the presence of complex fields in the ion chamber, the primary ion beam is always inhomogeneous in kinetic energy. Elements that select a narrow slice from this distribution are of limited value since the resulting signal attenuation would be too great. What is needed, therefore, is *energy focusing* (or *velocity focusing*). A sector electrostatic analyzer is a device that produces an energy *dispersion* in its image plane. The resultant aberration in the image can be combined with an equal and opposite aberration produced by a magnetic sector to give energy focusing. An instrument that combines an electric and magnetic sector therefore can give both direction focusing and energy focusing, hence the term *double-focusing* mass spectrometer. The chemist, however, is usually more interested in *what* an instrument can achieve than in *how* it is achieved. The requirement in this connection is accurate mass measurement rather than high resolution per se, although the two may be closely connected. The double-focusing arrangement is only one method of achieving high resolution and accurate mass measurement. Other techniques employ completely different means. For example, in quadrupole mass spectrometers high resolution is achieved by careful control of crossed rf and dc fields so that all ions, except those of some precisely defined mass, experience unstable trajectories through a rod assembly and are lost.

At this point it may be useful to emphasize that the types of focusing just discussed for sector fields represent only one aspect of ion focusing encountered in mass spectrometers. A series of slits can be used to form a crude lens to improve sensitivity by focusing the beam onto the source slit, provided the voltages on the slits are properly chosen. The ion beam emerging from the ion source is often focused at the source exit slit in this way in the course of its acceleration (Figure 3-1).

Energy analysis, as distinct from energy focusing, is important in the investigation of ion chemistry (see Chapter 5). It may be accomplished using an electric sector as the dispersive agent (Figure 3-3) or, for example, by using the parallel plate arrangement shown in Figure 3-4. In both procedures only those ions of correct energy are deflected through the correct angle to allow transmission. Both devices also possess direction-focusing properties.

One of the difficulties met in attempting accurate control of ion beams is space charge, the spreading of an ion beam caused by too high a charge density. In order to avoid deflection of the ion beam by surface charge, it must not be allowed to strike insulators. Even slits

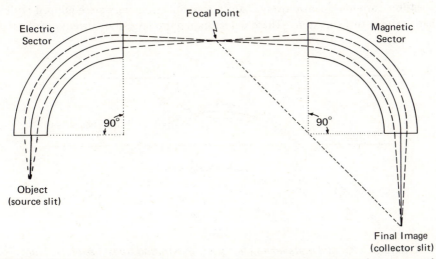

FIGURE 3-2. Direction-focusing properties of magnetic and electric sectors. All ions are brought to the same point of intermediate and final focus only if they are of identical mass/charge ratio and have identical kinetic energies.

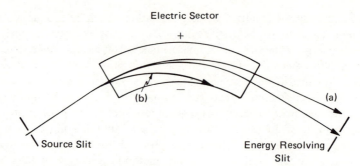

FIGURE 3-3. *Energy analysis using an electric sector. Beam (a) has too high and beam (b) too low a kinetic energy to pass centrally through the sector and to be transmitted by the energy resolving slit.*

and other metal objects in proximity to the ion beam may acquire a surface charge. Yet another problem is spurious signals due to ionized or neutral species formed by the reflection of the ion beam at the walls of the flight tube.

The use of mechanical slits plays an important role in the control of ion beams. Frequently, the ion beam is swept across a slit as the final step in mass analysis before ion detection. The width of this *collector slit* must be much narrower than that of the signal in order to record an undistorted peak shape. This effect is illustrated in Figure 3-5. It must be noted that wide slits are useful in measuring the total ion abundance. The *source slit* defines the width of the ion beam once it has been fully accelerated and before it enters the analyzer. Its dimensions limit the maximum theoretical resolution attainable.

3-4 Ion Detection

The most obvious method of detecting an ion beam is to use some type of electrometer that determines ion currents quantitatively. Most mass spectrometers are equipped with Faraday cups that perform this function. These detectors require large ion currents so their chief use

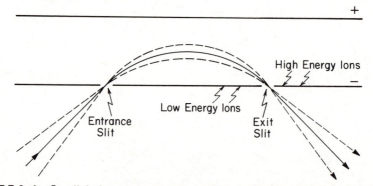

FIGURE 3-4. *Parallel plate energy analyzer, illustrating transmission and direction focusing of an ion beam of appropriate kinetic energy. Ions of lower or higher energy strike the walls at the positions shown and are not transmitted by the exit slit.*

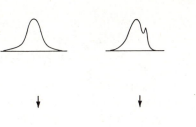

FIGURE 3-5. Effect of collector slit width upon recorded signal. Although the use of a wide slit masks details of peak shape, it has the advantage that the peak height is proportional to the total ion flux.

is to calibrate the gain of electron multipliers. The electron multiplier and the photographic plate, the most common detectors, respond with high sensitivity to all high energy particles, including charged species.

Electron multipliers function by a cascade effect in which the secondary electrons emitted when an energetic particle strikes a metal surface (such as copper–2% beryllium) are accelerated toward a second surface where more electrons are released, and so on through 10–20 stages. For maximum efficiency the secondary electron emission coefficient γ (the number of electrons released per incident ion) should be high. This number depends on surface properties, the kinetic energy of the ion, the mass of the ion, and the chemical nature of the ion. This last fact is of chemical significance since it may add considerably to the information usually obtained about a particular ion in the mass spectrometer. However, very little has been done in this connection and the experiments involved are rather difficult. The most that can be said at present is that γ can vary a great deal, being larger for polyatomic ions than for atomic ions.

The variation of electron multiplier response with ion mass, nature, and energy means that this device is intrinsically unsuitable for applications in which accurate ion abundances are sought. Such are its sensitivity (multiplication factors of 10^6 are normal, 10^7 quite common) and reliability, however, that these disadvantages are accepted and overcome by careful calibration and standardization.

Just as the primary events accompanying ion impact on a metal surface are still poorly understood, so a large number of events are involved in the ion-photoplate interaction. The chief process leading to reduction of silver ions, and hence image formation, is thought to involve liberation of conduction electrons in the solid on ion impact followed by electron capture with liberation of atomic silver. The dynamic range of the electron multiplier is far greater than that of the photoplate, but the latter has the advantage that the whole spectrum can be recorded simultaneously. Thus even though several hundred ions are required to give a perceptible photographic image in contrast to a single ion required in a high performance electron multiplier detector, a compensating gain in sensitivity (signal/noise) is achieved by the fact that no scanning is required. Parenthetically, response to a single ion allows an electron multiplier to be used as an *ion counter*, provided the sampling frequency is high enough to minimize the possibility of two ions reaching the detector in the same sampling interval.

Recent advances in electronics and data collection have made it possible to run very fast scans (10^{-1} sec per decade for magnetic instruments, 10^{-2} sec per decade for quadrupole instruments, and 10^{-4} sec per decade in time-of-flight instruments) or to hop back and forth between several peaks in very short times. Data retrieval from the photoplate tends to be more complex than from an electron multiplier system, the output of which is normally amplified and displayed using a fast oscillographic recorder. When complete high resolution spectra are required, however, the photoplate system again becomes competitive with the electron multiplier.

The technology behind ion detection is under vigorous development. Very small versions of the electron multiplier (channel multipliers) have appeared, and a unique arrangement now in use monitors the whole signal simultaneously rather than sweeping it past a slit. A rectilinear array of detector elements (multipliers) receives the signal, which is amplified in each multiplier to give electron beams that are allowed to impinge upon a phosphor, and the resulting visible signal is collected and digitized.

3-5 Mass Analysis

In this section the major methods of achieving mass analysis are discussed. Both the physical basis for each method and the properties of instruments based on the different methods are included.

Two properties of mass analyzers are of primary importance, (a) the transmission of the analyzer and (b) resolving power. As in other instrumentation, high signal and high resolution are mutually exclusive so the choice of mass analyzer will be dictated by the resolution requirements. The mass resolving power of a mass spectrometer is a measure of its ability to separate ions of different mass. An arbitrary definition of the term *separate peaks* must be established in terms of peak shapes. Several definitions are illustrated in Figure 3-6. The most widely accepted definition is the *10% valley*: two peaks of equal intensity are considered resolved when the valley between them is 10% of their (average) height. Since the separation of ions differing by one mass unit (i.e., the achievement of *unit resolution*) at low mass is a much easier task than at high mass, the definition of resolving power must take into account both mass m and mass difference Δm; hence resolving power $= m/\Delta m$.

In terms of practical usefulness there are certain thresholds that resolving power should reach in order to make worthwhile the trade-off against sensitivity. To illustrate: unit resolution up to mass 700 or so (i.e., resolving power 700) is generally required for obtaining low resolution spectra on organic compounds. If this resolving power is increased to say 5000,

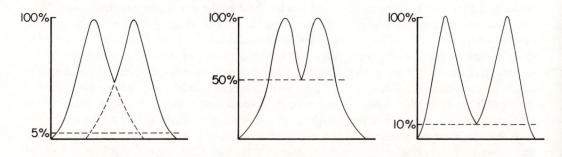

FIGURE 3-6. Two peaks representing ions of different masses that are just resolved using (a) 5% cross-contribution definition, (b) 50% valley definition, (c) 10% valley definition. The last is the preferred definition.

relatively little will be achieved since a resolving power in the vicinity of 10,000 is normally required in order to separate and identify the various ionic species that might constitute the signal at a given nominal mass in the spectrum of an organic compound. The next threshold in resolving power is reached at about 30,000-40,000 when $^{13}C/^{12}CH$ isotopic separations at m/e of about 100 become possible, and a final threshold occurs near 70,000, when $^{2}H/H_2$ assignments become possible. It must be emphasized that the primary interest is not in resolution per se but in accurate mass measurement, which is aided by good separation of the mass peaks. Deconvolution and multiple scan averaging of peaks that are not resolved on the basis of the above definitions may sometimes yield mass measurements of adequate accuracy. Indeed, the chief problem in using mass spectrometers of inadequate resolution for accurate mass measurement is not the stability or focusing of the ion beam but the possibility that multiplets will be taken for singlets, especially in the ions used as mass reference standards. Hence the position of a maximum in the overlapped peaks may be erroneously taken to correspond to one of the components when, in fact, it corresponds to neither.

An ion of mass m and charge e, which has been accelerated through a potential V and acquired a velocity v, will be transmitted by a sector magnet provided the force exerted on the ion at every point just balances the centrifugal force tending to drive the ion into the wall of the flight tube. The force exerted on the ion by a magnetic field of strength B[1] will be Bev, and hence we have eq. 3-1. The kinetic energy of the ion is given by eq. 3-2a.

$$Bev = \frac{mv^2}{r} \qquad\qquad 3\text{-}1$$

$$\frac{mv^2}{2} = eV \qquad\qquad 3\text{-}2a$$

Solution for v gives eq. 3-2b, which may be substituted into eq. 3-1 to give eq. 3-3. Rearrange-

$$v = \left(\frac{2eV}{m}\right)^{1/2} \qquad\qquad 3\text{-}2b$$

$$Be = \frac{m}{r}\left(\frac{2eV}{m}\right)^{1/2} \qquad\qquad 3\text{-}3$$

ment leads to eq. 3-4. Expressing V in volts, B in gauss, r in centimeters, m in atomic mass

$$\frac{m}{e} = \frac{B^2r^2}{2V} \qquad\qquad 3\text{-}4$$

units, and e in units of electronic charge, we obtain eq. 3-5.

$$\frac{m}{e} = \frac{B^2r^2}{20740V} \qquad\qquad 3\text{-}5$$

A magnetic sector acts as an ion momentum analyzer, as seen when eq. 3-1 is expressed

---

[1] See footnote 1, Chapter 1, Part One.

in the form of eq. 3-6. Thus, with an energetically homogeneous ion beam (section 3-2),

$$\frac{mv}{e} = Br \qquad\qquad 3\text{-}6$$

the magnetic sector can be used to analyze for the mass/charge ratio (m/e). Ions of different mass/charge ratio can be brought into focus by altering either the magnet field strength or the accelerating voltage. Because source conditions can be kept constant as the magnet strength is varied, this is the usual mode of operation. There are several reasons why mass spectra obtained using sector instruments are not recorded on precalibrated paper. The complicated relationship between magnet current and magnet field strength is affected by magnet hysteresis and saturation effects. In addition, m/e is proportional to B^2, not to B. Approximate position on the mass scale can be obtained by reading the magnet current, but a standard compound or, more commonly, specific known ions must be used to establish fixed positions on the mass scale. Because this procedure is somewhat inconvenient, the use of a mass marker, which depends on measurement of the field strength using a Hall probe (gaussmeter), has been adopted. If properly calibrated, these devices are accurate to a fraction of a mass unit.

The normal electron impact source yields an ion beam with a kinetic energy distribution that is sufficiently narrow to permit mass analysis at unit resolution through mass 500 or, in some instruments, considerably more. If an energy analyzer is not used, this then constitutes a single-focusing (low resolution) mass spectrometer. High resolution mass spectrometers are invariably sector instruments and are usually classified according to their ion path geometry. Two geometries that give both energy and direction focusing are commonly used: in the Nier–Johnson, there is only one point of double focus in the xy plane, and the collector or, more accurately, the collector slit is placed at this point. In the Mattauch–Herzog arrangement, there is a line along which the double focus is achieved. This fundamental difference has prompted the use of different ion detector systems in instruments of these geometries—the electron multiplier for the Nier–Johnson and the photoplate detector for the Mattauch–Herzog design.

Exact mass measurements using these two geometries are also accomplished by different means. A technique known as peak matching is used in Nier–Johnson instruments. The fractional increase in accelerating voltage required to bring an ion of unknown mass to a focus at the same point as an ion of reference mass is measured. In practice, the accelerating voltage is cycled between the two values and the peaks superimposed on an oscilloscope. The Mattauch–Herzog measurements are made by interpolation between two reference masses on the photoplate.

Methods of mass analysis that are proving increasingly valuable, especially in low resolution applications, are based on inducing cyclic motion in an ion. The two most important applications of this principle are the quadrupole mass filter and the crossed magnetic and electric fields employed in ion cyclotron resonance instruments. The quadrupole has a high ion transmission (~50% vs. 1% for sector instruments), as a result of the absence of narrow beam-defining slits. The ion drifts into an assembly of four parallel rods (Figure 3-7), which are charged by dc currents of opposite sign and are subjected to an oscillating rf field, also applied to opposite rods, that successively reinforces and then overwhelms the dc field. An ion moving in the z direction down the array experiences a force only in the plane normal to the rod length. Provided its oscillations in this plane are stable, the ion will drift down the rod assembly and reach the collector. Stable oscillations are only achieved by ions of given mass/charge ratio for a given rod assembly, oscillating frequency, rf voltage, and dc voltage.

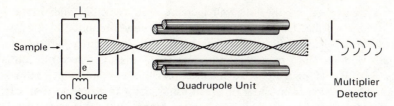

Sample →

e⁻

Ion Source

Quadrupole Unit

Multiplier
Detector

FIGURE 3-7. Quadrupole mass spectrometer. [Courtesy of Finnigan Corporation.]

Mass scanning is usually achieved by sweeping the dc and rf voltages, keeping their ratio and the oscillator frequency constant.

The ease with which electric currents (as opposed to magnetic fields) can be controlled is an important advantage of this instrument. The mass scale is linear, and after one calibration any mass can be dialed in directly during a working day. Computer control and data collection are much simplified. In addition, the absence of very high voltages simplifies the instrumentation and increases the reliability. Accurate construction of the rod assembly is the one factor critical to the success of the mass filter. Although a resolution of 20,000 has been reported with a quadrupole mass spectrometer, the high resolution condition requires extreme stabilization and control of the electrode voltages.

A disadvantage of this method of mass analysis is that metastable ion fragmentations cannot be followed, and breakdown schemes therefore are not subject to direct verification. This drawback is not important when mixtures of known compounds or of compounds of known types are being examined, as is often the case in gas chromatography–mass spectrometry (gc–ms; see section 3-6), which is perhaps the most successful way in which the quadrupole instrument has been applied. Another difference between sector and quadrupole instruments is that the transmission times are rather longer in the latter case, a difference shared by the other major mass analysis method in this group—ion cyclotron resonance.

Ion cyclotron resonance (icr) allows measurements other than mass analysis; its use in the study of ion-molecule reactions will be emphasized in section 5-7. The principles of ion cyclotron resonance can be derived from a consideration of the forces experienced by ions in magnetic and electric fields (eqs. 3-1 to 3-5). An ion of mass m and velocity v, in the plane normal to a magnetic field of strength B, experiences a force F ($= Bev$) normal to both the field direction and to its direction of motion (eq. 3-7). Hence the ion will describe a circle

$$\frac{mv^2}{r} = Bev \qquad\qquad\qquad 3\text{-}7$$

of radius r in the plane normal to the field, so that the cyclotron frequency is given by eq. 3-8.

$$\omega_{\mathrm{c}} = \frac{v}{r} = \frac{Be}{m} \qquad\qquad\qquad 3\text{-}8$$

Thus every ion has a characteristic cyclotron frequency that is inversely proportional to its mass/charge ratio and directly proportional to the applied magnetic field. While ω_{c} is independent of the velocity of the ion, the radius of the orbit is directly proportional to ion velocity.

If an oscillating electric field, angular frequency ω, is applied normal to B, then when

$\omega = \omega_c$, energy absorbed by the ions will increase their velocities and orbital radii. The absorption of energy will be signaled by a change in the power required by the oscillator. In order to achieve mass analysis the oscillator frequency is conveniently kept constant and the magnet strength varied. Since m/e is proportional to B, the mass spectrum obtained has a linear mass scale. If the pressure is such as to preclude ion-molecule reactions, the mass spectra do not differ much from those obtained using sector instruments. Mass resolution is typically satisfactory, but a limitation is the mass range (200–300) commonly attainable. Instruments capable of reaching higher mass are now appearing, and with the application of Fourier transform techniques improved resolution is also possible.

As in nuclear magnetic resonance, to which icr bears some resemblance, double resonance phenomena can be studied by irradiating at one frequency and observing at another frequency. If any signal is observed, it must be transmitted via interaction of the ionic species. Thus a mechanism exists for demonstrating that two ions are related by an ion-molecule reaction. In the most common case, irradiation at the cyclotron frequency of the reactant ion is used to increase its kinetic energy. The concentration of the product ion is thereby altered, with the result that the signal observed on irradiation at the resonance frequency of the product ion is changed.

Using special modifications, it is possible to trap ions in the icr cell for times of the order of seconds. Similar techniques allow ions to be trapped in the sources of quadrupole mass spectrometers. The extension of the mass spectrometer time scale in this way is of immense value in the study of both unimolecular and ion-molecule reactions.

Another method of mass analysis is based on ion flight time. The principle that ions of identical energy can be separated according to mass on the basis of differences in flight times has been applied with surprising success, considering the short times involved. Ions of mass m and charge e, which fall through a potential V and acquire a velocity v, will traverse the distance d from ion source to collector in time t (eq. 3-9). The ion kinetic energy expression (eq. 3-10)

$$t = \frac{d}{v} \qquad\qquad\qquad 3\text{-}9$$

$$\frac{mv^2}{2} = eV \qquad\qquad\qquad 3\text{-}10$$

gives the velocity (eq. 3-11), and substitution gives eq. 3-12 for the flight time. Hence, at a

$$v = \left(\frac{2\,eV}{m}\right)^{1/2} \qquad\qquad\qquad 3\text{-}11$$

$$t = \frac{dm^{1/2}}{(2\,eV)^{1/2}} \qquad\qquad\qquad 3\text{-}12$$

given ion-accelerating voltage, the flight time t is proportional to $(m/e)^{1/2}$.

In order to avoid the simultaneous collection of ions of different masses formed at different initial times, it is necessary to pulse the ion source and to employ a very rapid detection system. This requirement adds considerably to the complexity and cost of the instrument, which has only a limited mass range. For these reasons the time of flight does not provide an ideal general purpose method, although it has particular advantages in kinetic studies. A specialized application of this method of mass analysis is made in ionization employing high energy fission fragments. One fragment causes ionization, while the other serves to fix the exact time of ionization and hence allows accurate measurements of ion flight times.

3-6 Sample Inlet Systems

In the methods of primary interest, the sample, if not a gas, is converted to the vapor phase by heating to temperatures up to 400°. Alternative procedures include the (titanium) Knudsen cell used in high temperature studies, the methods used in spark source instruments where the sample is compacted to form a part of the electrode, and thermal emission methods by which ions are evaporated from a sample by direct thermolysis.

The ideal inlet system accepts samples in any physical form, including dilute solutions, and permits a small but constant amount of sample to be injected continuously into the ion source. This latter property necessitates transition from higher pressures to approximately 10^{-5} torr. As a result, the simplest inlet system consists of a flask that is filled to $\sim 10^{-2}$ torr with sample and connected to the ion source by a molecular leak, either ceramic or a finely perforated metal foil. Sample vapor is introduced into the reservoir flask via a system of glass or stainless steel valves. Volatile solid and liquid samples are introduced into the vacuum system and frozen down. The system is evacuated before the samples are vaporized. Liquids and solids can also be introduced directly into the reservoir by a very rapid method in which a Teflon plug containing the sample is introduced by displacement of a blank plug, as illustrated in Figure 3-8.

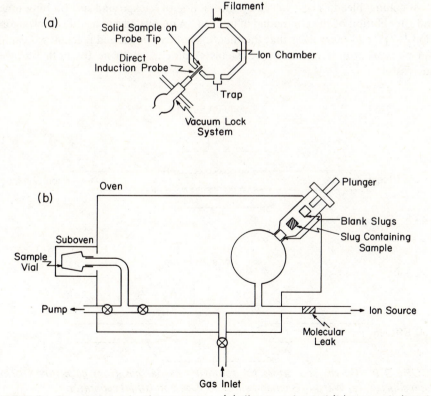

FIGURE 3-8. Sample introduction systems: (a) direct probe and (b) reservoir (batch) inlet.

For high-melting solids (generally mp $>100°$) the preferable method is the direct introduction (direct probe) system. This system (Figure 3–8) involves introduction of a small amount of sample on a probe tip directly into the ion chamber via a vacuum lock system. The probe temperature as well as the chamber temperature can normally be controlled so as to achieve a slow, steady evaporation rate. This method not only is suitable for samples of very low vapor pressure but also is particularly valuable for unstable solids and liquids, since probes can also operate below ambient temperatures. It is possible for a spectrum obtained using the reservoir inlet system to be complicated by the occurrence of thermal reactions in addition to electron-impact-induced reactions, with consequent confusion of the molecular ion. The problem of thermal reaction in the inlet system has, however, been considerably reduced by the development of all-glass and all-stainless-steel introduction systems. The direct introduction method is the only procedure applicable to very involatile solids, and it is by far the most sensitive sample introduction method since all of the sample reaches the ion chamber. Its chief disadvantage is that the sample evaporation rate, and hence the ion current, cannot be precisely controlled. This disadvantage can become significant in such applications as the determination of ionization efficiency curves (and hence ionization potentials).

The only way to introduce extremely dilute solutions is by coupling a gas chromatograph to the mass spectrometer, an arrangement that is also very valuable for the analysis of complex mixtures. Removal of carrier gas and carryover of column bleed into the mass spectrometer are the two major problems in gc–ms. The former is solved by a variety of separator devices that selectively enrich the eluant in sample and remove the carrier gas. Because this separation is never complete, the sensitivity of gc–ms to the sample is lowered. Two types of separators are shown in Figure 3–9. If capillary columns are used, separators are unnecessary. The other problem, that of column bleed, is best solved by subtraction of background spectra run immediately before and after elution of the compound of interest. A unique advantage of chemical ionization (see section 1–2) for gc–ms work is that the functions of carrier gas and reactant gas can be combined and no separator is necessary. Systems employing methane for both purposes have proved successful.

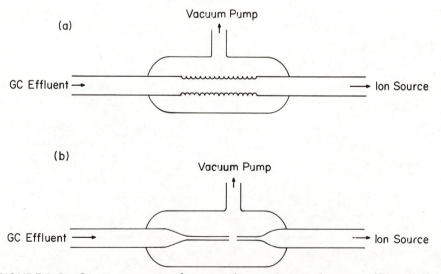

FIGURE 3–9. *Gc–ms separators for removing carrier gas from gc effluent prior to ion source: (a) Watson–Biemann sintered type and (b) jet separator.*

Some final comments on inlet systems are in order. All types feed into a metal ion source, the walls of which may absorb previously run compounds. Background scans are always taken, and if background peaks are too large, the source is baked to remove the contaminant. It may be noted that sometimes the compound of interest can itself be so strongly absorbed onto the source walls as to displace material not seen in the normal background scan in the absence of sample. Such cases are hard to predict, although strongly basic compounds appear to be prime offenders. This phenomenon can be missed if the associated change in spectrum with time is not seen.

3-7 Vacuum System

Probably a chief reason why many organic chemists still think of mass spectrometers as complex but nmr spectrometers as simple is the presence of the high vacuum system in the former. Here just a few lines are taken to note some basics. Mass spectrometer vacuum systems can be divided into two sections: a region that experiences a high gas load (the ion source and inlet region) and a region that does not (the analyzer region). To prevent sample from entering the analyzer there is a narrow orifice between the two regions, which are separately (differentially) pumped. The ultimate pressure in both regions in most analytical instruments is about 10^{-7} torr. This pressure is usually read using an ionization gauge (Bayard–Alpert gauge, Figure 3-10), which operates on principles that are familiar from our discussion of the ion source. Electrons are emitted from the tungsten filament and accelerated toward the anode or grid, which is about a hundred volts positive with respect to the filament. The energetic electrons collide with and ionize gas molecules. Ions formed inside the grid move toward the most negative region of the gauge, the collector, and the resulting ion current is a direct measure of the gas pressure for a given gas and electron emission current. The Bayard–Alpert ion gauge can be used to read pressures in the range 10^{-4} to 10^{-9} torr. It is suitable, therefore, for the high vacuum requirements of mass spectrometry, but a different device must also be incorporated into mass spectrometers to monitor higher pressures, including those in rotary pump lines. This measurement is frequently made with a thermocouple gauge. The temperature of a thermocouple junction in vacuum depends on the rate of heat loss from the metal to gas molecules and hence the gas pressure.

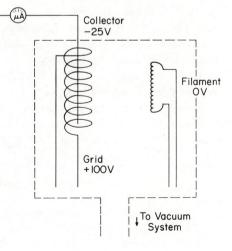

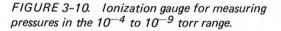

FIGURE 3-10. Ionization gauge for measuring pressures in the 10^{-4} to 10^{-9} torr range.

The vacuum needed in mass spectrometry can conveniently be obtained using ion pumps, mercury diffusion pumps, or oil diffusion pumps. All these methods require that pressures in the low micron range be attained by mechanical (rotary) pumps, and diffusion pumps require continuous backing by rotary pumps. Oil diffusion pumps are most commonly used in mass spectrometers, the oil being either a high-boiling hydrocarbon or polyphenyl ether. Diffusion pump oil vapor will to a very slight extent be present in the mass spectrometer and will contribute to the background. With cryogenic traps this contamination is negligible. A final aspect of vacuum technique is the practice of baking the system at several hundred degrees after breaking vacuum in order to avoid the slow outgassing of water vapor that would otherwise occur over a long period of time and severely limit the attainable vacuum.

3-8 Obtaining a Mass Spectrum

In many circumstances an instrument operator rather than the chemist will obtain mass spectra. A communication gap can arise unless the chemist is careful to provide certain items of information. First, the volatility of the sample (vapor pressure, or more frequently, melting point if a solid and boiling point if a liquid) will determine the type of inlet that can be used. On the other hand, use of the direct probe may be required even for relatively volatile samples if these are thermally unstable. Samples that are air or moisture sensitive can usually be run using the reservoir or probe inlet systems; a polyethylene glove bag usually suffices, but in extreme cases a break-seal container may be used with the reservoir inlet. Frequently, the chemist has some idea of the expected molecular weight, and this information assists the operator in deciding whether low energy scans are necessary. Particular structural features, such as the presence of alkyl bromide moieties, may make the observation of a molecular ion unlikely and if this information is available alternatives such as an $(M - Br)^+$ ion can be sought. Experienced operators will usually do a preliminary interpretation of the spectrum, and the more information they are provided by the chemist, the more they can assist in interpretation.

The form in which the mass spectrum is received will vary in different laboratories, but it is always important that the chemist examine the actual analog output. This chart contains information that may be lost as the data are further processed; this is particularly true of metastable peaks lying near integral mass peaks. Processed data will usually take the form of a list of masses vs. relative abundances, and in such cases a machine-plotted bar graph is often readily obtainable. Computer-processed high resolution spectra are described elsewhere (see section 4-2). However the mass spectrometric information is received, the process of interpretation remains largely a manual one, although automatic comparison of recorded spectra with libraries of thousands of mass spectra is becoming more common and the use of data-processing methods to identify all the reactions for which metastable peaks occur (in order to assist in molecular ion assignment) is now feasible. The use of computer techniques in specialized analyses, such as protein sequencing, also has made headway. In this type of approach specific groups of atoms are of interest, and their unique masses constitute the essential data needed to allow automated interpretation of spectra.

BIBLIOGRAPHY

1. General Treatments of Instrumentation: J. H. Beynon, *Mass Spectrometry and Its Applications to Organic Chemistry*, Elsevier, Amsterdam, 1960; R. W. Kiser, *Introduction to Mass Spectrometry and Its Applications*, Prentice-Hall, Englewood Cliffs, N.J., 1965, J. Roboz,

Introduction to Mass Spectrometry: Instrumentation and Techniques, Interscience, New York, 1968.

2. Electron Impact Ion Sources: A. O. C. Nier, *Rev. Sci. Instrum.*, **11**, 212 (1940); A. O. C. Nier, E. P. Ney, and M. G. Inghram, *Rev. Sci. Instrum.*, **18**, 294 (1947); R. M. Elliott in C. A. McDowell, ed., *Mass Spectrometry*, McGraw-Hill, New York, 1963, ch. 4.

3. Ion Optics: L. Kerwin in C. A. McDowell, ed., *Mass Spectrometry*, McGraw-Hill, New York, 1963, ch. 5; R. F. Herzog in A. J. Ahearn, ed., *Trace Analysis by Mass Spectrometry*, Academic, New York, 1972, ch. 3.

4. Ion Detection: D. M. Desiderio, Jr., in G. W. A. Milne, ed., *Mass Spectrometry: Techniques and Applications*, Wiley–Interscience, New York, 1971; C. La Lau in A. L. Burlingame, ed., *Topics in Organic Mass Spectrometry*, Wiley–Interscience, New York, 1970.

5. Vacuum Techniques: S. Dushman, *Scientific Foundations of Vacuum Technique*, 2nd ed., Wiley, New York, 1962.

4

PROBLEM SOLVING WITH THE MASS SPECTROMETER

4-1 Molecular Weight Determination

The single most valuable item of information that the mass spectrometer can provide is the molecular weight of a compound. The rapidity and accuracy of the method (nearest mass number for low resolution spectrometers) make it the preferred procedure for this determination. Not all compounds give molecular ions when the usual method of ionization, electron impact is employed. In addition, the usual requirement that a compound be vaporized before a mass spectrum can be obtained means that thermal degradation may preclude the observation of a molecular ion. This situation is met in relatively involatile compounds such as the sugars. The usual way to avoid the problem of facile mass spectral fragmentation and thermal degradation and to increase sample volatility is to derivatize the sample; hence the widespread use of acetylated, trimethylsilylated, and otherwise modified sugars, nucleosides, etc. These extra chemical transformations can be avoided by using chemical ionization (CI) or field ionization (FI) instead of electron impact (EI).

A practical problem of importance is the recognition of the molecular ion as such when it is present. Several tests can be used to check whether an ion may be the molecular ion:

(a) On lowering the electron energy to the region of the ionization potential, the molecular ion will be enhanced relative to fragment ions (compare Figure 1–5). This procedure forms the basis for the well-known test whereby the increase in the relative peak abundance on lowering E_e is used to recognize a molecular ion. It is important to remember that the increase in molecular ion abundance is relative. Rarely is it absolute.

(b) The assignment can be rapidly checked for its consistency with the fragment ions in the spectrum. For example, a fragment ion 7 mass units below the proposed molecular ion is impossible and points either to the presence of an impurity or to an incorrect molecular ion assignment.

438

(c) At least some of the metastable peaks in the high mass region of the spectrum are normally due to fragmentations of the molecular ion. Cases in which the molecular ion fails to exhibit metastable ion fragmentations of at least moderate abundance are rare. Occasionally, even when the molecular ion is absent, a metastable peak corresponding to its fragmentation may be present. This circumstance arises if the molecular ions fragment so fast that none survive the journey to the collector, but some survive long enough to fragment in the field free region.

(d) If the compound has an odd number of nitrogen atoms, its molecular weight will be an odd number.

We turn now to the use of "soft" techniques of ionization instead of electron impact. These are techniques in which the amount of excitation accompanying ionization is limited. The basic ionization processes have already been indicated in Chapter 1. *Chemical ionization* gives ions of low internal energy, but it still requires that the sample be vaporized. The standard *field ionization* technique also has this requirement, but a newer modification, *field desorption*, does not.

Chemical ionization spectra are generally characterized by a lower abundance of fragment ions than electron impact produces; moreover, the extent of fragmentation can be controlled by varying the ionizing gas. This is an important advantage since it means that attention can be focused on the question of molecular weight using an ionizing species that transfers very little energy, while other structural features can be explored by studying the fragmentation pattern obtained using a more reactive ionizing agent. The usual reactant ion is a protonated alkane ion RH_2^+, and the proton affinity of RH controls the energy transferred to the sample M in the reaction $RH_2^+ + M \rightarrow RH + MH^+$.

A disadvantage of CI is that proton transfer is by no means the only reaction occurring when a mixture of sample and reactive gas is subjected to electron bombardment at high pressure. Methane, a common reactant gas, gives CH_5^+ ions that are involved in proton transfer, $C_2H_5^+$ ions that abstract H^- and so form $(M - 1)^+$ ions, and other ionizing species. Commonly, therefore, several species will be formed in primary ionization reactions involving the sample, and the presence of $(2M + 1)^+$ and related ions indicates that further ion-molecule reactions are also possible. The complex mixture of bimolecular and unimolecular reactions that produces a CI spectrum is the chief drawback of the technique. The $(M + 1)^+$ ion, however, is almost always the most abundant ion in the high mass region of the spectrum, and the use of CI in molecular weight determination is straightforward. Figure 4-1 shows a case in which the molecular ion could barely be observed by EI, but the $(M + 1)^+$ species observed by the CI technique allowed molecular weight determination. This example also illustrates the use of mass spectrometry in peptide sequencing. Sequential fragmentation of the carbonyl bonds occurs from both the C-terminus and the N-terminus to give acylium ions and immonium ions, respectively. The appropriate cleavages are indicated in Figure 4-1.

Another example of the usefulness of chemical ionization in providing the molecular weight of a compound is shown in Figure 4-2. In this case the EI spectrum is the more valuable since, in addition to indicating the molecular weight, it allows deduction of the major structural features through its fragmentation pattern. (The losses of 15 and 43 mass units are readily related to the presence of the acetyl group.)

Field ionization has found important applications in the theory of mass spectra (see section 5-2) because the time scale involved is so much shorter than that for other techniques. This feature also contributes to less fragmentation and hence to more abundant molecular ions. The limited amount of internal energy transfer that accompanies field ionization also operates to increase molecular ion abundance. Thus FI is both a fast and a "soft" ionization technique.

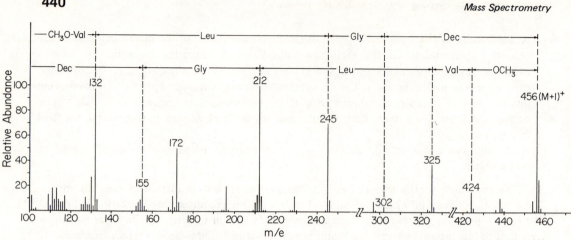

FIGURE 4-1. Chemical ionization mass spectrum (methane as reagent gas) of a tetrapeptide, showing an abundant (M + 1)⁺ ion and fragment ions from which the amino acid sequence can be deduced. Only the region above m/e 100 is shown. [Reproduced with the permission of Heyden and Son from A. A. Kiryushkin, H. M. Fales, T. Axenrod, E. J. Gilbert, and G. W. A. Milne, Org. Mass Spectrom., 5, 19 (1971).]

Figure 4-3 illustrates the application of field ionization to a problem in which EI was less suitable. It is particularly noteworthy that in spite of the high molecular weight and involatility of digitoxin, a good FI spectrum, which includes carbohydrate sequence information, is obtained.

Almost all fragment ions observed in a FI spectrum arise by simple cleavage, since the available time of $<10^{-10}$ sec is insufficient for rearrangements or eliminations, which require that the ion adopt some particular conformation. As in CI, there are a multitude of processes (usually minor) that can give rise to products in the FI spectrum. High local pressures of polar species at the emitter tip, for example, result in the formation of protonated or solvated ions.

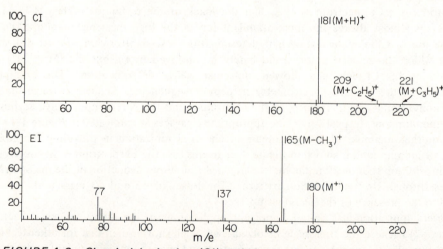

FIGURE 4-2. Chemical ionization (CI) and electron impact (EI) mass spectra of 3,4-dimethoxyacetophenone. Note the presence of low abundance (M + 29)⁺ and (M + 41)⁺ ions in the CI spectrum. [Courtesy of Finnigan Corporation.]

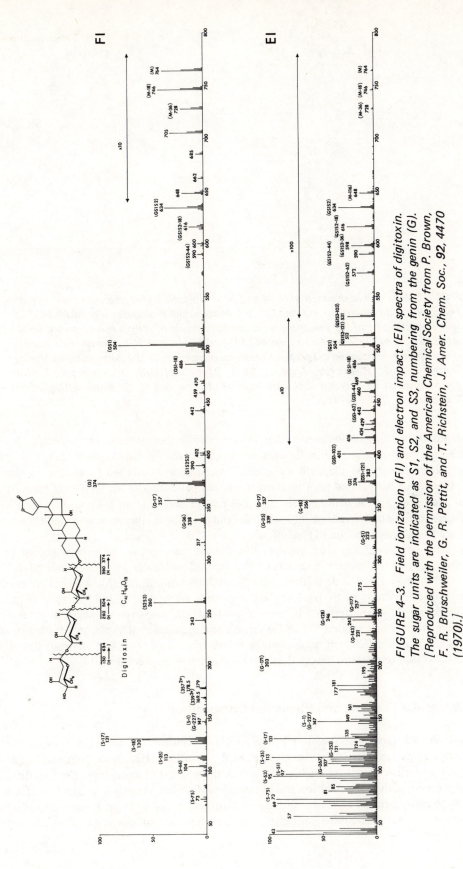

FIGURE 4-3. Field ionization (FI) and electron impact (EI) spectra of digitoxin. The sugar units are indicated as S1, S2, and S3, numbering from the genin (G). [Reproduced with the permission of the American Chemical Society from P. Brown, F. R. Bruschweiler, G. R. Pettit, and T. Richstein, J. Amer. Chem. Soc., 92, 4470 (1970).]

441

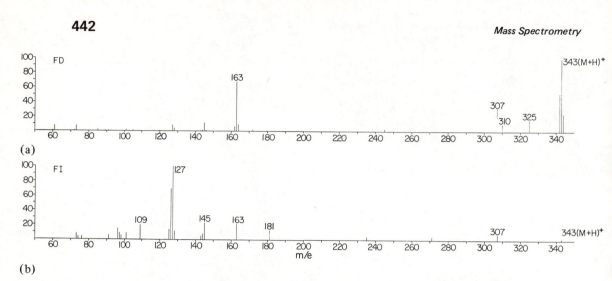

FIGURE 4-4.   *Field desorption (FD) and field ionization (FI) spectra of cellobiose. The greatly increased abundance of the protonated molecular ion (m/e 343) is characteristic of the FD spectrum, as is the occurrence of only a few fragment ions: m/e 325 (loss of H_2O), m/e 310 (loss of CH_3 and H_2O), m/e 307 (loss of H_2O twice), and m/e 163 (glycosidic cleavage). [(a) reproduced with the permission of Verlag Chemie from H. B. Beckey, Angew. Chem., 81, 662 (1969). (b) reproduced with the permission of Heyden and Son from H. Krone and H. D. Beckey, Org. Mass Spectrom., 5, 983 (1971).]*

A variation of field ionization that shows considerable promise for molecular weight determination on involatile, thermally unstable species is the field desorption technique. The solid is placed directly on the field ion emitter, and ions are desorbed by the strong electric field. The source is operated slightly above room temperature and the ions have virtually no internal energy, so predominantly molecular ions are formed. Figure 4-4 contrasts the FI and FD spectra of a nonderivatized disaccharide and highlights the increased molecular ion abundance characteristic of the FD technique.

It may be noted also that negative ion spectra have occasionally been shown to be superior to positive ion spectra in structural analysis and molecular weight determination. For example, riboflavin and its derivatives have been found unsuitable for analysis by normal electron impact methods partly because the compounds are very labile and ions of mass greater than the molecular weight appear. By way of contrast, the negative ion spectrum obtained at high source pressure (see section 5-1) is far simpler and includes a molecular ion of high relative abundance. The simplicity of such spectra belies the complexity and multiplicity of the processes that lead to negative ions in the mass spectrometer (see section 5-1).

4-2   Molecular Formula and Elemental Composition

Once a molecular ion is observed in a low resolution spectrum, the elemental formula can be determined by an exact mass measurement at high resolution. High resolution measurements can be done whether EI, CI, or FI is employed as the method of ionization. Frequently, only the composition of the molecular ion and of several of the more abundant fragment ions will be necessary to allow solution of the structural problem; in such a case the peak-matching technique (see section 3-5) would be most convenient. If the sample is available in very limited amounts and mass spectrometry is the only suitable method of structural analysis, the problem would justify a complete high resolution spectrum, either by electrical detection using computer-assisted data acquisition or by the photoplate method. Exact mass measurements frequently can be made to within 1 or 2 millimass units. Even with this accuracy, however, a unique fit is seldom obtained. The number of possibilities increases with the number of elements that may be present, with the number of atoms of each element possible, and with the molecular weight. For organic compounds that have only the common "heteroatoms" (O, N, F, Cl, Br, I, S) and molecular weights less than 500, only a few possibilities need be considered, and even cursory examination of the mass spectrum further reduces the number of possibilities.

To illustrate the application of complete high resolution measurements to the mass spectrum of a natural product available in small amounts, a portion of the spectrum of the ergot alkaloid dihydroelmyoclavine ($C_{16}H_{20}N_2O$) is reproduced in Table 4-1. This printout

Table 4-1. *Part of the High Resolution Mass Spectrum of an Alkaloid*

No.	Density	Distance	Measured	Calculated	Error (× 10³)	Elemental Composition(s)
126	−4.4	53.18786	235.98661	235.98722	−0.61	C7 F8
127	4.2	53.69972	237.13882	237.13917	−0.35	C16 H17 N2
128	3.2	54.14631	238.14644	238.14700	−0.55	C16 H18 N2
129	3.5	55.46508	241.13467	241.13409	0.59	C15 H17 N2 0
130	8.2	55.90821	242.14304	242.14191	1.13	C15 H18 N2 0
131	−7.6	56.27777	242.98564	242.98562	0.02	C6 F9
132	4.2	56.34733	243.14447	243.14527	−0.80	C14 (13C) H18 N2 0
133	−5.6	61.469.52	254.98564	254.98562	0.02	C7 F9
134	4.1	61.96979	256.15793	256.15756	0.37	C16 H20 N2 0

is obtained from a program that calculates peak centers in terms of atomic mass units and converts to elemental compositions listing all within a specified range (±3 millimass units). Peak positions are automatically measured from the photographic plate in terms of distance from a recognized peak in an internal standard (m/e 70 in perfluorokerosine). The mass scale is automatically calibrated using other perfluorokerosine peaks.

The first column is a number assigned to designate each peak. The second column, density, represents the abundance of the ion, a negative sign being assigned to ions from the perfluorokerosine standard. The distance given in column three is measured in millimeters from m/e 70 ($^{13}CF_3$) in perfluorokerosine. The next columns give the measured exact mass, the mass calculated from the formula shown, and the error in millimass units. The measured exact mass is seen to be accurate to 0.4 millimass unit for the molecular ion m/e 256. The presence of an ion $C_{15}H_{18}N_2O$ at m/e 242 shows that some of the demethyl homologue accompanies the alkaloid. The presence of a doublet at m/e 243 (lines 131 and 132) should be noted. The fragment ions due to methyl loss and H_2O loss, at m/e 241 and m/e 238, respectively, have the expected elemental compositions.

Even without high resolution equipment it is often possible to arrive at a molecular formula or at worst a few possible formulas before detailed interpretation of the low resolution mass spectrum. This information comes from analysis of isotopic clusters. This technique, if it is to be useful and not misleading, must be applied with some caution. The scan must be made slowly enough and with a wide enough collector slit that peak heights are proportional to ion abundances (see section 3-3). In addition, it is impossible to exclude completely the possibility of ion-molecule reactions contributing to the $(M + 1)^+$ species, so this abundance

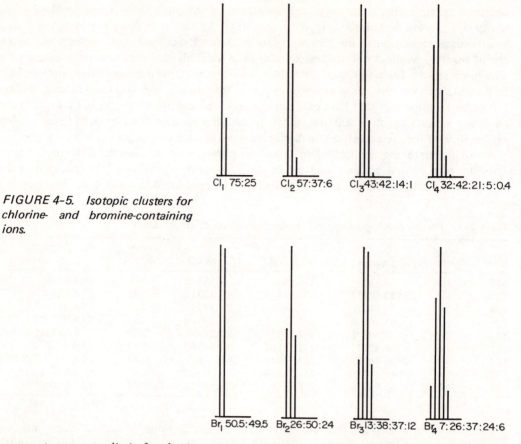

FIGURE 4-5. Isotopic clusters for chlorine- and bromine-containing ions.

represents an upper limit for the isotopic contributions. Finally the presence of $(M - H)^+$ ions can lead to confusion, and their removal by operating at low electron energies is sometimes desirable.

At this point the problem divides into two parts: (a) recognizing the presence and number of those atoms that have more than one isotope with significant (say >10%) abundance and (b) attempting the same analysis for species in which one isotope is dominant. The first problem is usually trivial when one is dealing with organic compounds or with organometallics if the metal(s) is specified. One tests the observed isotope cluster against calculated clusters for a given number of atoms of a given element. The most frequently met cases, Cl_x and Br_x $(x \geq 1)$, can be solved by inspection for small x (Figure 4–5). Note that the lines in the figure are separated by *two* mass units.

To illustrate the application of this method, consider the case of a chlorine-containing organic compound known not to contain any elements other than C, H, N, O, and Cl. To decide upon the number of chlorine atoms present it is only necessary to consider the $M^{+\cdot}$, $(M + 2)^+$, $(M + 4)^+$, . . . ,ions. The ratios (1 scan) found for this sequence were 53/100/78/33/8/1. The calculated ratios for Cl_6 are 51/100/82/35/9/1.

This type of calculation is based on simple statistics, as in eq. 4–1, where a_j is the number

$$\frac{[(M + 2)^+]}{[M^{+\cdot}]} = \sum_{j=1}^{n} a_j R_j \qquad 4\text{--}1$$

of atoms of a given element and R is the ratio of abundances of the isotopes leading to $(M + 2)^+$ and $M^{+\cdot}$ for each element.

The isotopic patterns for boron, sulfur, and silicon are also sufficiently distinctive to allow their ready determination in organic compounds, although boron and sulfur show quite large natural variations in isotopic content:

$$^{10}B/^{11}B \; = \; 19.6/80.4$$

$$^{28}Si/^{29}Si/^{30}Si \; = \; 92.2/4.7/3.1$$

$$^{32}S/^{33}S/^{34}S \; = \; 95/0.8/4.2$$

Iodine, phosphorus, and fluorine are monoisotopic and often can be recognized by the fact that the $(M + 1)^+$ ion has a much *lower* abundance relative to the molecular ion than can be explained on the basis of the presence of C, H, N, and O alone. Such a situation can be illustrated by the case of an unknown compound, molecular weight 322, which showed a $[(M + 1)^+]/[M^{+\cdot}]$ ion abundance ratio of only 17%. Since the natural abundance of ^{13}C is 1.1% of ^{12}C, the number of carbon atoms is $\leqslant 15$ and, as a maximum, the CH content can only account for 212 ($C_{15}H_{32}$) of the 322 mass units. A 6% $(M + 2)^+$ ion indicates 1 sulfur atom, but this still leaves $\geqslant 78$ mass units to be accounted for by heteroatoms. Such a situation suggests the possible presence of fluorine, iodine, or phosphorus atoms, and indeed in this case the actual molecular formula was $C_{14}H_8F_6S$.

The most difficult species to determine accurately by the isotope abundance procedure are the common elements C, H, O, and N. Since the presence of a single nitrogen atom can usually be recognized from the odd mass molecular weight, and that of two or more nitrogen atoms will usually be indicated by other than mass spectrometric evidence, the problem reduces most often to deciding between C and O. It has been suggested that O be recognized by the abundance of the $(M + 2)^+$ ion, but the method is not very satisfactory. It is preferable to estimate the upper limit of the number of carbon atoms from eq. 4-2 and to consider also

$$\frac{[(M + 1)^+]}{[M^{+\cdot}]} = 0.011(\text{no. of C atoms}) + 0.004(\text{no. of N atoms}) \qquad 4\text{-}2$$

the possibilities where one or two carbons are replaced by oxygen.

For example, a compound of molecular weight 162 showed an $M^{+\cdot}/(M + 1)^+/(M + 2)^+$ ratio of 100/13.0/1.2. The maximum possible number of carbon atoms is 11 since this figure requires $(M + 1)^+ \geqslant 12.1\%$; formulas with 10 carbon atoms, $(M + 1)^+ \geqslant 11.0\%$, should also be considered. Hence the probable molecular formulas are $C_{11}H_{14}O$, $C_{10}H_{14}N_2$, and $C_{10}H_{10}O_2$. The compound, in fact, was $C_{11}H_{14}O$, and the $(M + 2)^+$ ion abundance would have been of no value in indicating the number of oxygen atoms.

For organometallics the isotope cluster method can even become competitive with exact mass measurements for determining elemental compositions. As an illustration of the measure of agreement that can be expected, the molecular ion of triferrocenylborane ($C_{30}H_{27}BFe_3$) gave the calculated and observed isotope patterns shown in Table 4-2.

The equations on which these calculations are based are generalizations of those already given and can be found in many texts (e.g., Beynon's, reference 1, Chapter 3). Computer programs for these calculations are available.

Table 4–2. *Elemental Composition by the Isotope Cluster Method*
$$C_{30}H_{27}BFe_3$$

Mass, m/e	Relative Abundance (%)	
	Calculated	Observed
562	1.1	1.1
563	4.7	4.4
564	18.8	19.0
565	29.6	28.5
566	100	100
567	39.2	39.2
568	8.7	7.6
569	1.3	1.1

4–3 Isotopic Incorporation

The preceding discussion leads us naturally to the problem of determining the extent of incorporation of stable isotopes into molecules. Although recent improvements in nmr sensitivities have made the nmr method competitive in the case of some nuclei, mass spectrometry is still the standard technique for these measurements. In the following discussion, emphasis is placed on determination of the *extent* of incorporation, but it should be noted that the *site* of incorporation also can be determined from a knowledge of the fragmentation pattern of the compound. To be sure, any rearrangement of the label in the molecular ion will complicate this latter application, so it is fortunate that the types of molecules and the reaction conditions that promote label scrambling are becoming increasingly well understood (see section 5–3).

The most common stable labels are ^{2}H, ^{15}N, ^{13}C, and ^{18}O. The mass spectrometric method of determining isotopic incorporation has the following requirements.

(a) That the molecular ion (or fragment ion formed without label loss) be of reasonable abundance (a few percent of total ion current may suffice).

(b) That ion-molecule reactions leading to $(M + 1)^+$ be negligible or at least can be controlled so that they occur to a constant extent when the labeled and unlabeled samples are analyzed.

(c) That it be possible to correct for any $(M - 1)^+$ signal or, better still, that this process be removed by operation at low electron energy. This last requirement is less significant for ^{18}O determinations.

The analytical method involves obtaining the spectrum of the labeled compound and of the unlabeled analogue under as nearly identical conditions as possible. Since ion abundances are required, multiple slow scans are taken using wide collector slits. The spectrum of the unlabeled compound will include natural isotopic species, ion-molecule reaction products, and $(M - H)^+$ and $(M - H_2)^{+\cdot}$ ions. If these last can be completely removed, the determination becomes trivial; if not, the answer is still often accurate to 1% (if the enrichment is high). Deuterium incorporation determinations may have appreciable uncertainties if $(M - H)^+$ processes cannot be removed, since (1) accurate correction for primary isotope effects cannot be made and (2) the site of origin of the protium atom in the unlabeled ion is not always known.

We illustrate with two examples, one relatively simple, the other more complex.

Exchange of catechol in 50% NaOD–D_2O at 80° and subsequent methylation give 1,3-dimethoxybenzene-2,4,6-d_3. The problem is to determine the isotopic purity of this compound. The molecular ion is the base peak in the mass spectrum and the unlabeled compound shows no $(M - 1)^+$ ion at low ionizing energy. The molecular ion regions of the unlabeled (d_0) and labeled (d_3) compounds are

m/e	138	139	140	141	142	143
d_0	100	9.6	0.2			
d_3		0.5	2.6	100	9.6	0.2

Assigning compositions to the ions in the d_3 spectrum, starting from low mass and subtracting heavy isotope and protonated species in the ratio found for the d_0 compound, we have

	m/e	139	140	141	142	143
		0.5	2.6	100	9.6	0.2
m/e 139 = 0.5 unit d_1		0.5	0.05			
Subtracting		0.0	2.55	100	9.6	0.2
m/e 140 = 2.55 units d_2			2.55	0.24		
Subtracting			0	99.76	9.6	0.2
m/e 141 = 99.76 units d_3				99.76	9.58	0.20
Subtracting				0	0.02	0.00

Hence the labeled compound is 99.76 units d_3, 2.55 units d_2, and 0.5 unit d_1, i.e., 97.0% d_3, 2.5% d_2 and 0.5% d_1.

In our second example, an exact measurement of isotopic incorporation is not possible, although useful results are obtained. *n*-Butyl ethyl ether is generated from *n*-butanol, and the isotopic incorporation in the ether synthesized from butanol-1,1-d_2 is required. At 70 eV the mass spectrum of the unlabeled ether shows an $(M - H)^+$ ion that has ~13% of the abundance of the molecular ion. Even at 16 eV the $(M - H)^+$ ion is of substantial abundance, and the following experimental data were obtained for the labeled (d_2) and unlabeled (d_0) compounds:

m/e	100	101	102	103	104	105	106
d_0	0.2	4.6	100	6.5	0.3		
d_2			1.9	3.1	100	6.3	0.3

Analyzing the d_0 spectrum first, m/e 100 (0.2%) can only be due to loss of H_2 from the molecular ion. Its ^{13}C isotope will contribute 6.6% (<0.01 unit) to m/e 101, and this can be ignored. The signal at m/e 101 (4.6 units) must be entirely due to H˙ loss from the molecular ion, m/e 102, since H_2 loss from m/e 103 will be negligible, namely, 0.2% of 6.5 units or 0.01 unit. The ^{13}C isotope of $C_6H_{13}O^+$ (m/e 101) will contribute 6.6% of 4.6 units, or 0.3 unit, to m/e 102. Hence m/e 102 represents 99.7 units of $C_6H_{14}O$ and 0.3 unit of $^{13}CC_5H_{13}O$. The ions at masses m/e 103 and m/e 104 due to higher isotopes of $C_6H_{14}O$ have abundances of 6.6 units and 0.4 unit, respectively (considering ^{13}C and ^{18}O). Hence within experimental error (0.1 unit) m/e 103 and 104 are due entirely to isotopic contributions of $C_6H_{14}O$. Thus 99.7 units of $C_6H_{14}O^{+˙}$ gives 4.6 units of $C_6H_{13}O^+$ and 0.2 unit of $C_6H_{12}O^{+˙}$, or normalizing, 100 units of $C_6H_{14}O^{+˙}$ gives 4.6 units of $C_6H_{13}O^+$ and 0.2 unit of $C_6H_{12}O^{+˙}$.

In analyzing the d_2 ether, assumptions regarding the relative losses of H˙ and D˙ must be made. To a good approximation all H· loss from alkyl ethers occurs from the α position. Hence, if we assume a k_H/k_D isotope effect of unity for the fast source reactions under study (compare section 5-6), the labeled compound should suffer equal H· and D· loss.

Now if the labeled compound were 100% butyl-1,1-d_2 ethyl ether, the molecular ion region would show

m/e	102	103	104	105	106
	0.2 + 2.3	2.3	100	6.5	0.3

In this computation all molecular hydrogen loss is assigned to H_2 since loss of D_2 or HD is statistically much less likely, and $(M + 1)^+$ and $(M + 2)^+$ ion abundances are found experimentally for the d_0 compound rather than calculated (in this case the difference between the two is negligible). Agreement with experiment is fair but is optimized if the calculation includes some d_1 ether. The calculation for 99.0% d_2 1.0% d_1 gives

m/e	102	103	104	105	106
	2.4	3.1	100	6.1	0.3

which is in good agreement with the experimental data.

4-4 Compound Identification

A few remarks will suffice in connection with the problem of using a mass spectrum to finger-print a single unknown compound. (The ad hoc structure determination of an unknown is dealt with in the next section.) The general principles involved are similar to those applicable to the use of any type of spectrum to recognize a compound. One approach is to have a computer compare the spectrum of interest with all those in a library. Two points are particularly pertinent to mass spectrometry. First, there are usually too many peaks for all to be specified so only the most abundant are used. Eight-peak spectra and other abbreviated forms have therefore been introduced. Second, ion abundances taken in different laboratories reflect the source conditions and time scale appropriate to the particular experiment and instrument, so the limits allowable for a "fit" must be appropriately adjusted.

It is worth remarking that compounds may be fingerprinted by other spectra obtainable from a mass spectrometer, including doubly charged ion spectra, ion kinetic energy spectra, metastable maps, etc. (see Chapter 5).

4-5 Molecular Structure Elucidation

4-5a Introduction
It is our philosophy that structure elucidation rests on a few simple principles (see Chapter 1); the necessary operations can be illustrated but not taught, and the average organic chemist learns much faster when the spectra he is asked to examine are those of compounds he has prepared himself. It is also apparent that the topic of this section constitutes a much plowed field (see Bibliography). A highly selective approach, therefore, has been adopted in which two functional groups are treated in detail. The two functional groups chosen, ethers and ketones, are in many ways archetypical and their thorough study should prepare the reader for almost any problem short of those involving highly esoteric groups.

In what follows it has generally been assumed that high resolution data are not available. If, however, the compositions of the molecular and major fragment ions are known experimentally, the interpretation procedure is simplified, though basically unchanged. On the other hand, if a complete element map is available, a somewhat different approach to structure solving can be used in which groups in the molecule are recognized from homologous or other related series of ions given in the element map.

A final note: personal experience shows that the worst structure solvers are frequently the perfectionists. Success demands the ability to ignore some information temporarily, however distasteful this may be. This approach does not imply that certain ions are impossible to explain, just that a particular structure can cause an increased contribution from a normally unlikely reaction.

4-5b Ethers

The major primary fragmentation occurring in dialkyl ether molecular ions is α cleavage[1] (eq. 4-3). If α cleavage can occur at several alternative positions, the process leading to the

$$R-\overset{|}{\underset{|}{C}}-\overset{+\cdot}{O}-\overset{|}{\underset{|}{C}}-R_1 \quad \rightarrow \quad R-\overset{|}{\underset{|}{C}}-\overset{+}{O}=\overset{|}{\underset{|}{C}} \quad + \quad R_1\cdot \qquad 4\text{-}3$$

most stable carbonium ion will occur preferentially. If different radicals can be lost by α cleavage, then the more stable radical (tertiary > secondary > primary) will preferentially be lost, provided the rule concerning product ion stabilities is not contradicted. Thus ethyl isobutyl ether loses the $C_3H_7\cdot$ radical to give an ion of 8% relative abundance and the $CH_3\cdot$ radical to give an ion of less than 1% relative abundance. If formation of a stable ion and formation of a stable radical are competitive processes, then ion stabilization is more important. Thus isobutyl isopropyl ether loses a secondary propyl radical to give ion **1** to a smaller extent than it loses a primary methyl radical to give ion **2**. All of these generalizations cover other functional groups as well as ethers.

$$\overset{+}{C}H_2-O-i\text{-}Pr \leftrightarrow CH_2=\overset{+}{O}-i\text{-}Pr$$
$$\mathbf{1}$$

$$i\text{-}Bu-O-\overset{+}{C}H-CH_3 \leftrightarrow i\text{-}Bu-\overset{+}{O}=CH-CH_3$$
$$\mathbf{2}$$

In cases where alternative fragmentation reactions give radical and ionic products that have identical degrees of substitution, ion abundance is controlled by the size of the alkyl radical lost: the loss of the larger alkyl radical gives the more abundant fragment ion in 70 eV spectra. The operation of this rule is seen in the fact that *n*-butyl *n*-propyl ether reacts by loss of the larger radical ($C_3H_7\cdot$) to give an ion of 24% relative abundance, while loss of the smaller ethyl radical yields an ion of 13% relative abundance. The rule is also applicable to other functional groups. The most probable explanation is that the loss of the smaller radical gives the larger ion, which has more available fragmentation modes and more internal energy than the smaller ion. It therefore undergoes more extensive secondary fragmentation. At low electron energies or when metastable ions are studied, the order is reversed and indeed the

---

[1] The first carbon–carbon bond, numbering from the functionality, is designated as the α bond in all functional groups.

appearance potential for loss of the smaller radical is lower than that for loss of the larger radical.

The even-electron α cleavage product from an ether is formed with a distribution of internal energies such that some ions will fragment further, both in the source and in the analyzer region of the mass spectrometer. This further fragmentation is governed by two considerations. First, formation of an even-electron product ion with elimination of an even-electron neutral will be favored on energetic grounds (see section 1-3). Second, the tendency to form the most stable possible product will usually mean retention of the charge on the heteroatom. As so often happens when these two considerations obtain, a neutral hydrocarbon molecule is lost. Thus ethyl isobutyl ether forms an ion m/e 31 of 76% relative abundance, by the sequences shown in eq. 4-4. As an alternative mode of secondary fragmentation,

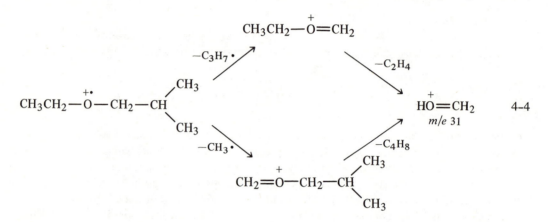

$$4\text{-}4$$

neutral formaldehyde may be eliminated in the process shown in eq. 4-5. Naturally, this

$$\overset{+}{R-O}=CH_2 \;\rightarrow\; R^+ + H_2CO \qquad\qquad 4\text{-}5$$

pathway will be most important when R^+ is a stable ion, such as a benzyl cation. (Direct formation of R^+ from the molecular ion also occurs.)

Those secondary fragment ions that retain enough energy for further fragmentation will again expel a neutral molecule. Thus $CH_2{=}O^+{-}H$ (m/e 31) loses H_2 to give $H{-}C{\equiv}O^+$ (m/e 29). Usually, however, it is unnecessary to consider these higher order fragmentations because of the time and energy limitations imposed upon ionic fragmentation and because the low mass region of the spectrum provides little specific information on molecular structure.

The second *primary* mode of fragmentation which is important in dialkyl ethers is formation of an alkyl cation by C–O bond cleavage. The relative importance of ions with and without heteroatoms in ethers is intermediate between the situation for more electropositive heteroatoms such as nitrogen, where hydrocarbon ions are of low total abundance (the fragmentation patterns of amines therefore resemble those of ethers, but are simpler), and less electropositive atoms such as the halogens (but also sulfur), where there is a strong tendency to form hydrocarbon ions since fragmentations are less strongly directed by the heteroatom. Naturally, C–O cleavage is most significant when a tertiary, benzylic, allylic, or other stabilized carbocation can be formed. Further fragmentation of the cation follows a pattern typical of the spectra of alkanes and other species in which alkyl cations are formed: H_2 loss is ubiquitous, and CH_4 and larger alkane losses also can be observed. Highly unsaturated alkyl

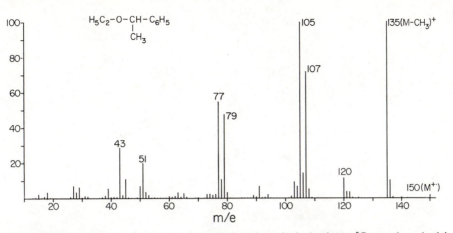

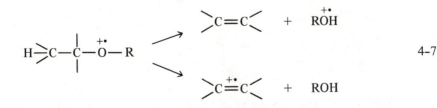

FIGURE 4-6.   *The mass spectrum of ethyl 1-phenylethyl ether. [Reproduced with the permission of John Wiley & Sons, Inc., from E. Stenhagen, S. Abrahamsson, and F. W. McLafferty, eds., Atlas of Mass Spectral Data,* © *1969.]*

cations, like $C_7H_7{}^+$ tend to lose carbon-containing fragments (eq. 4-6) rather than H_2.   The

$$C_7H_7{}^+ \longrightarrow C_5H_5{}^+ + C_2H_2$$
$$m/e\ 91 \qquad\quad m/e\ 65 \searrow$$
$$C_3H_3{}^+ + C_2H_2 \qquad\qquad 4\text{-}6$$
$$m/e\ 39$$

spectrum of ethyl 1-phenylethyl ether (Figure 4-6) shows many features from hydrocarbon ions.   The base peak is due to α-cleavage (methyl loss) to give the 1-ethoxybenzyl cation (m/e 135).   The expected alkene elimination then occurs to give m/e 107.   Cleavage of the C–O bond gives a stable 1-methylbenzyl cation (m/e 105) in high abundance, and further decomposition of this ion occurs via loss of C_2H_4 and then C_2H_2.

Both primary fragmentations of ethers considered so far involve simple cleavage.   Further cleavages are far less likely since only special substitution features would give stable product ions.   Hence the remaining primary fragmentations all involve elimination reactions or rearrangements.   As usual, these processes give rather low abundance ions in the 70 eV spectra, but become much more important at lower electron energies (see section 5-3).

Some bond-forming reactions, particularly simple eliminations occurring in the molecular ion, are of analytical value.   In ethers there are two such reactions: loss of an alkene and loss of an alcohol.   The reactions are complementary, only the location of the charge being changed (eq. 4-7).   Retention of the charge by the alcohol is usually favored (compare section 4-5d).

$$4\text{-}7$$

Both reactions are evident in the butyl ethyl ether spectrum (Figure 4-7), which shows an ion at m/e 56 that has the composition $C_4H_8{}^{+\cdot}$ and an ion $C_2H_5OH^{+\cdot}$ at m/e 46.

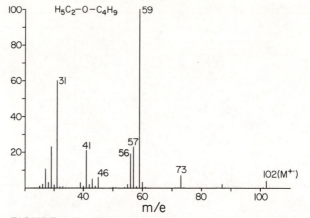

FIGURE 4–7. *Mass spectrum of n-butyl ethyl ether.*

The species $R\overset{+}{O}H_2$ is another fragment ion that should be mentioned, not because it is abundant in 70 eV spectra of ethers (it is usually barely detectable), but because it is the dominant fragment ion at low electron energies. Obviously, several bonds must be cleaved to form this ion, and the absolute value of the entropy of activation is large. On the other hand, the ion is very stable. Similar stable ions formed by multiple hydrogen transfers occur for other functional groups. The spectrum of di-*n*-hexyl ether taken at 70 and 12 eV is given in Figure 4–8. The 70 eV spectrum is dominated by formation and further fragmentation of

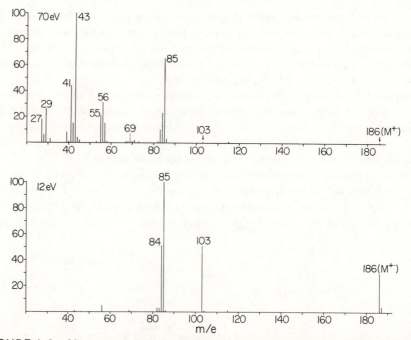

FIGURE 4–8. *Mass spectra of di-n-hexyl ether at 70 and 12 eV. Note the double hydrogen rearrangement leading to m/e 103, which becomes important at low energy. [Reproduced with the permission of United Trade Press from G. Spiteller and M. Spiteller-Friedmann in R. Bonnett and J. G. Davis, eds., Some Newer Physical Methods in Structural Chemistry, 1967.]*

the alkyl cation (*m/e* 85) and of the ionized alkene (*m/e* 84). Oxygen-containing ions are almost absent. At low energy further fragmentation of *m/e* 84 and 85 is virtually eliminated, but the relative abundances of the molecular ion and the $\overset{+}{R}OH_2$ rearrangement ion (*m/e* 103) have increased dramatically.

Dialkyl ethers illustrate so very well the simple principles governing mass spectrometric fragmentation that one apparent contradiction should be noted. An ion that is apparently due to simple β cleavage is observed with low to moderate abundance. For example, an ion $C_4H_9O^+$ (8% relative abundance) appears at *m/e* 73 in the spectrum of *n*-butyl ethyl ether (Figure 4-7). It is unreasonable to believe that this peak is due to the unstable alkoxy cation **3** or to the

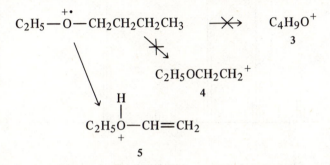

primary carbonium ion **4**. In fact, the ion apparently has the stable structure **5** as indicated by deuterium labeling experiments. It is only by coincidence that this rearrangement ion can be confused with the product of the unfavorable β cleavage reaction.

Alkyl aryl ethers differ from dialkyl ethers in that they typically show molecular ions of much greater abundance. This stabilizing role of the aryl group is a general phenomenon. *n*-Butyl phenyl ether can be compared to di-*n*-butyl ether for purposes of illustration. The molecular ion abundance of the aromatic compound is 20%, that of the dialkyl ether ~1%. Aryl methyl ethers differ from dialkyl ethers in two of their most important primary fragmentations. One of these reactions is loss of the methyl radical to give the aroxy ion (ArO^+), a far more stable species than alkoxy cations. Further fragmentation of this even-electron cation occurs by elimination of CO. The second new reaction—loss of the heteroatom as formaldehyde to give the molecular ion of the parent aromatic compound—is an illustration of the fact that simple elimination reactions can lead to relatively abundant ions in 70 eV spectra. The mechanism of the reaction is considered to be the simple four-centered transition (eq. 4-8),

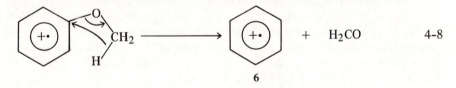

as indicated by substituent-effect evidence (see section 5-5). More recently the observation that the methyl and the ortho hydrogens in anisoles undergo mutual exchange prior to metastable reactions suggested that the five-membered transition state mechanism of eq. 4-9

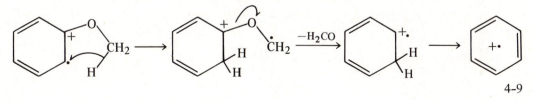

4-9

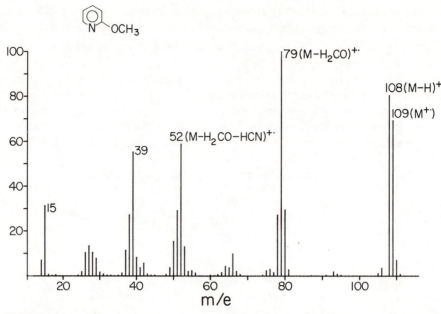

FIGURE 4-9. *Mass spectrum of 2-methoxypyridine. [Reproduced with the permission of John Wiley & Sons, Inc., from E. Stenhagen, S. Abrahamsson, and F. W. McLafferty, eds., Atlas of Mass Spectral Data, © 1969.]*

is also possible, as did measurements on the mode of energy partitioning (see section 5-4). The possible existence of the five-membered cyclic mechanism is one of numerous instances that emphasize the importance of radical sites in initiating rearrangements that lead to fragmentation.

However formed, the product ion **6** undergoes the same fragmentation reactions as do $ArH^{+\cdot}$ ions formed by direct ionization. However, since it does not show exactly the same rate characteristics, it probably has a different internal energy distribution. The subsequent reactions (ignoring those due to substituents that may be present) are $H^{\cdot}$ loss, C_2H_2 loss, and $C_3H_3^{\cdot}$ loss. The spectrum of 1-methoxynaphthalene shows ions at m/e 127, 102, and 101 corresponding to analogous reactions. That of 2-methoxypyridine (Figure 4-9) shows the sequence $M^{+\cdot} \rightarrow (M - H_2CO)^{+\cdot} \rightarrow (M - H_2CO - HCN)^{+\cdot}$ in which the $(M - H_2CO)^{+\cdot}$ ion apparently has the same structure as the pyridine molecular ion. This spectrum is also important because it provides our first example of an ortho effect. The molecular ion loses $H^{\cdot}$ to a far greater extent than do most methyl ethers; here $(M - 1)^+ > M^{+\cdot}$, whereas in anisole $(M - 1)^+$ is 3% of $M^{+\cdot}$. Direct interaction of the amino and methoxyl groups occurs to form a unique stable ion, perhaps **7**. In benzene (and anisole), ions due to $ArH^{+\cdot}$

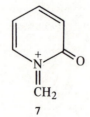

7

fragmentation occur at *m/e* 77, 52, and 39. The benzene molecular ion also shows losses of two and three hydrogen atoms to give *m/e* 76 and 75, further fragmentation of *m/e* 77 to give *m/e* 51 and of *m/e* 76 to give *m/e* 50, as well as other minor reactions. An illustration of the type of feature that may worry the student and is best ignored during the structure-solving operation is a low abundance ion at *m/e* 63 in the anisole and benzene spectra. Since exact mass measurements confirm that the responsible ion is $C_5H_3^+$, how does benzene lose a methyl radical? There is no metastable peak for this process, but there is a metastable for the reaction $C_6H_6^{++} \rightarrow C_5H_3^+ + CH_3^+$ in the doubly charged ion spectrum of benzene. Indeed the energy release accompanying this metastable reaction has been used to demonstrate that the charges in $C_6H_6^{++}$ are ~5.4 Å apart and that the species is linear, at least in this reactive form.

Another primary reaction of anisoles, this one also observed in dialkyl ethers, is loss of the alkoxyl radical. This process is important because the ion Ar^+ (*m/e* 77 in phenyl ethers) provides the quickest method of recognizing the aromatic group.

In certain substituted anisoles the loss of the formyl radical occurs to yield the species ArH_2^+. Why should the ion undergo a double hydrogen transfer in order to yield this product? As always, product stability is the answer. The process is restricted to anisoles that bear oxygen- or nitrogen-containing substituents to which a hydrogen can be transferred. It is also restricted to anisoles in which the competitive methyl elimination process is not dominant. These points are well illustrated by comparing the mass spectra of the *m*- and *p*-dimethoxy-benzenes (Figure 4-10). The $(M - CH_3)^+$ ion in the para isomer is resonance stabilized so that it gives the base peak in the spectrum (37% of the total ion current); formyl radical loss is not observed. In the meta isomer methyl radical loss barely occurs, giving rise to an ion

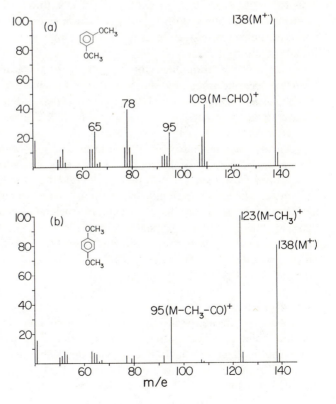

FIGURE 4-10. The mass spectra of (a) m-dimethoxybenzene and (b) p-dimethoxybenzene. The differences between these spectra are pronounced; see especially m/e 109 and 123. [Reproduced with the permission of Holden-Day from H. Budzikiewicz, C. Djerassi, and D. H. Williams, Mass Spectrometry of Organic Compounds, 1967.]

carrying <1% of the total ion current; formyl radical loss is the dominant fragmentation with 11% of the total ion current. The formyl loss product probably has either of the stable structures **8** or **9**. Either pathway requires a second heteroatom and accounts for the absence

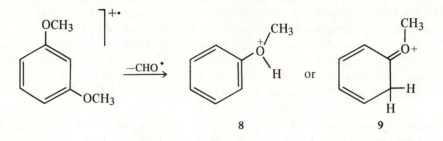

of this fragmentation in anisole itself. This process is comparable to $R\overset{+}{O}H_2$ formation from long chain ethers.

The third isomer, *o*-dimethoxybenzene, exhibits a mass spectrum that is different from both the others. In particular, interaction of the ortho substituents occurs. Again loss of CHO˙ from the molecular ion does not occur. Therefore, in this particular case mass spectrometry can be used to distinguish isomeric disubstituted benzenes. Other isomeric substituted anisoles give very nearly identical mass spectra. Once again we have encountered a general situation that is far more widely applicable than just to this class of compounds. Mass spectrometry serves to identify isomeric substituted benzenes in perhaps half of all cases.

It is an open question whether or not substituent isomerization is responsible for the similarities observed in spectra of some disubstituted aromatic compounds. Evidence for substituent scrambling, both by valence tautomerism and through radical migration, has been found in some systems (see section 5-3). In other instances reversible ring expansion can lead to isomerization.

The fragmentation behavior of a disubstituted aromatic compound will frequently be dominated by one group, and substituents can be ranked in order of their reactivity. This order also applies to monosubstituted benzenes, in which it determines the extent of fragmentation relative to the molecular ion abundance. Table 4-3 gives this crude reactivity order and also the major reaction of each functional group. The activation energies (AP-IP) for each process are seen to correlate approximately with the reactivity of the substituent. An exact correlation is not expected because of differences in the frequency factors for the different reactions.

Turning from anisoles to other alkyl aryl ethers, one finds that alkene elimination to give the ionized phenol is the dominant reaction. Diaryl ethers are notable for their stable molecular ions, and for the occurrence of a complex skeletal rearrangement resulting in CO elimination (compare the spectrum of diphenyl ether, Figure 5-4). While this particular rearrangement could hardly have been predicted, guidelines have been recognized (see section 5-3) that rationalize most skeletal rearrangements. Let us not forget, however, that these are minor processes in analytical 70 eV spectra.

4-5c Ketones

Ketones are representative of the second major functional type, those possessing a multiple bond to a heteroatom. A major primary process is acylium ion formation, and the rule regarding large vs. small radical loss is applicable here too. In other words, R_LCOR_S will give R_SCO^+ with the greater abundance at 70 eV, even though R_LCO^+ has a slightly lower appearance potential. If R_S is secondary and R_L primary, then of course, this factor becomes

Table 4-3. *Reactivity Order in Substituted Benzenes, C_6H_5Y*

Y	Fragmentation[a]	IP[b]	AP[c]	ϵ_0[d]	$[D^+]/[M^+]$[e]
$\begin{matrix} O \\ \| \\ -CCH_3 \end{matrix}$	CH_3	9.27	10.0	0.7	2.2
$-C(CH_3)_3$	CH_3	8.68	10.3	1.6	2.1
$\begin{matrix} O \\ \| \\ -CH_3 \end{matrix}$	OCH_3	9.35	10.8	1.5	1.5
$-Et$	CH_3	8.76	11.3	2.5	0.95
$-NO_2$	NO_2	9.92	12.2	2.3	0.72
$-CH_3$	H	8.81	11.8	3.0	0.53
$-I$	I	8.73	11.5	2.8	0.43
$-Br$	Br	8.98	12.0	3.0	0.36
$-Cl$	Cl	9.07	13.2	3.1	0.16

[a] Neutral lost in major reaction.

[b] From J. L. Franklin, J. G. Dillard, H. M. Rosenstock, J. T. Herron, K. Draxl, and F. H. Field, *Ionization Potentials and the Heats of Formation of Gaseous Ions*, NSRDS–NBS 26, 1969.

[c] From D. H. Williams and I. Howe, *Principles of Organic Mass Spectrometry*, McGraw–Hill, London, 1972, p. 60.

[d] Activation energy, AP–IP.

[e] Relative fragment ion/molecular ion abundance at 18 eV where secondary reactions are largely eliminated.

dominant. Further fragmentation of the acylium ions is facile, and the ions $R_L{}^+$, $R_S{}^+$, $R_L CO^+$, $R_S CO^+$ are easily recognized in almost all ketone spectra.

The second major fragmentation process is the six-membered cyclic hydrogen transfer and associated β cleavage leading to loss of an alkene molecule. The reaction has been termed the McLafferty rearrangement. Although some hydrogen exchange occurs in ketone molecular ions even at 70 eV, the reaction has repeatedly been shown to be highly specific. It is by far the most studied rearrangement in mass spectrometry, and the weight of evidence now favors a stepwise reaction initiated by the radical site on oxygen (eq. 4–10) rather than

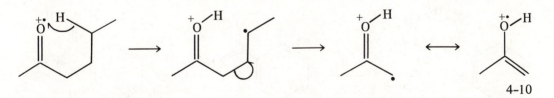

4-10

a fully concerted process. The product ions are unusually abundant for a rearrangement fragmentation. The McLafferty rearrangement product is the enolic species shown. (Evidence for the structure comes from icr, isotopic labeling, and metastable studies.) Further fragmentation of the ionic product occurs by α cleavage after ketonization or by alkene elimination.

Several other notable features of the McLafferty rearrangement are

(i) A methyl group does not rearrange when substituted for the γ hydrogen.

(ii) A formally analogous process can occur in even-electron ions.

(iii) Related reactions occur in other C=X systems, for example, oximes.

(iv) Secondary hydrogen atoms in the γ position are more readily abstracted than are primary.

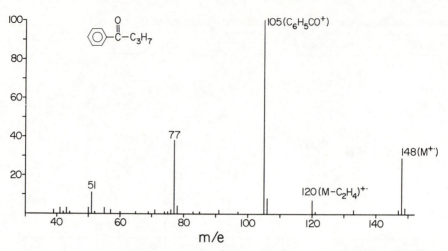

FIGURE 4-11. Mass spectrum of n-butyrophenone. Note the McLafferty rearrangement product, m/e 120.

(v)   If both alkyl groups can undergo McLafferty rearrangement, the more abundant product is due to reaction from the larger group. In addition, the product of two successive alkene eliminations (double McLafferty rearrangement) will usually appear as an abundant ion.

Many of the major features of ketone mass spectra are illustrated in the spectra of *n*-butyrophenone (Figure 4-11) and 4-nonanone (Figure 4-12).

The *n*-butyrophenone spectrum is very simple, the base peak m/e 105 ($C_6H_5CO^+$), being formed by methyl radical loss from the McLafferty rearrangement product, m/e 120, as well as by α cleavage in the molecular ion. It is interesting to note that the methyl loss process involves the CH_2 group of the enol ion and an ortho ring hydrogen, *not* the hydroxyl hydrogen. In other words, a hydrogen atom is abstracted via a five- rather than a four-membered cyclic transition state (eq. 4-11).

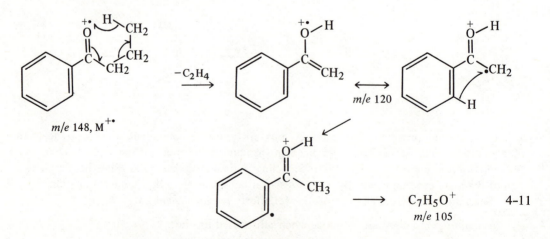

4-Nonanone fragments by α cleavage to give the acylium ions at m/e 71 and 99. These ions fragment by CO loss to give the alkyl cations at m/e 43 and 71 (the latter is composed of

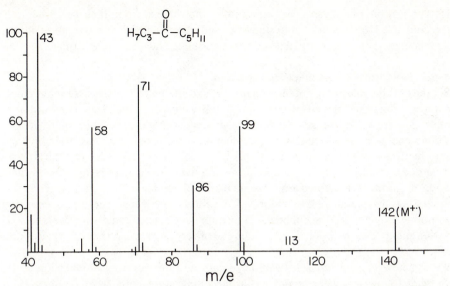

FIGURE 4-12.   Mass spectrum of 4-nonanone.   [Reproduced with the permission of Heyden and Son from G. Eadon, C. Djerassi, J. H. Beynon, and R. M. Caprioli, Org. Mass Spectrom., 5, 917 (1971).]

both $C_5H_{11}^+$ and $C_3H_7CO^+$).   Formation of the alkyl cations directly from the molecular ion is unimportant.   Of the three possible odd-electron ions due to McLafferty rearrangement, that associated with elimination of the smaller alkene is not observed (loss of C_2H_4 would give m/e 114), but loss of C_4H_8 gives the ion at m/e 86 and the double McLafferty rearrangement gives m/e 58.   Both of these ions fragment by α cleavage and then CO elimination and so contribute to the acylium and alkyl ions already mentioned.   The only other ion that appears in the spectrum besides hydrocarbon fragments at m/e 55, 41, etc., is a low abundance ion, m/e 113, due to loss of $C_2H_5 \cdot$ from the molecular ion.   This product is due to a rearrangement reaction and probably has the stable structure 10, although a cyclized structure cannot be

$$\overset{+OH}{\overset{\|}{C_3H_7-C-CH=CH-CH_3}}$$

10

excluded.   Although not evident in the 70 eV mass spectrum of 4-nonanone, several other low energy rearrangement processes do appear in ketone spectra.   One of these is analogous to the McLafferty rearrangement, except that two hydrogen atoms are transferred to the carbonyl group, probably to give the stable ion **11**.   It will be recognized that this is analogous to $R\overset{+}{O}H_2$

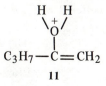

11

formation from ethers.   Thus these rather complex rearrangements do not interfere with the analytical applications of mass spectrometry, nor are they haphazard in their occurrence. Rather, they are controlled by product ion stability, and they increase the structural information available in a mass spectrum, particularly when metastable reactions are examined.

4–5d Basic Rules Governing Fragmentation

Several of these rules were noted in Chapter 1, and the preceding discussion has emphasized their validity. Thus the roles of product ion stability, odd- vs even-electron nature of the ion, and simple cleavage vs. rearrangement have been adequately covered. The value of the charge localization concept and the tendency in fragmentations not to lose the heteroatom (especially if it can effectively stabilize charge) have also been well illustrated. Several indications of the control exerted by transition state size also have been given, and it may be noted here that in HX elimination reactions, the favored ring size increases as the size of X decreases. Thus for sulfur the favored size is five-membered, for chlorine five- and six-membered, for oxygen six-membered, and for fluorine six- and seven-membered. Transition state ring size may, however, be secondary to activation energy considerations, as the examples of very large cyclic transition states in section 5–3 show.

We can now consider in a little more detail the rules covering product ion abundance. Much of what follows can be summarized in one simple statement: when fragmentation takes place, the positive charge remains on the fragment with the lowest ionization potential. Expressed more cautiously, the lower IP fragment will be more abundant. This rule has been shown to be true both for simple bond cleavages, where radical IP's are compared, and for elimination reactions, where the fragments are even-electron species. Consider, for example, the rearrangement of **12**. The IP of 2-butene is 9.13 eV, that of acetaldehyde 10.22 eV.

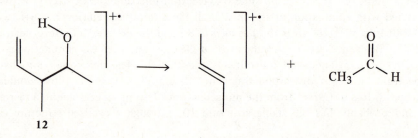

12

In keeping with the rule, the relative abundances of the corresponding ions formed by fragmentation of **12** are 100% and ~5%, respectively. Similarly, simple C–C cleavage of 2-hydroxyethylamine (eq. 4–12) can result in charge retention by either the N- or the O-con-

$$ HO-CH_2-CH_2-NH_2 \Big]^{+\cdot} \longrightarrow HO-CH_2^{\cdot} + CH_2 {=} \overset{+}{N}H_2 \qquad 4\text{-}12 $$

taining fragment. Charge retention by nitrogen is dominant (50/1), and the ionization potentials of the radicals $\cdot CH_2NH_2$ and $\cdot CH_2OH$ are 6.2 and 7.6 eV, respectively. The simple statement that product ion stability (enthalpy) governs fragmentation certainly covers this behavior. Cases are occasionally encountered where differences in the stabilities of the neutral species are sufficient to reverse this order.

The rule regarding the control exerted by fragment ionization potentials can be derived as follows. Consider the fragmentation of $R_1R_2^{+\cdot}$ to give either R_1^+ or R_2^+. Ion abundances of R_1^+ and R_2^+ will depend on the relative activation energies of the two reactions, provided that further fragmentation occurs to a similar extent. The implication that frequency factors (entropies of activation) are comparable for the two processes is justified by the fact that the same bond(s) is broken in both reactions. Since the activation energy for any fragmentation $AB^{+\cdot} \to A^+ + B^\cdot$ can be equated to the difference in the appearance potential of A^+ and the ionization potential of $AB^{+\cdot}$, the difference in the activation energies for R_1^+ and R_2^+ formation is equal to the difference in their appearance potentials. Hence, ignoring excess energy terms (see section 4-9), we develop eqs. 4-13 and 4-14. Thus the more abundant

$$AP(R_1^+) = \Delta H_f(R_1^+) + \Delta H_f(R_2^\cdot) - \Delta H_f(R_1 R_2)$$

and $$AP(R_2^+) = \Delta H_f(R_2^+) + \Delta H_f(R_1^\cdot) - \Delta H_f(R_1 R_2)$$

therefore $$\Delta AP = \Delta H_f(R_1^+) - \Delta H_f(R_1^\cdot) - [\Delta H_f(R_2^+) - \Delta H_f(R_2^\cdot)] \qquad \text{4-13}$$

but $$IP(R_1^\cdot) = \Delta H_f(R_1^+) - \Delta H_f(R_1^\cdot)$$

therefore $$\Delta AP = IP(R_1^\cdot) - IP(R_2^\cdot) \qquad \text{4-14}$$

fragment ion will be that associated with the radical of lower ionization potential.

4-6 Stereochemical Assignments by Mass Spectrometry

The persistence of the myth that reactions occurring in the mass spectrometer have little relation to the normal behavior of organic compounds has been unfortunate. A major concern of this Part has been to demonstrate the fallaciousness of this viewpoint. Important aspects of the argument involve the recognition of the close mechanistic relationships between electron impact reactions and those encountered in photochemistry, thermochemistry, and ordinary solution chemistry. Lack of space and the availability of other treatments preclude extensive discussion of the photochemical and thermochemical analogies. The relationship to solution organic chemistry is met implicitly or explicitly in many places in these pages (see especially section 5-5) and again in this section. A major feature of the results that are discussed below is the *delicacy* of the mass spectrometric method.

A mass spectrum can be used to deduce the relative orientation of two groups in a molecule by studying a fragmentation reaction that involves both groups. Such a process will typically involve the elimination of some neutral molecule. For example, the elimination of acetic acid from an acetate involves the acetoxy group and a hydrogen on the adjacent carbon. If two epimers are compared, in only one of which a hydrogen atom can closely approach the carbonyl group, then the relative abundances of the $(M - CH_3CO_2H)^{+\cdot}$ ions will differ sharply. To enhance the differences between the behavior of the epimers, the comparison might be better made using the corresponding metastable peaks or the fragment ion abundances in the low electron energy spectrum. Alternatively, the appearance potentials for the reactions in question could be compared.

This general technique (illustrated in Figure 4-13a) has had considerable practical value in allowing stereochemical assignments in complex molecules, including steroids and sugars, where nmr methods have been difficult to apply. Systematic study of particular classes of compounds has also shown that it is frequently not necessary to obtain data on both epimers. The technique does require, however, that care be used in selecting the probe section. Such elementary steps as avoiding contributions from thermal elimination reactions and using only reactions with specific geometrical requirements naturally obtain.

By way of illustration, consider the following example. The epimeric 3-methoxy-17-ethyl-$\Delta^{1,3,5(10)}$-estratrien-17-ols both lose water upon electron impact, but the ratio of the

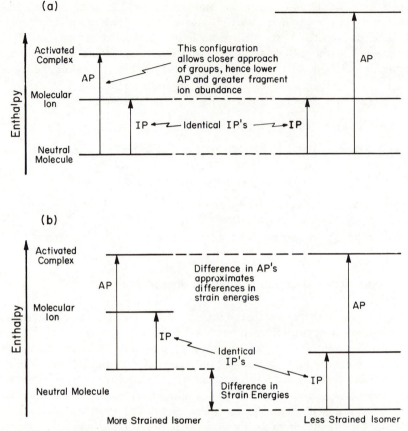

FIGURE 4–13. Stereochemical determinations by mass spectrometry: (a) determination of configuration and (b) determination of conformational energy.

$(M - H_2O)^{+\cdot}$ ion abundance to that of the molecular ion is 0.41 for the 17 α-hydroxyl compound and 0.21 for the 17 β (equatorial) analogue. Although the difference in these ratios is not large in 70 eV spectra, it can be accentuated by using lower energy conditions. The fact that only the axial hydroxyl group can closely approach a hydrogen atom (that at C-12) accounts for the more efficient ionic dehydration and provides a basis for the stereochemical assignment.

    The sensitivity of the mass spectrometric method is best illustrated by the fact that it can be used to assign the stereochemistry of diastereomers that differ only in their H/D substitution. HCl loss by gaseous ions is normally a 1,3-elimination involving a five-membered cyclic transition state. Hence 2-chloropentane should undergo preferential loss of H_α relative to H_β. This reaction has been used to confirm the stereochemistry of the diastereomers **13**

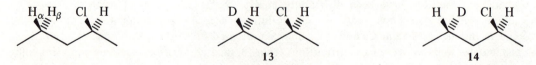

and **14**, the latter undergoing preferential HCl elimination and the former losing DCl preferentially. Analogous results were found for the 1,4-elimination of H_2O and DHO from the 2-hexanols **15** and **16**. More significantly still, the alkoxy radical formed under Barton

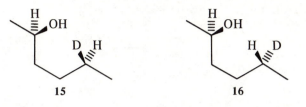

15   16

reaction (liquid phase) conditions showed the same specificity in abstracting a C-5 hydrogen atom. Two important conclusions follow:

(a) The electron impact process is sensitive to what must be very minor differences in activated complex enthalpies (estimated as ~1 kcal mole^{-1}).
(b) The structure of these gaseous ions can differ but little from those of the corresponding neutral species.

A similar conclusion is required in the case of the dicarbomethoxydienes **17** and **18**. Cis-trans

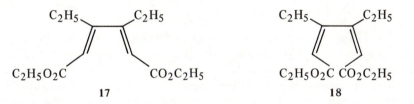

17   18

isomerization accompanies some mass spectral reactions, notably those studied using high energy ions, particularly if the double bond is the only functional group present. Compounds **17** and **18** are easily distinguished, however, even in their 70 eV mass spectra where the $(M - MeOH)^{+\cdot}/M^{+\cdot}$ ion abundance ratio is 11 in compound **17** but only 0.3 in **18**.

Stereochemical assignments have also been made from measurements of the kinetic energy release that accompanies metastable ion fragmentation (see section 5-11). The energy release is determined by the nature of the activated complex and hence is stereochemically controlled.

A rather different application of mass spectrometry to stereochemical problems allows the assignment of relative conformational strain in isomers. The requirement here is that there exist a mass spectral reaction that involves a transition state in which crowding has been relieved. Frequently a simple cleavage reaction will be appropriate. The abundance of the fragment ion formed from the sterically strained molecule (and hence molecular ion) is found to be greater than that due to the less strained isomer. In Figure 4–13(b) this result is rationalized by an illustration of the energetic situation, which contrasts with that appropriate to the elimination reactions discussed above. This illustration also emphasizes that while ion abundances are most conveniently measured, appearance potential data would serve at least as well. Indeed it has proved possible with appearance potential measurements to assign strain energies quantitatively. For example, the enthalpies of *cis*-4,6-dimethyl-1,3-dioxane and the trans isomer differ by some 3.4 kcal mole^{-1} as determined by standard calorimetric methods. The cis isomer is the more stable, since interaction of the C-2 hydrogen and the axial C-4 and C-6 methyls raises the enthalpy of the trans compound. The mass spectrometric method makes use of $(M - H)^+$ formation since all strain should be released in the transition state leading to the $(M - H)^+$ ion. This reaction is known to involve the C-2 hydrogen exclusively. Hence the appearance potential for the cis isomer (**19**) should be 3.4 kcal mole^{-1} higher than that for the trans compound (**20**). The measured AP's were $9.69_3 \pm 0.005$ and $9.54_0 \pm 0.003$ eV, respectively, i.e., they differ by 0.15_3 eV, or 3.5 kcal mole^{-1}.

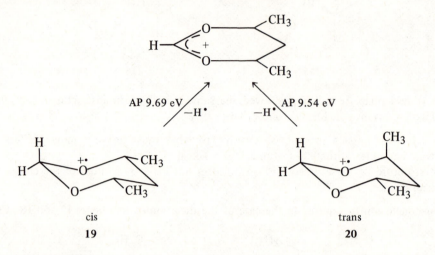

cis   trans

19   **20**

Yet another type of mass spectral study on molecular conformation involves the determination of tautomeric forms in the gas phase. The methods generally used to assign tautomeric structure in solution (nmr, ir, uv, etc.) are not easily applied to gaseous species. The essence of the mass spectral method is the assignment of certain fragment ions exclusively to one or the other tautomer. For example, ring chain tautomerism in the oxazolidine (**21**)

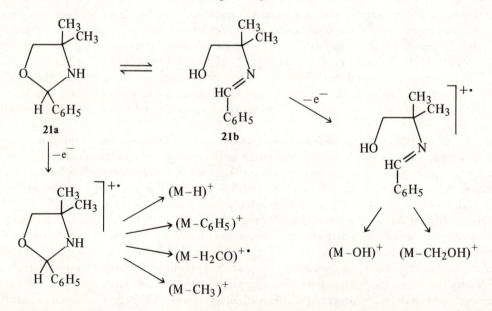

can be studied by this method. The cyclic molecular ion is expected to lose H$^{\cdot}$ and C$_6$H$_5$$^{\cdot}$ with great ease, since the processes represent α-cleavage to both an ether and an amine functionality. This molecular ion should also undergo methyl loss (α cleavage relative to the amino group) and H$_2$CO elimination. These reactions are all observed in addition to HO$^{\cdot}$ loss and CH$_2$OH$^{\cdot}$ loss, processes that are reasonably ascribed to the acyclic form. The mass spectral method gives values for the enthalpy of tautomerism that are in reasonable agreement with nonpolar solution values.

4-7 Sequence Determinations

The analysis of compounds of biochemical origin has not so far been singled out for special mention since the considerations involved are similar to those for any organic compound. The quantities of material available, however, are often more limited, and problems of thermal instability and involatility may be acute. Complex mixtures are also common and hence gc–ms (see section 4–8) has special importance in this area. A unique feature of some of the most important compounds of biological origin is the fact that they contain sequences of chemically related groups. Proteins, peptides, sugars, and nucleotides fall into this group, and sequence analysis in these systems is of considerable importance.

The potential of mass spectrometry in sequence analysis lies in the fact that the sequence may be recognized from the presence of a set (or sets) of fragment ions that contain varying numbers of the groups forming the sequence. In the ideal case, the only fragment ions formed would arise by cleavage between the groups forming the sequence and each such group would have a unique mass. Automated sequence determination would be simple in such a situation. In practice, complications do arise, although considerable progress has been made, especially in peptide sequencing. Much attention has been devoted to finding derivatives that have the appropriate thermal and mass spectrometric properties, especially in regard to increasing the abundance of the less recognizable sequence ions. Some investigators have used high resolution spectra for sequencing peptides and considerable automation has been achieved in this area.

For further discussion of peptide sequencing, the reader is referred to the treatments listed in the Bibliography (reference 6). Here the principles of sequencing by mass spectrometry will be illustrated by an oligosaccharide example. The difficulties presented by these samples in terms of involatility, thermal instability, and variety of stereoisomers are considerable. Nevertheless, traditional methods of sequencing are laborious (not so for peptides where the Edman degradation is a routine operation) and mass spectrometry appears to have merit as a supplementary analytical technique.

In one approach to oligosaccharide sequencing, an aromatic group is incorporated into the sugar in order to provide a charge-stabilizing site. This moiety increases the molecular ion abundance and favors the formation of fragment ions that retain the aromatic group, together with 1, 2, 3,. . ., monosaccharide units. Thus the 1-phenylflavazole derivative of maltopentose (22) lends itself to sequence determination. The compound was peracetylated to increase

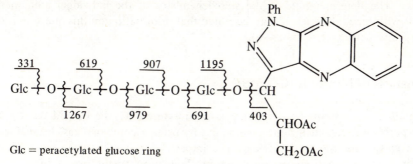

Glc = peracetylated glucose ring

22

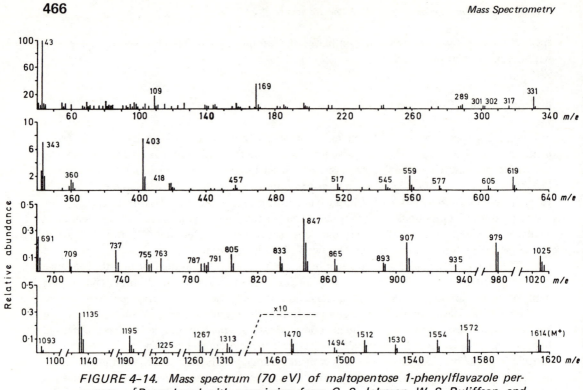

FIGURE 4-14. *Mass spectrum (70 eV) of maltopentose 1-phenylflavazole per-acetate. [Reproduced with permission from G. S. Johnson, W. S. Ruliffson, and R. G. Cooks, Chem. Commun., 587 (1970).]*

volatility, and the 70 eV mass spectrum shown in Figure 4-14 was obtained. The presence of the molecular ion at m/e 1614 is noteworthy. Even more significant, however, is the occurrence of fragment ions belonging to two *sequences*, either of which can be used to sequence the molecule. These ions, with their daughter ions due to acetic acid and/or ketene loss, constitute virtually the only fragment ions of significant abundance above m/e 350. They arise by glycosidic bond cleavages starting from either end of the molecule.

In addition to the molecular weight of the oligosaccharide and the sequence of masses of its constituent monosaccharides, the mass spectra of peracetylated 1-phenylflavazoles also allow the stereochemistry of each glycosidic linkage to be determined, at least in the cases so far examined. For example, the fragment ions belonging to the sequence which does not retain the flavazole undergo ready loss of acetic acid if the linkage is 1–4 but not if it is 1–6.

The above example illustrates the potential of mass spectrometry in sequencing biological molecules. The determination of the stereochemistry of the individual monomers has not been dealt with here, although it may be noted that information on this question is obtainable by mass spectrometry (see section 4–6).

4-8   Analysis of Mixtures (GC-MS)

The separation step essential to any analysis of mixtures may be carried out by controlled sample vaporization or, much more commonly, by using a combined gas chromatograph–mass spectrometer (gc-ms, see section 3–6). The impact of gc-ms has been tremendous, adding the power of mass spectrometry in structure elucidation to the power of gc to deal with

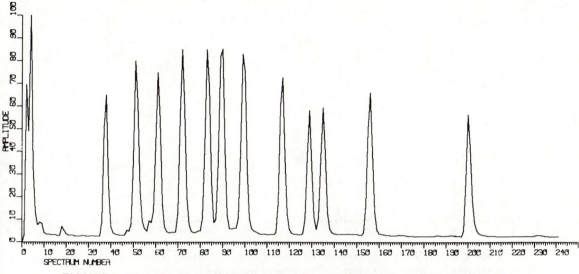

FIGURE 4-15.   Simulated gas chromatogram of a barbiturate mixture (0.2 μg each) obtained on a gc-ms instrument. [Courtesy of Finnigan Corporation.]

small sizes and multicomponent mixtures of complex compounds. Applications include flavor chemistry where coffee aroma, for example, was found to consist of a mixture of several hundred relatively simple compounds.   Applications to medicine (drug metabolites in body fluids) and to environmental science (pesticides, water quality) are rapidly expanding. In typical cases it is possible to identify all components present in concentrations of more than 1% in submicrogram quantities of sample.   Before describing an application of gc-ms, some relatively minor limitations of the technique should be noted. First, column bleed is always transmitted into the instrument and, although its spectrum can be taken and subtracted, it corresponds essentially to having a moderate background always present. Second, undetected reactions, such as isomerization, may occur on the column; this is a perennial gc problem. Third, there are sample volatility limitations.   Fourth the carrier gas not removed by the separator causes higher source pressures than are optimal.

A mixture of twelve barbiturates (0.2 μg each), *N*-methylated to improve their gc properties, provides a suitable case to illustrate the power of the combined gc-ms technique. Over 200 mass spectra were taken during the gc elution and stored in a computer. Figure 4-15 shows a gas chromatogram obtained using the mass spectrometer as detector; the chromatogram is a plot of total ion intensity vs. spectrum number. A mass spectrum of a component with one of the shorter retention times, for example, is obtained by displaying spectrum number 62 (Figure 4-16). With the knowledge that the unit **23** is present, this spectrum

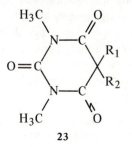

23

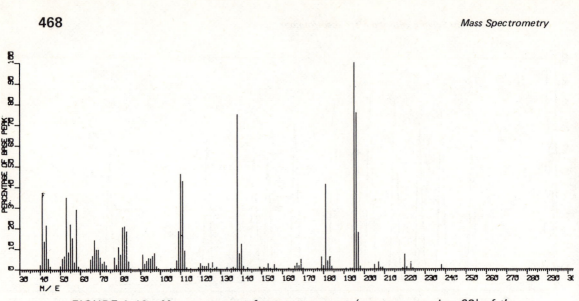

FIGURE 4-16. Mass spectrum of one component (spectrum number 62) of the barbiturate mixture. [Courtesy of Finnigan Corporation.]

is easy to interpret. The masses of R_1 and R_2 must total 106 in order to account for the molecular ion at m/e 238. This observation suggests that $R_1 + R_2 = C_6H_{12}$, i.e., that there is just one ring or double bond in the two substituents. Alurate (R_1 = isopropyl, R_2 = allyl) is the only common barbiturate that fits these data. The fragmentation pattern, especially the base peak at m/e 195 due to allylic cleavage with loss of the isopropyl group, confirms the assignment.

The gc–ms computer system allows another method of determining the compositions of the components in the mixture. It is possible to display a mass chromatogram—a plot of the abundance of a particular ion as a function of spectrum number. In some of the barbiturates one of the substituents is the allyl group. A mass chromatogram for m/e 195 (molecular weight less the other C-6 substituent) then indicates which fractions contain these barbiturates. Figure 4-17 shows this mass chromatogram. It is seen that m/e 195 is concentrated in just three of the twelve gc components, including that due to alurate.

4-9 Thermochemical Determinations[2]

Mass spectrometric techniques allow the determination of many relatively inaccessible thermochemical quantities. Heats of formation, ionization and appearance potentials, and electron affinities of ions and radicals can be measured and used to obtain bond dissociation energies, gas phase acidities and basicities, heats of hydration, etc.

The applicable experimental measurements divide into two types, those in which ion-molecule reactions (see section 5–7) are involved and those in which they are not. The most frequently and easily measured thermodynamic properties are ionization and appearance potentials. These terms refer to the minimum energies required to form molecular and fragment ions, respectively, from the neutral molecule. The use of the term minimum energy

---

[2]This section covers an application of mass spectrometry and properly belongs here in Chapter 4. However, a prior reading of section 5–1 may be helpful in understanding the basis for this application.

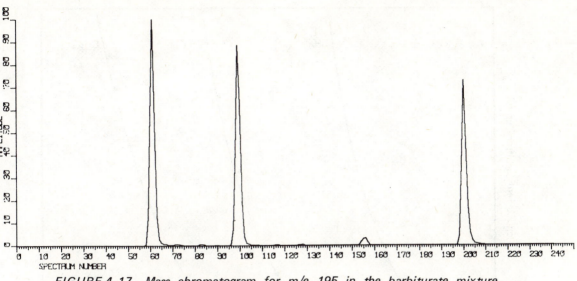

FIGURE 4-17. *Mass chromatogram for m/e 195 in the barbiturate mixture.*
[Courtesy of Finnigan Corporation.]

in these definitions needs to be qualified. Electron impact and photoionization are vertical rather than adiabatic processes, so that the lowest vibrational level of the molecular ion will not necessarily be reached (see Figure 2-1). In practical terms the AP is not the minimum energy required to form an ion by a given route using a given ionizing agent since there is the additional specification that the reactions occur within the ion source, i.e., in a time of the order of a microsecond. As a result extra energy, referred to as the *kinetic shift*, is required to effect reaction. Furthermore, if the fragment ion of interest is not the product of the lowest energy fragmentation reaction, another excess energy term, sometimes called the *competitive shift*, is involved. This term represents the extra energy necessary to reach rate constants $> 10^6$ sec^{-1} such that the process of interest gives a detectable signal in its competition with the lowest activation energy reaction.

The measurements with which we are concerned are threshold measurements—the minimum energy required to detect a process. In addition, at least in large molecules, the nature of the relationship between ion current and ionizing energy is not well defined. Therefore it is not surprising that empirical or semiempirical methods of fixing the signal onset have to be used. Typical ionization efficiency curves are shown in Figure 4-18 for the threshold region (a few eV above the onset of the ion.) The almost asymptotic behavior at threshold is due in large measure to the distribution of energies of the electron beam itself, and much sharper onsets can be obtained by using photoionization or by using the retarding potential difference method (Figure 4-19). In this procedure a modified electron impact ion source is employed, and the signal due to a narrow electron energy segment ΔE is measured as the difference between that measured for electrons of energy ($E_e + \Delta E$) and that measured for electrons of energy E_e. Accuracies within a few hundreths of an eV are possible. The same results can be achieved directly by using an experimental arrangement in which the electrons are energy selected (using a electric sector) prior to entering the ion source.

It is possible to use an unmodified electron impact source to get excellent reproducibility and fair accuracy (0.1-0.2 eV) by the following simple procedure (or variants on it). Stray fields in the source are minimized by operating the repeller(s) at 0 volt, and the electron current

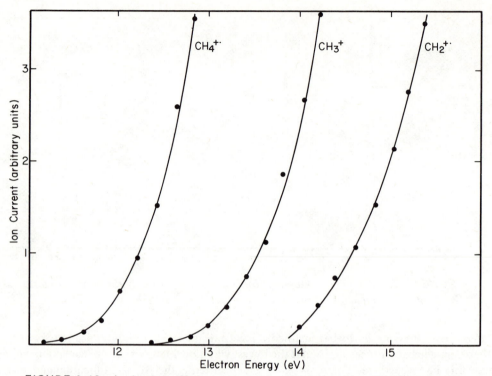

FIGURE 4-18. Ionization efficiency curves in the threshold region for some ions in methane. [Reproduced with the permission of The Faraday Society from C. A. McDowell and J. W. Warren, Disc. Faraday Soc., **10**, 53 (1951).]

is decreased (typically from 100–200 μamp to 10–20 μamp) so that it can be accurately controlled at low energies. Each sample is run mixed with a calibrant chosen to give an ion of well known (spectroscopic or photoionization) IP. Several calibrants are used for each sample because the potential applied between ion chamber and filament is not usually a reliable measure of the electron energy since stray fields cause some electron acceleration or deceleration. The region of the ionization efficiency curve of interest is that at threshold, in practice, that in

FIGURE 4-19. Retarding potential difference method of determining ionization efficiency curves. A retarding electrode system admits into the ion chamber only electrons of energy greater than a specified quantity. If this quantity is changed from E to $(E + \Delta E)$, the resulting difference in total ionization represents the ionization effected by electrons with energies in the narrow range E to $(E + \Delta E)$ (hatched area).

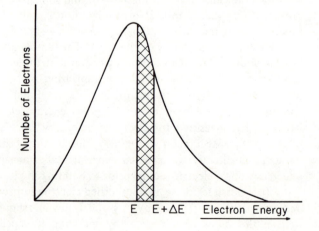

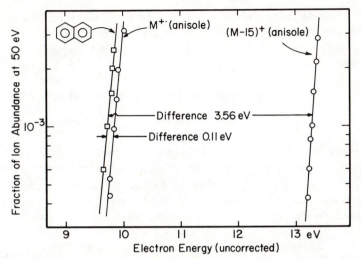

FIGURE 4-20.   Determination of IP (anisole) and AP (M − CH₃)⁺ by the semilog
plot method using naphthalene, IP 8.12 eV, as standard.   Adding the measured
differences to the IP of the standard gives an IP of 8.23 eV for anisole and an AP
of 11.68 eV for the (M − CH₃)⁺ ion.

which the ion current is 10^{-4} to 10^{-2} of its value at 70 eV. A semilog plot over this region
gives straight and essentially parallel lines for most organic ions. The results obtained by this
technique are illustrated in Figure 4-20. Naphthalene is used as the standard since its ionization potential (8.12 eV) is accurately known from photoionization experiments. The naphthalene molecular ion and the two ions of interest, the anisole molecular ion and the $(M - CH_3)^+$
fragment ion, give good straight lines on the semilog plot. Naphthalene is an excellent
standard for the anisole molecular ion, but its IP is too low for it to serve as the most accurate
standard for the $(M - CH_3)^+$ fragment ion. The measured value of the anisole IP (8.23 eV)
is in good agreement with literature photoionization values (8.20 and 8.22 eV).

Ionization and appearance potential values can be used to determine the heats of formation of gaseous ions, from which can be calculated bond dissociation energies, heats of formation of free radicals, and other relatively inaccessible thermochemical quantities. The usefulness
of mass spectrometry as a route to thermochemical parameters is best illustrated by some
examples. All heats of formation refer to the gas phase.

Acetonitrile gives CH_3^+ ions with a measured AP of 14.6 eV (eq. 4-15). Making the

$$CH_3CN + e^- \rightarrow CH_3^+ + CN^\bullet + 2e^- \qquad \Delta H \text{ reaction} = 14.6 \text{ eV (336.7 kcal mole}^{-1}) \qquad 4\text{-}15$$

usual assumption that no excess energy is involved in this reaction, we write eq. 4-16. The

$$\Delta H \text{ reaction} = \Delta H_f(CH_3^+) + \Delta H_f(CN^\circ) - \Delta H_f(CH_3CN) = 336.7 \text{ kcal mole}^{-1} \qquad 4\text{-}16$$

quantity $\Delta H_f(CH_3CN)$ is known from calorimetric methods to be 20.9 kcal mole^{-1}, and the
heat of formation of CH_3^+ is accurately known (260 kcal mole^{-1}) from mass spectrometric
methods (e.g., photoionization of methane) and from spectroscopic measurements on $CH_3^\bullet$.
Hence, from eq. 4-17, we obtain a quantity that compares well with the value of about

$$\Delta H_f(CN^\bullet) = 336.7 + 20.9 - 260 = 97.6 \text{ kcal mole}^{-1} \qquad 4\text{-}17$$

100 kcal mole^{-1} obtained by photoionization of HCN, by shock tube experiments, and by electronic absorption studies of the reaction $CN^{\cdot} + H_2 \rightarrow HCN + H^{\cdot}$.

After calculation of $\Delta H_f(CN^{\cdot})$, the bond dissociation energy of the C—C bond in acetonitrile is easily obtained, (eq. 4-18).

$$D(CH_3\text{—}CN) = \Delta H_f(CH_3^{\cdot}) + \Delta H_f(CN^{\cdot}) - \Delta H_f(CH_3CN)$$

$$= 33.2 + 97.6 - 20.9 = 109.9 \text{ kcal mole}^{-1} \qquad \text{4-18}$$

The measurement of the ionization potentials of free radicals is a relatively difficult task, yet it provides direct data on heats of formation of radicals. For example, the ethyl radical formed by thermolysis of tetraethyllead has an IP (electron impact) of 8.4 eV (194 kcal mole^{-1}). If the heat of formation of $C_2H_5^+$ is taken as 223 kcal mole^{-1} (obtained from the AP of the ethyl ion formed from ethane), then the result of eq. 4-19 is close to the accepted

$$\Delta H \text{ (reaction)} = 194 = \Delta H_f(C_2H_5^+) - \Delta H_f(C_2H_5^{\cdot})$$

$$\therefore \Delta H_f(C_2H_5^{\cdot}) = 223 - 194 = 29 \text{ kcal mole}^{-1} \qquad \text{4-19}$$

value (25 kcal mole^{-1}), the difference probably arising because the ion formed from ethane has some excess energy at threshold.

Determination of the heats of formation of gaseous ions is an important means of characterizing ions. For example, an ion $C_2H_5O^+$ can be generated from methyl ethers and secondary alcohols, but it is not known whether these ions have the same structure. For dimethyl ether $AP(C_2H_5O^+) = 10.70 \pm 0.1$ eV (247 kcal mole^{-1}). Hence from $\Delta H_f(CH_3OCH_3) = -44$ kcal mole^{-1} and $\Delta H_f(H^{\cdot}) = 52$ kcal mole^{-1}, we get eq. 4-20.

$$\Delta H \text{ reaction} = 247 = \Delta H_f(C_2H_5O^+) + \Delta H_f(H^{\cdot}) - \Delta H_f(CH_3OCH_3)$$

$$\therefore \Delta H_f(C_2H_5O^+) = 247 - 52 - 44 = 151 \text{ kcal mole}^{-1} \qquad \text{4-20}$$

Other determinations, including one in which the electrons are energy selected prior to entering the ion chamber, gave slightly higher values, the best value being approximately 155 kcal mole^{-1}. This figure is to be contrasted with a value of 141 kcal mole^{-1} for the ion generated from a series of secondary alcohols. It is thought that the ion generated from dimethyl ether does not possess appreciable excess energy at threshold since the kinetic energy release accompanying the fragmentation of the metastable ion is very small, 0.1 kcal mole^{-1}. Hence the ions are indeed isomeric and the structures $CH_3\overset{+}{\text{—}}O{=}CH_2$ and $CH_3\text{—}CH{=}\overset{+}{O}\text{—}H$ are suggested. Independent results on ion-molecule reactions confirm these conclusions.

BIBLIOGRAPHY

1. Molecular Weight Determination (Alternative Methods of Ionization): H. D. Beckey in A. L. Burlingame, ed., *Topics in Organic Mass Spectrometry*, Wiley–Interscience, New York, 1970; H. M. Fales in G. W. A. Milne, ed., *Mass Spectrometry: Techniques and Applications*, Wiley–Interscience, New York, 1971; J. M. Wilson, *Mass Spectrom. (Spec. Period. Report)*, **1** (1971); H. M. Fales, G. W. A. Milne, H. U. Winkler, H. D. Beckey, J. N. Damico, and R. Barron, *Anal. Chem.*, **47**, 207 (1974).

2. Exact Mass Measurements: G. W. A. Milne, *Quart. Rev.*, **22**, 75 (1968); K. Biemann, *Advan. Mass Spectrom.*, **4**, 139 (1968); R. Venkataraghvan, R. J. Klimowski, and F. W. McLafferty, *Accounts Chem. Res.*, **3**, 158 (1970).
3. Isotopic Incorporation Determinations: K. Biemann, *Mass Spectrometry: Organic Chemical Applications*, McGraw-Hill, New York, 1962.
4. Molecular Structure Elucidation: See reference (1) of Chapter 1.
5. Stereochemical Assignments: S. Meyerson and A. W. Weitkamp, *Org. Mass Spectrom.*, **1**, 659 (1968); K. Pihlaja and J. Jalonen, *Org. Mass Spectrom.*, **5**, 1363 (1971); M. M. Green, R. J. Cook, J. M. Schwab, and R. B. Roy, *J. Amer. Chem. Soc.*, **92**, 3076 (1970); K. C. Kim and R. G. Cooks, *J. Org. Chem.*, **40**, 511 (1975).
6. Sequence Determinations: B. C. Das and E. Lederer in A. Niederwiess and G. Pataki, ed., *New Techniques in Amino Acid, Peptide and Protein Analysis*, Ann Arbor Science Publishers, Ann Arbor, Mich., 1971; K. Biemann in G. R. Waller, ed., *Biochemical Applications of Mass Spectrometry*, Wiley–Interscience, New York, 1972; G. S. Johnson, W. S. Ruliffson, and R. G. Cooks, *Carbohydrate Res.*, **18**, 233 243 (1971); T. J. Mead, H. R. Morris, J. H., Bowie, and I. Howe, *Mass Spectrom.(Spec. Period. Report)*, **2** (1973).
7. GC–MS: S. Stallberg-Stenhagen and E. Stenhagen in A. L. Burlingame, ed., *Topics in Organic Mass Spectrometry*, Wiley–Interscience, New York, 1970; W. H. McFadden, *Techniques in Combined Gas Chromatography/Mass Spectrometry: Applications in Organic Analysis*, Wiley, New York, 1973.
8. Thermochemical Determinations: J. L. Franklin in G. A. Olah and P. von R. Schleyer, eds., *Carbonium Ions*, vol. 1, Interscience, New York, 1968, ch. 2; A. G. Harrison in A. L. Burlingame, ed., *Topics in Organic Mass Spectrometry*, Wiley–Interscience, New York, 1970.

ION CHEMISTRY

5-1 Ionization Phenomena

5-1 Positive Ions

We emphasize here those processes of positive ion formation that can be observed in an analytical sector mass spectrometer equipped with an electron impact source. Other methods of ionization are also discussed and, in the second part of the section, we remark on negative ion formation. The basic ionization reactions have already been given in section 1-2.

The electron-impact-induced vertical transition from the neutral species to some charged state is discussed in section 1-2. The separate consideration of electronic and nuclear motion (the Born–Oppenheimer approximation) is justified for electron impact ionization by the fact that a 70 eV electron moves 5 Å in 10^{-16} sec. Since nuclear vibrations are at least several orders of magnitude slower, an ionizing electron will have completed its interaction with a molecule before the nuclei can adjust appreciably. Electron impact therefore is characterized by vertical (Franck–Condon) excitation, and the lowest vibrational level of the ground state will not necessarily be populated preferentially.

Important properties of the ionization process are its efficiency (cross section) and the manner in which this property varies with the energy of the ionizing electrons. Energies in the neighborhood of the ionization potential—that is, in the threshold region—are of particular interest. Electron impact obeys a linear threshold law of the form $[M^{+\cdot}] \propto E_e - \epsilon_0$. In other words, the abundance of the molecular ion is linearly dependent upon the energy available in excess of the minimum energy ϵ_0 required to cause ionization. Cross sections for electron impact excitation reach a maximum at about 70 eV and then slowly decrease. Plots of ion abundance vs. electron energy (E_e) are known as ionization efficiency curves. An upward break in an ionization efficiency curve (Figure 5-1a) indicates that a new process starts to contribute to the formation of the ion, again with a linear dependence of ion current upon

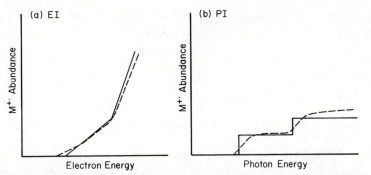

FIGURE 5-1. Ionization efficiency curves for molecular ions generated by (a) electron impact and (b) photoionization. The full lines represent an idealized situation, the broken lines the actual situation in large molecules.

excess energy.   The positions of upward breaks in the curve can sometimes be related to the energies of excited states of the ion in question. These features are seen far more readily in photoionization efficiency curves since the threshold law for photoionization is a step function (Figure 5-1b).

The ionization efficiency curves of large organic ions (discussed in connection with IP and AP determinations, section 4-9) usually show no detectable breaks when electron impact is used because the states lie close together. The importance of observing and defining the energies of excited states of ions, however, has led to much careful study of these curves. Their finer features become much more apparent if the second derivative of the electron impact curve (Figure 5-2) or the first derivative of the photoionization curve is plotted. Further detail becomes apparent if the curve is deconvoluted with respect to the energy distribution of the ionizing radiation.   Indeed, for a compound that shows no breaks in the ionization efficiency curve, the second derivative gives directly the energy distribution of the ionizing electrons. Measurements of this type tell us that the average energy spread at half height for electrons emitted from a rhenium filament is of the order of several tenths of an electron volt.

One of the factors that complicate ionization efficiency curves is the occurrence of autoionization (sometimes termed preionization), in which neutrals are excited to states with energies above the ionization potential and the resulting superexcited species then undergo

FIGURE 5-2. Ionization efficiency curves for electron impact together with their first and second derivatives. Curve (i) represents the idealized case in which there is no energy spread in the electron beam; curve (ii) represents the actual situation where the electron energy spread is several tenths of an electron volt; curve (iii) represents a case where two states of the ion are populated.

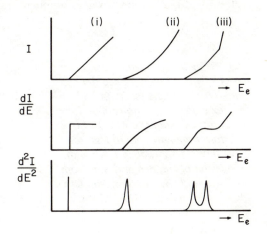

curve crossing to yield the molecular ion. The initial step is a resonance process for photo-ionization (eq. 5–1), but not for electron bombardment (eq. 5–2) where excess energy can

$$AB + h\nu \longrightarrow AB^* \qquad\qquad 5\text{--}1$$

$$AB + e^- \longrightarrow AB^* + e^- \qquad\qquad 5\text{--}2$$

$$AB^* \longrightarrow AB^{+\,\cdot} + e^- \qquad\qquad 5\text{--}3$$

be carried away by the electron. These autoionization processes produce sharp peaks in photoionization efficiency curves, and the ions formed (eq. 5–3) are readily recognized.

A related phenomenon is predissociation, in which an excited state of the ion may cross with a nonbonding state so that ionization occurs to give fragment ions directly. If predissociation occurs slowly, because it is spin or symmetry forbidden, then it will give rise to a metastable peak. Little is known regarding the occurrence of curve-crossing processes in complex organic ions, but a growing number of cases are being identified in simple species.

A related consideration in the general area of ionization phenomena concerns the existence of isolated electronic states of an ion that are not in equilibrium with highly excited ground state ions of the same energy. Again this situation is relatively common in simple ions, but for complex organic ions it has generally been considered unimportant. Such states would be recognizable by the fact that they behave as if they possess far less internal energy than they actually do. Their detection is difficult because the internal energy distribution of an ion is a most difficult quantity to measure. The charge exchange method of ionization (see section 1–2), however, does give ions of closely defined internal energy, and, if several reactant ions are used, molecular ions ($M^{+\,\cdot}$) can be formed with each of several different internal energies. The fragmentation behavior of the molecular ion can be followed, and the results plotted in the form of a breakdown pattern as a function of internal energy. Analogous data can be obtained from photoionization *if* the energy of the released electron is measured in coincidence with that of the ion. Breakdown curves are a direct measure of the relationship between k (the unimolecular rate constant) and ϵ (the internal energy of the reactant) and thus are of primary importance in kinetics. (Breakdown patterns are essentially a more fundamental form of mass spectra; see section 5–2.) A breakdown pattern for methyl *n*-propyl ketone is shown in Figure 5–3. Two points deserve special notice. First, the appearance potential for loss of the smaller methyl radical is the lower (compare section 4–5b). The crossover to dominance of propyl loss is in agreement with the observed

FIGURE 5–3. Idealized breakdown pattern for 2-pentanone derived from charge exchange data. Note the existence of molecular ions with internal energies $\epsilon(M^{+\,\cdot})$ of 6–7 eV that fragment much less than expected. [Reproduced with the permission of Heyden and Son from J. Turk and R. H. Shapiro, Org. Mass Spectrom., 5, 1373 (1971).]

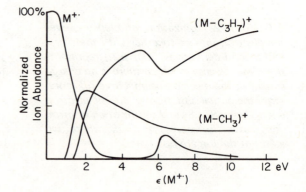

$(M - CH_3)^+/(M - C_3H_7)^+$ ratio in the 70 eV mass spectrum (which weights energies in the range of several eV most heavily). The second point is the occurrence of a maximum at high energy in the molecular ion curve. These high energy molecular ions obviously do not behave normally since they fragment far less than their internal energies suggest that they should. They are, therefore, probably indicative of the presence of an isolated state(s) of $M^{+\cdot}$ in this energy range. An alternative explanation is that the ions do not in fact possess the energy indicated because they have suffered energy loss by radiation. Experiments have been done in which the radiation emitted from the ion chamber is detected and monitored, but little is yet known about this phenomenon.

We turn now from electron impact to alternative methods of ionization. Photoionization has been partly covered in the preceding pages. The apparatus is complex; radiation in the vacuum ultraviolet is needed so a gas discharge lamp is usually attached directly to the ion source. The lack of transparent window materials in this energy range means that differential pumping and high pumping speeds have to be handled. Spectra resemble electron impact spectra at comparable energies (20 eV or less). The sensitivity of the electron bombardment technique is the greater, especially when 70 eV electrons are used. The great advantage of photoionization, however, is obtained near threshold; the sharp onsets lead to better sensitivity at threshold, and the accuracy with which ionization potentials can be measured is high (typically ± 0.01 eV, i.e., 0.2 kcal mole^{-1}).

In field ionization a sharp edge such as a razor blade (radius of curvature ~ 1000 Å) is subjected to a high positive potential (~ 8 kV) and surface fields of the order of 10^8 V cm^{-1} develop. As a molecule approaches the surface, its Coulomb potential is deformed and the probability that an electron will tunnel from the molecule to the tip becomes large. This process occurs without excitation of the molecule. Immediately after ionization occurs, the ion is accelerated from the anode by the intense field of several volts per Ångstrom. Almost all fragmentation therefore occurs some distance from the anode and corresponds to metastable fragmentations in electron impact spectra. If fragmentation occurs during acceleration, the daughter ion can still be detected but will not be sharply focused. These processes with lifetimes of 10^{-11} to 10^{-8} sec are referred to as fast metastable decompositions, in contrast to the slow 10^{-8} to 10^{-6} sec decompositions that occur in the field free region reached after complete acceleration. The relation between ion lifetime and fragment ion kinetic energy can be established experimentally for fast metastable decompositions, so that the kinetics of these very fast unimolecular reactions can be studied.

The above considerations explain why field ionization frequently has been characterized as both a "soft" and a "fast" ionization technique. Fragment ions are observed at their normal m/e positions, provided that they are generated within $\sim 10^{-13}$ sec. Under these conditions the fragment ions are of low abundance and result predominately from simple cleavage. The reactions that do give rise to normal daughter ions may also be field induced: the presence of the very intense field deforms the potential energy surface of the ion sufficiently to allow low energy fragmentation routes not otherwise possible.

A comparison of the field ionization spectrum of diphenyl ether with the electron impact spectrum highlights several of the major characteristics of FI spectra (Figure 5-4; see also Figures 4-3 and 4-4). The molecular ion comprising $\sim 90\%$ of the total singly charged ion current is much more abundant in the FI spectrum. Most of the remainder of the ion current is due to its ^{13}C isotope. The most abundant fragment ion is m/e 77, formed by simple cleavage. The skeletal rearrangement ion $(M - CO)^{+\cdot}$, so important in the electron impact spectrum, cannot be observed at all.

The intricate ion chemistry of FI is illustrated further by the occurrence of ion-molecule

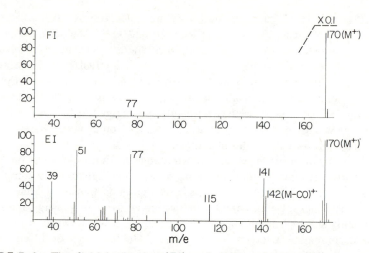

FIGURE 5-4. *The field ionization (FI) and electron impact (EI) mass spectra of diphenyl ether. Fragment ions are barely visible in the FI spectrum, in spite of being plotted at ten times the molecular ion sensitivity. The most abundant fragments in the EI spectrum, m/e 39, 51, 77, 115, and 141, occur commonly in the spectra of aromatic hydrocarbons. [Reproduced with the permission of Elsevier Publishing Company from H. D. Beckey, H. Hey, K. Levsen, and G. Tenschert, Int. J. Mass Spectrom. Ion Phys., **2**, 101 (1969).]*

reactions that yield species with masses greater than the molecular weight. In particular high local concentrations of polar species can accumulate near the tip and can be responsible for these effects. The sensitivity of the method is rather low compared to electron impact, and there are practical problems in maintaining the requisite strong fields. Nevertheless, the exceptional abundance of the molecular ion makes this a valuable ionization technique for molecular weight determination.

In chemical ionization (see sections 1-2 and 4-1), the basic ion-molecule reactions (proton transfer, hydride abstraction) that lead to ionization of the sample are relatively simple processes individually, although complex mixtures of ion-molecule and unimolecular reactions may occur in the ion source. The chemical ionization spectra of 4-methylbenzophenone with isobutane and methane as reactants (Figure 5-5) illustrate (i) the typical high abundance of the $(M + 1)^+$ species; (ii) the simple fragmentation pattern including the structurally diagnostic Ar^+ and $ArCO^+$ ions; (iii) the presence of ions of mass greater than $(M + 1)^+$; and (iv) the degree of control that can be exercised over the extent to which fragmentation occurs.

One of the less well-known methods of ionization, Penning ionization, involves reaction of the compound of interest with an electronically excited neutral species. Metastable rare gas atoms are frequently used as reagents (eq. 5-4). Figure 5-6 compares the 70 eV electron

$$Ne^* + C_2H_5OH \longrightarrow Ne + C_2H_5OH^{+\cdot} + e^- \qquad\qquad 5\text{-}4$$

impact spectrum of ethanol with that obtained by Penning ionization using Ne metastables in the $^3P_2$ and $^3P_0$ states. These metastables have internal energies of 16.6 and 16.7 eV relative to the Ne ground state. Since the ionization potential of ethanol is 10.5 eV, the maximum energy available for excitation of the ethanol molecular ion is 6.2 eV. The actual excitation energy will be somewhat less than this amount since the ejected electron may carry off some kinetic energy.

The striking similarities between the 70 eV electron impact and the neon Penning

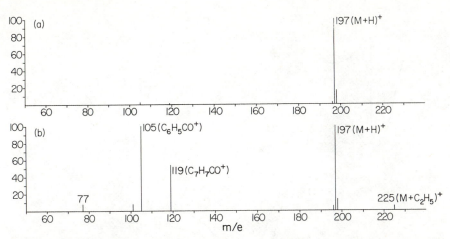

FIGURE 5-5. Chemical ionization spectra of 4-methylbenzophenone using (a) isobutane and (b) methane as reagents. The $(M + H)^+$ ion formed from CH_5^+ (the
reactive ion when methane is the reagent) has much more internal energy than that
formed from $t\text{-}C_4H_9^+$ (reactive ion in the isobutane mixture); consequently more
fragmentation is observed in spectrum (b) than in (a). [Reproduced with the permission of Heyden and Son from J. Michnowitz and B. Munson, Org. Mass
Spectrom., **4**, 481 (1970).]

ionization spectra provide strong support for the contention that the average internal energy of molecular ions formed from 70 eV electrons is only a few electron volts. This fact deserves emphasis because of the misapprehension that much higher internal energies are involved. The reasons why so little of the electron energy is transferred to the ion are twofold.

(i) Many electron–neutral molecule interactions are relatively weak, so they cause ionization but little excitation; stronger interactions are statistically less probable.

(ii) Secondary electrons, which have lower energies, are probably responsible for some fraction of the ionization events in an electron impact source.

In this discussion of ionization phenomena we have so far dealt entirely with the

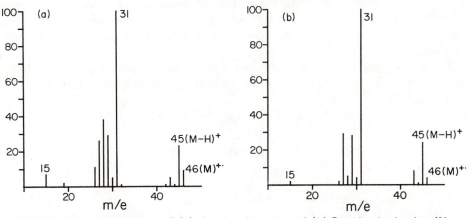

FIGURE 5-6. Comparison of (a) electron impact and (b) Penning ionization (Ne;
$^3P_2$ and $^3P_0$) spectra of ethanol. [(b) reproduced with the permission of Elsevier
Publishing Company from E. G. Jones and A. G. Harrison, Int. J. Mass Spectrom.
Ion Phys., **5**, 137 (1970).]

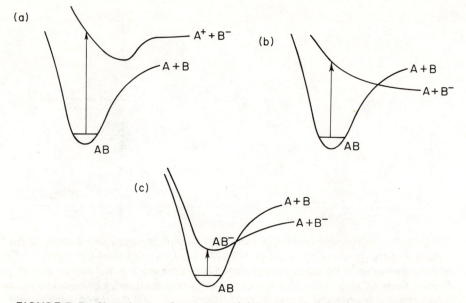

FIGURE 5-7. Negative ion formation: (a) ion pair production (excitation by an energetic electron or by other means); (b) dissociative electron attachment; (c) electron capture, a resonance process.

yield and properties of the molecular ion and have ignored the electrons produced simultaneously. The energies of these electrons, however, provide a complementary source of information on the energy transfer process. Photoelectron spectroscopy and the photoelectron–photoion coincidence method are aspects of this field. Electron energy spectroscopy also has been applied to electron impact and to Penning ionization. Electron energy analysis constitutes one of the few available experimental approaches to the internal energy distribution $P(\epsilon)$ and is an invaluable source of accurate thermochemical data on ions.

5-1b Negative Ions

For two primary reasons, studies on negative ion mass spectra have lagged very much behind those on positive ions. First, negative ion abundances are two or three orders of magnitude lower than those of positive ions. Second, the reactions by which negative ions are generated in the mass spectrometer are both more various and less well understood than those that lead to positively charged ions.

The major reactions leading to negative ions are illustrated in Figure 5-7. In ion-pair production (eq. 5-5) AB undergoes a vertical excitation on interaction with an electron or

$$AB + e^- \longrightarrow A^+ + B^- + e^- \qquad 5\text{-}5$$

photon to give the species AB*, which correlates with $A^+ + B^-$. This process is important under normal mass spectrometer source conditions (i.e., $\sim 10^{-5}$ torr pressure, >20 eV electrons). It cannot give negatively charged molecular ions.

At lower electron energies the electron may become attached to the molecule AB. Frequently such attachment will be followed, within a molecular vibration, by dissociation, giving rise to the dissociative attachment mechanism (eq. 5-6). As an alternative to dissocia-

$$AB + e^- \longrightarrow (AB^{-\cdot}) \longrightarrow A^{\cdot} + B^- \qquad 5\text{-}6$$

tion, the captured electron can be given up in an Auger process. If the AB⁻· state reached is
bonding, the lowest electronic state of AB⁻· often will be repulsive (unlike AB⁺·, which
will normally be bonding in the ground state, although less strongly so than AB) and, just
as for positively charged ions, radiationless transitions will tend to convert the internal energy
of the ion into vibrational energy of the ground electronic state. Hence dissociation will follow
immediately if the ground state is reached.

Negatively charged molecular ions are observed for some compounds. They arise by the
electron attachment reaction (eq. 5-7), which is a reasonance process. If the molecular ion is

$$AB + e^- \longrightarrow AB^{-\cdot} \qquad\qquad 5\text{-}7$$

formed in an excited state, it may reach the ground state by photon emission.

The negative ion spectrum of anisole obtained under normal conditions (70 eV, source
pressure $\sim 10^{-5}$ to 10^{-6} torr) is shown in Figure 5-8. This spectrum illustrates the principle
that fragmentation patterns in negative as well as positive ion spectra may be predicted from
relative ionic stabilities known from solution studies. No molecular ion is observed, but the
fragment of highest mass is the phenoxyl anion, $C_6H_5O^-$. The most abundant fragment ions
belong to the series C_nH^-. It is eminently reasonable that the acetylide anion should be the
most stable hydrocarbon fragment as is also the case in the spectra of aliphatic compounds.
The presence of a moderately abundant $O^{-\cdot}$ ion should also be noted.

Abundant negative molecular ions (molecular anions) can be produced if a large number
of very low energy electrons are made available. This situation can be effected by mixing
the sample with a much larger quantity of a gas such as argon in a low voltage gas discharge
ion source (10^{-2} to 10^{-3} torr). A high density of thermal energy electrons is produced
in the plasma, and resonance electron capture by the sample is promoted. An example of
structural analysis using this method of negative ion spectrometry is quoted in section 4-1.
Abundant ions in the molecular ion region $[(M^{-\cdot}), (M + H)^-, (M - H)^-,$ etc.] are a feature
of these spectra.

With certain types of electron impact ion sources, negative ions can be formed in good
abundance using the experimental conditions normally associated with positive ion mass
spectra, but with reversed polarities of the ion-accelerating field, the electric sector field,
and the magnetic field. The fact that molecular anions ($M^{-\cdot}$) are observed indicates that
low energy electrons must be responsible for ionization even though the ionizing electron
beam has an energy of 70 eV. The mechanism of formation of the slow electrons is not
known, although they may be generated in the course of positive ion formation. Molecular
anion formation requires that there be stabilization of both a radical and an anion. Of the
isomeric diacetylbenzenes only the ortho and para compounds give molecular anions using
the normal conditions for obtaining positive ion mass spectra but with reversed polarities.
Resonance stabilization of the anion, e.g., **1**, is possible for these isomers, but not for *m*-diacetyl-

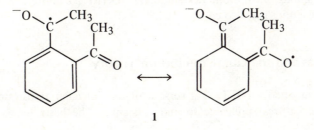

1

benzene. Fragmentation of molecular anions is generally limited because of the low internal

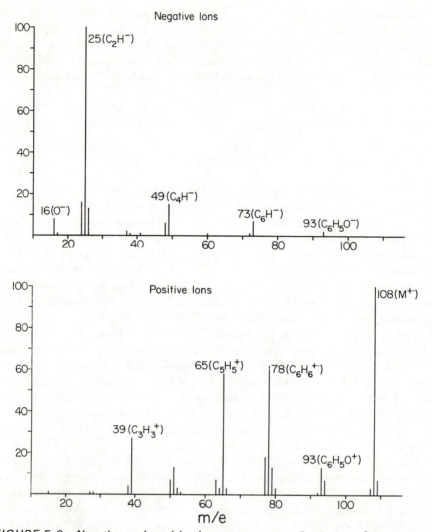

FIGURE 5–8. *Negative and positive ion mass spectra of anisole.* *[Upper repro-duced with the permission of the American Chemical Society from R. T. Aplin, H. Budzikiewicz, and C. Djerassi, J. Amer. Chem. Soc., 87, 3180 (1965).]*

energies of the molecular ions. The processes that do occur, however, are relatively simple to rationalize. For example, carboxylic acids lose H˙ to give carboxylate anions, ethers give alkoxyl or aroxyl anions, nitro aromatics eliminate NO˙ to give aryloxyl anions, and hydrogen atom abstraction reactions can also be observed.

5–2 Ion Kinetics and the Quasiequilibrium Theory

The quasiequilibrium theory (QET) is a version of unimolecular transition state theory pos-sessing some extra features applicable to ionic systems. The basic tenets of the theory are

(a) A mass spectrum is the result of a network of competing and consecutive unimolecular reactions of energized ions that represent isolated systems.

(b) The reaction rate depends on the concentration of activated complexes.

(c) Activated complexes are in equilibrium with reactant ions.

(d) The fraction of ions in the activated complex configuration is equal to the fraction of states that represent activated complexes.

In more immediate terms, energetic ions undergo internal energy transfer and return to the ground electronic state, from which fragmentation takes place. The rate of any fragmentation reaction depends on the probability that an ion will adopt the activated complex configuration and consequently depends on the energy and entropy of activation. The collection of ions present at a certain time after ionization (the ion source residence time, ~1 μsec) will comprise the mass spectrum.

The rate constant formulation used in QET is derived from absolute reaction rate theory. Early QET calculations employed approximate rate expressions; more recently, there has been a tendency to employ exact counting of states and to use the formulation of Rice, Ramsperger, Kassel, and Marcus (RRKM). In this and other statistical theories, the reaction rate $k(\epsilon)$ at a given ion internal energy ϵ depends on the ratio of the number of states representing activated complexes to all other states with energy ϵ. Obviously, reaction rate theory is only absolute if the molecular parameters of the energized reactants and the activated complex are known—a formidable task indeed. The usual procedure is to assume a "reasonable" activated complex configuration and to compare theory with experiment. Since the total number of states of a given energy ϵ is required, the density of states (number of states per unit of energy) is an important parameter in rate calculations. This density can be determined directly by counting states, or approximately by other methods. Both techniques, however, rely on prior assumptions of transition state properties.

In the most general form, the reaction rate is given by eq. 5-8, where $[AB^{+\ddagger}]$ is the

$$k(\epsilon) = [AB^{+\ddagger}] \times (\text{rate of passage over barrier}) \qquad 5\text{-}8$$

density of activated complex states. Since equilibrium between reactant and activated complexes is maintained, we have eq. 5-9. The equilibrium constant in eq. 5-9 is given by

$$[AB^{+\ddagger}] = K^{\ddagger} [AB^{+}] \qquad 5\text{-}9$$

eq. 5-10, where $\rho(\epsilon)$ is the density of states of reactant ions with energy ϵ and $\rho^{\ddagger}(\epsilon^{\ddagger})$ is the

$$K^{\ddagger} = \frac{\rho^{\ddagger}(\epsilon^{\ddagger})}{\rho(\epsilon)} \qquad 5\text{-}10$$

density of activated complex states with nonfixed energies, $\epsilon^{\ddagger}$, between 0 and $(\epsilon - \epsilon_0)$. The summation over $\epsilon^{\ddagger}$ from 0 to $(\epsilon - \epsilon_0)$ arises because one degree of freedom in the reactants represents translation (i.e., fragmentation) in the activated complex and can take any energy between zero and the total nonfixed energy of the activated complex.

Standard activated complex theory gives eq. 5-11, and by substitution we get eq. 5-12.

$$K^{\ddagger} = k(\epsilon)h \qquad 5\text{-}11$$

$$k(\epsilon) = \left(\frac{1}{h}\right)\left(\frac{\rho^{\ddagger}(\epsilon^{\ddagger})}{\rho(\epsilon)}\right) \qquad\qquad 5\text{-}12$$

A fundamental fact emerging from the theory is that for any given energy ϵ, the rate constant represents an average rate of passage over the barrier (depending on the value of ϵ), and hence $k(\epsilon)$ represents a dissociation probability, not a precise ion lifetime.

The determination of $k(\epsilon)$ using RRKM theory cannot be detailed further here. A good feel for the principles involved in mass spectral kinetics is provided by a consideration of a simplified $k(\epsilon)$ relation in which the density of states is not exactly counted but is mathematically approximated using the harmonic oscillator model (eq. 5-13, where ν represents

$$k(\epsilon) = \nu\left(\frac{\epsilon - \epsilon_0}{\epsilon}\right)^{s-1} \qquad\qquad 5\text{-}13$$

a frequency (entropy) factor and s is the total number of degrees of freedom of the reactant ion). To obtain agreement with experiment and still use reasonable values of ν ($\sim 10^{14}$ sec^{-1} for simple cleavages, smaller for rearrangements), it is found that the exponential term in this approximate expression must be reduced by a factor of 2 or 3. The major properties and consequences of the $k(\epsilon)$ relationship are as follows (compare the introductory discussion, section 1-3).

(a) At high internal energy, $k(\epsilon)$ tends toward ν, i.e., the reaction rate is controlled by the frequency factor.

(b) Consequently, reactions with high frequency factors such as simple bond cleavages should predominate over low frequency factor reactions such as rearrangement and elimination reactions in high energy ions.

(c) At low internal energies, i.e., in the threshold region, the rate constant will be controlled by the activation energy. A low activation energy will increase the nonfixed energy, and hence the density of activated complex states, thereby increasing the rate constant.

(d) Consequently, two competitive low energy reactions (such as metastable fragmentations) will only occur from the same precursor ion if the activation energies of the two processes are similar. Otherwise only the process with lower activation energy will be observed.

(e) The larger the reacting ion is, the greater are the number of degrees of freedom and, hence, the smaller the $k(\epsilon)$ for a given ϵ.

(f) It follows that the rate of increase of $k(\epsilon)$ with increasing ϵ will be smaller for large ions. This factor, together with the frequency factor ν, determines the magnitudes of the kinetic shift,[1] which is small for small ions and for simple cleavages and larger for large, ions and for rearrangements.

The consequences of these facts have been met repeatedly in previous chapters. The facts themselves can be adduced from the $k(\epsilon)$ of eq. 5-12 instead of eq. 5-13, if frequency factors are replaced by the actual vibrational frequencies.

In order to calculate mass spectra, it is necessary to know (a) ϵ_0 and $k(\epsilon)$ for each

---

[1] The kinetic shift is the internal energy, in excess of the activation energy, that must be supplied to a reactant ion in order to raise the rate of reaction so that product formation can be observed on the time scale appropriate to the experiment. Thus fragmentation in the ion source must occur at a rate of $\sim 10^6$ sec^{-1} (the source residence time is $\sim 10^{-6}$ sec) and the energy required to increase the reaction rate from the minimum rate to $k = 10^6$ is the kinetic shift.

reaction involved, (b) $P(\epsilon)$, and (c) the characteristic times of the instrument. The determination of $P(\epsilon)$ has been discussed in section 5-1a. The internal energy distribution in a fragment ion is usually derived from $P(\epsilon)$ for molecular ions, assuming a statistical distribution of excess energy between the ionic and neutral fragments. The instrumental parameters necessary are the source residence time, if daughter ion abundances are to be calculated, or the range of lifetimes during which the ion traverses a field-free region, if metastable ion abundances are required. An excellent method of calculating the mass spectrum of a compound involves prior calculation of the breakdown curve, which is a plot of the ionic distribution as a function of the internal energy of the molecular ion. The breakdown pattern for methyl propyl ketone is shown in Figure 5-3. Each position on the energy axis represents the mass spectrum that would be observed if molecular ions received the specified internal energy. At low internal energies no fragmentation occurs and only the molecular ion is observed; as the internal energy is raised other processes begin to compete. A breakdown curve is not entirely instrument independent, since the time available for reaction must be specified. It is, however, more fundamental than a mass spectrum that is merely a weighted sum of points on the breakdown curve.

The breakdown pattern for a compound is calculated from the known fragmentation pattern of the compound, the activation energies ϵ_0, and the $k(\epsilon)$ relationships derived from QET. It can then be transformed into a mass spectrum using an internal energy distribution $P(\epsilon)$ obtained from photoelectron spectroscopy or by other means. Comparison of the resulting mass spectrum with that found experimentally constitutes the classical and most rigorous test of the QET theory. Breakdown curves are also accessible experimentally: a charge-exchange mass spectrum obtained with a given reagent constitutes one set of points on the breakdown curve, as discussed in section 5-1.

Calculations of the mass spectra of large organic ions are still in a rudimentary stage. The procedure just outlined can successfully reproduce the major features of the spectra of such molecules as propane and thiabutane, but even in these cases adjustments of the activated complex parameters to fit the results cannot be rigorously justified. Fortunately, it is possible in many cases to predict semiquantitatively the major features of the mass spectra of large organic molecules. This is the case when the reactions involved are under thermodynamic control, i.e., when ion abundances reflect relative appearance potentials. This condition is approximately true when only simple bond cleavages occur. Figure 5-9 illustrates the situation for a case in which only two reactions are possible, $AB^{+\cdot} \rightarrow A^+ + B^\cdot$ and $AB^{+\cdot} \rightarrow B^+ + A^\cdot$. Estimation of the relative abundances of $[AB^{+\cdot}]$, $[A^+]$, and $[B^+]$ requires that the ionization potential of AB, the appearance potentials of A^+ and B^+, and the electron energy E_e be known. $P(\epsilon)$ may be measured or a model distribution may be used. To a first approximation, area **1** will represent the molecular ion abundance (ions with insufficient energy to

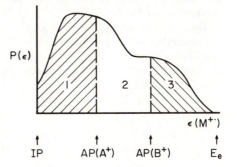

FIGURE 5-9. *Relationship between the internal energy distribution $P(\epsilon)$ of molecular ions $AB^{+\cdot}$, the appearance potentials of the fragments A^+ and B^+, and the relative abundances of $AB^{+\cdot}$, A^+, and B^+. Note that A^+ and B^+ formation is energetically possible in area 3, but for simple bond cleavages B^+ will usually not compete successfully.*

undergo any reaction), and areas **2** and **3** the abundance of A^+. The abundance of B^+ will be approximately zero. In a crude approximation, ions with sufficient energy to fragment in the source ($k \geqslant 10^6$ sec^{-1}) *all* do so, and of those with smaller rate constants *none* do so.

It should be apparent that this approach could be improved to include consecutive fragmentations and extended to allow for differences in frequency factors, but it will be illustrated in its most primitive form. Table 4-3 provides thermochemical data for a series of monosubstituted benzenes. Each compound undergoes one major primary reaction, and a rough correlation is seen between the activation energy for this process and the fragment ion/ molecular ion abundance ratio at 18 eV. This information allows one to predict the spectra of disubstituted benzenes. For example, substituted bromobenzenes should show relatively little Br$^\bullet$ loss when this process is in competition with methyl loss from an acetyl group, but successively more Br$^\bullet$ loss as the series (Table 4-3) is descended. In agreement with this prediction, *p*-bromoacetophenone gives a $(M - Br)^+/(M - CH_3)^+$ ratio of 0.004, *p*-bromo-toluene a $(M - Br)^+/(M - H)^+$ ratio of 7.2, and *p*-bromochlorobenzene a $(M - Br)^+/(M - Cl)^+$ ratio of 63.

Recently $k(\epsilon)$ measurements have been made directly for the first time using an instrument capable of selecting the lifetimes of the ions to be examined. The measured curves for H$^\bullet$ and for C_2H_2 loss from benzene are shown in Figure 5-10. These reactions have similar (and unusually high) activation energies, ~ 4 eV; both give metastable peaks and both give daughter ions of similar relative abundances in normal 70 eV mass spectra. These facts are consistent with the experimental $k(\epsilon)$ curves only if the two reactions $C_6H_6^{+\bullet} \rightarrow C_6H_5^+ + H^\bullet$ and $C_6H_6^{+\bullet} \rightarrow C_4H_4^{+\bullet} + C_2H_2$ are not in competition with each other. If they were competitive, $(M - H)^+$ formation always would occur much more slowly than $(M - C_2H_2)^{+\bullet}$.

The occurrence of reactions from isolated states does not require that QET be abandoned, merely that it be generalized to cover such cases. The usual model—that the electronic energy acquired during the primary ionization step is rapidly converted to vibrational energy of the reactive (ground) electronic state—must now allow for several vibrationally excited but non-interacting reactive states. Accumulation of vibrational energy in the bond (degree of freedom) undergoing fragmentation occurs as before, but separately for the isolated states. Isolated electronic states may undergo the same reaction as the ground state with different $k(\epsilon)$ parameters, they may undergo a different reaction, as in the $C_6H_6^{+\bullet}$ case, or they may not react. This latter possibility is apparently the case in some simple ketones. For example, charge

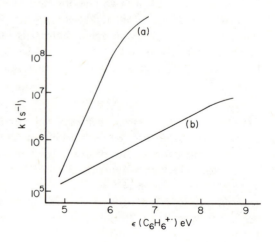

FIGURE 5-10. Experimentally determined relationship between rate of fragmentation and internal energy of the reactive benzene molecular ion for (a) $C_6H_6^{+\bullet} \longrightarrow C_4H_4^{+\bullet} + C_2H_2$ and (b) $C_6H_6^{+\bullet} \longrightarrow C_6H_5^+ + H^\bullet$. [Reproduced with the permission of Verlag der Zeitschrift für Naturforschung from B. Andlauer and C. Ottinger, Z. Naturforsch., 27a, 293 (1972).]

exchange of 2-pentanone gives a breakdown curve (see Figure 5-3) that shows high energy nonreacting molecular ions.

A central principle of QET, and of all statistical unimolecular reaction theories, is that the behavior of the reactant is determined only by its total energy, not by the manner in which it is generated. (This is the hypothesis that the activated complex is in equilibrium with reactants. Instances in which it breaks down at high energy have recently been found.) The principle is difficult to test satisfactorily since different methods of preparation usually result in differences in $\epsilon^{\ddagger}$, the internal energy of the activated complex. For example, the ion $C_4H_9^+$ generated from various *n*-alkanes always releases $\sim$80 meV of kinetic energy (compare section 5-4) in fragmenting to yield $C_3H_5^+ + CH_4$. The small but real differences that do exist between the energy released on changing the source of the ion appear to be due to differences in $\epsilon^{\ddagger}$. Differences in internal energies of course have marked effects upon fragmentation rates and hence upon ion abundances even when structurally identical ions are compared. (See sections 5-4 and 5-8 for further discussion of ion structure.)

5-3   Rearrangements

Reactions that result in bonds between previously nonbonded atoms (*skeletal rearrangements* if neither atom is hydrogen, *hydrogen rearrangements* otherwise) are important because of the limitations they cause in the use of mass spectrometry in structural studies, particularly in obscuring the positions of labeled atoms. They also contribute greatly to the richness of unimolecular ion chemistry. The subject acquired a certain mystical quality when early observations could not immediately be ordered within a simple framework. Although unexpected processes still turn up, a few simple principles now cover most of this field. Our emphasis here is on reaction mechanisms, not on the unique energetics and kinetics of rearrangements discussed in sections 1–3 and 5–2. It is worth reiterating that since rearrangements have relatively unfavorable entropies of activation, they must have low activation energies in order to compete with simple cleavages.

Since only the ionic fragment is collected in mass spectrometry, only processes in which bond formation has occurred in this fragment are of interest here. Bond formation in the neutral is also common (cf. elimination of HOH, HCl, HOAc, MeOMe, etc.), but unless there is clear evidence for concomitant bond formation in the ionic product, such processes are not considered here.

5-3a   Skeletal Rearrangements

Skeletal rearrangements can be divided into two major mechanistic types, according to whether the bond formation step is initiated by a free radical or by a nucleophile. Usually odd-electron ions that possess heteroatoms are preferentially ionized at the heteroatom, and this free radical site initiates radical rearrangement reactions (including, of course, H˙ abstraction, discussed in section 5-3b). Odd-electron ions that do not include heteroatoms rearrange under the influence of either the charged site or the radical site. Sometimes it is not clear which is responsible and in some systems concerted electrocyclic rearrangements, including valence tautomerism, occur. Even-electron ions such as carbonium ions possess electrophilic centers and may readily undergo intramolecular nucleophilic rearrangements. We now illustrate these two major types of skeletal rearrangement (radical and nucleophilic) and discuss some reactions, including ring expansion, that are more appropriately classified separately.

Simple examples of radical-induced skeletal rearrangement are provided by ϵ cleavage in amines (2) and δ cleavage in alkyl bromides (3). These reactions are instructive, even though

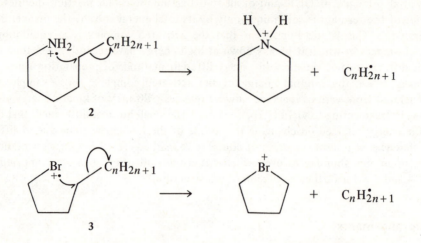

they are somewhat atypical in that bond formation is not accompanied by skeletal reorganization and hence is "hidden." Both reactions are driven by the stability of the product ions. The size of the heteroatom accounts for the fact that the five-membered ion is favored in the bromine case, whereas the six-membered ammonium ion is most stable. Another difference between the two classes of compounds is that the illustrated process is of major importance in the spectra of alkyl bromides, but the amine reaction does not lead to abundant product ions. The fact that simple α cleavage in amines can lead to very stable product ions explains this difference. A final general point is that reactions of the type shown always have to compete with hydrogen abstraction processes, which provide an alternative means of satisfying the valence requirements of the heteroatom and often lead to prominent fragment ions. For example, dialkyl thioethers generally give abundant ions, $RSH^{+\cdot}$, because of hydrogen radical abstraction by the heteroatom and C–S bond cleavage.

An even more reactive heteroatom-centered free radical may be generated by ionization of sulfoxides, sulfones, nitro compounds, phosphine oxides, and other $-\overset{+}{X}-\overset{-}{Y}$ groups. For example, methyl phenyl sulfoxide (4) shows the rearrangement ions $C_6H_5O^+$ and CH_3S^+ in its

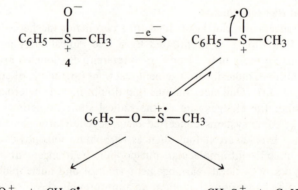

mass spectrum. Molecular ion isomerization occurs prior to fragmentation and results in a more stable ion. The lower enthalpy of the sulfenate ion means that it has a greater vibrational energy than the ionized sulfoxide. This fact, plus the accessibility of more stable products, leads to the result that fragmentation proceeds almost entirely from the rearranged molecular ion. This general rearrangement type is not restricted to 1,2 shifts. The reaction of phenyl dimethylthiophosphinate (**5**) is an example of a 1,3 shift. Isomerizations such as these are

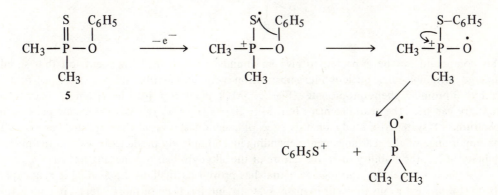

favored when aryl rather than alkyl migration is involved, in agreement with the general observation that aryl groups have the greater migratory aptitudes.

One of the most common skeletal rearrangement types can be represented in the generalized form $A-B-C^{+\circ} \rightarrow A-C^{+\cdot} + B$. The rearrangement occurs with many groups, especially in unsaturated compounds. Normally, B is a stable neutral molecule such as N_2, CO, C_2H_2, and SO_2. The detailed electronic nature of the reaction cannot be firmly established, but the role of radical sites in initiating many other reactions implicates them here. In numerous cases a three-centered reaction is believed to be involved. To quote just one example out of many, benzophenone loses CO to give a $C_{12}H_{10}^{+\cdot}$ ion.

Another instructive case of radical-induced bond formation is provided by the cyclohexenol (**6**). Clearly the charged site plays a passive role in this reaction, which is driven by

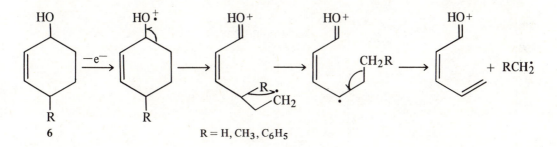

R = H, CH$_3$, C$_6$H$_5$

the formation of a very stable product. The fact that hydrogen, alkyl, and aryl groups all migrate reinforces the proposed radical mechanism.

The second class of skeletal rearrangements includes those that involve nucleophilic rearrangements. The reactive site may be a carbonium ion or another electron-deficient atom. Reaction may occur through a three-, four-, five-, or six-membered (or even larger) cyclic transition state, and it may result in the elimination of almost any stable even-electron neutral species. The principles involved can be illustrated by a single example in which a molecule is tailored to undergo this reaction in the mass spectrometer. First, a suitable carbonium ion must be generated from the molecular ion, for instance by loss of Br· from an alkyl carbon atom. Next, a suitable nucleophile (Nu) must be incorporated in the molecule, and finally the group to be eliminated (N) must be a stable entity. Thus in general terms the molecule required is **7**.

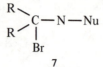

7

This compound can be expected to give an abundant $(M - Br)^+$ fragment ion that should undergo intramolecular nucleophilic substitution with elimination of N. The spectrum of methyl 3-bromo-3-phenylpropionate (Figure 5-11) illustrates just how simple mass spectrometry can be. The rearrangement ion $(M - Br - CH_2CO)^+$ represents the base peak in the spectrum. A systematic study on this type of reaction shows all the expected results, with the importance of the rearrangement depending in a completely predictable way on the nucleophilicity of Nu, the stability of N, the nature of the alkyl groups, and the ring size.

A facet of skeletal rearrangement that has proved particularly fascinating is ring expansion. This process seems to occur in many systems, but has been of most interest in connection with the problem of the structure of the $C_7H_7^+$ ion, which may be the benzyl or the tropylium ion. Extensive labeling data, using both ^2H and ^{13}C, point to the fact that the hydrogens and the carbons are all equivalent in the high energy ions that fragment by acetylene elimination. These results strongly suggest the tropylium structure for the reactive $C_7H_7^+$ ion. The structure of the $C_7H_7^+$ ion at threshold is not necessarily the same. The heats of formation of the benzyl and the tropylium ions are 220 and 209 kcal mole^{-1}, respectively, as determined from the ionization potentials of the corresponding free radicals. The $C_7H_7^·$ ion formed from toluene has a heat of formation of 232 kcal mole^{-1}, which indicates that excess energy is required in its formation. This observation suggests that rearrangement, rather than simple cleavage, is necessary for $C_7H_7^+$ formation from toluene. (By way of contrast, the much lower enthalpy of $C_7H_7^+$ from benzyl bromide suggests simple cleavage and formation of the benzyl ion at threshold in this case.)

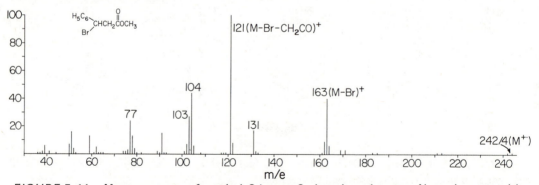

FIGURE 5-11. *Mass spectrum of methyl 3-bromo-3-phenylpropionate. Note the ease with which the benzylic bromine is lost and the fragmentation of the (M − Br)$^+$ ion by skeletal rearrangement with ketene loss to give m/e 121 or by loss of methanol to give m/e 131. [Reproduced with permission from R. G. Cooks, J. Ronayne, and D. H. Williams, J. Chem. Soc., C, 2601 (1967).]*

There is also strong evidence that ring expansion can occur in the molecular ion ($C_7H_8{}^{+\cdot}$) in the case of toluene. Thus labeling data show that $H^\cdot$ loss to give $C_7H_7{}^\cdot$ occurs with equal probability from both the ring and the methyl group. As the internal energy of the fragmenting $C_7H_8{}^{+\cdot}$ ions is increased, direct $H^\cdot$ loss from the side chain to give the benzyl cation becomes increasingly likely, but the resulting benzyl cation rapidly equilibrates with the tropylium structure before further C_2H_2 loss.

Still unanswered is the question of whether the toluene molecular ion has the toluene structure or a rearranged structure in the case of those ions with insufficient energy to undergo any unimolecular fragmentation. Preliminary icr results, in which the ease of nitration of various $C_7H_8{}^{+\cdot}$ ions is examined, show that the toluene radical cation (**8**) undergoes nitration

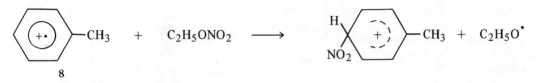

8

readily but cycloheptatriene does not. This result strongly indicates that the molecular ions are structurally distinct.

Among the preceding cases of skeletal rearrangement, there have been examples of rearrangement occurring directly from the molecular ion as well as instances of molecular ion rearrangement prior to fragmentation. It is because of their exceptionally low activation energies that molecular ion isomerization can sometimes compete with direct fragmentation mechanisms. Valence tautomerism reactions provide a particularly striking example of this phenomenon. The carbon atoms of benzene all become equivalent prior to loss of C_2H_2 or $C_3H_3{}^\cdot$ from the metastable molecular ion. This was shown by studying benzene-1,2,$^{13}C_2$-3, 4,5,6-d_4 and it was concluded that valence tautomerism is rapid on the time scale corresponding to metastable ion reaction. (The co-occurrence of a direct hydrogen scrambling mechanism in benzene is discussed in section 5–3b.)

5–3b   Hydrogen Rearrangements

Hydrogen rearrangements vary from simple radical abstractions to complex processes involving several sites in the molecule. The ease with which hydrogen is abstracted as $H^\cdot$ by a radical provides the major reason why hydrogen rearrangements are so common in mass spectrometry. There are few new principles involved, so the following discussion will be brief.

Many elimination reactions involve hydrogen rearrangement. It has been usual to consider these as concerted processes, but there is growing evidence that stepwise mechanisms apply, inter alia, to the McLafferty rearrangement and to some 1,5 eliminations of water. Many elimination reactions also occur in even-electron ions, in which case a radical type of mechanism is unlikely. Moreover, in these cases—for example, alkene loss from the α cleavage product in ethers—there is no evidence that the reactions occur in a stepwise fashion.

Numerous other examples of hydrogen rearrangement appear in Chapter 4. Reciprocal hydrogen transfer is observed in the fragmentation of some odd-electron ions. Hydrogen abstraction by the radical site gives an isomeric form of the molecular ion, which fragments by a pathway involving transfer of some other hydrogen atom back to the site from which the original hydrogen atom was derived. The γ cleavage reaction in ketones (section 4–5c) provides an example. Another more complex case, multiple hydrogen rearrangement, has also been met in the ketone and ether studies of section 4–5, where the control of product ion stability is

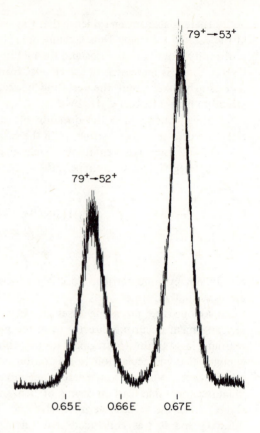

FIGURE 5-12. Metastable peaks for acetylene loss from benzene-d_1 showing complete scrambling of the hydrogens and deuteriums in the molecular ion. The ratio of C_2H_2 to C_2HD loss is 2/1.

shown to be the driving force for these low entropy reactions. Even triple hydrogen rearrangements have occasionally been encountered.

In some ions low energy pathways are available whereby all the hydrogen atoms of a given structural type can become equivalent. This phenomenon is termed *scrambling*. It is a low activation energy, low frequency factor process and, in consequence, the degree to which scrambling is complete prior to a given fragmentation reaction generally increases as ions of longer lifetime (lower internal energy) are examined. Such ions usually are studied by lowering the ionizing electron energy or by using metastable rather than daughter ions. Figure 5-12 illustrates the fact that complete H-D scrambling occurs in benzene-d_1 prior to elimination of C_2H_2 in a metastable reaction. These metastable peaks were recorded using a new type of mass spectrometer (mike spectrometer) described in section 5-4.

Since scrambling and related isomerizations must be detected by isotopic labeling, it is not always obvious whether (i) hydrogen is the mobile agent and the molecular skeleton is unaffected or (ii) the skeletal atoms, usually carbon, are mobile and the hydrogens are merely carried along with them. Carbon labeling experiments have been carried out in only a few cases. The benzene experiment described in section 5-3a showed ring atom scrambling to occur, but it showed extensive hydrogen scrambling, since loss of $^{13}C_2D_2$ was observed.

Hydrogen scrambling, or at least the tendency toward complete scrambling, is not restricted to aromatics, but occurs in virtually all types of systems. It is important to realize that the phenomena underlying hydrogen scrambling are just those responsible for specific hydrogen rearrangements. Thus the stepwise hydrogen transfer that allows H_2CO loss from the anisole molecular ion (see section 4-5b) is a reversible reaction and may not lead directly to products.   In this way scrambling can occur.   Similarly, dialkyl ketone molecular ions (**9**) undergo hydrogen scrambling both *between* and *within* the alkyl groups (to a very small

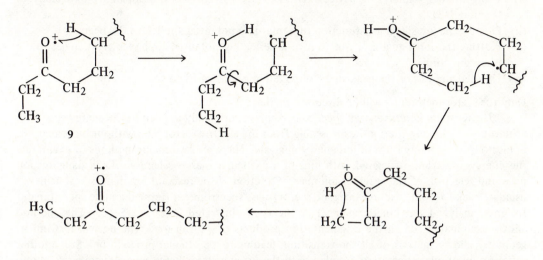

extent in 70 eV spectra, more extensively in metastable ions).   These processes occur via cyclic hydrogen transfers similar to those that are responsible for the McLafferty rearrangement.

If radical abstraction reactions are indeed the basis for the mobility of hydrogen in molecular ions, then replacing hydrogen by other groups that are easily abstracted as radicals should be possible.   Bromine, for example, has been shown to interchange positions on the thiophene nucleus with hydrogen by a mechanism that involves Br$^{\cdot}$ abstraction by a carbon radical.   Similar processes were observed for $C_6H_5{}^{\cdot}$, although complete scrambling did not ensue.   On the other hand, alkyl groups will not substitute for H$^{\cdot}$ in the McLafferty rearrangement, although trimethylsilyl groups apparently will.

In summary, it is the existence of rearrangements that usually has been claimed as evidence for the view that chthonic forces are at work in mass spectrometry.   On the preceding pages and elsewhere in this treatment, we have shown that this is not the case.   Rearrangement reactions, like simple cleavages, occur in accordance with the basic principles of organic chemistry and, indeed, startlingly close relationships with solution chemistry are common (see also section 5-5).

5-4   Ion Kinetic Energy Spectra (IKES) and High Energy Ion–Molecule Reactions

5-4a   Unimolecular Reactions

Not surprisingly, *mass measurement* has traditionally been the principal subject of mass spectrometry.   Recently, however, vigorous development of *kinetic energy measurements* and their application to problems of gaseous ion chemistry has taken place.   The traditional double-focusing sector instrument allows energy analysis as a prelude to mass analysis (see Chapter 3). Recent developments make mass analysis secondary to energy analysis in several types of mass spectrometric study.   These studies involve both unimolecular reactions occurring in a field-free region, that is, metastable ion fragmentations, and collisional spectroscopy, that is, ion-molecule reactions in the keV energy range.   Such high energy ion-molecule reactions are to be contrasted with those occurring at thermal energies or at relative kinetic energies of a few electron volts in icr and tandem mass spectrometers.

Four methods may be used for determining the energy of an ion beam.

(i) Using an energy analyzing device such as an electrostatic sector or a parallel plate analyzer (section 3-3).

(ii) Plotting the ion signal as a function of a retarding potential applied to the beam.

(iii) Plotting the ion signal as a function of a deflecting potential applied at right angles to the beam direction.

(iv) Detailed analysis of the peak shape observed in a time-of-flight spectrometer.

Only the sector method (i) will be discussed further.

The term *ion kinetic energy spectrum* (ikes) was originally applied to the energy spectrum achieved by detecting the ion beam issuing from the electric sector, i.e., without mass analysis, but is now used for any form of ion energy analysis. There are two major methods of examining the energy spectrum associated with ions of a particular mass: selection of the daughter ion mass and selection of the reactant ion mass. Selection of the reactant ion mass is best achieved using a "back-to-front" (source–magnet–sector) mass spectrometer (termed a mike spectrometer for mass-analyzed ion kinetic energy) of the type illustrated in Figure 5-13. This method allows examination of all the fragmentation products of a given ion. The normal instrument geometry allows analysis of all the transitions leading to a particular product ion. Still another variation, photoplate detection coupled with the usual source–sector–magnet geometry, allows the recognition of all transitions that involve a certain kinetic energy change, whatever the masses of reactant and daughter ions.

Establishing the position of a peak in an energy spectrum allows one to define (unequivocally if mass analysis is also performed) the reaction involved. Figure 5-14 shows the ion kinetic energy spectrum of aniline, without mass analysis, at low analyzer pressure. The spectrum was obtained by scanning the electric sector voltage from zero to that value E required to transmit stable ions generated in the ion source. Each peak in the spectrum is due to a

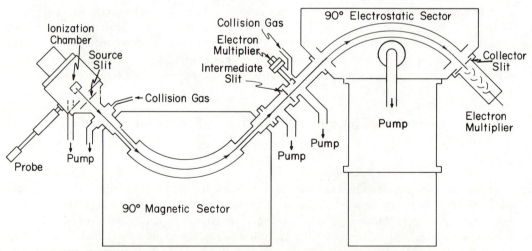

FIGURE 5-13. A mike spectrometer is a reversed sector mass spectrometer in which reactant ions are selected by mass and their unimolecular and collision-induced reactions occurring in the intersector region are then followed by kinetic energy analysis. [Reproduced with the permission of Elsevier Publishing Company from R. G. Cooks, J. H. Beynon, R. M. Caprioli, and G. R. Lester, Metastable Ions, 1973.]

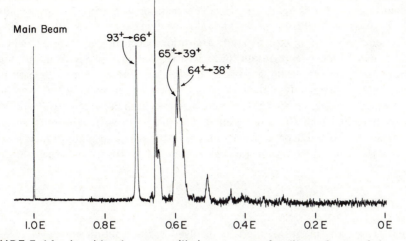

FIGURE 5-14. Ion kinetic energy (ike) spectrum of aniline. Some of the metastable ion reactions that give rise to the peaks in this spectrum are indicated.

unimolecular reaction occurring in the field-free region that precedes the electric sector. These reactions are the counterparts of those occurring in the second (premagnet) field-free region that give rise to diffuse (metastable) peaks in the normal mass spectrum. The processes occurring in the first field-free region are also due to reactions of metastable ions.

In the ikes technique all the metastable reactions of the molecular ion and of all fragment ions formed in the ion source are recorded at high sensitivity. The value of ike spectra lies in the following factors.

(i) They allow the detection of parent-daughter ion relationships, an essential step in structure elucidation.

(ii) They are much more sensitive to differences in molecular structure than are 70 eV mass spectra because the reactant ions have only small excess energies; consequently, isomers are distinguishable by their ikes spectra even when they give identical mass spectra.

(iii) The determination of isotopic label incorporation has been a major advantage of conventional mass spectrometry. In ^{2}H-incorporation analysis, however, this method is limited by the requirement that the molecular ion not undergo H$^\bullet$ loss or at least that this loss be insignificant at low ionizing electron energies (see section 4-3). In many cases, ikes overcomes this problem, since the degree of labeling of the ion can be determined from the relative abundance of the signal for any convenient transition that does not involve a primary isotope effect.

(iv) Many studies on ionic rearrangements and other processes are facilitated by the high sensitivity achievable. The detailed nature of the information attainable is illustrated by the fact that the rate of carboxyl group rotation in the benzoic acid molecular ion has been estimated from the rate of H-D isomerization between the ortho and carboxyl hydrogens.

The energy spread of the product ion beam resulting from a metastable reaction is important since it can be used to determine the kinetic energy release (T), i.e., the amount of internal energy of the reactant ion that has been converted into translational energy of the products. This quantity is of considerable thermochemical significance.

In order to determine T, the measurement must be done under conditions of good energy resolution. For this reason metastable reactions occurring in the second field-free region which give rise to diffuse peaks at nonintegral masses, are usually unsuitable. Some technique is required such as ikes to study transitions occurring in the first field-free region under conditions of high energy resolution. This requirement stems from the fact that the energy spread of the main beam is usually appreciable, but the energy release nonetheless must be detected. Fortunately, for reactions in which loss of mass occurs, an amplification effect operates on conversion from the center-of-mass system, in which T eV is released, to the laboratory system, in which the kinetic energy of the resulting ion beam is analyzed. The exact amplification involved depends on the masses of the reactants and products and is implicit in eq. 5-14, the general equation for calculation of T from ikes measurements, where

$$T = \left(\frac{m_2 V}{16 m_3}\right)\left(\frac{\Delta E}{E}\right)^2 \qquad\qquad 5\text{-}14$$

m_1, m_2, and m_3 are the masses of the ions involved in the reaction $m_1^+ \rightarrow m_2^+ + m_3$; V is the ion-accelerating voltage; and $\Delta E/E$ is the fractional spread in kinetic energy over which the product of the metastable ion reaction is transmitted. The effect may be illustrated by considering the case of $H^{\cdot}$ loss from the propane molecular ion. The half-width of the main beam of stable ions under given conditions was 0.95 eV; the energy spread of the product ions of the metastable transition was 3.15 eV, i.e., 2.2 eV greater. The energy release in this case, however, is only 2.0×10^{-3} eV, and only because of the amplification effect can a quantity as small as this be determined using instrumentation that is capable of a resolution of 1 in 10^4 (1.0 eV in 10,000 eV).

By way of contrast, kinetic energy releases of several electron volts have also been observed, particularly in the fragmentations of doubly charged ions. Although the behavior of doubly charged ions in leading to kinetic energy release is not basically different from that of singly charged ions, it is convenient to discuss charge-separating fragmentation reactions of doubly charged ions as a distinct case. In the course of the reaction, $AB^{++} \rightarrow A^+ + B^+$, energy is released equivalent to that required to bring two charges from a distance of infinity to that which exists in the activated complex structure. If all this energy appears as kinetic energy of separation of the product ions, then the kinetic energy release provides a direct measure of the intercharge distance in the activated complex. (Other contributions to the stabilization of products relative to the transition state will be much smaller than that due to coulombic repulsion.) The measured kinetic energy release in the reaction $C_6H_6^{++} \rightarrow C_5H_3^+ + CH_3^+$ in benzene, $T = 2.5$ eV, corresponds to an intercharge distance in the transition state of 5.5 Å. This value is in agreement with that estimated for fragmentation from a linear structure with terminal methyl groups.

The foregoing principles have formed the basis for extended investigations of the reactions of doubly charged ions. In addition to information on ion structures, the following are the most important results to emerge so far.

(i) Charges tend to be localized at the termini of the linear chains formed when cyclic ions undergo ring scission. This factor, however, is balanced by a tendency for heteroatoms, if present, to carry a charge.

(ii) Isomeric compounds such as substituted heteroaromatics frequently isomerize before fragmentation.

(iii) Crude ion structural information, such as distinction between a cyclic and a ring-opened molecular ion, is frequently possible. For example, *p*-xylene, gives 106^{++} (the molecular ion), 104^{++}, and 102^{++}, all of which undergo loss of CH_3^+. The intercharge distances, 5.6, 5.5, and 7.7 Å respectively, indicate that removal of four hydrogen atoms is accompanied by ring opening.

We turn now to the best-studied aspect of this topic, kinetic energy release accompanying unimolecular fragmentation of singly charged ions in the field-free region. The range of T values encountered is very large, varying from 2×10^{-4} eV to over 1 eV (a range of nearly 10^4). These quantities constitute an excellent means of characterizing ionic structures. Thus they are only slightly dependent on the reaction used to form the ion in question. For example, the ion $C_{12}H_{10}O^{+\cdot}$ formed by direct ionization of diphenyl ether undergoes CO loss accompanied by an energy release of 0.435 eV, while the $C_{12}H_{10}O^{+\cdot}$ ion formed by fragmentation of diphenylcarbonate undergoes the same reaction with an energy release of 0.438 eV. This result strongly indicates that the reactive forms of the two ions have identical structures; conversely, pronounced differences in values of T provide evidence against identical structures. Unlike some parameters that have been used to characterize ion structures, especially metastable ion abundances, the internal energy distribution of the reactant ion does not have a major effect on observed T values. Moreover, these quantities are also relatively insensitive to electron energy, ion-accelerating voltage, and exact source conditions. Many of these desirable properties of kinetic energy release measurements arise because (1) the ions that fragment in the first field-free region have internal energies of only 0.1 or 0.2 eV in excess of the activation energy for the reaction in question and (2) ions fragmenting in the same time interval will have similar internal energies, given that the ions are generated with a broad distribution of internal energies. The nonfixed energy of the activated complex $\epsilon^{\ddagger}$ is statistically partitioned between that degree of freedom which corresponds to movement along the potential energy surface toward products and all the others. Hence, in organic ions of even moderate complexity, $\epsilon^{\ddagger}$ provides only a small kinetic energy release. The remainder is derived from $\epsilon_0{}^r$, the reverse activation energy (if present). Thus we have eq. 5-15, where T^e and $T^{\ddagger}$ represent the contribu-

$$T = T^e + T^{\ddagger} \qquad\qquad 5\text{-}15$$

tions of the reverse activation energy (electronic energy) and the internal energy of the activated complex, respectively.

When $\epsilon_0{}^r$ is large, $T \cong T^e$ and measured T values provide direct information on the potential energy surface for the reaction. Not only can T then be used to characterize the reaction, but it is also possible to determine $\epsilon_0{}^r$ independently and hence to study the energy partitioning in these reactions. In addition to the fundamental importance of such an endeavor in physical chemistry, knowledge of energy partitioning provides a new and powerful test of ionic reaction mechanisms. Let us illustrate this point with an example, the elimination of of formaldehyde from benzyl methyl ether. This reaction might occur by either the four-membered (**10**) or the six-membered (**11**) cyclic intermediate. Reactions analogous to both

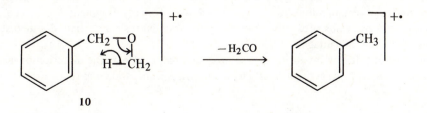

10

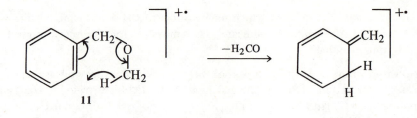

these mechanisms are known, and further fragmentation of the product ion does not clarify the mechanism since ring expansion can be expected to intervene in both alternatives. It is known, however, that the proportion of the reverse activation energy partitioned into kinetic energy of separation of the products of fragmentation of a metastable ion is large for reactions that proceed through cyclic activated complexes of small ring size. In the present case, $T = 0.20$ eV, while the reverse activation energy for formation of ionized toluene is 0.34 eV. The value of $T/\epsilon_0{}^r$ (~60%) is high, as expected for the first mechanism. It is difficult to estimate $\epsilon_0{}^r$ for the alternative process, but it must be smaller (the product ion is less stable); therefore $T/\epsilon_0{}^r$ must be even higher than 60%, a situation that is inconsistent with a six-membered cyclic rearrangement. The loss of C_3H_6 from *n*-butylbenzene, an example of a reaction that does occur via a six-membered cyclic rearrangement, is accompanied by a kinetic energy release of 0.02 eV and a $T/\epsilon_0{}^r$ value of approximately 3%. It is concluded that formaldehyde loss from benzyl methyl ether occurs via a four-centered cyclic complex.

Our final topic in this section concerns measurements made at high energy resolution, $E/\Delta E \geqslant 1000$. It is possible under these circumstances to measure not just T values but also the actual distribution of kinetic energy releases. Moreover, *fine structure* can be discerned in metastable peaks. For example, the unimolecular metastable fragmentation of $NO^+ \rightarrow O^{+\cdot} + N^\cdot$ shows vibrational fine structure; in addition, two separate electronic states are seen to contribute to the observed metastable peak. Composite metastable peaks (Figure 5–15) are being observed with increasing frequency in organic molecules. The loss of H_2CO from anisoles and substituted anisoles constitutes such a case. Independent evidence including energy-partitioning data has been offered suggesting two competitive mechanisms, four-membered and five-membered cyclic hydrogen transfers, to account for this particular observation.

5–4b Ion-Molecule Reactions

One of the many high energy ion-molecule reactions that can be studied by mass spectrometry is collision-induced dissociation. This process involves fragmentation in a field-free region, just as do reactions of metastable ions. Excitation of stable ions occurs on collision with subsequent rather rapid fragmentation. Collision-induced transitions have been little used in analytical mass spectrometry. This situation should change, however, since these transitions provide similar information to unimolecular metastable reactions. They seem to be particularly useful in identifying the molecular ion; typically far more collision-induced reactions of the molecular ions can be observed than for any fragment ions. The fact that collision-induced reactions often give much more abundant signals than do metastable ion reactions is a further advantage.

The type of collision involved in these processes barely affects the direction or momentum of the ion. It does, however, cause an electronic transition paralleling that caused by electron

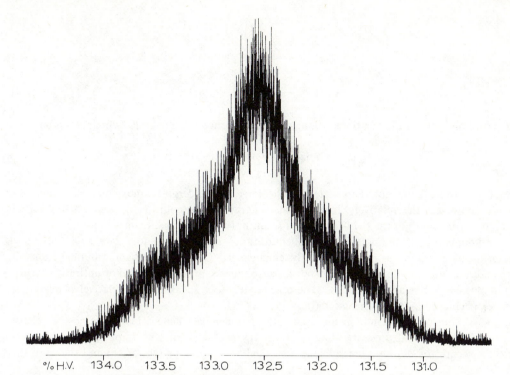

FIGURE 5-15.   Composite metastable peak for H_2CO loss from p-methylanisole. The overlying peaks are centered at exactly the same position; one corresponds to a large kinetic energy release, the other to a small energy release.   Two competitive mechanisms of H_2CO loss are responsible for this behavior.   The energy scale is given in terms of the accelerating voltage (HV) required to transmit the product ion.   [Reproduced with the permission of the American Chemical Society from R. G. Cooks, M. Bertrand, J. H. Beynon, M. E. Rennekamp, and D. W. Setser, J. Amer. Chem. Soc., 95, 1732 (1973).]

impact in the ion source. Most ions passing through the field-free region are stable and possess small internal energies, but energy transfer occurs in the course of collisional excitation. Although only sparse evidence is available, it seems that the internal energies of the resulting excited ions are, on average, intermediate between those of ions that fragment within the source and those of ions that fragment unimolecularly in the field-free region. This result is useful, since defining and selecting the internal energies of ions is one of the most important exercises in ionic chemistry and especially kinetics. It has been used in the study of the internal energy dependence of isotope effects and of rearrangement vs. cleavage processes.

Another property of collision-induced reactions is that consecutive reactions may occur in the same field-free region. By contrast such an event is rare in unimolecular processes since the higher energy ions that can undergo several successive reactions will have fragmented before the ions reach the field-free region. To illustrate this property of collision-induced dissociations, benzene shows the unimolecular metastable transitions given in eqs. 5-16 to 5-19.

$$C_6H_6^{+\cdot} \longrightarrow C_6H_5^+ + H^\cdot \qquad\qquad 5\text{-}16$$

$$C_6H_6^{+\cdot} \longrightarrow C_4H_4^{+\cdot} + C_2H_2 \qquad\qquad 5\text{-}17$$

$$C_4H_4^{+\cdot} \longrightarrow C_4H_3^{+} + H^{\cdot} \qquad\qquad 5\text{-}18$$

$$C_6H_5^{+} \longrightarrow C_4H_3^{+} + C_2H_2 \qquad\qquad 5\text{-}19$$

Its collision-induced dissociations include the process shown in eq. 5-20, which could be due to

$$C_6H_6^{+\cdot} + N \longrightarrow C_4H_3^{+} + C_2H_3^{\cdot} + N \qquad\qquad 5\text{-}20$$

reaction from a state not represented in the ions that undergo unimolecular reaction. More likely, however, the relatively high internal energy of the collisionally excited ion and the length of time spent in the field-free region result in two successive fragmentations.

Frequently, if unimolecular fragmentation of a metastable ion gives an intense peak, addition of collision gas will not greatly enhance it. On the other hand, abundant ions due to collision-induced dissociations often correspond to processes that are entirely absent as unimolecular metastable reactions. These generalizations speak for the relatively large energy transfer achieved by collisional excitation.

We now turn to ion-molecule reactions that involve changes in the number of charges carried by an ion and thus alteration of its energy/charge ratio. Two important reaction types are conversion of doubly charged ions into singly charged ions (eq. 5-21) and the ionization

$$AB^{++} + N \longrightarrow AB^{+} + N^{+} + e^{-} \qquad\qquad 5\text{-}21$$

of singly charged ions (eq. 5-22). Each of these reactions forms the basis for a new type of

$$AB^{+} + N \longrightarrow AB^{++} + N + e^{-} \qquad\qquad 5\text{-}22$$

mass spectrum. Thus, by operating a double-focusing mass spectrometer at accelerating voltage V and electric sector voltage $2E$ (where E is the electric sector voltage necessary to transmit stable ions formed in the ion source), only products from reactions of the type 5-21 are collected, and mass analysis will provide a spectrum of doubly charged ions. Singly charged ions formed in the ion source are not transmitted under these conditions. This spectrum is unique in that analysis is by way of the charge exchange reaction (eq. 5-21). It has been shown that doubly charged mass spectra are useful adjuncts to singly charged spectra in analytical, particularly structural, studies. The so-called $2E$ spectrum is really just a spectrum of the doubly charged ions, although the lower internal energies of the doubly charged ions sampled in the first field-free region and the method of analysis make it differ somewhat from the spectrum obtained by the tedious alternative of exact mass measurement and subtraction of all singly charged ions. The sensitivity with which $2E$ spectra can be plotted is good, perhaps two or three orders of magnitude less than normal singly charged ion spectra.

The use of $2E$ spectra is still in its infancy. In hydrocarbons they have been found to be very similar for isomeric compounds, while in amines they are structure specific. Their chief value in the hydrocarbon series is the extra avenue of approach to ion structures that they provide. A most striking result is that the *n*-decane doubly charged molecular ion loses in succession eight molecules of H_2 in the ion source (i.e., within ~ 1 μsec), driven by the tendency to form the very stable species $C_{10}H_6^{++}$. The study of doubly charged ion mass spectra reveals several particularly stable ions, such as $C_nH_2^{++}$, $C_nH_6^{++}$, $C_nH_7^{++\cdot}$, and possibly structures of the cumulene type $H-\overset{+}{C}\mkern-8mu=\mkern-8mu C=C\overset{+}{\underset{n}{=}}C-H$.

A doubly charged ($2E$) spectrum is given in Figure 5-16. Interpretation of this spectrum can be based on the principles made familiar from normal mass spectra. The ion of mass 93 is apparently the molecular ion, and the abundance of the associated isotopic species suggests that there are six or fewer carbon atoms in the molecule. The odd molecular weight indicates an odd number of nitrogen atoms while the loss of various numbers of hydrogen atoms from the molecular ion is reminiscent of the behavior of benzene. These facts suggest the molecular formula C_6H_7N, so the compound is probably aniline. The abundant ion of mass 62 is of the type $C_nH_2^{++}$.

If the electric sector voltage is set at $E/2$, a spectrum representing all the singly charged ions that undergo the stripping reaction (eq. 5-22) can be obtained. This spectrum also will include all daughter ions formed by the fragmentation reaction (eq. 5-23), but these processes

$$m_1^+ \longrightarrow \left(\frac{m_1}{2}\right)^+ + \left(\frac{m_1}{2}\right)^\cdot \qquad\qquad 5\text{-}23$$

can be distinguished by the fact that they occur with loss of mass. Hence they always involve some kinetic energy release and give broadened peaks (see section 5-4a). Their subtraction from ions due to eq. 5-22 is easily accomplished. The analytical value of these processes is just now being explored. What has energed, however, is that they can be useful in defining the internal energies of reactant ions.

In any high energy ion-molecule reaction, some of the energy necessary for reaction may be supplied from the kinetic energy of the reactants. This is always the case in collision-induced fragmentation processes, although the kinetic energy lost by the reactant ion is generally a very small fraction of its total energy. In charge transfer reactions kinetic energy loss can be

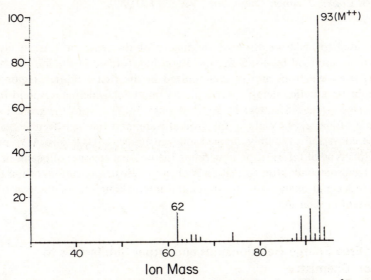

FIGURE 5-16. *Doubly charged (2E) mass spectrum of aniline. [Reproduced with the permission of the American Chemical Society from R. G. Cooks, J. H. Beynon, and T. Ast, J. Amer. Chem. Soc., 94, 1004 (1972).]*

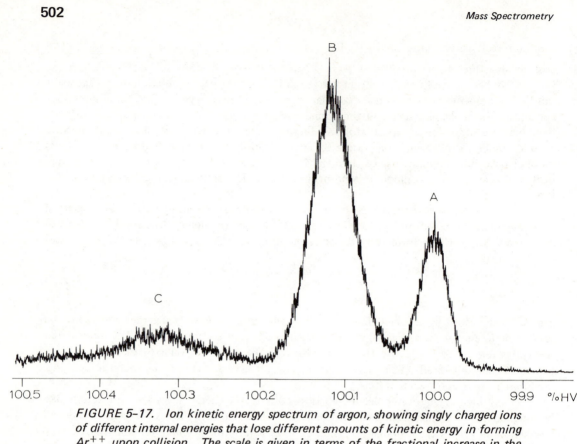

FIGURE 5-17.   Ion kinetic energy spectrum of argon, showing singly charged ions of different internal energies that lose different amounts of kinetic energy in forming Ar^{++} upon collision.   The scale is given in terms of the fractional increase in the accelerating voltage (HV) above that necessary to transmit Ar^+ ions.   [Reproduced with the permission of the American Chemical Society from T. Ast, J. H. Beynon, and R. G. Cooks, J. Amer. Chem. Soc., 94, 6611 (1972).]

measured and used to calculate the endothermicity of the reaction. As an illustration of a stripping reaction described by eq. 5-22, the argon ion, formed in the ion source by electron impact using 70 eV electrons and further ionized in the first field-free region, shows three distinct peaks in its kinetic energy spectrum, the onsets of which correspond to three states (or groups of closely spaced states) of $Ar^{+\cdot}$ (Figure 5-17). Thus the ground state ion, the metastable state of energy 9 eV above the ground state, and the high Rydberg states are each observed to be converted to doubly charged ions, giving peaks C, B, and A, respectively. The usefulness of this type of information in defining the internal energies of beams of ions at times of the order of microseconds after formation is obvious. Most applications of ion beams involve times of this order and analysis of the states and internal energies of the constituents of ion beams is important in their use.

5-5   Linear Free Energy Relationships and Other Similarities to Solution Chemistry

The theme of the unity of organic chemical phenomena, whether they occur in condensed phase or in the gas phase, runs through much of the present discussion of mass spectrometry (see especially section 4–6 on stereochemistry and section 5–7 on ion cyclotron resonance). This section contains further examples.

5–5a Linear Free Energy Relationships

Benzophenone fragments upon electron impact to give the benzoyl ion, $C_6H_5CO^+$ (eq. 5–24), and a substituted benzophenone will undergo an analogous reaction (eq. 5–25).

$$C_6H_5COC_6H_5^{+\cdot} \xrightarrow{k_0} C_6H_5CO^+ + C_6H_5^\cdot \qquad \text{5–24}$$

$$Y-C_6H_4COC_6H_5^{+\cdot} \xrightarrow{k} C_6H_5CO^+ + Y-C_6H_4^\cdot \qquad \text{5–25}$$

Under steady state conditions (eqs. 5–26 and 5–27), the concentration ratio of ions is given

$$\frac{d\,[C_6H_5CO^+]}{dt} = 0 \qquad \text{5–26}$$

$$\frac{d\,[C_6H_5CO^+]}{dt} = k_0\,[C_6H_5COC_6H_5^{+\cdot}] - \sum k_{\text{removal}}\,[C_6H_5CO^+] \qquad \text{5–27}$$

by eq. 5–28.

$$\frac{[C_6H_5CO^+]}{[C_6H_5COC_6H_5^{+\cdot}]} = \frac{k_0}{\sum k_{\text{removal}}} \qquad \text{5–28}$$

Writing the concentration ratio (eq. 5–28) as Z_0 for the unsubstituted case and as Z for the substituted case and assuming that $\Sigma k_{\text{removal}}$ is independent of the nature and internal energy of the molecular ion, we have eq. 5–29. Hammett relationships take the form of

$$\frac{Z}{Z_0} = \frac{k}{k_0} \qquad \text{5–29}$$

eq. 5–30, and so eq. 5–29 can be expressed as eq. 5–31.

$$\log \frac{k}{k_0} = \rho\sigma \qquad \text{5–30}$$

$$\log \frac{Z}{Z_0} = \rho\sigma \qquad \text{5–31}$$

Some of the simplifications used in deriving eqs. 5–24 to 5–31 will already be evident from a reading of earlier sections in this chapter. (1) The substituent effect on $P(\epsilon)$, the internal energy distribution of the molecular ions, is neglected. (2) The use of a single rate constant instead of a $k(\epsilon)$ function is a further simplification. (3) The occurrence of competitive and consecutive reactions is assumed to be independent of substituent. Therefore it is not expected that an overall correlation of ion abundance with σ will always be observed; however, there may well be cases in which the simplified treatment is applicable.

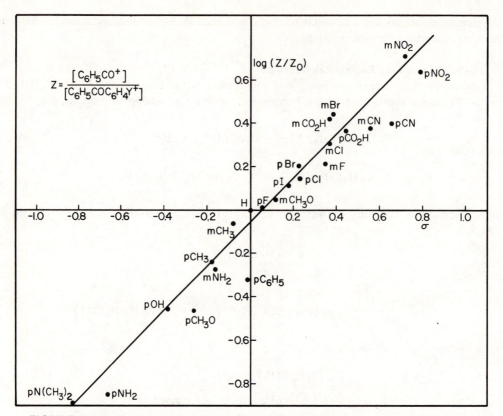

$$Z = \frac{[C_6H_5CO^+]}{[C_6H_5COC_6H_4Y^+]}$$

FIGURE 5-18. Correlation of log Z/Z_0 with the Hammett σ constant for the benzoyl ion, $C_6H_5CO^+$, generated from monosubstituted benzophenones. [Reproduced with the permission of Heyden and Son from M. M. Bursey, Org. Mass Spectrom., 1, 31 (1968).]

Figure 5-18 shows a plot of Z/Z_0 vs. Hammett σ constants for reaction 5-25. It is apparent that a linear free energy relationship does exist in this case. This result implies that each of the many factors that control ion abundance correlates with σ and that, in general, correlations of ion abundance with σ may result when one factor is dominant or when different factors reinforce each other. There is direct evidence that some of these individual factors do correlate with σ (or σ^+). For example, Figure 5-19 shows that ionization potentials for a series of substituted benzenes are linearly dependent on σ^+. In addition, activation energies (AP-IP) in several systems have also shown to be proportional to σ^+. We have already emphasized (see Figure 5-9) the control of activation energies over ion abundances. Because ion abundances depend on several different and sometimes opposing substituent effects, linear free energy relationships in the gas phase are not simply related to reaction mechanism as is often the case in solution.

5–5b Chemical Ionization and Acid-Catalyzed Reactions

Chemical ionization (CI) is one of several techniques in which the compound of interest is ionized by an ion-molecule reaction (eqs. 5-32 and 5-33). It is distinguished from charge

$$XH^+ + AB \longrightarrow X + ABH^+ \quad (H^+ \text{ transfer})$$ 5-32

$$XH^+ + ABH \longrightarrow XH_2 + AB^+ \quad (H^- \text{ transfer})$$ 5-33

exchange (eq. 5-34) by the fact that mass transfer is involved.

$$X^{+\cdot} + AB \longrightarrow X + AB^{+\cdot} \quad (e^- \text{ transfer})$$ 5-34

In the present context our interest is largely limited to chemical ionization reactions that involve proton transfer to the species of interest. The protonated product, ABH^+, will possess more or less internal energy depending on the exothermicity of the ionization reaction, that is, on the relative basicities of AB and the conjugate base X of the reagent XH^+ (equation 5-32). The presence of this internal energy can cause rearrangements and fragmentations that may resemble solution reactions catalyzed by Brønsted acids. The fact that the protonated molecular ion is an even-electron species whereas the molecular ion formed by electron impact is a free radical explains the tendency for elimination of simple neutral molecules in CI spectra and the lower incidence of complex rearrangements. The energy transferred to the ion is also normally smaller than that deposited on electron impact.

In the gas phase, as in solution, the most basic site in the molecule is preferentially

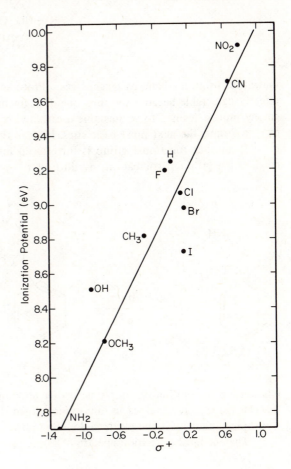

FIGURE 5-19. Correlation of the ionization potentials (measured by photoionization) of substituted benzenes with σ^+.

protonated and fragmentation usually proceeds from the resulting species. If fragmentation of this protonated species is unfavorable, however, reactions proceeding from ions protonated in other positions will predominate. The similarities between gas phase and solution acid-catalyzed processes are illustrated by the examples that follow.

Simple amino acids (12) undergo loss of H_2O and HCO_2H, but not NH_3, on chemical

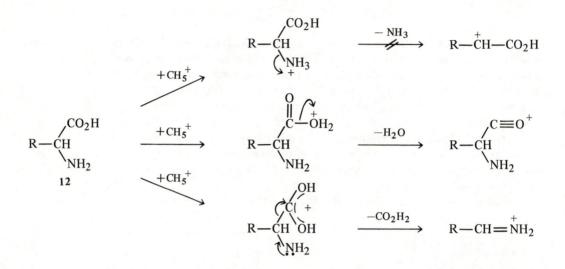

ionization using methane as reagent gas. Protonation at the most basic site, nitrogen, is not directly detectable because the subsequent fragmentation of this ion by loss of NH_3 is a high energy process leading to an unstable α-carbonyl carbonium ion. Protonation at either of the oxygen atoms, the next most basic sites, leads to the favorable fragmentation reactions shown.

If another functional group is introduced into the amino acid, NH_3 loss can occur via neighboring group participation, as illustrated by the reaction of phenylalanine (13). The

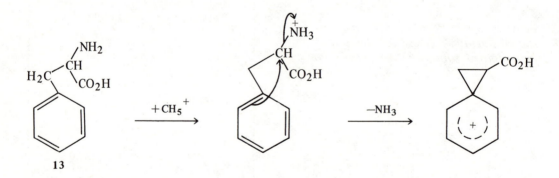

relationship to solution chemistry is further underlined by the fact that the loss of NH_3 from the $(M + 1)^+$ ion increases in importance down the series phenylalanine, histidine, tyrosine, tryptophan—that is, as the electron-donating power of the ring increases.

The chemical ionization spectra of esters provide further links with reactions in solution.

An ester such as benzyl acetate under CI conditions forms the ion **14**, which fragments to the

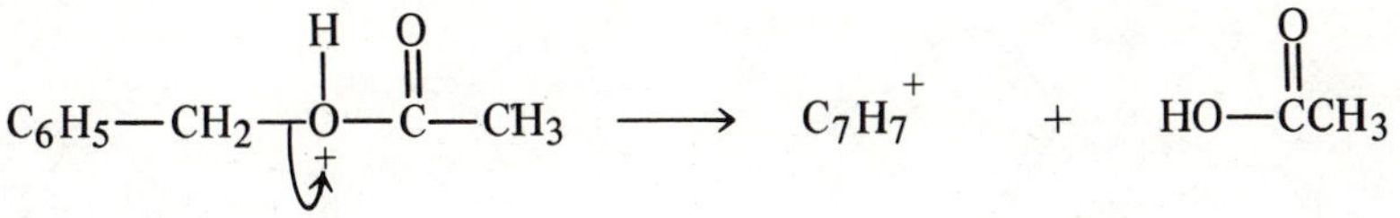

$$C_6H_5-CH_2 \underset{\overset{+}{\wedge}}{\overset{H \;\; O}{\underset{|}{\overset{||}{\underset{O-C-CH_3}{}}}}} \longrightarrow C_7H_7^+ \quad + \quad HO-\overset{O}{\overset{||}{C}}CH_3$$

14

ion $C_7H_7^+$ by the reaction shown. The relationship of this reaction to the acid-catalyzed alkyl cleavage mechanism ($A_{AL}1$) of ester hydrolysis is evident.

An alternative CI fragmentation mode is formation of the acylium ion (**15**), a reaction

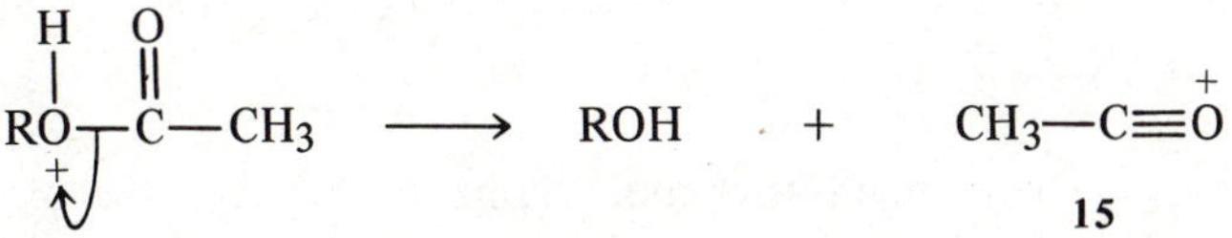

$$RO \underset{\overset{+}{\wedge}}{\overset{H \;\; O}{\underset{|}{\overset{||}{\underset{C-CH_3}{}}}}} \longrightarrow ROH \quad + \quad CH_3-\overset{+}{C}{\equiv}O$$

15

that is more important in alkyl than in benzyl esters. This process is the analogue of the acid-catalyzed acyl cleavage mechanism ($A_{AC}1$) of ester hydrolysis.

Acid-catalyzed ester exchange also has its analogue in the gas phase. If methane is used as the chemical ionization reagent gas, $C_2H_5^+$ ions are formed as well as CH_5^+. Addition of this electrophile ($C_2H_5^+$) to the carbonyl group gives an ion (**16**) that can eliminate ethene via a six-membered hydrogen transfer to give the alkoxy-protonated form (**17**) of the original

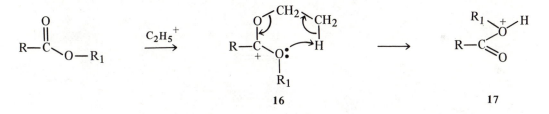

16   **17**

ester. Alternatively, an alkene molecule can be eliminated in the course of transferring a β hydrogen from the R_1 group to the (original) carbonyl oxygen. This process yields the protonated form (**18**) of the ester in which R_1 has been replaced by C_2H_5.

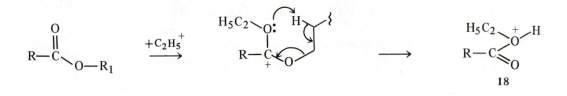

18

5–5c Photochemical–Mass Spectral Correlations

To illustrate this very extensive subject we will discuss the N^+-O^- group in azoxy compounds and nitrones. These compounds exhibit a notable tendency for photorearrangement, which frequently occurs via the three-membered cyclic isomer. Azoxy compounds (19)

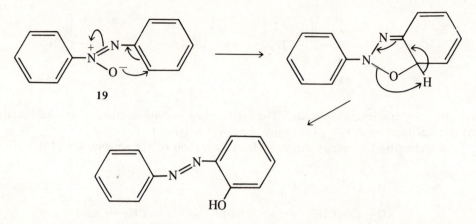

19

rearrange to hydroxyazo compounds by bond formation between the oxygen atom of the azo group and the ortho position of the more distant ring. Upon electron impact an analogous oxygen–ortho-carbon bond formation reaction occurs, although this step is followed by expulsion of the ortho substituent. This latter process leads to moderately abundant ions (20)

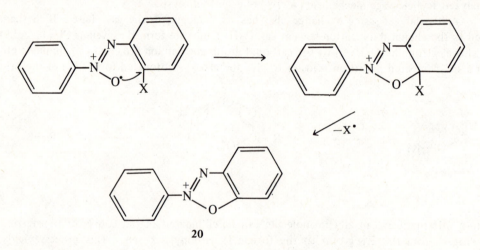

20

when the substituent is H• and to much more abundant ions when it is Cl•. (It should be noted that this rearrangement represents an example of free radical intramolecular aromatic substitution; compare the nucleophilic counterpart in section 5–3).

Nitrones also displace ortho groups in the course of their mass spectral fragmentation, but the formation of *o*-hydroxyanils corresponding to *o*-hydroxyazo compounds apparently has not been observed photochemically. Instead, isomerization of the nitrone (21) to the

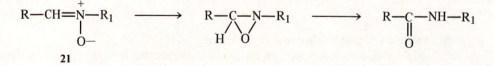

21

oxaziridine is followed by amide formation. This reaction also has its mass spectral analogue, since much of the fragmentation of aromatic nitrones, including the formation of abundant benzoyl ions, can only be explained if molecular ion isomerization to the amide occurs.

5-5d Solvolysis–Mass Spectral Relationships

The stability of certain intermediates in the mass spectrometer and the instability of others finds parallels in solution. In some of these cases, the mass spectral data were obtained long before the corresponding solution data, but were generally ignored or ascribed to the unique conditions that pertain in the ion source.

A case in point concerns the cyclic five-membered bromonium ion. This species was proposed to be relatively stable (considerably more so than the corresponding four- and six-membered cyclic ions) to account for the fact that alkyl bromides show unusually pronounced δ cleavage. For example, in *n*-octyl bromide this ion is the most abundant in the mass spectrum; by way of contrast γ and ϵ cleavages give ions with 2% and 10% relative abundance, respectively. In agreement with the mass spectral data, the tetramethylenebromonium ion has recently been prepared in superacid media, and it has been shown to be far more stable than the four-membered analogue.

Intramolecular or neighboring group displacement reactions should also have their counterparts in the gas phase. Thus phenonium ions (three-membered rings) should be particularly stable, and hence abundant, whereas the less stable spiro[5.3] cations (four-membered rings) should be of much lower abundance. The results for numerous functional groups support the hypothesis that aryl participation occurs in the gas phase and leads to phenonium ion formation. The simplest argument supporting this view comes from the abundance of the corresponding fragment ions compared to model systems. Substituent effects and thermochemical data also provide indirect support for anchimeric assistance in the gas phase. In certain nonbenzenoid systems more direct evidence for aryl participation has been forthcoming. Indeed, in the azulene system, mass spectral results have been used to predict solvolytic behavior, and subsequent solvolysis studies have confirmed the predictions. For example, it was predicted that 4-azulylpropyl esters (**22**) should undergo solvolysis by a k_Δ (participation) mechanism and therefore should be much more labile than the 6-substituted isomers, where only a k_s (solvent displacement) mechanism was expected to be important.

Solvolysis:

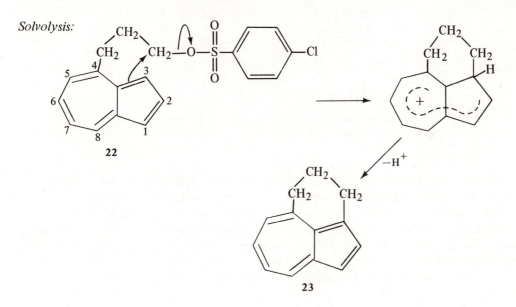

Electron impact:

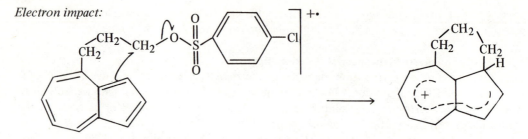

The isolation of the tricyclic product (23) from the acetolysis mixture, as well as the kinetic and thermodynamic data for the acetolysis, confirmed the mass spectrometric predictions. These and related results indicate the potential application of mass spectrometry as a rapid technique to survey systems of potential solvolytic interest.

The participation reactions just discussed emphasize the tendency for isolated ions to stabilize themselves by internal solvation. A large number of bond-forming reactions in mass spectra can be rationalized on the basis of this model, in which coiling of the ion allows solvation, and hence stabilization. Among numerous other examples, consider methyl 12-hydroxystearate (24). Solvation of the ionized ester by the alcoholic group stabilizes the molecular

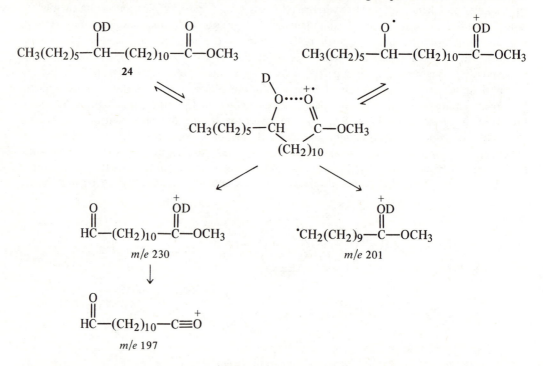

ion. However, since the internal energy of the molecular ion is fixed, any such stabilization must promote reaction by releasing as internal energy the potential energy of solvation. Hence it is not surprising that reactions tend to proceed from a form of the molecular ion in which the two functional groups interact, in spite of the large entropy requirements for such interaction. The three major fragment ions in the spectrum of methyl 12-hydroxystearate (24), with hydroxyl deuteration, all appear to involve such a functional group interaction although one, cleavage to give m/e 230, can be explained without this hypothesis.

The point that emerges from the above discussion is that stabilization of a gaseous ion by internal solvation *increases* its reactivity, whereas such stabilization decreases reactivity in solution. This result, which is something of an extention of the charge localization hypothesis of mass spectra (see section 1-3d), arises because a molecular ion is an energetically isolated system.

5-6 Isotope Effects

As in other branches of chemistry, the study of isotope effects has much to contribute to an understanding of reaction mechanisms in the gaseous ion milieu. Only a beginning has been made, however, in a field that should provide a wealth of new data on ionic reactions in the near future. The fact that nonequilibrium isotope effects are involved in low pressure unimolecular reactions is of cardinal importance. As a result, upper limits are not imposed upon kinetic isotope effects for mass spectral fragmentations, and values in the thousands have been measured. In addition to kinetic isotope effects, denoted k_H/k_D in the common hydrogen-deuterium case and measured from relative ion abundances, there are several other mass spectrometric measurements that show the operation of isotope effects. These all concern thermochemical parameters and arise because of the origin of isotope effects in zero point energy differences. Hence the zero point energy difference between isotopically substituted neutral species and the difference between the corresponding ions will seldom be exactly equal and will give rise to an isotope effect on the ionization potential. Similarly, appearance potentials have associated isotope effects, as of course do enthalpies of formation of ions. These effects, however, are all very small.

Major interest focuses on isotope effects on ionic reaction rates, i.e., ion abundances. This discussion will be restricted to one isotopic pair, H–D, and to primary isotope effects. The rate of a unimolecular mass spectral reaction depends on the internal energy of the reactant ion, i.e., the nonfixed energy $\epsilon^{\ddagger}$ of the activated complex. For a molecular ion $M^{+\cdot}$ that can lose either $H^{\cdot}$ or $D^{\cdot}$, the potential energy diagram shown in Figure 5-20 applies. The activation energies for the two reactions differ, that for loss of $H^{\cdot}$ being smaller. Now, if an ion of the total energy shown is examined, it will have a greater $\epsilon^{\ddagger}$ for loss of $H^{\cdot}$ than for loss of $D^{\cdot}$ and hence $k_H > k_D$. By selecting ions of lower and lower total energy, the difference between the activation energies becomes relatively more important and the rate difference larger. In principle, ions with energy in excess of the activation energy for $H^{\cdot}$ loss, but with insufficient energy to undergo $D^{\cdot}$ loss could be chosen. An infinitely large isotope effect (k_H/k_D) would result. The inverse relationship between size of isotope effect and internal energy has repeatedly been confirmed. Fast reactions occurring in the ion source often show small isotope effects, but these effects increase as the electron energy is lowered and especially as metastable ions are sampled. For example, k_H/k_D for $H^{\cdot}$ loss from toluene is ~1.5 in the source region of a typical instrument and 2.4 in the metastable region. More striking differences are observed for $H^{\cdot}$ loss from isobutane, which shows a small isotope effect in ion source reactions but a k_H/k_D value of more than 1000 for metastable ions. The observation of very large isotope effects in the reactions of low energy ions appears, from the limited data available, to be associated with simple cleavage reactions rather than with bond-forming reactions. Isotope effects on H–D scrambling rates in the metastable range, for example, do not appear to be very large, since the same difference in $\epsilon^{\ddagger}$ has a much greater effect on the rate constant for a simple cleavage than for a rearrangement (compare the shapes of the k vs. ϵ curves in Figure 1-4).

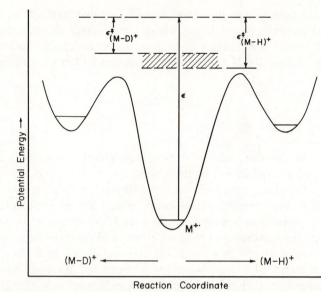

FIGURE 5-20. Potential energy diagram for competitive H· and D· loss from a partially labeled molecular ion $M^{+\cdot}$. Differences in the zero point energies of the activated complexes for the two reactions give rise to differences in the activation energies. For an ion of a given internal energy ϵ the excess energy of the activated complex appropriate to H· loss, i.e., $\epsilon^{\ddagger}_{(M-H)^{+}}$, is greater than that for D· loss. Hence, H· loss is the faster reaction. [Reproduced with the permission of Elsevier Publishing Company from M. Bertrand, J. H. Beynon, and R. G. Cooks, Int. J. Mass. Spectrom. Ion Phys., **9**, 346 (1972).]

In illustrating the application of isotope effect measurements to reaction mechanisms, just two examples can be cited. One involves an isotope effect on kinetic energy release, and the other an effect on reaction rate. *s*-Triazole undergoes N_2 loss from the molecular ion **25**

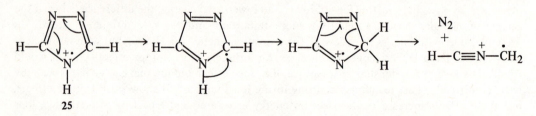

with $T = 1.41 \pm 0.02$ eV, while the N-d_1 analogue shows $T = 1.57 \pm 0.02$ eV. This result suggests that hydrogen bond cleavage is involved in the rate-controlling step. The second example concerns whether or not the loss of C_2H_4 from phenetole occurs to give the phenol molecular ion (e.g., **26**); if it does, there may be an isotope effect on loss of CO since H transfer must precede fragmentation. The alternative is direct formation of a keto ion (**27**), in which case no isotope effect would be expected. Ion abundances measured for metastable ions could conveniently be expressed relative to Br· loss, which is a competitive reaction not expected to show an isotope effect. It was found that the $(M - Br)^{+\cdot}/(M - CO)^{+\cdot}$ ratio for bromophenetole-d_5 was three times as great as that for the bromophenetole itself. The con-

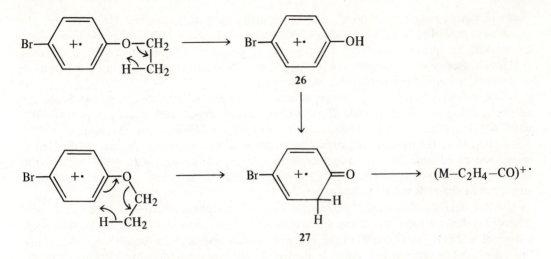

clusion—that the rearrangement leads to the phenol molecular ion—is strictly applicable only to those phenetole ions that have enough energy to yield $(M - C_2H_4)^{+ \cdot}$ ions that fragment further.

5-7 Ion Cyclotron Resonance

This section is devoted to a brief account of some of the ion chemistry now being done with the ion cyclotron resonance spectrometer. Because bimolecular processes, which occur at low (thermal) kinetic energies, are the chief concern of icr, these reactions come closest to resembling ionic processes in condensed phase. This relationship is currently being exploited in an attempt to assess the role of solvent in affecting the course and rates of solution reactions and the relative stabilities of intermediates. It is perhaps worth noting that some of the icr results, such as the order of gas phase acidities and basicities, have also been found using other ion-molecule reaction techniques, such as the tandem mass spectrometer and the chemical ionization method.

Topics of interest to organic chemists are gas phase acidities and basicities, gas phase nucleophilic substitutions, and gas phase qualitative organic analysis. The experimental methods (see section 3–5) involve monitoring product ion abundance, sometimes as a function of pressure or ionizing electron energy. Double resonance is also possible and serves to identify the reactant ion. In the usual experiment, irradiation at the resonance frequency of the reactant ion results in an increase in its kinetic energy and hence in a change in the rate constant for product formation and a change in signal strength. If a pulsed irradiating field is used, then a phase-sensitive detector allows the determination of the masses of all reactant ions contributing to a particular product ion.

Ion cyclotron resonance methods allow the determination of relative gas phase acidities and basicities. Other methods must be used to provide reference values in order to establish absolute magnitudes. The icr method depends on examination of proton transfer reactions of the general types shown in eqs. 5–35 and 5–36. A fundamental assumption is that ion-molecule

$$A^- + BH \longrightarrow AH + B^- \qquad \text{5-35}$$

$$A + BH^+ \longrightarrow AH^+ + B \qquad \text{5-36}$$

reactions having reasonable rates all involve negligible activation energies. Hence if eq. 5–35 proceeds to the right, the proton affinity of A^- is greater than that of B^-. If eq. 5–36 proceeds to the right, the proton affinity of A is greater than that of B. The proton affinity of a species is a direct measure of its basicity, while the proton affinity of a base is a quantitative measure of the acid strength of the conjugate acid.

As a further check on the proton affinity orders derived from eqs. 5–35 and 5–36, the reverse reaction can be examined. If a double-resonance experiment shows a decrease in B^- when A^- is irradiated, this is evidence for an exothermic reaction in the direction $A^- \rightarrow B^-$. If, in addition, irradiation of B^- causes an increase in the A^- signal, the reverse reaction is considered to be endothermic. Together these data provide strong evidence that A^- has the greater proton affinity, i.e., $PA(A^-) > PA(B^-)$. Hence HA must be a weaker acid and A^- a stronger base than HB and B^-, respectively.

These relationships form the basis for ordering gas phase acidities and basicities. The observed orders are most interesting when they differ from the order found in solution. Such a reversal is found, for example, in the order of alcohol acidities. In the gas phase the acidity order is t-butyl > isopropyl > ethyl > methyl > H_2O, whereas inductive and hyperconjugative effects have been interpreted as yielding the opposite order in solution. The order of amine basicities is an even more complex problem in solution, and all explanations of the experimental order have had to rely upon combinations of effects. The gas phase basicity is much simpler and points up the complications that may be introduced by solvent. When the same substituents are compared, the order of basicity in the gas phase is tertiary > secondary > primary. In addition, within any series (primary, secondary, or tertiary), the basicity increases as the size of the alkyl substituent(s) increases. This effect of substituent size is exactly analogous to that observed for acids. Hence the stabilizing role of alkyl groups in the gas phase is believed to reside in their polarizability. Unlike the inductive effect, this property can account for the stabilization of both positively and negatively charged ions.

The basicity order of alkyl halides found in the gas phase is also much simpler than in solution, again because the complications of solvent do not mask electronic effects. Thus the proton affinity order $C_2H_5I > C_2H_5Br > C_2H_5Cl > C_2H_5F$ is observed. In addition, $C_2H_5Cl > CH_3Cl > HCl$.

Electrophilic aromatic addition, apparently analogous to the first step in solution electrophilic aromatic substitution reactions, has been observed in the gas phase in several systems. For example, a Friedel-Crafts type of addition of CH_2Cl^+ to benzene is observed, but after electrophilic attack, the gaseous ion (**28**) eliminates HCl to give the stable $C_7H_7^+$ ion, rather

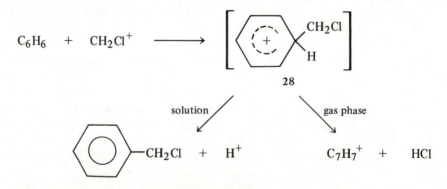

than eliminating H^+ as occurs in solution. The reaction of benzene with CH_3^+ in the gas phase also leads to $C_7H_7^+$ ions with the elimination of H_2.

The use of icr in qualitative organic analysis may be illustrated by the nitration of aromatic substrates. In studies of this type, either the ion or the neutral may be the "reagent," depending on whether the inquiry concerns the structure of a neutral or an ionic species. A future laboratory exercise in qualitative organic analysis might proceed along the following lines.

You are provided with a liquid unknown sample. (1) Run its mass spectrum. The observed molecular weight of 92 suggests the molecular formula C_7H_8. (2) Consider the fragmentation pattern. One cannot distinguish toluene, cycloheptatriene, and norbornadiene. (3) React the unknown with ionized ethyl nitrate in the icr cell. If nitration occurs to give the addition product $C_7H_8NO_2^+$, then the compound must be toluene since the other isomers are not aromatic and are unreactive under these conditions. (4) Carry out the reverse experiment, reacting the ionized C_7H_8 compound with neutral ethyl nitrate. The same result is obtained; only the toluene ion reacts. The conclusion that nondecomposing toluene molecular ions do not ring expand is interesting in the light of the problem of the structure—benzylic or tropylium—of $C_7H_7^+$ in mass spectrometry.

Functional group analysis by icr is of course in its infancy, but it holds considerable promise. An example is the reactivity of various compounds with the acetylating ionic reagent $CH_3COCOCH_3(COCH_3)^+$. This ion is formed from 2,3-butadione by self-acetylation, and it acts as a source of CH_3CO^+. The nucleophilicity of an unknown determines the ease with which it undergoes acetylation. Thus, the order of decreasing reactivities, amines > alcohols > alkyl halides > hydrocarbons, that has been observed can be used to classify unknowns. With slightly stronger or slightly weaker acetylating reagents, more refined structural conclusions should become possible. The advantages of this approach to qualitative analysis are that impurities do not interfere and extremely small quantities of sample can be used.

Ion cyclotron resonance results have shed interesting new light upon nucleophilic substitution reactions. As in solution, the nucleophile may be an anion (eq. 5–37) or a neutral molecule

$$Y^- + RX \longrightarrow YR + X^- \qquad\qquad 5\text{-}37$$

(eq. 5–38). With reagent ions of the type CH_3ClH^+, the order of *C*-nucleophilicites observed in

$$Y + RX^+ \longrightarrow YR^+ + X \qquad\qquad 5\text{-}38$$

icr experiments is NH_3 > CO > H_2S > H_2CO > HI > H_2O > HBr > HCl > N_2 > HF. This is essentially the same order as observed in solution under S_N1 conditions.

Proton affinities determined in the gas phase by icr can be compared with methyl cation affinities (defined as the negative of the enthalpy change for the reaction $M + CH_3^+ \rightarrow MCH_3^+$). This comparison offers a quantitive measure of the "hardness" or "softness" of a given nucleophile. This HI, a "soft" nucleophile in solution, can be compared with HCl, CH_3F, and HF, all "harder" nucleophiles. The methyl cation affinities of these nucleophiles are 67, 51, 44 and 36 kcal mole^{-1}, while the proton affinities are 145, 142, 151, and 137 kcal mole^{-1}. The softest nucleophile thus has the greatest cation affinity, but it does not have the largest proton affinity.

Results of gas phase acetylation studies have also provided information on steric effects. These effects operate in some unimolecular fragmentations, too, but they have not been extensively studied. Gas phase steric effects are potentially very important since they represent steric factors due only to molecular sizes without the complication of solvent steric effects.

Experiments using negative ions in the gas phase have revealed that solvent effects are mainly responsible for the fact that the S_N2 mechanism involves attack from the rear. In the gas phase attack occurs from the front, i.e., by an S_Ni process. This result is shown most dramatically by the fact that 1-bromoadamantane reacts with Cl^- at approximately the same rate as do other tertiary bromides.

5–8 Ion Structures

In terms of practical importance and intellectual challenge, ion structure is probably the most important problem in mass spectrometry today. Knowledge of structure is fundamental to the development of any chemical science. The importance of organic ions is growing in such diverse fields as isotope separation, ionospheric processes (including those resulting from the presence of pollutants), plasma chemistry, cosmochemistry, radiation processes, high energy beams in therapeutics, and electronic devices that depend on ion-surface interactions such as ion pumps and ion gauges. It is therefore imperative that ionic structural information be obtained. The determination of complex organic structures within a time of about a microsecond is an intellectual challenge of the first order.

Throughout the text there have been many instances in which ion structural data were explicitly obtained. Before these threads are gathered together, a few general comments are pertinent. The term ion structure really covers two situations that must not be confused. The structure with which an ion is formed at threshold may or may not be the same as that which it adopts when it has sufficient energy to undergo some particular fragmentation reaction. Thus "reacting" and "nonreacting" ion structures, in general, have to be distinguished. Frequently, a given method of probing ion structure will be applicable to only one of these classes. A further observation is that many of these methods give information that is relevant not to a particular point on a potential energy surface, but to the whole reactant–product system. A final feature of ionic structure is the fact that mixtures of structures may occur. The difficulties that this complexity can cause are largely unexplored.

Turning now to the individual methods of obtaining information of ion structure, we consider first those applicable to nonreacting (threshold) ions. An important method of distinguishing ions is by determination of their enthalpies of formation, but considerable care is necessary to ensure that the excess energy terms in the thermochemical equations are properly treated (see section 4–9). This method, like many others, characterizes an ion without directly defining its structure. Other methods appropriate to nonreacting ions concentrate on the properties of reactions leading to the ion in question. The properties used include kinetic energy release and energy partitioning (section 5–4), label incorporation into the product ion, and substituent effect studies (section 5–5). Ion-molecule reactions can also be used to probe the structures of low energy ions (section 5–7). In particular, collision-induced dissociation can provide a mass spectrum of any selected ion when a reversed sector (mikes) spectrometer is employed. Photodissociation of ions is also possible.

In deducing the structures of reacting ions, one must not confuse differences in behavior resulting from differences in internal energy with those behavioral differences that have a structural origin. The best procedure is to employ metastable ions, which have low internal energies. The parameters used in this type of study include ion abundances (which approximate reaction rates), kinetic energy releases, and isotope effects on ion abundances and on kinetic energy release. Generation of a given ion in turn by direct ionization, by a fragmentation route, and by charge exchange of a doubly charged ion provides an important method of assessing the relative effects of structure and internal energy on ionic properties. In a very few cases experimental determinations of the rate constant as a function of internal energy have been made. In the case of the $C_6H_6^{\cdot+}$ ion formed from benzene, these data have suggested that this ion has at least two structures, one that fragments by $H^{\cdot}$ loss and the other by C_2H_2 loss.

Non-mass spectrometric methods of deducing gaseous ion structures have also made some headway. In particular, small ions have been the subject of theoretical studies and their emission and absorption spectra have been measured. In addition, icr and other experiments on the effects of irradiation on ionic behavior have commenced, and in related work the radiation emitted during ionic reactions has been studied.

BIBLIOGRAPHY

1. Photoionization: N. W. Reid, *Int. J. Mass Spectrom. Ion Phys.*, **6**, 1 (1971).
2. Field Ionization: H. D. Beckey in A. L. Burlingame, ed., *Topics in Organic Mass Spectrometry*, Wiley–Interscience, New York, 1970; H. D. Beckey, *Field Ionization Mass Spectrometry*, Pergamon, Oxford, 1971.
3. Negative Ions: C. E. Melton in F. W. McLafferty, ed., *Mass Spectrometry of Organic Ions*, Academic, New York, 1963, ch. 4; C. E. Melton, *Principles of Mass Spectrometry and Negative Ions*, Marcel Dekker, New York, 1970; J. H. Bowie and B. D. Williams in *MTP Review of Science*, Physical Chemistry Section, Mass Spectrometry, Vol. 5, 1975.
4. Quasiequilibrium Theory: H. M. Rosenstock and M. Krauss in F. W. McLafferty, ed., *Mass Spectrometry of Organic Ions*, Academic, New York, 1963, ch. 1; M. Vestal in P. Ausloos, ed., *Fundamental Processes in Radiation Chemistry*, Interscience, New York, 1968, ch. 2.
5. Rearrangements: P. Brown and C. Djerassi, *Angew. Chem. Int. Ed. Engl.*, **6**, 477 (1967); R. G. Cooks, *Org. Mass Spectrom.*, **2**, 481 (1969).
6. Ion Kinetic Energy Spectrometry: R. G. Cooks and J. H. Beynon, *J. Chem. Educ.*, **51**, 437 (1974); R. G. Cooks, J. H. Beynon, R. M. Caprioli, and G. R. Lester, *Metastable Ions*, Elsevier, Amsterdam, 1973.
7. Hammett Relations: M. M. Bursey, *Org. Mass Spectrom.*, **1**, 31 (1968).
8. Chemical Ionization: F. H. Field, *Accounts Chem. Res.*, **1**, 42 (1968).
9. Relation to Photochemistry: R. A. W. Johnstone, *Advan. Phys. Org. Chem.*, **8**, 151 (1970).
10. Relation to Solvolysis: R. G. Cooks, N. L. Wolfe, J. R. Curtis, H. E. Petty, and R. N. McDonald, *J. Org. Chem.*, **35**, 4048 (1970).
11. Isotope Effects: I. Howe and F. W. McLafferty, *J. Amer. Chem. Soc.*, **93**, 99 (1971); M. Bertrand, J. H. Beynon, and R. G. Cooks, *Int. J. Mass Spectrom. Ion Phys.*, **9**, 346 (1972).
12. Ion Cyclotron Resonance: J. D. Baldeschwieler and S. S. Woodgate, *Accounts Chem. Res.*, **4**, 114 (1971); J. I. Brauman and L. K. Blair, *J. Amer. Chem. Soc.*, **92**, 5986 (1970); M. M. Bursey, T. A. Elwood, M. K. Hoffman, T. A. Lehman, and J. M. Tesarek, *Anal. Chem.*, **42**, 1370 (1970).

WORKED STRUCTURAL PROBLEMS

The problems presented here (Figures A–1 to A–10) illustrate the lines of argument that can be followed in deducing molecular structures from mass spectra, although the thought processes involved are more efficiently conveyed verbally than in writing. The steps listed were not necessarily the only ones taken in solving the problem. Blind alleys were sometimes followed, but it is not feasible to set out a complete scheme covering all possible approaches to each structure. It should also be noted that the set of problems used is relatively small and compilations of mass spectral data or books on spectral problems should be consulted for more comprehensive collections.

The examples used, while simple, all represent real structural problems, and the mass spectra were determined to provide structural information. Very limited use of metastable peaks is made here, despite their immense value in determining fragmentation patterns, because these peaks are not readily observed with some instruments. No source of information besides mass spectrometry was employed in solving these problems. Of course, this is seldom the case in practice, so that much more complex structures than these examples are actually amenable to analysis.

Molecular ions are assigned since the actual spectrum would otherwise have had to be reporduced. This assignment was trivial in almost all cases. It was, nevertheless, checked by the methods discussed in section 4-1. Finally, the spectra are restricted to ions with greater than 1% relative abundance.

The material of Chapters 1 and 4 is particularly pertinent to these problems, and many of the arguments derive directly from those chapters.

518

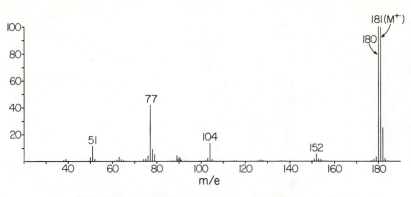

FIGURE A-1. *Mass spectrum of unknown A-1.*

1. This is a very simple spectrum, with the characteristics of an aromatic compound, namely, an abundant molecular ion and only a few abundant fragment ions.
2. In particular, the abundances of the ions at *m/e* 77 and 51 indicate a mono-substituted benzene derivative. The loss of 77 mass units to give *m/e* 104 tallies with this assignment.
3. The odd molecular weight indicates the presence of an odd number of nitrogen atoms.
4. Identification of a phenyl group and a nitrogen atom still leaves a massive portion of the molecule unaccounted for. Most significantly, there seems to be no indication of the nature of this fragment in the spectrum. Such a circumstance often suggests that a unit is repeated, in this case the phenyl ring.
5. A clue that this diagnosis is correct is provided by the low abundance ions at *m/e* 152–154, which correspond to ionized biphenyl (*m/e* 154) and biphenylene (*m/e* 152). The former species could be generated by the common rearrangement in which two aryl groups are bonded together with the elimination of a stable neutral molecule. In the present case the neutral has mass 27 and includes a nitrogen atom. Hence the unknown is the imine $C_6H_5CH{=}NC_6H_5$.
6. It is noteworthy that the $(M + 1)^+$ ion is far more abundant than the ^{13}C isotopic contribution requires. This is a frequent observation for nitrogen-containing compounds and is due to protonation of the molecular ion.

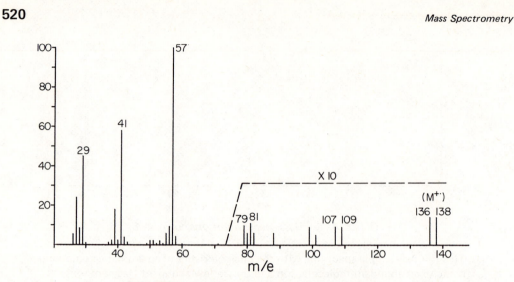

FIGURE A-2. Mass spectrum of unknown A-2.

1. The molecular ion abundance is very low, but the isotope distribution clearly indicates the presence of a single bromine atom in the molecule. The *m/e* 79/81 and 80/82 doublets confirm this conclusion.

2. Loss of bromine from the molecular ion occurs with great ease to give the base peak *m/e* 57. This ion and the presence of other hydrocarbon ions at *m/e* 27, 29, 39, and 41 indicate a molecular formula C_4H_9Br.

3. The molecular ion loses an ethyl radical to give the bromine-containing ions at *m/e* 107 and 109. Hence the compound is 2-bromobutane. Methyl radical loss from the molecular ion does occur, but gives a product of very much lower abundance (not recorded) than that due to loss of the larger radical (compare section 4-5).

4. The presence of ions at *m/e* 88, 99, and 101 due to impurities may be noted.

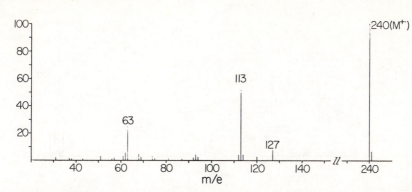

FIGURE A-3. Mass spectrum of unknown A-3. Metastable peaks were observed at m/e 35.2 and m/e 53.2.

1. This spectrum shows the behavior typical of aromatic compounds in that the molecular ion and two or three fragment ions carry the bulk of the ion current.

2. It is striking that in spite of the high molecular weight (240) the ^{13}C isotope of the molecular ion has an abundance of only 7% of $M^{+\cdot}$, limiting the number of carbon atoms present to a maximum of six. Clearly, a considerable proportion of the mass of the molecule is contributed by monoisotopic elements. Iodine and fluorine are the prime candidates.

3. The atomic weight of iodine is 127; hence the ion at *m/e* 127 corresponds to I^+. In addition, *m/e* 113 is due to M − 127.

4. The only other fragment ion having even moderate abundance is *m/e* 63, and this is formed, at least in part, from *m/e* 113, as shown by the presence of a metastable peak at *m/e* 35.2. The loss of 50 mass units as a neutral species is unusual except in fluorine-containing compounds where CF_2 elimination is common.

5. Assuming that *m/e* 63 is $C_5H_3^+$, the formula $C_6H_3F_2I$ is obtained. This formula requires four rings and/or double bonds. A trisubstituted benzene fits all the above data. In fact, the compound is 1,3-difluoro-5-iodobenzene, but the positions of the substituents cannot be determined from the mass spectrum.

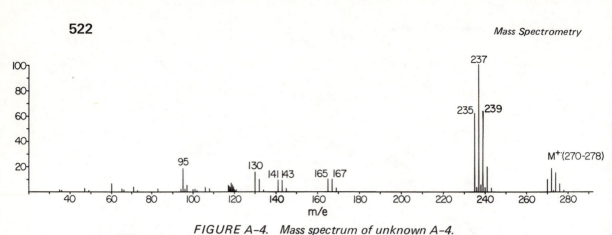

FIGURE A-4. Mass spectrum of unknown A-4.

1. In spite of complex isotope patterns, this spectrum is rather simple, just a few ions carrying the bulk of the ion current.

2. At first glance an organometallic might be suspected, but the isotope pattern of the molecular ion differs from that of the major fragment ions. Indeed, the pattern becomes more and more simple as one moves to low mass.

3. This behavior suggests the presence of several atoms of the same multi-isotopic element, and the characteristic chlorine 3/1 pattern with a separation of two mass units is seen in several low mass ions.

4. One concludes that this is a polychlorinated compound. The isotope patterns indicate the number of chlorine atoms in each fragment: for example, *m/e* 71/73 has one chlorine; 130/132/134 has two chlorines; 141/143/145/147 has three chlorines; 235/237/239/241/243 has five chlorines and the molecular ion, 270/272/274/276/278/280 has six chlorines (the peak at *m/e* 280 is not recorded in the spectrum since its relative abundance is <1%).

5. Given a molecular weight of 270 (^{35}Cl) and the presence of six chlorine atoms, only 60 mass units remain, possibly C_5. The complete absence of any hydrocarbon ions (*m/e* 27, 29, 39, 41, 43, etc.) makes the perchlorinated formula C_5Cl_6 seem reasonable.

6. This formula requires three rings and/or double bonds, implying a cyclopentadiene structure, a vinyl cyclopropene formula, or a pentenyne structure. A decision among these alternatives is not readily made from the mass spectrum. The compound, in fact, is perchlorocyclopentadiene.

7. The extent to which Cl˙ loss predominates over ring fragmentation is shown by the presence of abundant ions in the series $C_5Cl_5^+$ (*m/e* 235), $C_5Cl_3^+$ (165), $C_5Cl_2^{+\cdot}$ (130), and C_5Cl^+ (95) where the mass numbers refer to the ^{35}Cl isotope. The only major ions that involve carbon–carbon bond cleavage are $C_3Cl_3^+$ (141) and $C_3Cl_2^{+\cdot}$ (106).

8. The abundance of the doubly charged $(M - Cl)^{++}$ ion (in the cluster from *m/e* 117 to 121) is notable.

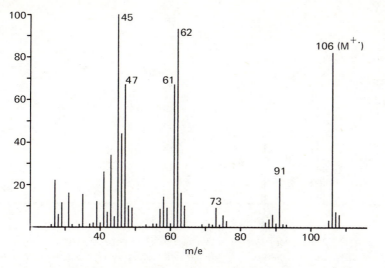

FIGURE A-5. *Mass spectrum of unknown A-5.*

1. The presence of one sulfur atom is suggested by the abundance of the $(M + 2)^{+\cdot}$ ion.
2. The ion *m/e* 47 is unusual since this region of the spectra for compounds containing C, H, N, and O is invariably blank. The ion CH_3S^+ is probably responsible for this peak.
3. The proposed *S*-methyl group may also account for the observed loss of a methyl radical from the molecular ion.
4. The ion *m/e* 61 is apparently the higher homologue of *m/e* 47. This result suggests the partial structure CH_3SCH_2-.
5. The remainder of the molecule has mass 45 and indeed an ion *m/e* 45 forms the base peak in the spectrum. This ion must have the composition CO_2H^+ or $C_2H_5O^+$.
6. The presence of ions due to loss of 17, 18, and 19 mass units from the molecular ion suggests an alcohol. In conjunction with the abundance of *m/e* 45, this observation requires the structure $CH_3SCH_2CH(OH)CH_3$.
7. The propensity of sulfur to engage in bond-forming reactions in mass spectrometry is exemplified by two ions in this spectrum. First, the loss of SH˙ occurs to give *m/e* 73, even though a thiol group is not present in the molecule. As is usual for such complex rearrangements, this ion is of low relative abundance. Second, the formation of *m/e* 62 by elimination of acetaldehyde is due to the ease with which the ionized sulfur atom can abstract a hydrogen atom via a five-membered cyclic transition state.

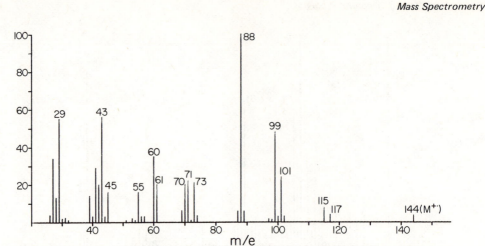

FIGURE A-6. *Mass spectrum of unknown A-6. Note that the composition of the molecular ion is $C_8H_{16}O_2$. Metastable peaks were observed at m/e 55.7, 53.7, and 50.9.*

1. Exact mass measurement establishes the formula of this unknown as $C_8H_{16}O_2$. Hence there is just one ring or double bond.

2. The base peak, *m/e* 88, corresponds to loss of 56 mass units from the molecular ion. This observation indicates that a molecule of butene is eliminated from the molecular ion; the presence of a metastable peak at *m/e* 53.7 confirms the transition (but not the composition of the neutral).

3. Since the molecule has but one ring or double bond and two oxygen atoms and since the base peak arises by alkene elimination from the molecular ion, it is reasonable to assume that McLafferty rearrangement to a carbonyl group is responsible for butene loss.

4. The second oxygen atom might be present as several functionalities, but the abundant ion at mass 99 (M − 45) and the presence of an ion *m/e* 45 suggest the ethoxy group, and in all probability the unknown is an ethyl ester. An ethyl ester would be expected to give an ion $EtOC\equiv O^+$, *m/e* 73, and this ion is observed.

5. The loss of butene from an ester $C_5H_{11}CO_2C_2H_5$ in a McLafferty rearrangement requires that the α carbon be unsubstituted.

6. The only remaining question concerns whether or not the C_4H_9 unit in $C_4H_9CH_2CO_2C_2H_5$ is branched. The presence of the $(M − 43)^+$ ion, which may at first sight seem puzzling, helps to answer this question. Just as ethers undergo a rearrangement reaction that appears to be simple β cleavage (section 4-5b), so carbonyl compounds undergo a reaction that appears to be simple γ cleavage. This reaction would give the $(M − 43)^+$ ion, *m/e* 101, if both the α and β carbons were unsubstituted. The ion *m/e* 115 is the higher homologue of *m/e* 101, and its formation requires that the entire amyl chain be linear, so the unknown is ethyl hexanoate.

7. The origin of several other ions may be briefly noted: *m/e* 71 is the $C_5H_{11}^+$ ion, formed in part by CO loss from the acylium ion; *m/e* 60 is ionized acetic acid, generated in part by C_2H_4 loss from the McLafferty rearrangement product. The ion at *m/e* 117 due to $C_2H_3{}^{\cdot}$ loss is unexpectedly abundant for an ethyl ester, although the product ion, $RCO_2H_2{}^+$, is particularly stable.

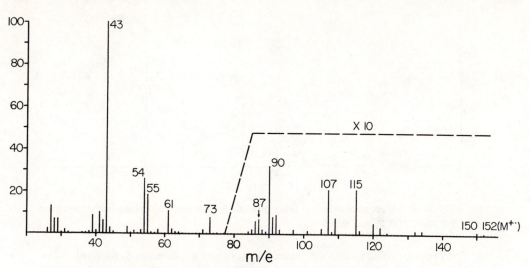

FIGURE A-7.   *Mass spectrum of unknown A-7.*

1. Although the molecular ion is of very low abundance, the two isotopic forms suggest that one chlorine atom is present; this conclusion is confirmed by the isotope pattern seen in several other ions, notably *m/e* 107/109.
2. The fact that the chlorine atom is readily lost (to give *m/e* 115) suggests that it is present as an alkyl or acyl chloride. The presence of the ion CH_2Cl^+ (*m/e* 49/51) and the loss of $CH_2Cl^\cdot$ from the molecular ion to give *m/e* 101 clarify the nature of the chloro group.
3. The abundance of *m/e* 43 suggests the presence of an acetyl group; the loss of acetic acid from the molecular ion to give *m/e* 90/92 confirms that the compound is an acetate.
4. The units $ClCH_2-$ and $CH_3C(O)O-$ leave 42 mass units to be accounted for. Only C_3H_6 and C_2H_2O are reasonable possibilities. The structure $Cl(CH_2)_4O$-$COCH_3$ is suggested by the presence of homologous ions belonging to the series 115, 101, 87, and 73.
5. The ion at *m/e* 54 corresponds to ionized butadiene and is apparently formed by loss of protonated acetic acid from the $(M - Cl)^+$ species. There are several other ions of low abundance that arise by rearrangement reactions.

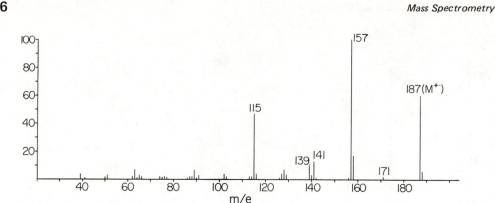

FIGURE A-8. *Mass spectrum of unknown A-8. Metastable peaks were observed at m/e 131.8, 106.2, and 93.7.*

1. The presence of just a few dominant ions suggests an aromatic compound.
2. The odd molecular weight implies that there is an odd number of nitrogen atoms (probably one rather than three or more).
3. There is no indication from isotope patterns of the presence of any elements besides C, H, N, and O.
4. The major fragment ion in the high mass region (and coincidentally the base peak) is m/e 157, which corresponds to loss of 30 mass units from the molecular ion (a reaction that is substantiated by the observation of a metastable peak at m/e 131.8). This observation is good evidence for the presence of a nitro group. The low abundance $(M - 16)^+$ ion is diagnostic of an N—O group. The loss of 46 mass units from the molecular ion (to give m/e 141) provides further evidence for the presence of the nitro group.
5. The unknown is therefore an aromatic nitro compound, X—NO$_2$, where X has mass 141 and could have zero, one, or more oxygen atoms. Hence the formulas that must be considered for this unit are $C_{11}H_9$, $C_{10}H_5O$, and C_9HO_2. Only $C_{11}H_9$ seems at all likely, and it is strongly indicated by the fact that m/e 115 (loss of 26, i.e., C_2H_2, from 141) is abundant. In fact, both m/e 141 and 115 are common ions in larger aromatic compounds.
6. The molecular formula is therefore $C_{11}H_9NO_2$. The structure must exclude ring-substituted methyl groups since an $(M - H)^+$ ion is not observed. The actual compound is p-(2H-cyclopentadienyl)nitrobenzene.

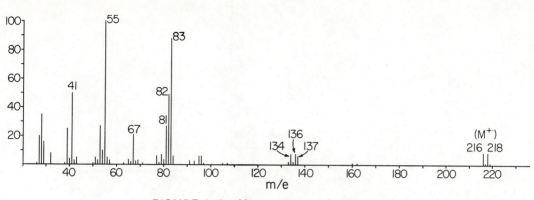

FIGURE A-9. Mass spectrum of unknown A-9.

1. The presence of a single bromine atom is immediately evident from the isotope distribution in the molecular ion.
2. The spectrum shows the typical features of aliphatic compounds, including a molecular ion of relatively low abundance and abundant fragment ions of low mass.
3. The loss of bromine does not give rise to an abundant fragment ion; hence bromine is not bonded to a saturated carbon, but rather to an alkenic carbon.
4. The low abundance of *m/e* 43, 57, and 71 indicates that the compound does not contain a saturated alkyl chain of more than two atoms.
5. The abundant ion at *m/e* 83 (and its fragment ion *m/e* 55) provides the key to this structural determination. Recognition that the cyclohexyl group has mass 83 allows one to rationalize all the abundant ions in the spectrum. Thus, *m/e* 82 is ionized cyclohexene, the major fragmentation of which is methyl radical loss to give *m/e* 67. Further loss of C_2H_2 should give rise to *m/e* 41.
6. The cyclohexyl group and the bromine atom leave 54 mass units to be accounted for. Possible formulas are C_4H_6 and C_3H_2O. Reasonable structures can be deduced on the basis of both possibilities, but an exact mass measurement proved the hydrocarbon formula C_4H_6.
7. The formula $C_6H_{11}-C_4H_6-Br$ requires one ring or double bond in addition to the cyclohexyl ring. The bromine must be attached to the C_4H_6 unit, and it has already been suggested that it is an alkenic bromide. The absence of an $(M-15)^+$ ion precludes an allylic methyl group.
8. A unique structural assignment can be made on the basis of the formation of *m/e* 134/6. This fragment contains bromine and is due to elimination of a neutral molecule of cyclohexene from the molecular ion. This process may occur via a six-membered cyclic rearrangement only for 1-bromo-2-methyl-3-cyclohexyl-1-propene if the above-mentioned structural requirements are also to be satisfied.

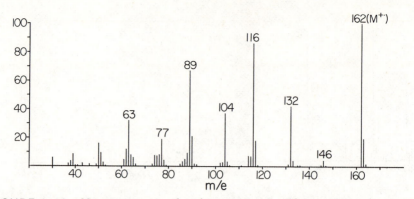

FIGURE A-10. Mass spectrum of unknown A-10. Metastable peaks were observed at m/e 81.9 and 68.3.

1. The fragment ion of highest mass, *m/e* 146, which corresponds to loss of 16 mass units from the molecular ion, probably results from loss of an oxygen atom. Such loss is diagnostic of an N—O group.
2. This analysis is confirmed by the fact that the molecular ion also loses fragments having masses of 30 and 46. An aryl nitro group is clearly indicated by these processes.
3. Since the molecular weight of the compound is even, it must contain an even number of nitrogen atoms, probably two.
4. Metastable peaks show that, at least in part, the ion at *m/e* 104 arises from 132 and that at *m/e* 89 from 116. The subsequent loss of CO after NO$^{\cdot}$ loss to give *m/e* 132 is an expected process. The further loss of 27 mass units after loss of $NO_2^{\cdot}$ almost certainly means that HCN is the neutral.
5. One knows, therefore, that the compound contains an aromatic nitro group and a cyano group, and the presence of *m/e* 77 indicates a benzene ring (although not necessarily monosubstituted). The molecular weight of 162 leaves only 14 mass units unaccounted for, which suggests a methylene group.
6. As there is no loss of H$^{\cdot}$ from the molecular ion, there can be no ring methyl group. Hence the compound must be a cyanomethylnitrobenzene. The mass spectrum does not allow one to choose among the ortho, meta, and para isomers, but the compound is *p*-(cyanomethyl)nitrobenzene.

Part Five

X-RAY CRYSTALLOGRAPHY

1

CRYSTALS AND CRYSTALLOGRAPHY

1-1 Introduction

Although X-ray crystallography differs in many ways from the spectroscopic techniques discussed in the other parts of this volume, its application to organic chemistry is now undergoing the explosive growth shown earlier by ultraviolet, infrared, nmr, and mass spectral methods. In all these cases the cause has been the same: the appearance of commercial instrumentation that can permit the chemist to make measurements and to analyze the results with a minimum of effort, as an adjunct to other problems rather than as a research subject in itself. In the case of crystallography this instrumentation took two forms, neither of which would have been adequate alone. First, there occurred the development and distribution of large scale computing facilities that, together with programs prepared by professional crystallographers, replaced the slow and tedious computation of an earlier day with rapid, cheap, and effortless solutions of most of the routine crystallographic problems. Second, there appeared on the market automatic diffractometers that almost entirely eliminated the tedium of collecting a set of data for analysis, reducing the time required to only a few days. Thus, in principle, an entire analysis can be carried out in less than two weeks, and the investments in time and effort that may be lost in case of failure is relatively small.

The material that follows is not intended to enable the reader to undertake structure analyses singlehandedly. To achieve this ability requires a greater understanding of the principles and techniques than can be provided in this text. Rather the material serves as an introduction for those who wish to proceed to more comprehensive works, and as a survey of what is possible, reasonable, and correct for those who wish to interpret the growing literature of applied structure analysis.

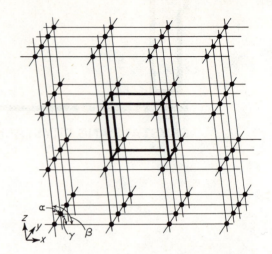

FIGURE 1-1. A three-dimensional lattice showing a unit cell (heavy lines).

The process of determining a crystal structure can be divided into two parts, one experimental and the other computational. The first involves choosing a crystal, determining its lattice geometry and symmetry, and measuring the relative intensities of a large number of diffracted rays. The second transforms the crude data into a representation of the electronic distribution in the crystal, from which the desired information about the molecules composing the structure can be deduced. Each of these parts is associated with a corresponding body of theory, at least some of which must be comprehended in order to avoid pitfalls and to provide meaning to the operations. In the material that follows, the theory is held to a minimum and is intentionally nonrigorous, but what appears is needed to illuminate the experimental methods.

1-2 Crystals in Principle

1-2a The Crystal Lattice

Crystals are solid bodies characterized by a microscopic structure in which identical points are repeated at regular intervals in three dimensions. These points can be connected to form a three-dimensional grid or lattice. The positions of lattice points in the crystal are arbitrary, although the spacings between them are not. Similarly, the manner in which the lattice is chosen on the set of points is arbitrary, although convention and convenience often favor one arrangement strongly. Once chosen, the lattice defines volume elements, called *unit cells*, which are the parallelepipeds whose sides are successive grid lines (Figure 1-1).

Every unit cell is characterized by six parameters—the three cell edges, a, b, and c, and the three angles, α (between b and c), β (a and c), and γ (a and b). In the least symmetric crystals (*triclinic*) these may assume any arbitrary values. As the crystals become more regular, identities appear in the edge lengths, and special values (90°, 120°) in the angles. Consequently, the number of independent parameters decreases. The crystal systems (classes) so generated are listed in Table 1-1, together with the specifications for their unit cell parameters.

In general, crystals made up from small, low symmetry molecules, both organic and inorganic, tend to fall in the first three classes. Ionic crystals and those in which the repeat is a subunit of a polymer more commonly show higher symmetry.

Table 1-1. *The Crystal Classes*

Class	Independent Parameters	Special Parameter Values		Lattice Symmetry
Triclinic	6	$a \neq b \neq c$;	$\alpha \neq \beta \neq \gamma$	$\bar{1}$
Monoclinic	4	$a \neq b \neq c$;	$\alpha = \gamma = 90°$; $\beta \neq 90°$	$2/m$
Orthorhombic	3	$a \neq b \neq c$;	$\alpha = \beta = \gamma = 90°$	mmm
Tetragonal	2	$a = b \neq c$;	$\alpha = \beta = \gamma = 90°$	$4/mmm$
Rhombohedral	2	$a = b = c$;	$\alpha = \beta = \gamma \neq 90°$	$\bar{3}m$
Hexagonal	2	$a = b \neq c$;	$\alpha = \beta = 90°$; $\gamma = 120°$	$6/mmm$
Cubic	1	$a = b = c$;	$\alpha = \beta = \gamma = 90°$	$m3m$

1-2b Symmetry

The significant feature in determining the crystal system is not, for example, the experimental equality of two cell edges, but the presence of the symmetry that requires they be identical. Thus each system is characterized by a lattice symmetry as given in Table 1-1, and the lattice directions are chosen to take the greatest advantage of this symmetry.

Symmetry in crystals is best described in terms of two operations (also see Part Two, section 3-10): rotation and reflection.[1] *Rotation* occurs about an axis and is described as *n*-fold if rotation by $360°/n$ gives a structure indistinguishable from the original. Thus a rectangular solid has a twofold axis perpendicular to the center of each face (Figure 1-2); if the edges are equal, the resulting cube has fourfold axes. Owing to the requirements of unit cell packing, lattices occur only with one-, two-, three-, four-, and sixfold axes.

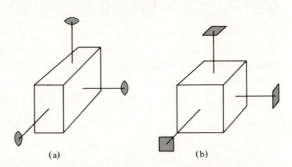

FIGURE 1-2. (a) Rectangular solid with twofold axes (●). (b) Cube with fourfold axes (■).

A *reflection* or mirror plane exists if a structure can be divided into two equal parts such that one half is the mirror image of the other. A reflection plane requires that, for every point *A* on one side, there be an identical point on the other side located on the perpendicular from *A* to the plane and at the same distance from the plane as *A* (Figure 1-3). Mirror planes are designated as *m*.

Reflection can occur not only in a plane but also in a point, a *center of symmetry*. In this case the image of *A* is on the line from *A* through the center and at the same distance beyond the center. The symbol for a crystallographic center of symmetry is $\bar{1}$ (Figure 1-4).

[1]There are two sets of nomenclature commonly used for describing symmetry, the Schoenflies and the Hermann–Maugin. The former is widely used by specroscopists and others; the latter (which is simpler) is used by crystallographers and will be considered here.

FIGURE 1-3.   A mirror plane.

These symmetry elements often combine to form groups[2] so that the presence of two generates a third. In a common case, a twofold axis perpendicular to a mirror plane (symbol $2/m$, the slash indicating perpendicularity) generates a center of symmetry at their junction. Alternatively a twofold axis through a center produces a mirror plane, or a center placed in a mirror generates a perpendicular twofold axis. The identity of these three cases is clear, since a single asymmetric object gives the same three additional images, as shown in Figure 1-5, when acted on by any two of the symmetry operations. Thus the twofold axis relates (2) to (1) and the mirror relates (4) to (1). The last point (3) derives from (1) either by the action of the center ($\bar{1}$) *or* by the combined action of the axis and mirror, i.e.,

$$(1) \xrightarrow{2} (2) \xrightarrow{m} (3)$$

An additional kind of crystallographic symmetry element combines rotation and reflection through a point. These *rotary inversion axes* relate identical points that are obtained from the initial one by *n*-fold rotation followed by reflection in a center on the axis. Such axes are symbolized by a bar over the number indicating the multiplicity of the rotation axis, e.g., $\bar{2}, \bar{4}$. Figure 1-6 shows four points related by a fourfold rotary inversion axis. As with simple

FIGURE 1-4.   A center of symmetry.

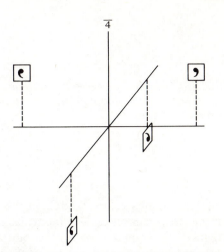

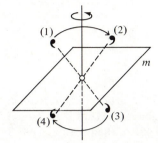

FIGURE 1-5. The combined operations of a twofold axis, a mirror, and a center.

FIGURE 1-6. A fourfold rotary inversion axis.

rotations, only $\bar{1}, \bar{2}, \bar{3}, \bar{4}$, and $\bar{6}$ axes occur in crystal symmetry, but of theses $\bar{1}$ is merely a center of symmetry and $\bar{2}$ is equivalent to a mirror plane perpendicular to the axis.

Figure 1-7 shows the symmetry elements that occur in the unit cells of the three crystal lattices of lowest symmetry. Since all lattice points are identical, a cell with lattice points at the corners is inherently centrosymmetric and the lowest symmetry possible is $\bar{1}$. Addition of either a twofold axis or a mirror generates the $2/m$ combination, which is characteristic of the monoclinic system. One edge of the unit cell is placed along the axis and the other two are laid in the mirror. Since the axis must be perpendicular to the mirror, two of the angles of the cell are constrained to be 90°. If a second axis or plane is added perpendicular to the first, the combination leads to an orthorhombic unit cell with three right angles, three twofold axes along the cell edges, and three mirror planes at the faces. Such symmetry is $2/m\ 2/m\ 2/m$, but is generally called *mmm* since the intersection of two mirrors at right angles generates a twofold

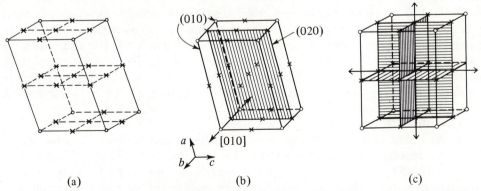

(a)   (b)   (c)

FIGURE 1-7. The symmetry elements of one unit cell in different crystal systems: (a) triclinic, (b) monoclinic, and (c) orthorhombic. Note: ○ = lattice point; x = center of symmetry. In (b) there are mirror planes midway between and parallel to those coincident with (010). In (c) the faces of the unit cell together with the planes shown constitute the full set of mirror planes.

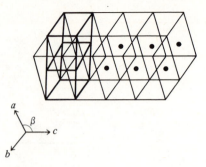

FIGURE 1–8.   A C-centered monoclinic lattice (light lines) with a superimposed alternative primitive assignment of axes.

axis along the intersection.[3] Addition of higher-fold axes to the lattice leads to more constraints on the cell parameters and to the remaining systems of Table 1–1. Thus the tetragonal system resembles the orthorhombic with a fourfold axis replacing one of the twofolds. As a result one of the cell faces is a square and two of the axes are identical. Rhombohedral and hexagonal cells have three- or sixfold axes, and the cubic system has a complicated array of symmetry elements that make all cell edges the same.

The combinations of symmetry elements given so far are *point groups*, which describe the symmetry of the empty unit cell with lattice points at the corners. If actual molecules or ions are placed in the cell, the arrangement need not be centrosymmetric; so other symmetry combinations are possible. In fact, there are 32 *crystallographic point groups* that are compatible with the inherent symmetry of the various lattices. Each of these belongs to one of the seven crystal systems. The symmetry of the empty lattice is the most symmetric point group of the system. Groups of lesser symmetry can also be fitted to the system lattice, providing their effect on the cell axes and angles agrees with the system. Thus the point groups 2, *m*, and 2/*m* all require one axis to be perpendicular to the plane of the other two and so fall into the monoclinic system.

The lattices so far described possess lattice points only at the corners of the cell and thus have only one lattice point per cell (8 corners with each corner point shared among 8 cells). Such lattices are called *primitive* and are symbolized by the symbol *P* before the symmetry designation. More complicated arrangements that contain two or four points per cell also exist and lead to a total of 14 unique crystal lattices (Bravais lattices). Although the nonprimitive lattices can always be described in terms of a primitive triclinic system, to do so ignores the additional symmetry present in the nonprimitive cell (Figure 1–8). If the additional points occur in the centers of two opposing faces, the cell is *side-centered* and designated *A*, *B*, or *C* depending on the pairs of faces involved. If all faces are centered, it is *face-centered* (*F*), while if there is a single additional point in the center of the unit cell, it is *body-centered* (*I*).

1–2c Space Groups

A real crystal consists neither of an ideal lattice of isolated points nor of a single unit cell with its contents; rather, it is a repeating collection of unit cells that may for practical purposes be considered infinite. The symmetry possibilities of such an array would appear to be those produced by the combination of the lattice symmetries with the appropriate point groups, which describe the unit cells. In fact, there are more such *space groups* than would be predicted

---

[3]Note that the intersection of two perpendicular two-fold axes generates a third at right angles to their plane, but does not produce any mirrors (unless a center of symmetry is also present). Thus 222 is also a symmetry group, but unlike *mmm* it is not equivalent to 2/*m* 2/*m* 2/*m*.

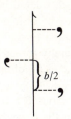

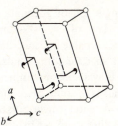

FIGURE 1-9.   A 2_1 screw axis along b.   FIGURE 1-10.   An a glide plane.

from a simple combination of this kind because the repetition of the lattice allows the addition of two new types of symmetry elements. As a result, 230 space groups are needed to describe all of the symmetries that three-dimensional objects can show in their crystalline arrangements.

The new symmetry elements are *screw axes* and *glide planes*. A screw axis carries one object into its symmetry image by a combination of a rotation about the axis *and* a shift along it. A n_m-fold screw combines a rotation of $360°/n$ with a translation along the axis of m/n times the lattice translation in the axial direction. Thus a 2_1 screw axis produces a rotation of $180°$, and a translation of $1/2$ the axial repeat (Figure 1-9). Two such operations simply serve to translate an object in one unit cell to the corresponding position in another. It is for this reason that these symmetry elements can exist only in the extended lattice. As with rotation axes, the crystallographic screw axes are restricted, and only $2_1, 3_1, 3_2, 4_1, 4_2, 4_3,$ $6_1, 6_2, 6_3, 6_4,$ and 6_5 occur. Of these only 2_1 is common for small molecules of low symmetry.

In much the same way that screws are related to rotation axes, glide planes produce images by a combination of mirror reflection and translation. In the simplest form ($a, b,$ or c glides), reflection as in a mirror plane is combined with a shift of $1/2$ the repeat distance along the $a, b,$ or c axis (Figure 1-10). In an n glide the shift is $1/2$ along the two cell axes in the plane, i.e., along the face diagonal of the cell. Thus an n glide perpendicular to the b axis of a cell produces a translation of $a/2 + c/2$. The less common d glide occurs only in some centered lattices and has translations of $1/4$ on two axes.

A space group is designated by a capital letter identifying the lattice type (P, A, B, C, F, I), followed by symbols describing the symmetry elements present. Usually only a partial set of symmetry operations is given, sufficient to characterize the group uniquely, but often omitting extraneous, though important, elements. Thus the common space group $P2_1/c$ involves a primitive lattice and a twofold screw axis perpendicular (symbolized by the slash) to a c glide. That it also contains a center of symmetry must be discovered from a tabulated description or by relating it to the point group $2/m$ by converting the translational symmetry elements (screws, glides) to their corresponding nontranslational forms (axes, mirrors). In this way all space groups may be reduced to the point groups from which they derive, e.g., $P2/m,$ $P2_1/m, C2/m, P2/c, P2_1/c,$ and $C2/c$ all derive from $2/m$. From the point groups the crystal system may be deduced.

1-3   Crystals in Practice

1-3a   Choosing a Crystal

It is clear that in order for a crystallographic analysis to be possible, the compound under study must be crystalline. Mere crystallinity is not sufficient, however; the crystals must also be single and of adequate size to give satisfactory intensities for the diffracted beams. In the early days of crystallography a suitable research problem could often be obtained by finding any reasonable crystal, but present interests are almost always directed to specific questions about specific compounds. Thus the chemist-crystallographer is generally forced to work with what is available or within a limited set of chemical modifications. As a result, obtaining suitable crystals is in some cases the most difficult part of a structure analysis.

A good single crystal is roughly equidimensional with edges about 0.1–0.3 mm. The upper limits are set by the need to keep the crystal totally bathed in the X-ray beam, and the lower by the proportionality between crystal volume and diffracted intensity. Many crystals are not ideally shaped, and much work has been done with needles or plates in which one or two dimensions are smaller than desired, but usually at a cost in the number of observable reflections and so in the accuracy of the final determination.

If a good single crystal is easily available, excellent! If not, recrystallizations with various solvents and conditions may yield good results. If these fail, the usual solution is to prepare a derivative, often one of noticeably different polarity, in the hope that different forces will dominate the crystal structure and lead to a more favorable outer form.

Derivatives are often prepared with an eye to simplifying the later stages of structure solution. This process is usually easier if the molecule contains one atom that is markedly heavier than the others. Thus for many years most organic problems *began* by searching for a suitable "heavy-atom derivative," without regard to the crystalline nature of the starting material. Recent advances in computation have made this approach much less necessary, but it still may represent the easiest route. Bromine and iodine are the commonest heavy atoms, either covalently bound or as the anions in salts, but any element with $Z > 30$ will serve well in the presence of C, H, O, and N.

One disadvantage of heavy atoms is that their presence in a crystal increases its absorption coefficient μ for X-rays. Since they also increase the scattering power, the average loss in intensity is not usually serious. Real problems can arise, however, since individual reflections will have their intensities affected differently by absorption if they travel different average path distances through the crystal. If the shape of the crystal is accurately known, *absorption corrections* can be computed, but the calculations are lengthy and complicated even on modern computers. More commonly, an attempt is made to keep the corrections as small as possible by the use of the lightest effective heavy atom and sometimes by shaping the crystal into some approximation of a sphere, for which the variations in intensity due to absorption are usually tolerable. The residual corrections are then ignored.

1–3b Mounting a Crystals

The standard mount for a crystal to be studied by X-ray diffraction consists of a short metal pin with a small axial hole in which a glass fiber is held in place by sealing wax. The crystal is fastened to the end of the glass fiber with a suitable adhesive (Duco, epoxy resin, Eastman 910, or shellac may be used), and the pin is clamped in a *goniometer* head (Figure 1–11). This device, which screws onto the X-ray camera or diffractometer, has slides and axes that serve to center the crystal and tip it within a limited range ($\pm 20°$) to obtain the exact orientation desired. This manipulation may be performed either by adjusting the head arcs while observing light reflections from the crystal faces in an optical goniometer or with the aid of X-ray diffraction techniques.

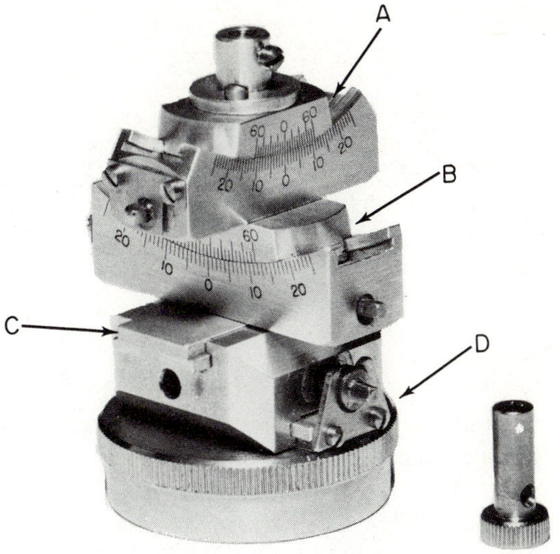

FIGURE 1-11.   A standard goniometer head.
[Courtesy of Charles Supper Co.]

1-4   Diffraction

1-4a   Planes and Indices

The process of X-ray diffraction is often visualized by analogy with the more common one of reflection from mirrors. Such an analogy, which can be justified rigorously, requires the development of suitable reflecting planes in the crystal. Actually the concept of lattice planes, which antedates X-ray diffraction, arose as a consequence of the need for a description of the surface faces of a crystal in terms of a crystal lattice.

Lattice planes are sets of parallel planes placed in a lattice so that every lattice point lies on some member of the set. Figure 1-12 demonstrates corresponding lattice lines in a two-dimensional lattice. A set of planes is characterized by three indices (hkl) (by two, hk, in the two-dimensional cases), that give the number of planes encountered in passing from one lattice point to the next along the a, b, and c cell edges. Thus for the lines in Figure 1-12, the first has the indices (11), the second (13), and the third ($2\bar{1}$). The $\bar{1}$ indicates "−1" and implies that a plane that cuts the lattice on the +x axis from a given lattice point also intersects the

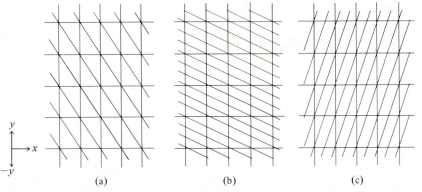

FIGURE 1-12.   Three families of lattice "planes" in a two-dimensional lattice: (a) (11), (b) (13), (c) ($2\bar{1}$).

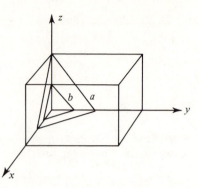

FIGURE 1-13. A lattice plane (221) (a) and an interleaving plane (442) (b).

−y rather than the +y axis. In the same way the three-dimensional plane of Figure 1-13 has the indices (221).

One of the characteristics of lattice planes is that their indices have no common divisor. Indices of this type (*Miller indices*) are used in geometrical crystallography to describe the faces of real and idealized crystals. In X-ray diffractions, however, it is advantageous to imagine further sets of regularly spaced planes, some of which do not pass through the lattice points. These planes have indices that are multiples of the Miller indices of the planes they parallel. Thus in Figure 1-13 the planes described by (442) are parallel with (221) but occur twice as often.

One special case occurs for planes parallel with an axis. These planes clearly never encounter that axis and so have corresponding indices of 0.[4] Thus the faces of the unit cell are the three sets of planes (100), (010), and (001).

1–4b X-Ray Diffraction

The process of X-ray diffraction was originally, and correctly, described by von Laue in terms of diffraction from a three-dimensional lattice. Very shortly thereafter Bragg proposed a simpler, if less rigorous, derivation in terms of reflection from lattice planes. Thus for a series of lattice planes (Figure 1-14) with a perpendicular separation d and assuming that the incident and reflected beams make equal angles θ with the planes as in ordinary reflection, it may easily be shown that reinforcement occurs between waves reflected from successive planes of the set only for values of θ given by the Bragg equation, $n\lambda = 2d\sin\theta$. This reinforcement

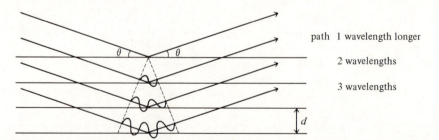

path 1 wavelength longer

2 wavelengths

3 wavelengths

FIGURE 1-14. Diffraction resulting from in-phase addition of waves reflected from successive planes.

[4]Indices are often defined as reciprocals of the fractional intercepts made on an axis by a family of planes. Thus planes that divide the a axis into segments of length $\frac{1}{3}a$ have h indices of 3. Planes parallel to axes have infinite intercepts and again lead to 0 indices.

occurs because at the angle θ the pathlengths for rays reflecting from adjacent planes differ by exactly an integer number n of wavelengths λ. Since the spacing between the planes is of the order of 10^{-8} mm, very many such planes participate in a reflection and it is observed only at very specific values of θ. A real crystal usually shows a spread of reflection over an angular range of 0.1–0.5°, but this spread arises principally from the existence in the crystal of slightly misaligned sections of perfect lattice (*mosaic blocks*), since the reflection from a single mosaic block is very sharp indeed.

1-4c The Reciprocal Lattice

Although the process of diffraction can be explained in terms of reflection, it is difficult to visualize the relationships of the various families of planes and the manner in which the planes move into reflecting positions as the crystal is rotated. A less obvious but extremely useful description of the process involves the concept of the *reciprocal lattice* (rl).

The reciprocal lattice may be constructed by considering the normals from some lattice point to each of the sets of planes (hkl). On each of these normals, the distance $1/d_{hkl}$ is marked off, where d_{hkl} is the perpendicular distance between the planes in the set (hkl) (Figure 1-15). The array of points so generated may be shown to form a lattice in which each point corresponds to a particular set of planes in the direct (crystal) lattice and is referred to by the same index. The parameters of the reciprocal lattice are symbolized by the addition of an asterisk to the corresponding symbols for the direct lattice. Thus the reciprocal cell constants are a^*, b^*, c^*, and α^*, β^*, γ^*, and the volume is V^*. The relative shapes of the direct and reciprocal cells depend on the angles involved. For example, in an orthorhombic lattice the direct axes a, b, and c are perpendicular to the cell faces (100), (010), and (001). Therefore the normals to these planes lie along the axes and so a and a^* coincide, as do b and b^*, and c and c^*. The distances between the planes are a, b, and c, respectively, so the corresponding reciprocal axial lengths are $a^* = 1/a$, $b^* = 1/b$, and $c^* = 1/c$, that is, the longer the direct axial length, the shorter the reciprocal axis and the more closely spaced are the rl points in that direction.

If the cell angles are not 90°, the plane normals do not coincide with the direct axes and the direct and reciprocal axes do not point in the same direction. Similarly, the distance $d_{100} \neq 1/a$, etc., but is a function of the cell angles. The simplest and commonest case of this sort occurs with monoclinic crystals, in which the a and c axes meet at an obtuse

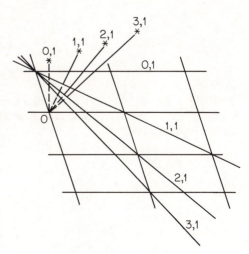

FIGURE 1-15. A two-dimensional projection of planes in direct space represented by points in reciprocal space.

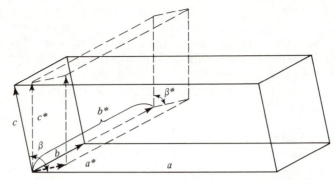

FIGURE 1-16.   *Monoclinic direct and reciprocal cells.*

angle β, while b is perpendicular to both. The normal to (010) (the ac plane) again lies along b and the rl spacing in this direction is $b^* = 1/b$. In the ac plane, however, the two normals are separated by an angle $\beta^* = 180 - \beta$ and the interplanar distances are $d_{100} = a \sin \beta$, $d_{001} = c \sin \beta$. Thus $a^* = 1/a \sin \beta$ and $c^* = 1/c \sin \beta$. The result is a reciprocal cell that is a right parallelepiped similar to the direct cell, but with an acute angle at the origin (Figure 1-16).

In triclinic crystals the directions and lengths of the reciprocal axes depend simultaneously on all three angles. Consequently, both the formulas and the visualization become quite complicated and we shall not consider them.

One important generalization about the reciprocal lattice is that regardless of the lattice type any pair of *rl* axes, e.g., a^* and b^*, is always perpendicular to the third *direct* axis, here c. As a consequence the plane of rl points containing these axes, i.e., the plane of points with the indices $hk0$, and the parallel planes containing the points $hk1$, $hk2$, etc., are always perpendicular to c.

To connect the rl with Bragg's law, consider a crystal in a beam of X-rays of wavelength λ, oriented so that the beam lies in the a^*c^* section of the rl and passes along the line $x0$ through the rl origin 0 (Figure 1-17). Describe a circle of radius $1/\lambda$ about the point C on $x0$ so that 0 lies on its circumference. Now consider some other rl point P also lying on the circumference. The angle $0PB$ is inscribed in a semicircle and is a right angle so that we have eq. 1-1.

FIGURE 1-17. *Diffraction in terms of reciprocal lattice: the reciprocal lattice and the sphere of reflection.*

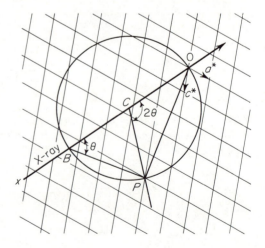

$$\sin 0BP = \sin \theta = \frac{OP}{OB} = \frac{OP}{2/\lambda} \qquad \text{1-1}$$

Since P is a rl point, the length $0P$ is defined to be $1/d_{hkl}$. Substitution and rearrangement give eq. 1-2, which is clearly Bragg's law with $n = 1$.

$$\sin \theta = \left(\frac{1}{2}\right)\left(\frac{\lambda}{d_{hkl}}\right)$$
$$\lambda = 2d \sin \theta \qquad \text{1-2}$$

This result is quite general and implies that whenever a rl point coincides with a circle constructed as described, Bragg's law is satisfied and reflection occurs from the corresponding direct plane. Since CP makes an angle of 2θ with the direct beam, it defines the direction of the diffracted beam. The construction is not limited to one plane, but will hold for any points on the surface of the sphere (the *sphere of reflection*) produced by rotating the circle about $0x$. If the cyrstal is rotated, the rl rotates with it about 0 so that points can be brought onto the surface of the sphere and the corresponding reflection observed. Clearly, only those points less than $2/\lambda$ (the diameter of the sphere) from the rl origin can be made to touch the sphere since the origin is also constrained to remain on the surface. One consequence of this restriction is that reducing the wavelength of the X-rays used increases markedly the number of reflections that may theoretically be observed.

The relationship of the reciprocal lattice and the sphere of reflection often is derived in slightly different terms. The rl is defined for a particular radiation in terms of λ/d_{hkl} rather than $1/d_{hkl}$. The lattice then expands or contracts with the radiation used and is expressed in the dimensionless quantities of rl units. The sphere of reflection then has a constant radius of 1 rlu. The two views always lead to the same predictions, but one may be more convenient than the other for a particular discussion.

BIBLIOGRAPHY

1. General Works: G. H. Stout and L. H. Jensen, *X-Ray Structure Determination*, Macmillan, New York, 1968; J. P. Glusker and K. N. Trueblood, *Crystal Structure Analysis*, Oxford University Press, London, 1972; D. E. Sands, *Introduction to Crystallography*, W. A. Benjamin, New York, 1969; G. B. Carpenter, *Principles of Crystal Structure Determination*, W. A. Benjamin, New York, 1969; M. M. Woolfson, *X-Ray Crystallography*, Cambridge University Press, Cambridge, 1970.
2. Symmetry and Crystallography: F. C. Phillips, *An Introduction to Crystallography*, 3rd ed., Wiley, New York, 1964; Sands (reference 1); Carpenter (reference 1).
3. Space Groups: A listing of the 230 space groups, their equivalent positions, and standardized drawings showing the relative positions of the symmetry elements is to be found in N. F. M. Henry and K. Lonsdale, eds., *International Tables for X-Ray Crystallography*, Vol. I, Kynoch Press, Birmingham, England, 1952. Discussions of the symbolism used can be found in Woolfson (reference 1), pp. 24–33.

DATA COLLECTION

2–1 Introduction

The apparatus used for data collection in X-ray diffraction consists of a source of X-rays, a device for rotating the crystal relative to the X-ray beam, and a detection system for the diffracted rays. This last component may use either film, in a diffraction camera, or a photon-counting system, in a diffractometer. Historically, photographic methods have mainly been used, but recently they have been largely supplanted by more or less automated diffractometers.

X-rays are generated by the bombardment of a metal anode with high voltage electrons. The rapid deceleration of the electrons produces a continuum of X-radiation (*white radiation*) on which are superimposed intense sharp lines of *characteristic radiation* with wavelengths specific to the target element. These lines arise from the ejection of inner shell electrons from the target atoms, followed by the descent of an outer electron and the emission of a photon of sharply defined energy. Since the characteristic lines are essentially monochromatic and since they account for about half the total X-ray intensity, they are used universally for diffraction measurements. For most common anode materials the ejected electron arises from the K shell. If the vacancy is filled by descent of an electron from the L shell, K_α radiation (actually a close doublet, $K_{\alpha 1}$ and $K_{\alpha 2}$) is produced. If the emitting electron comes from the M shell, the emitted quantum K_β, has higher energy and a shorter wavelength.

The K_α radiation is sufficiently close to being monochromatic that it is usually treated as such and assigned an average wavelength $\lambda_{K\bar{\alpha}}$. The K_β emission represents a serious contaminant, however, and is removed either by means of a crystal monochromator or by selective absorption by an element (β filter) with atomic number one or two less than that of the anode.

A crystal monochromator selects a specific wavelength from the incident beam by diffraction from a crystal set at a specific angle. The geometry of the monochromator is such that the beam from this crystal follows the normal path of the incoming X-rays and strikes the crystal being studied. A very highly monochromatic beam can be obtained in this way, with quartz

544

and highly oriented graphite being the commonest monochromator crystals, but there are difficulties with loss of intensity and inhomogeneities in the resulting beam. The β filter is less effective but simpler and widely used. It depends on the markedly greater absorption of any element for X-rays that are just energetic enough to eject an electron from its K shell as compared with those that cannot be absorbed by this process. This effect results in a sharp absorption edge, and by the proper choice of the filter element this edge can be placed between the K_α and K_β wavelengths of the target element. Thus there is high absorption of the more energetic K_β radiation and weak absorption of the K_α, resulting in a more nearly monochromatic beam.

The usual X-ray tubes for diffraction work contain Cu or Mo anodes and produce the wavelengths shown in Table 2-1. Other anodes are sometimes used when other wavelengths are

Table 2-1. *Wavelengths of Cu and Mo Characteristic Radiations*

	Cu	Mo
α_1	1.5405	0.70926
α_2	1.5443	0.71354
$\bar{\alpha}$	1.5418	0.71069
β_1	1.3922	0.63225
β filter	Ni	Nb

needed for special purposes. Since these tubes run at 10–50 kV and 10–20 mamp, considerable energy is contained in the electron beam. Much of this energy appears as heat and is dissipated by circulating cold water through the anode. The energy content of the emitted beam is also high, however, and careful safety precautions should be taken against both direct and scattered radiation.

X-rays, like visible light, activate silver halides for reduction. Thus photographic film has long been used as a means of detecting diffracted beams. In addition, the amount of darkening produced under standardized conditions is directly proportional to the number of X-ray quanta striking the film (not true for light) and can be used for quantifying the beam intensity. X-ray films usually have a heavier coating of silver halide than do ordinary photographic emulsions and may even be coated on both sides. The greater density of coating increases the probability of an X-ray photon being absorbed and recorded. The resulting attenuation of the beam enables a stack of several films to be used, the first one or several reducing a strong reflection by some factor until the intensity falls within the linear response range of some one of the pack.

Generally photographic methods are used for preliminary crystal studies intended to reveal the unit cell parameters and symmetry, even if the final intensity data are to be collected on a diffractometer. Film cameras are relatively inexpensive and have the great virtue that they show a section through the reciprocal lattice rather than merely one point in it. Thus one can be made aware of anomalies, curious regularities, symmetry, or signs of trouble that are not easily detected in the output from a diffractometer. The parameters from film are less accurate than those from a diffractometer, but with care are good to $\sim 0.1\%$, which is more than adequate for many purposes.

The principal distinction of a diffractometer is its use of a quantum-counting device instead of film for detecting and measuring the diffracted beams. The commonest such device is a scintillation counter, in which the X-ray quantum produces a flash of light upon striking a sodium iodide crystal. This light is detected and amplified by a photomultiplier system, and finally counted electronically. One great advantage of this system is that it is much more sensitive than film. Thus fewer than 1% as many quanta are required to give a detectable peak as compared with those needed for a just detectable blackening on film. Moreover, the efficiency of detection is nearly 100% for both CuK_α and MoK_α, whereas film allows much of the highly penetrating MoK_α to pass undetected. A certain amount of wavelength discrimination can be achieved by requiring the magnitude of the light pulse to fall within certain set limits (*pulse height discrimination*), and in this way the effects of the white radiation can be reduced. The resolution achieved is not very good, however, and certainly cannot replace filtering or monochromatization to isolate the characteristic wavelength.

The disadvantage of a diffractometer, aside from its very considerable cost ($50,000–$100,000 or more), arises from its examining only a single point in the reciprocal lattice at a time. Thus each reflection is studied in turn, and any other diffracted beam that may occur during the process is not detected and is wasted. With film, on the other hand, a section of the reciprocal lattice is studied at one time and reflections accumulate intensity whenever they occur. For this reason the saving in time from a diffractometer, although real, is not so great as might be supposed from the relative sensitivities of the detection systems.

Some diffractometers have a geometry similar to an X-ray camera, but most now are "four-circle" instruments. In such a device angular adjustments are made in four arcs or circles to bring the desired plane to the proper reflecting angle and orient it so that the emitted ray is in the plane perpendicular to the main axis of the instrument. At the same time the detector is rotated about this axis to the proper position to receive the ray. Figure 2–1 shows schematic drawings of a four-circle orienter and gives the symbols by which the various arcs are commonly designated.

Once the approximate lattice constants are known from photographs and the orientation of the crystal on the diffractometer found by hand, the proper values for setting the arcs may be calculated by a computer. More sophisticated computer-controlled diffractometers often have programs that enable them to search for three reflections and work out from these the cell parameters and the crystal orientation. Since the arcs may be set to $\sim 0.01°$, and since the values are related to the unit cell parameters, precise measurement of the optimum angles for a number of reflections allows a more accurate determination of these parameters than that normally possible with film. On the other hand, the same sensitivity to orientation demands accurate setting even for routine data collection.

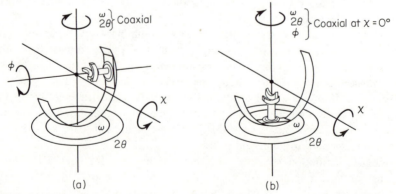

(a)   (b)

FIGURE 2-1. Schematic of a four-circle diffractometer (a) at $\chi = 90°$ and (b) at $\chi = 0°$.

In the early diffractometers the angular values were cranked into the instrument by hand. Later electromechanical devices were used to read precalculated setting values from punched cards or tape. Most recently small computers have been used as controllers, and the proper values are calculated in turn for each reflection as needed. The computer provides great flexibility for other parts of the measurement process and can even carry out preliminary calculations on the data if desired.

2-2   Geometry of Diffraction

2-2a   Rotation and Oscillation Methods

In the simplest use of the X-ray camera, a crystal is rotated about a direct axis while exposed to an X-ray beam perpendicular to this axis. A sheet of film is located as a cylinder around the axis (Figure 2-2). The rl points lie in planes perpendicular to the direct axis, regardless of the crystal symmetry; consequently, as the crystal and rl are rotated, each plane will cut the sphere of reflection in a circle. The rays passing from the crystal through the rl points intersecting on these circles will form a nested family of cones which intersect the cylinder of film. When the film is developed and unrolled, the intersections appear as lines of spots (Figures 2-2 and 2-3). The spacing between the layers of spots is simply related to the separation between the rl planes and thence to the length of the direct cell axis lying along the rotation axis.

A similar photograph taken by oscillating the crystal through a small angle (5-15°) can show the presence of symmetrically related reflections and serve to limit the possible point groups of the crystal.

2-2b   Weissenberg Methods

Rotation and (to some extent) oscillation photographs suffer from the disadvantage that they project the points in one layer of the rl onto a single line in a photograph. Thus it may be very difficult to determine the indices belonging to a particular spot. For a clearer view, and particularly for the measurement of intensities, it is convenient to map one rl level onto a whole sheet of film. The Weissenberg and precession methods represent the most useful solutions to this problem.

The Weissenberg camera resembles the rotation camera in structure; in fact, rotation photographs are usually taken on a Weissenberg camera with the film translation deactivated. The rotation of the crystal spindle, however, is geared to a drive screw that moves the film

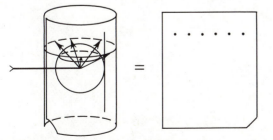

FIGURE 2-2.   Recording reflections from one level of the reciprocal lattice as a single layer line on cylindrical film.

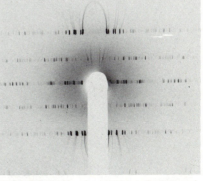

FIGURE 2-3.   A rotation photograph.

carrier along the spindle axis in unison with the rotation. At the end of the film travel, a reversing switch is struck and a rotation in the opposite sense is started. A slotted screen placed between the crystal and the film allows only the rays from a selected rl level to pass to the film. The reflections that would appear in a single line on a rotation photograph are spread over the entire film as a consequence of the varying amounts of crystal rotation needed to produce them.

Weissenberg photographs may be obtained in turn from each level that appears on the rotation photograph. Zero level photographs (those containing reflections for which the

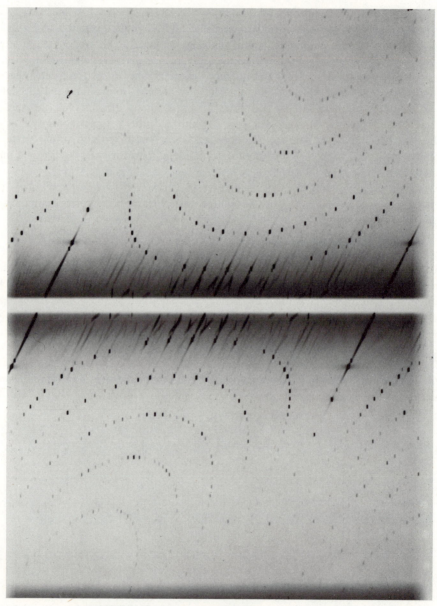

FIGURE 2–4.   Zero-level Weissenberg photograph.

index associated with the axis of crystal rotation is zero) are taken with the beam perpendicular to the camera. Upper level photographs require other beam–camera angles.

A Weissenberg photograph (e.g., Figure 2-4) is a rather distorted view of a rl net, but with a little practice it is easy to interpret. Points that lie on a line passing through the rotation axis in the rl (e.g., axial reflections) give rise to film spots that lie on a straight line extending across the top and bottom halves of the film at an angle of ~63.5°. Lines of points that do not pass through the rotation axis fall on curves (*festoons*). Every spot lies on the intersection of two festoons (or a festoon and an axis) and serves to identify the indices of the reflection.

Since a Weissenberg photograph displays a rl section normal to the rotation axis, it can provide unit cell data that complement those from rotation photographs. The distance traveled by the film from the appearance of one axis to the appearance of the other is simply related to the angle between the axes. Measurement of the separation of spots on the axial lines permits the calculation of the corresponding rl parameters, and thus the entire rl net can be reconstructed.

2–2c Precession Methods

The precession camera provides an alternative method of photographing sections of the rl. In this apparatus the normal to the section (e.g., a direct axis) is aligned with the X-ray beam rather than with the spindle axis. The normal is then tipped 5–20° off the beam by an angle μ, with the result that the section cuts the sphere of reflection in a circle (Figure 2-5). Movement of the crystal in a complicated fashion causes the normal to describe a circle (to *precess*)

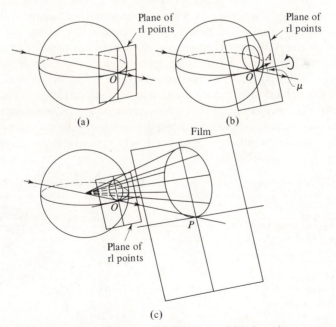

FIGURE 2-5. Precession geometry. (a) Reciprocal lattice plane perpendicular to X-ray beam (tangent to sphere of reflection). (b) Reciprocal lattice plane in (a) rotated slightly about horizontal axis showing circular intersection with sphere of reflection. (c) Film parallel to reciprocal lattice plane showing circular trace on film of circular intersection with sphere of reflection.

about the beam while the flat sheet of film is tipped and turned so that it remains parallel to the rl section. During this motion the circle marking the intersection of the rl section and the sphere of reflection appears to rotate about the beam, with the result that the reflections strike the film so as to form an undistorted image of the rl. The rl cell constants can then be obtained directly from the film measurements and the known geometry of the camera.

A great advantage of the precession method is that it provides photographs of sections perpendicular to those viewed by the Weissenberg method. Thus by using both techniques it is possible to get views of all three reciprocal axes and the angles between them with a single crystal mounting and with a minimum of realignment of the head arcs.

2-2d Reflection Symmetry, Systematic Absences, and Space Group Determination

The importance of being able to view the various rl axes and sections arises from the relationship between the symmetry of the reciprocal lattice as seen in the photographs and that of the direct lattice. If the points of the reciprocal lattice are weighted in proportion to the intensity of the corresponding reflection, the rl has the symmetry of the crystal point group plus a center of symmetry produced by the diffraction process. Thus all triclinic crystals give a rl with $\bar{1}$ symmetry, all monoclinic $2/m$, and all orthorhombic mmm. This symmetry appears in the photographs in terms of *equivalent reflections*, whose indices are related by symmetry (e.g., hkl, $\overline{hkl}$, $h\bar{k}l$, $\bar{h}\,\bar{k}l$) and whose intensities are the same. With practice it is easy to recognize the repetitions of pattern that point to this symmetry and so to deduce the crystal system.

Further information about the space group may often be obtained from special effects produced by certain symmetry elements. In particular, those elements that contain a translational component (glides, screws, lattice centering) cause the systematic extinction of sets of reflections. Conversely, if a pattern of systematic absences is recognized, the presence of the corresponding symmetry element is shown and the choice of possible space groups is much restricted. Thus if a crystal gives photographs that show the characteristic mmm symmetry of the orthorhombic lattice and at the same time systematic absence of the axial reflections of odd index (e.g., 100, 010, 001), the presence of three perpendicular twofold screw axes is indicated. The only orthorhombic space group that contains just these translational symmetry elements is $P2_12_12_1$, a very common noncentrosymmetric group belonging to the point group 222, one of the less than maximally symmetric point groups allowed in the orthorhombic class. Table 2-2 shows the absences associated with various common symmetry elements,

Table 2-2. *Partial Set of Translational Symmetry Elements and Their Extinctions*

Symmetry Element	Reflections Affected	Conditions for Systematic Absence
2-fold screw along a (2_1)	$h00$	$h = 2n + 1 = $ odd
b	$0k0$	$k = 2n + 1$
c	$00l$	$l = 2n + 1$
3-fold screw along c ($3_1, 3_2$)	$00l$	$l = 3n + 1, 3n + 2 = $ not evenly divisible by 3
4-fold screw along c ($4_1, 4_3$)	$00l$	$l = 4n + 1, 2,$ or 3
Glide plane perpendicular to b	$h0l$	
translation $a/2$ (a glide)		$h = 2n + 1$
$c/2$ (c glide)		$l = 2n + 1$
Glide plane perpendicular to c	$hk0$	
translation $a/2$ (a glide)		$h = 2n + 1$
C-centered lattice (C)	hkl	$h + k = 2n + 1$
Body-centered lattice (I)	hkl	$h + k + l = 2n + 1$

and it may be seen that screws extinguish some fraction of the reflections on a rl axis; glide planes, those in a section; and centering, a set throughout the rl.

Not all of the possible space groups are defined by their systematic absences, but two of the commonest, $P2_12_12_1$ and $P2_1/c$ (monoclinic centrosymmetric), are. Knowledge about the compound may sometimes be used to restrict the choices. In particular, optically active compounds contain only right-handed (or left-handed) molecules and so cannot crystallize in space groups that contain reflection elements such as mirror (or glide) planes or centers of symmetry. Thus while a monoclinic crystal showing only absences of odd $0k0$ reflections could be either $P2_1$ or $P2_1/m$ (since the mirror plane causes no additional extinctions), knowledge that the sample is optically active would restrict the possibility to $P2_1$.

2–3 Intensity

2–3a Choice of Intensities

Once the crystal has been selected and mounted, the cell parameters determined, and the systematic absences characterized, the principal operation of the data collection process remains—the measurement of the intensities of a sufficient set of reflections to permit the accurate determination of the structure. Usually a unique set, i.e., one of each set of symmetry-equivalent reflections, is measured in a sphere around the origin of the reciprocal lattice. The radius[1] of this sphere is limited by the restriction that only rl points closer than $2/\lambda$ from the origin can be brought onto the sphere of reflection. Fewer data may be used as determined by the accuracy desired, or by the instrumental parameters, or by the absence of any significant number of observable reflections outside a smaller radius. A good data set usually extends to values of $\sin\theta/\lambda > 0.5$. More than one unique set may be measured and the result for equivalent reflections averaged to reduce the random error, but the process is time consuming and usually limited to cases in which the highest accuracy is needed.

The number of reflections needed depends on the number of adjustable parameters that will be used to describe the final structure, and on the accuracy desired. An analysis that uses three positional and one temperature parameter per atom will usually allow reasonable refinement and will provide all the information necessary for answering most chemical questions. More refined treatments invoke six parameters instead of one to describe the thermal motions of atoms. For most purposes three to five reflections per parameter will serve to define a structure reliably. Additional numbers of reflections will increase the precision of the results, but the effort involved increases sharply.

2–3b Photographic Methods

Film methods for measuring intensities have the virtues of being inexpensive, especially in terms of capital investment, and of being easily fitted into small units of working time. On

---

[1]Various descriptions of the extent of the data in reciprocal space (the space containing the rl) are used. The geometry of diffractometers is such that 2θ (θ = the Bragg angle) is a natural measure. The value of 2θ for a specific reflection varies with the radiation, and $(\sin\theta)/\lambda$ (eq. 2–1) is often used as

$$\frac{\sin\theta}{\lambda} = \left(\frac{1}{2}\right)\left(\frac{1}{d}\right) \qquad \text{2–1}$$

a radiation-independent measure of the distance from the rl origin. The maximum value of $(\sin\theta)/\lambda$ that can be reached with CuK_α is 0.65, with MoK_α 1.4.

the other hand, they are less accurate, more tedious, slower, and more demanding of time and knowledge on the part of the worker than are counter methods.

It is not particularly difficult to obtain diffraction photographs in which the film blackening at any point is directly proportional to the radiation received. The blackening cannot be too intense or proportionality is lost, but most films will allow measurement of a spot at least fifty times as dark as one just visibly darker than the background. A much more serious problem occurs because the reflections to be measured do not produce spots of uniform density on the film. Rather they are darkest near the center and fall off with increasing steepness near the edge of the spot. Thus in order to determine the total amount of radiation received, an integrated measure of blackening over the whole spot is needed.

Many methods have been devised for obtaining such a measure. The oldest, and still a useful technique, is to compare the spot visually with a calibrated gray scale or preferably with a series of spots made by timed exposures of a reflection from the crystal being studied. Since the eye is a reasonably good integrator, such comparisons are surprisingly accurate ($\pm$8–12% in favorable cases) if spots of similar size and shape are compared. Although the method is not much used now, it still has much to recommend it to an investigator with limited equipment and a relatively small problem in gross structure determination. A particular advantage is that photographs may be obtained whenever equipment is available and the data worked up at a later time.

The limiting inaccuracies with visual estimation arise not from the imprecision of eye comparison but from the marked variations in spot size that occur in a data set. Some of these result from differences in projected dimensions of irregular crystals and can be minimized by selection and shaping. Others arise from the gradual separation of the reflections of $K_{\alpha 1}$ and $K_{\alpha 2}$ at high values of sin θ. Most important and troublesome, however, are the expansion and contraction of spots on Weissenberg photographs of rl nets that do not pass through the rl origin (upper level photographs). The change in area can be large and is difficult to correct for, either empirically or by calculation.

A more sophisticated and somewhat more accurate approach is to measure the densities of the reflections photometrically. A point of light is projected on the film, and the absorption is proportional to the amount of blackening. Here the problem of integration becomes acute, however, since photometers measure only a very tiny area at a time. Measurement at a single point (peak height measurement) has been used but is inferior to good visual estimation. Instead a scanning process that averages over the entire spot area is needed. This required integration also solves the difficulties of varying spot size. A scanning photometer may be used in which the light is moved rapidly in a zigzag path over the film spot and the integrated absorption measured. More recently there has been increasing interest in computer-driven photometers in which the blackening of the film is measured point by point and digitized; the integrated intensities, background correction, etc., are then calculated from the digitized data. Such an approach, although expensive, is attractive for studies on large molecules such as proteins, for which photographic methods theoretically are faster than a diffractometer and have additional advantages because they integrate over any changes in intensity with time.

Integrating photometers, regardless of their details, are expensive and specialized pieces of equipment, and an alternative is to modify the camera so that it performs the integrating function. If the film holder is shifted systematically between each occurrence of a reflection, the various parts of a spot can be superimposed on a particular location. The film darkening measured at this location is then proportional to the integrated intensity and can be determined by a single measurement. The disadvantage of such integration is that the spot area

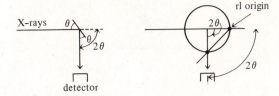

FIGURE 2-6. Geometry of diffraction: (a) Bragg reflection and (b) reciprocal lattice view.

increases at least fourfold and many weak reflections are lost in the background. A useful compromise involves camera integration in only one direction, followed by a single scan with a nonintegrating photometer and mechanical integration of the area under the photometer output trace. Regardless of the details of the method, good integrated photometric measurements are more accurate (± 5–8%) than visually estimated ones, and are less subject to variations that depend on the measurer.

2-3c Counter Methods

It is true that the majority of structure determinations are now being carried out with data collected on diffractometers. Although the capital investment is high, the instruments are now widely available and provide in general the simplest and most accurate means of measuring intensities.

Three basic techniques (plus some intergrades) have been used. The conceptually simplest is to orient the crystal so that a desired set of lattice planes is in the reflection position, making an angle θ with the X-ray beam. The detector is placed at an angle of 2θ to the beam so that it receives and counts the diffracted beam (Figure 2-6a). The corresponding reciprocal lattice view is that a rl point is placed on the sphere of reflection and the resulting beam measured (Figure 2-6b). This technique (the stationary crystal, stationary counter method) suffers from a number of disadvantages that reduce its accuracy and is rarely used despite its simplicity and speed. Instead one of two scanning methods is generally employed.

In the ω scan, the crystal is rotated through a small angle $\Delta\omega$ from $\theta - \delta$ to $\theta + \delta$. The desired rl point then passes through the sphere of reflection while the counter remains fixed at a position (2θ) that will receive the beam. In the θ–2θ scan, the crystal rotates in the same way, but in addition the counter moves at twice the angular velocity. Thus while the crystal is scanned through $\Delta\omega = \Delta\theta$ (usually 1-2°) about θ, the counter moves through $2\Delta\theta$ about 2θ.

The effective difference in the two techniques may be seen in Figures 2-7 and 2-8. The ω scan examines an arc in reciprocal space that lies at a constant distance from the rl origin (Figure 2-7). The θ–2θ method, however, scans a straight line passing through the

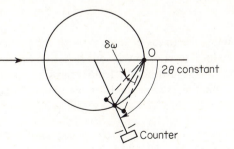

FIGURE 2-7. An ω scan.

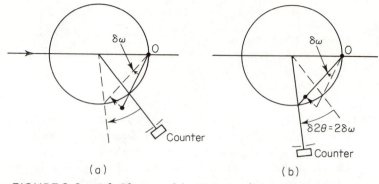

FIGURE 2-8. A θ-2θ scan: (a) starting position; (b) final position.

rl origin (Figure 2-8). For perfectly monochromatic X-rays the two methods are interchangeable provided that the accuracy of setting and the size of the counter receiving window are sufficient to capture all of the photons diffracted as the rl point passes from entirely outside to entirely inside the sphere of reflection.[2] In practice, however, perfectly monochromatic X-rays are rarely available (although crystal monochromators come sufficiently close for most purposes), and the two methods are then not equivalent. The difference arises because the presence of X-rays of wavelengths other than the characteristic one causes the rl point to appear stretched out along the line joining it to the rl origin. (Recall that the distance of the rl points *hkl* from the origin may be defined as λ/d_{hkl}.) Thus the desired reflection appears embedded in a streak of diffraction of other wavelengths (see Figure 2-4). The ω scan cuts across this streak while the θ-2θ scan runs along it. Measurements of the noncharacteristic background at the two ends of the scan will detect the presence of the streak in a θ-2θ scan but will overlook it in an ω scan. This omission can be serious since it is possible for the streak from one reflection to overlap the next one out on the radial line. Thus an erroneously high intensity can be obtained from an ω scan. For this reason the majority of experimenters elect the θ-2θ technique.

The normal process of data collection involves orienting the crystal by means of the various goniometer adjustments so that the desired rl point lies in the plane of the X-ray source and the detector, and then scanning through an angle Δθ (or Δω) that carries the rl point entirely through the sphere of reflection. The size of this angle depends on the geometry of the X-ray beam, the value of 2θ, and particularly the intrinsic mosaic spread of the crystal, i.e., the extent to which the perfectly ordered blocks are slightly out of alignment. The scan range is often calculated from eq. 2-2, in which A and B are constants empirically determined

$$\Delta 2\theta = A + B \tan \theta \qquad\qquad 2\text{-}2$$

by examining the traces of scans through the particular crystal being studied.

The local level of extraneous radiation (incoherent scatter from the crystal and its mount, noncharacteristic diffraction from nearby rl points) is estimated by taking nonscanning background counts at the two ends of the scan. These counts, multiplied by the ratio of scan time

[2]In terms of the Bragg picture, as the plane passes from a state of no reflection, through the reflecting position, and beyond to no reflection again.

to background time, are subtracted from the scan counts to give a net value that is the experimental integrated intensity for the reflection.

It is extremely useful to have some idea of the precision of the observed intensities, and one of the virtues of the diffractometer is that it permits a quantitative estimation of this precision. Because of the statistics of the counting process, the standard deviation (see section 3-4c) of the intensity is given by eq. 2-3, in which N is the *total* number of counts, scan *plus*

$$\sigma_I = \sqrt{N} \qquad\qquad 2\text{-}3$$

background.[3] The fractional uncertainty in I, σ_I/I, decreases as the total number of scan counts increases, and in principle could be made as small as desired. In practice there appears to be a lower limit imposed by instrumental variations, and intense reflections usually show a limiting precision of 0.5–2% of I for the standard deviation.

An estimate of the instrumental precision and hints of problems are usually obtained by measuring a few standard reflections repeatedly during the course of data collection. Changes in the observed intensities may arise from movement of the crystal, changes in the X-ray source or counter, or, most commonly, from crystal decomposition under the X-ray bombardment. The changes in the standards can be used to estimate a scale factor to adjust intensities measured after decomposition begins to the values they would have had in the undegraded crystal. The various standards rarely behave in a consistent fashion, however, and the scaled intensities are always less reliable than those obtained from stable crystals.

BIBLIOGRAPHY

1. Data Collection—General: E. W. Nuffield, *X-Ray Diffraction Methods*, Wiley, New York, 1966, pp. 29–45, 172–201.
2. X-Ray Tubes and Wavelength Selection: C. H. Macgillavry and G. D. Rieck, eds., *International Tables for X-Ray Crystallography*, Vol. III, Kynoch Press, Birmingham, England, 1962, pp. 71–78.
3. Diffractometry: U. W. Arndt and B. T. M. Willis, *Single Crystal Diffractometry*, Cambridge University Press, Cambridge, 1966.
4. Geometry of Diffraction: G. H. Stout and L. H. Jensen, *X-Ray Structure Determination*, Macmillan, New York, 1968, pp. 83–133; M. J. Buerger, *The Precession Method*, Wiley, New York, 1964; Nuffield (reference 1), pp. 250–341; J. W. Jeffery, *Methods in X-Ray Crystallography*, Academic, London, 1971, pp. 175–237; M. M. Woolfson, *X-Ray Crystallography*, Cambridge University Press, Cambridge, 1970, pp. 125–55.
5. Intensity: Stout and Jensen (reference 4), pp. 165–94; Arndt and Willis (reference 3); Jeffery (reference 4), pp. 256–83; M. J. Buerger, *Crystal Structure Analysis*, Wiley, New York, 1960, pp. 77–151.

[3] This approach assumes that the backgrounds were counted for the same total time as the scan. If not, the standard deviation increases somewhat because of greater uncertainty in the background value.

3

SOLUTION OF THE STRUCTURE

3–1 Data Analysis

3–1a Data Reduction

Once a set of raw intensities has been obtained, the next task is to convert it to a codified set of *structure factor magnitudes (structure amplitudes)* that will be used for all subsequent calculations. These parameters are related to the observed data by eq. 3–1, in which K is a

$$|F_o| = \sqrt{\frac{KI_o}{Lp}} \qquad \qquad 3\text{–}1$$

scaling factor, I_o is the observed intensity, and Lp represents the Lorentz and polarization corrections. The polarization term p, given by eq. 3-2, arises from the partial polarization of

$$p = \frac{1 + \cos^2 2\theta}{2} \qquad \qquad 3\text{–}2$$

the X-rays on reflection. The Lorentz correction L depends on the geometry of the measurement method used and on the manner in which a rl point passes through the sphere of reflection. For zero level Weissenberg photographs and for four-circle diffractometers, it is given by eq. 3-3, but for other techniques, especially the precession camera, it is more complicated.

$$L = \frac{1}{\sin 2\theta} \qquad \qquad 3\text{–}3$$

The scaling factor may in principle be determined absolutely, but in practice it is

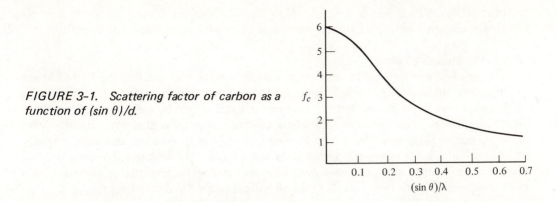

FIGURE 3-1.   Scattering factor of carbon as a
function of (sin θ)/d.

always obtained by scaling the set of observed structure factors to those calculated for the final model. Preliminary scaling will be necessary, however, to combine data obtained from different films, e.g., from zero and higher level Weissenberg photographs. Most commonly, one or more *cross level photographs* are taken of the crystal rotating about another axis, and the data from these (after conversion to $|F_o|$'s) are used to scale together the other films. These scale factors are then refined later by comparison with calculated values. Alternatively, precession photographs can be used to give a view of the rl at right angles to the sections measured by the Weissenberg technique. One of the advantages of most diffractometers is that all of the reflections are measured in exactly the same way, and so a single scale factor suffices for the entire data set.

Data reduction is now always carried out by computer. The raw intensities (relative intensities from film, scan and background counts from a diffractometer) and their indices are punched onto cards or paper tape or written on magnetic tape and used as input to the program, together with the cell parameters and information on the data measurement method. The computer calculates the structure amplitudes and may scale and average various multiply measured reflections to obtain a unified data set that is put out on cards or tape for later use.

It is usual at the same time to obtain for each reflection suitable values of the *scattering factor f* for each kind of atom in the structure. The scattering factor describes the scattering power of an atom in terms of an equivalent number of electrons concentrated at one point. It decreases with increasing (sin θ)/λ after starting with a value equal to the atomic number of the element (or to the total number of electrons in an ion) (Figure 3-1). The decrease results from the finite size of the electron cloud about an atom and from the increasing interference at high (sin θ)/λ of X-rays scattered by electrons in different parts of the cloud. The assumption is usually made that the electrons are spherically distributed about the nucleus and thus that the scattering factor depends only on (sin θ)/λ, i.e., on the indices of a given reflection. Since the proper values of the scattering factors are needed for later calculation of structure factors, they are generally obtained during data reduction by tabular interpolation for each reflection and then put out as part of the information accompanying each $|F_o|$.

The scattering factors in the literature have been calculated quantum mechanically for atoms at rest. In fact, atoms in real crystals are always in thermal motion about their nominal locations, and this motion has the effect of enlarging the apparent electron cloud. As a result the scattering factors decline more rapidly than for a stationary atom. The usual correction for this effect (eq. 3-4) consists of an exponential term that depends on $(\sin^2 \theta)/\lambda^2$ and a

$$f_{\text{real}} = f_{\text{stationary}}\, e^{-B(\sin^2 \theta)/\lambda^2} \qquad\qquad 3\text{-}4$$

temperature factor B, generally 2–6 Å². More detailed analyses replace the single *B* by six parameters, allowing the moving atom to assume an ellipsoidal rather than a spherical form.

3–1b Structure Factors

Bragg's law and its usual derivation define the direction of a diffracted beam but say nothing about its intensity. Furthermore, the concept of reflection from lattice planes is easily seen for an ideal lattice, but is less clear for a real crystal in which the lattice is filled with molecular electron density. More refined treatments show that even though X-ray scattering occurs from each electron in the cell regardless of its position, diffracted intensity is observed only in the directions predicted by Bragg's law. The intensity, however, is determined by the manner in which the individually scattered waves reinforce or interfere. As a result it depends on both the positions of the electrons and the direction of scattering.

The *structure factor F* is a vector quantity that describes the results of the combined scattering in terms of a magnitude $|F|$ and a phase α. The magnitude $|F|$ is expressed in terms of an equivalent number of electrons all scattering in phase, and α is the phase difference of the resultant wave from that which would be produced by one electron at the origin of the unit cell. In general *F* is a complex number and may be expressed by eqs. 3–5 (see also Figure 3–2).

$$F = |F|e^{2\pi i\alpha} \qquad\qquad 3\text{-}5a$$

$$F = A \cos 2\pi\alpha + iB \sin 2\pi\alpha \qquad\qquad 3\text{-}5b$$

The computation of the structure factor, once the coordinates of the atoms in the cell are known, is straightforward and routine—albeit lengthy without the aid of a computer. The assumption of a spherical distribution of electrons about an atom renders the combined scattering power of an atom independent of the direction of the diffracted beam. As a result, a scattering factor may be used that depends only on the kind of atom and the spacing (but not the orientation) of the reflecting planes. Thus, rather than considering the position of every electron in the cell, it is necessary only to consider the positions of the atomic centers. In these terms the structure factor for a reflection *hkl* is calculated from eqs. 3–6, in which the

$$F_{hkl} = \sum_j^N f_j\, e^{2\pi i(hx_j + ky_j + lz_j)} = \sum_j^N f_j e^{2\pi i\delta} \qquad\qquad 3\text{-}6a$$

$$F_{hkl} = \sum_j^N f_j\, [\cos 2\pi\,(hx_j + ky_j + lz_j) + i \sin 2\pi\,(hx_j + ky_j + lz_j)] \qquad\qquad 3\text{-}6b$$

FIGURE 3–2. Representation of a vector as a complex number.

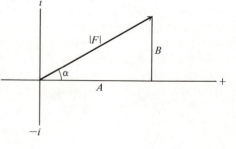

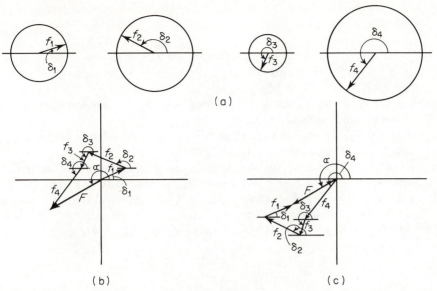

FIGURE 3-3. (a) Vectors representing four waves of different amplitude f_j and phase δ_j; (b) resultant; (c) resultant after addition in a different order.

summation is over the N atoms of the cell, f_j is the scattering factor (corrected for thermal motion) for the jth atom, and the exponential term gives the phase difference between the wave scattered at the atom with coordinates x_j, y_j, z_j[1] and the wave hypothetically scattered at the origin. The summation corresponds to the addition of a set of vectors, each representing the scattering by one atom, and the resultant is the structure factor (Figure 3-3).

Various interesting characteristics of the structure factor expression may be visualized with diagrams similar to Figure 3-3. One of the most significant concerns the effect of a crystallographic center of symmetry on the phase angles. If the center of symmetry is taken as the origin of the lattice, it requires that every atom at position x, y, z must be accompanied by another at $-x, -y, -z$. The structure factor may then be expressed by eq. 3-7. Thus

$$F = \sum_{j=1}^{N/2} f_j e^{2\pi i(hx_j + ky_j + lz_j)} + \sum_{N/2+1}^{N} f_j e^{2\pi i(-hx_j - ky_j - lz_j)} \qquad 3\text{-}7$$

each pair has phase contributions $e^{2\pi i\alpha_j}$ and $e^{-2\pi i\alpha_j}$, so that the combined vectors always add as shown in Figure 3-4. The resultant always lies along the horizontal (real) axis, and the

FIGURE 3-4. The effect of a crystallographic center of symmetry on the phase resulting from the addition of $e^{i\alpha}$ and $e^{-i\alpha}$.

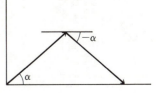

[1]The coordinates used in crystallography are always expressed as fractions of the corresponding cell edge, and so generally run from 0.0 to 1.0. The change from the origin of one cell to that of another therefore causes a phase shift of an integral number of cycles and so all cells scatter in phase.

structure factor expression reduces to eq. 3–8. Here the phase angles are restricted to 0 and

$$F = 2 \sum_{j=1}^{N/2} f_j \cos 2\pi \, (hx_j + ky_j + lz_j) \qquad\qquad 3\text{–}8$$

$180°$ (0 and π radians, 0 and $\frac{1}{2}$ cycle). This restriction causes the exponential term of eq. 3–5 to have the values ± 1 and F to be simply $\pm |F|$.

By a similar analysis it may be shown that for the reflections hkl and $\overline{hkl}$, $|F|_{hkl} = |F|_{\overline{hkl}}$ but $\alpha_{hkl} = -\alpha_{\overline{hkl}}$. This result is known as Friedel's law.

In practice structure factors are computed for each reflection in the data set using a list of the atoms in the structure, their x, y, z parameters, and the scattering factor lists prepared during data reduction. The magnitude of the phase of F_c (F calculated) is listed for comparison with F_o for each reflection, and usually $\Delta F (= |F_o| - |F_c|)$ is calculated as well. The overall agreement is usually described (roughly) by the residual index R (eq. 3-9), which may vary from

$$R = \frac{\sum |\Delta F|}{\sum |F_o|} \qquad\qquad 3\text{–}9$$

~3% for the best refined correct structures to ~60-80% for totally incorrect ones. The comparison of observed and calculated structure magnitudes can also be used to rescale the observed values and bring them to the absolute scale of the calculated $|F_c|$'s.

3-1c Fourier Synthesis

As has been described, it is routine to compute structure factors from a given electron distribution, usually in its atomic approximation. For crystal structure analysis it is also

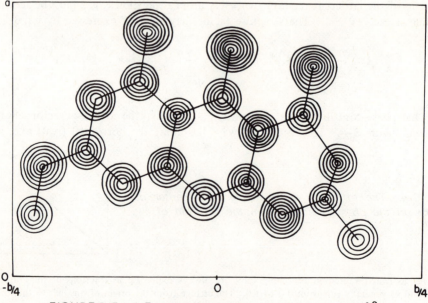

FIGURE 3-5. A Fourier map with contours at 1 electron/Å³.

necessary to be able to perform the reverse, i.e., to calculate an electron distribution from a given set of structure factors. Such a process takes the form of a Fourier synthesis, a summation that unravels the contribution made to each structure factor by the scattering ability (expressed as *electron density* ρ) at some point x, y, z in the cell. The summation is expressed by eq. 3–10 and is normally calculated for a grid of points spaced 0.3–0.5 Å apart throughout

$$\rho_{xyz} = \frac{1}{V} \sum_h \sum_k \sum_l F_{hkl}\, e^{-2\pi i (hx + ky + lz)} \qquad\qquad 3\text{--}10$$

the unique volume of the unit cell. Such a grid provides enough information to allow one to draw lines (*contours*) through points of equal electron density and to observe the maxima that represent the location of atomic centers (Figure 3–5).

Fourier calculations are now routinely carried out in three dimensions to represent the density in the volume of the unit cell. The values are calculated point by point in sections that represent slices through the cell parallel to one face. These sections are contoured, the contours are traced onto plastic sheets, and the sheets are stacked in a frame. The result is a spatial representation of the cell in which atoms appear as roughly spherical contour regions (Figure 3–6).

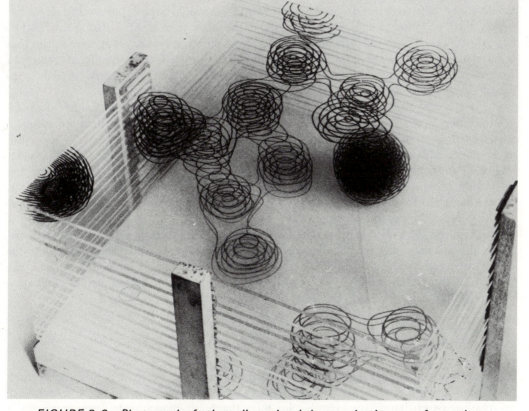

FIGURE 3–6. Photograph of a three-dimensional electron density map of potassium benzyl penicillin, showing use of stacked plastic sheets to represent sections. [By permission from D. Crowfoot, C. W. Burn, B. W. Rogers-Low, and A. Turner-Jones in The Chemistry of Penicillin, H. T. Clarke, J. R. Johnson, and R. Robinson, eds., National Academy of Sciences, Princeton University Press, Princeton, 1949, p. 327.]

Equation 3–10 resembles very much that used for calculating structure factors (eq. 3–6), differing from it mainly in the quantity $1/V$ and in the minus sign in the exponential term. In fact, the relationship is a general one and the density and structure factors are *Fourier transforms* of each other. Indeed it is possible to calculate a continuous Fourier transform throughout reciprocal space that corresponds to the electron density distribution in a single unit cell in real space. The collection of many unit cells into a lattice reduces the observable values of this transform to only those occurring at the usual reciprocal lattice points.

As will be seen later, various modifications of Fourier syntheses in which the coefficients F_{hkl} are replaced by other quantities serve a number of important functions in crystal structure analysis. In particular, the Patterson function, in which the coefficients are $|F_{hkl}|^2$, provides a map of the vectors between atoms rather than of the atoms themselves. The difference map, calculated with coefficients $\Delta F = |F_o| - |F_c|$, shows differences between an assumed model and the one implied by the actual data, and is useful in finding unplaced atoms and in refining the positions of others.

3–1d The Phase Problem

The previous discussion indicated that it is straightforward to calculate structure factors from atomic coordinates, and an electron density map from structure factors. Thus the solution of crystal structures would appear to be relatively routine. Would it were so! Unfortunately, the data available from the experimental intensities are not structure factors but merely their magnitudes, and information about the phase term of eq. 3–5 is lost. This loss occurs because the observed intensities are squares of the structure factors and the square root is extracted during data reduction. Thus it is impossible to tell directly whether an observed I corresponds to a positive or negative structure factor in a centrosymmetric crystal or to measure the phase angle α in the noncentrosymmetric case.

As a result of this uncertainty, the greatest single stumbling block to the successful determination of a structure is the need to acquire a suitable set of phases that are adequate approximations of the true values. Given such a set, a Fourier synthesis will show—with uncertainties and inaccuracies—an identifiable representation of the molecule or of a major fragment. Interpretation of the irregular electron density map in terms of spherical atoms and calculation of structure factors based on these will give better phases, which may be used as the starting point for another such cycle. In this way the approximate phase may be refined to true values as the structure develops.

The difficulty with this process is clearly the need for a starting set of phases. Alternatively, if a suitable portion of the structure (a *phasing model*) can be placed properly in the cell, it can serve as an equivalent starting point since structure factors and phases can be calculated from the partial model. As we shall see, the various techniques used to initiate a structure solution involve one or the other of these routes.

3–2 Solutions to the Phase Problem: Heavy Atom Method

In the heavy atom method the phasing model for the structure determination is a single atom (or at most a very few atoms) that is significantly heavier than the others in the structure. Such an atom has two striking virtues in this role. First, the power of an atom in determining phases is roughly proportional to the *square* of its atomic number, so the concentration of electron density in a single atom rapidly increases its utility as a model. Second, it is possible to locate a suitably heavy atom in the cell by means of the Patterson function with no prior knowledge of the phases. For these reasons the heavy atom method has historically been the method of choice for phase determination, and much effort has been spent in chemical modification of interesting molecules in order to prepare a suitable derivative.

The Patterson function is a Fourier series in which the usual coefficients have been replaced by $|F|^2$'s. These quantities are the intensities after correction for various geometric factors and are directly available from experiment. It was shown by Patterson in 1935 that the results of such a calculation could be derived from the electron density by the relationship that two points in the crystal, x_1, y_1, z_1 and x_2, y_2, z_2, with densities ρ_1 and ρ_2, lead to a point in the $|F|^2$ map with the coordinates $u = (x_1 - x_2)$, $v = (y_1 - y_2)$, and $w = (z_1 - z_2)$ and a value proportional to $\rho_1 \rho_2$. In other words, the Patterson function gives a weighted map of the vectors between regions of electron density. Two atoms in the crystal give rise in the Patterson to a peak corresponding to the vector between their centers (Figure 3-7).

Two major problems complicate the use of the Patterson function. Since there is a vector between every pair of atoms, for N atoms in the cell there are N^2 vector peaks in the same (or smaller) Patterson volume. Of these, N are vectors of zero length from an atom to itself and appear as a very large peak at the origin of the map, but there are still $(N^2 - N) = N(N - 1)$ other peaks to be interpreted. Also, the Patterson peaks are larger and more diffuse than the electron density peaks because they represent a span of vectors between all parts of the electron density (Figure 3-8). As a result Patterson maps from crystals containing a moderate number of

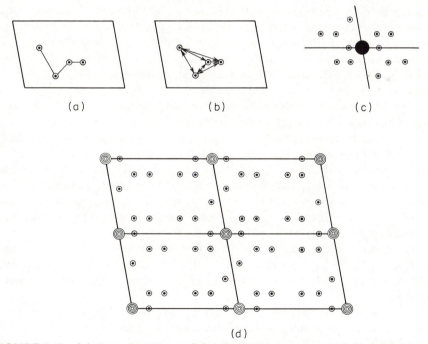

FIGURE 3-7. (a) Set of points.   (b) Interatomic vectors.   (c) Patterson peaks about the origin from one set (● = origin peak).   (d) Patterson peaks in four unit cells.

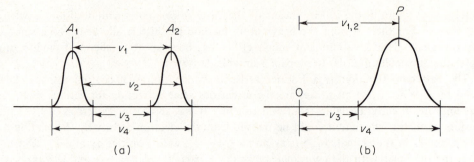

FIGURE 3–8. (a) Two atoms. (b) Patterson peak resulting from atoms in (a), showing spread of vector peak (arbitrary height). Origin peak not shown.

atoms of comparable atomic number tend to have broad high and low regions but no sign of discrete vector peaks.

It is possible to reduce the effects of peak broadening by a process known as *sharpening*. In this method the coefficients of the Patterson are modified to approximate those that would have been found if the atoms were actually points rather than extended distributions. The resulting map then shows many more maxima, but in practice these generally still represent the coincidence of a number of individual vectors. The modification of the coefficients involves multiplying $|F|$'s by the ratio of the atomic number of an average atom of the structure (the scattering factor of a point atom) to the average scattering factor appropriate to the reflection. It is not perfect, partly because of the approximations involved in averaging scattering factors, but principally because the correct representation of a point atom or vector peak requires an infinite series. If the summation is finite, as is necessarily true in practice, the peak is not a true point and is surrounded by false ripples that may overlap to give spurious maxima. During the sharpening process it is also possible to remove the origin peak arising from the self-vectors. This removal has some computational advantages but, unlike sharpening, does not generally increase the interpretable information significantly.

Locating a heavy atom or other molecular fragment with the aid of the Patterson involves identifying its position with respect to the symmetry elements of the real cell. This process uses the vectors that arise between an atom and its symmetry-related mates. For example, an atom at x, y, z in a centrosymmetric space group has a related atom at $\overline{x}, \overline{y}, \overline{z}$, and this pair gives rise to a vector at $u = 2x, v = 2y, w = 2z$. If the atoms are sufficiently heavier than the rest of the structure, this vector peak will be identifiable in the Patterson in the presence of all of the other vectors, and the positions of the heavy atoms in the real cell may be deduced. In a similar fashion atoms related by other symmetry operations give rise to vector peaks from which the atomic positions may be calculated.

The condition that the heavy atom → heavy atom vectors be readily distinguishable is easily met if $Z^2_{\text{heavy}}/\Sigma Z^2_{\text{light}} \sim 1$, where Z is the atomic number. In practice the ratio may be significantly less than 1, but at the cost of more difficulties in interpretation and poorer starting phases.

Once the heavy atom has been located, it is used as the basis for a structure factor calculation. The phases from this calculation are combined with the observed structure amplitudes, and a suitable Fourier series is computed. Normally some or all of a reasonable chemical structure will be identifiable in the resulting map, and addition of these atoms to the list and recalculation of structure factors will allow the process to be repeated as needed until all of the structure is revealed. In some cases, especially when centrosymmetric space groups are involved, the process will be very rapid. In others a sizable number of cycles will be needed, and it may be necessary at times to backtrack and remove incorrectly placed atoms.

3–3 Solutions to the Phase Problem: Direct Methods

3–3a Phase Relationships

Undoubtedly the most striking change in X-ray crystallography since 1970 has been the meteoritic growth of structure solutions based on the direct estimation of phases from the observed intensities. Analyses that would have been regarded a few years earlier as spectacular achievements have become commonplace. The success of these methods suggests that in the near future the structure of nearly any crystalline material will be determinable in a purely routine fashion, supplanting both the arts of deductive organic chemical analysis and of non-mechanical X-ray interpretation.

Although the mathematical and conceptual roots reach back farther in time, perhaps the easiest approach to direct methods is through the work of Sayre. In a classic 1952 paper, he demonstrated simply that a knowledge of the magnitudes and phases of all reflections but one would permit the calculation of the missing quantities through a series of the form given by eq. 3–11. Thus any reflection *hkl* is related to those pairs of reflections whose indices add

$$F_{hkl} = K \sum_{h'} \sum_{k'} \sum_{l'} F_{h'k'l'} \cdot F_{h-h',k-k',l-l'} \qquad 3\text{-}11$$

to give *hkl*. While this equation does not appear intuitively useful, Sayre pointed out that the summation is dominated by products of two large $|F|$'s, and that a large sum can be obtained only if the majority of these products indicate the same phase and add rather than cancel. Thus for centrosymmetric crystals, in which the phase problem reduces to a question of sign, it is probably true that if all three reflections are large, then eq. 3–12 holds.

$$\text{sign } F_{hkl} = \text{sign } F_{h'k'l'} \cdot \text{sign } F_{h-h',k-k',l-l'} \qquad 3\text{-}12$$

It had been shown earlier that if the reflections are sufficiently strong relationship 3–12 is necessarily true. The Sayre equation, however, provided an argument for a probability statement even in the absence of a certainty. Subsequent workers provided mathematical expressions for the probability as a function of the magnitude of the $|F|$'s involved.

A fundamental difficulty with using F's in such a relation is their marked decline in average magnitude with increasing indices, or $(\sin \theta)/\lambda$. It may be shown that for a cell of N atoms eq. 3–13 holds. Thus since the scattering power of the atom decreases at higher

$$\langle I \rangle_{\text{av}} = \sum_{i=1}^{N} (f_i \, e^{-B(\sin^2\theta)/\lambda^2})2 \qquad 3\text{-}13$$

values of $(\sin \theta)/\lambda$, the intensities fall off. However, the derivations of the Sayre and related equations in no way require them to be restricted to real atoms of finite size and subject to thermal motion. Instead it is possible to modify the $|F|$'s to approximate those that would be obtained for fixed point atoms (compare the sharpening of the Patterson function, section 3–3). These quantities, which do not suffer a general decline as $(\sin \theta)/\lambda$ increases, can then be used in phasing relationships.

The modified F's in common use now are symbolized as E's, and are obtained from eq. 3-14. Thus $|E_{hkl}|$ is a measure of the actual value of $|F_{hkl}|$ as compared with the average

$$|E_{hkl}|^2 = \frac{|F_{hkl}|^2}{\epsilon \sum\limits_{i}^{N} (f_i e^{-B(\sin^2\theta)/\lambda^2})^2} \qquad 3\text{-}14$$

or expected value.[2] Those reflections with large $|E|$ values (>1.5) are relatively improbable, contain much information about the structure, and are the ones of importance in direct phasing.

The distribution of $|E|$ values depends on whether or not a crystal is centrosymmetric. Table 3-1 gives some theoretical values and shows the uncommon nature of large $|E|$'s.

Table 3-1. *Theoretical Distributions of $|E|$ Values*

	Centrosymmetric	*Noncentrosymmetric*
Av $|E|^2$	1.0	1.0
Av $|E|$	0.798	0.886
% $|E| > 1.0$	32.0	36.8
% $|E| > 2.0$	5.0	1.8
% $|E| > 3.0$	0.3	0.01

3-3b Centrosymmetric Phasing

The problem of phase determination in centrosymmetric crystals is much simpler than that in noncentrosymmetric ones, both because of the two-valued ($+, -$) nature of the phase and because there tend to be more reflections with high $|E|$ values. As a result of both effects the phase predictions have greater reliability.

The classic approach, now known as the *symbolic addition method*, was used independently by many workers, but came to its fullest development in the hands of the Karles. This method consists of constructing for those $|E|$'s greater than some value (e.g., 1.5) a list of all the triples in which two indices add to give a third. Thus a possible triple would be $E_{421} E_{\bar{2}33} E_{254}$. The starting point consists of this list and a small set of reflections (usually 3) whose phases may be set arbitrarily to fix the origin of the unit cell with respect to the symmetry elements. These reflections cannot be chosen at random, but must meet certain specified conditions. If two of these are, say, $E_{421} = +|E_{421}|$ and $E_{\bar{2}33} = -|E_{\bar{2}33}|$, then the relationship of eq. 3-11 predicts that $E_{254} = -|E_{254}|$. The probability that this statement is true depends on the magnitudes of the $|E|$'s involved and on the number of atoms in the structure. For a structure containing N equal atoms the probability may be calculated approximately from eq. 3-15, and if it is high, the implication is assumed to be true.

$$P = 1/2 + 1/2 \tanh \left\{ \frac{1}{\sqrt{N}} |E_{hkl} E_{h'k'l'} E_{h-h',k-k',l-l'}| \right\} \qquad 3\text{-}15$$

In principle such a process could provide step-by-step a route to phases for all the large

---

[2] The quantity ϵ is an integer, usually 1, but sometimes greater, that corrects for the fact that for certain classes of reflections $\langle I \rangle$ is a multiple of that given by eq. 3-13.

E's. In practice, however, it is unlikely that a complete set can be built up with high probabilities starting from just the origin-defining phases. Instead it is usually necessary to feed in further unknown additional phases by assigning letter symbols to reflections that will prove useful in the phase development. Thus if $E_{2\bar{1}2}$ is assigned the symbolic phase a (which may be either + or −), then the triple $E_{2\bar{1}2}$ $E_{\bar{2}33}$ E_{025} suggests that E_{025} will have the phase −a, i.e., will be opposite in sign to $E_{2\bar{1}2}$. In this way a large list of phases can be built up in terms of symbols and combinations of symbols.

As the process continues, several triples may be found that indicate phases for a single E. In general, these will agree and will serve to reinforce the probability that the phase is correct. It may also occur that indications appear in terms of two symbols, and the expectation of agreement can lead to a relationship between the symbols. Thus, if a large $|E|$ has several indications of a and several of −b for its phase, it is likely that these correspond to the same + or − phase. Consequently, it is probable that a is the same sign as −b. In this way the list of variables can be reduced.

After the process of symbolic addition is complete, most of the large E's will have phases determined in terms of either signs or a few symbolic variables. The simplest approach, and one often used, is merely to calculate a Fourier map corresponding to each of the possible combinations of sign assignments to each of the variables. The calculation of 16 or even 32 such maps is not a major undertaking and generally costs only a few hundred dollars in computer charges. The E's rather than the corresponding F's are generally used as the coefficients for the Fourier summations, since they lead to much more sharply defined atomic peaks, even though only a relatively few reflections are used for the calculation.

In general, the symbolic addition procedure has proved very powerful and easy to apply to centrosymmetric structures. Difficulties can arise if one of the early relationships fails to give the correct phase (i.e., if the improbable phase is actually right), and it may be necessary to try a different set of origin phases and a different pattern of phase development to avoid troublesome triples. Various computer techniques have been devised to assist the process of finding the triples and developing the phases, but any more complicated methods are generally special cases of those described below for the noncentrosymmetric problem.

3–3c Noncentrosymmetric Phasing

Direct phasing in the noncentrosymmetric cases is significantly more difficult in principle than that for centrosymmetric. It was not, in fact, until the 1960's that the first successful analyses were reported. The principal difficulty is the continuous variable nature of the phase, which increases immensely the number of possible solutions. Fortunately, it is not necessary for the calculated phases to be exactly correct; in a number of successful solutions the average error appears to be about $20°$ and a few large errors can be tolerated.

The basic probability equation for centrosymmetric phasing, which corresponds to eq. 3-12, is eq. 3-16, in which ϕ is the phase angle.[3] The relationship between the two equa-

$$\phi_{hkl} = \phi_{h'k'l'} + \phi_{h-h',k-k',l-l'} \qquad \text{3-16}$$

tions is seen by noting that the sign of $|F_{hkl}|$ is $e^{i\phi_{hkl}}$ and thus eq. 3-12 may be written as eq. 3-17. Taking logarithms of both sides gives eq. 3-16.

$$e^{i\phi_{hkl}} = e^{i\phi_{h'k'l'}} \cdot e^{i\phi_{h-h',k-k',l-l'}} \qquad \text{3-17}$$

---

[3]This is the same quantity symbolized above and in most other crystallographic discussions as $2\pi\alpha$.

An equivalent expression that is often used rewrites eq. 3-16 as eq. 3-18, in which the

$$\phi_{\overline{hkl}} + \phi_{h'k'l'} + \phi_{h-h', k-k', l-l'} = C \qquad\qquad 3\text{-}18$$

constant C has values ranging around zero. For triples involving strong reflections the probability of a value close to zero is high, and thus knowledge of two of the phases can lead to a value of the third. It is possible to take advantage of these relationships and carry out symbolic addition with noncentrosymmetric phases, but the process is more difficult than in the centrosymmetric case and the chances of having to make more than one trial are greater.

Once more than single indications begin to appear for phases, a method is needed to combine the various values to a most probable value. This operation is carried out by means of the *tangent formula* (eq. 3-19). This formula, like eq. 3-12, relates the phase of one *hkl*

$$\tan \phi_{hkl} = \frac{\displaystyle\sum_{h'k'l'} |E_{h'k'l'} E_{h-h', k-k', l-l'}| \sin (\phi_{h'k'l'} + \phi_{h-h', k-k', l-l'})}{\displaystyle\sum_{h'k'l'} |E_{h'k'l'} E_{h-h', k-k', l-l'}| \cos (\phi_{h'k'l'} + \phi_{h-h', k-k', l-l'})} \qquad 3\text{-}19$$

reflection to all the pairs of others whose indices add to *hkl*. It can be used with a starting set of phases, derived for example by symbolic addition, both to extend the determination to new reflections and to provide a method of refining all of the phases to the most consistent set.

One of the most interesting variations on this approach at present consists of attempts to calculate a priori the values of C in eq. 3-18. Given even an approximate value for this quantity, it would be possible to identify those triples that were likely to provide wrong answers in the usual symbolic addition method (i.e., those for which C is far from zero) and to avoid their use. Although some large structures have been solved using such calculations, they are quite complicated and are still in the experimental stages.

Perhaps the most routinely applicable and the most powerful method is the multisolution approach of Woolfson and his collaborators. This method is in many ways similar to symbolic addition, but attacks the problems in a somewhat different order. First the list of important triples is made as usual. Then a "convergence map" is constructed by dropping out one at a time the reflections (and the triples in which they appear) for which the phase-indicating power of the triples is the least. Thus at any point in the process one has remaining those reflections for which there is the highest probability of being able to deduce the phases correctly. As the process draws to an end, it is manipulated to arrive at a legitimate set of reflections that may be used as an origin-determining starting set. In addition there are usually a few others that are important in the later phase development but whose phases are too weakly (or not at all) indicated from the starting set. In symbolic addition these would be assigned letter symbols and used as variables. In the multisolution technique they are simply given reasonable phase values and a complete set of phases developed for each combination with the aid of the tangent formula.

For centrosymmetric problems and for certain special reflections in noncentrosymmetric ones the possible phases are only two, $+$ and $-$, and the choices are clear. For general phase angles, it has been found that if four phases are tried, so that the maximum error is $45°$ and the probable error less, the tangent formula will refine to the correct values. Thus an unknown general phase is normally tested at 45, 135, 225, and $315°$. Since the number of combinations to be examined is the product of the number of trials for each of the unknown phases, a few general phases can rapidly lead to a very large number of phase sets. Usually a reasonable compromise between accepting indications that are too weak and having too many sets leads to between 16 and 64 possibilities. Fortunately it is not necessary to calculate and examine all of the corresponding Fourier maps because correct solutions can often be recognized from various internal tests. In particular, it is possible to use a formula similar to the Sayre series to calculate values of E for each reflection in the set, and then to calculate a "Karle R" (eq. 3-20) for the set. Normally the value of this agreement measure will be significantly

$$R = \frac{\sum \big| \, |E_c| - |E_o| \, \big|}{\sum |E_o|} \qquad\qquad 3\text{-}20$$

lower for the correct solution (or solutions, more than one often being found) than for the others.

The widespread application of multisolution methods is fairly recent and is largely based on the use of a particular program, MULTAN. A number of laboratories have reported its use, and although a certain amount of experimentation is sometimes necessary, the method appears to approach a routine structure-solving technique. It, as well as symbolic addition and other variations, has been applied to noncentrosymmetric structures containing up to 50 light atoms (C, N, O), with a remarkable record of success. Thus it appears that the phase problem is almost entirely under control.

3-4 Finishing the Job

3-4a Structure Completion

The general process by which a phasing model is extended to a complete structure by repeated cycles of structure factor and Fourier calculation has already been described. A number of related techniques, employing modifications of the simple Fourier series, have been devised to simplify finding new atoms. The most useful of these utilizes the difference (or ΔF) *Fourier synthesis*; the coefficients are given by eq. 3-21. The map resulting from such

$$\Delta F = (|F_o| - |F_c|)e^{2\pi i \alpha_c} \qquad\qquad 3\text{-}21$$

a calculation is the same as would be obtained by subtracting point by point an F_c Fourier from the normal F_O Fourier. In other words, it shows both positive and negative regions corresponding to too little and too much electron density in the model. Such a map does not show peaks corresponding to atoms already included in the model unless these are misplaced somewhat.

If entire atoms are missing from the structure, they appear in the ΔF map as pronounced peaks, more prominent than those in the normal Fourier because they are not overshadowed by larger ones arising from atoms included in the phasing model. Conversely, incorrect atoms (when these are few in number) appear in well-defined negative regions and are marked for removal.

By the use of such maps it is usually relatively easy to extend a phasing model to a complete structure. Indeed, once the heavier atoms are placed with reasonable accuracy, it is even possible to observe hydrogen atoms in a difference map, although this is not usually done until after a certain amount of refinement.

3-4b Refinement

The process of refinement converts a structure in which all of the nonhydrogen atoms have been placed with a precision of at worst 0.1 or 0.2 Å into one that give the best possible agreement between $|F_o|$ and $|F_c|$. The extent to which this agreement can be achieved depends on the accuracy of the data, the perfection of the crystal, and the sophistication of the model used to describe the crystal. The most widely used measure of agreement is the residual R, defined in eq. 3-22. A structure that is grossly correct will give $R \sim 0.2$-0.3, and careful

$$ R = \frac{\sum \left| |F_o| - |F_c| \right|}{\sum |F_o|} = \frac{\sum |\Delta F|}{\sum |F_o|} \qquad \text{3-22} $$

refinement in the best cases can lower this quantity to 0.02. From a theoretical view, R is not an ideal measure of agreement since it cannot be related directly to the accuracy of the structure. It has been long used, however, and crystallographers have developed a feeling for its implications.

Crystallographic refinement normally includes more adjustments than merely the position of the skeleton atoms. One or more scale factors are used to relate the various sets of data (e.g., levels from Weissenberg photographs) to the absolute scale provided by the $|F_c|$'s. Occasionally, a few other parameters such as the fraction of atomic locations actually occupied by atoms may be needed for disordered structures. The principal variation in refinements of different degrees of accuracy, however, lies in the alternative descriptions used for the thermal motion of the atoms.

As has already been discussed, the scattering factors calculated for an ideal stationary atom must be modified to account for the additional spreading of the electron cloud caused by atomic motion. The simplest correction, and one general used during the early stages of structure solution, assumes that all atoms in the molecule move equally and spherically This motion is described by a single overall isotropic thermal parameter B. Clearly, the assumption that the molecule moves as a whole is a doubtful one, and a more reasonable approximation assigns an individual isotropic thermal parameter to each atom to describe its unique motion. In this way each atom is characterized by four parameters rather than three.

A still more sophisticated description of the atomic motion, which is used for most modern refinements, recognizes that the motion may not be the same in all directions (anisotropic). Thus the atom sweeps out an ellipsoid rather than a sphere. Such motion is described by six parameters, making a total of nine per atom. The use of these extra parameters increases the complexity and the cost of the refinement very noticeably, but it does also permit a significantly better fit to the data.

Refinement may be performed using a variety of techniques. The oldest, but now little used, method involves the careful fitting of atomic centers to the maxima of electron density in F_o Fourier maps. Various interpolation methods have been used to locate the maximum density, but even so it was difficult to obtain very high precision in this way and refinement of thermal parameters was rarely attempted.

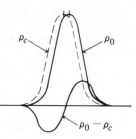

FIGURE 3-9.   The gradient produced in a ΔF Fourier as a result
of differences in observed and calculated density functions.

    The most widely used Fourier technique for refinement involves the difference (ΔF)
Fourier (section 3-4a).   Although a correctly placed and identified atom is invisible in a
difference map, any residual differences between the true structure and the model are clearly
visible.   Thus, if a nitrogen atom has been called carbon and assigned carbon scattering factors,
there will be a positive region reflecting the missing electron.   If the scattering factor is correct,
but there is a difference between the assumed and true position, a very characteristic gradient
appears at the assumed site (Figure 3-9).   This gradient indicates both the magnitude and the
direction of the needed positional changes.   Similarly, failure of a spherical model to match an
ellipsoidal density distribution leads to a pattern of positive and negative regions that can, with
experience, be used as a basis for roughly adjusting anisotropic temperature parameters.

    With the advent of large computers, however, Fourier methods for refinement have been
largely replaced by various analytic processes that may be lumped under the head of *least
squares methods*.   In these methods a systematic procedure is applied to minimize the sum of
the weighted $|\Delta F|^2$.   If a set of approximate values is available for all the $p_1, p_2, \ldots, p_n$
parameters (positional, scale, and thermal) characterizing the structure, shifts ($\Delta p_1, \Delta p_2,
\ldots, \Delta p_n$) that tend to the desired minimization may be found by the solution of a set of
linear simultaneous equations, the *normal equations* (eq. 3-23).   In these equations the

$$\sum w\left(\frac{\partial F}{\partial p_1}\right)^2 \Delta p_1 + \sum w\left(\frac{\partial F}{\partial p_1}\right)\left(\frac{\partial F}{\partial p_2}\right)\Delta p_2 \cdots \sum w\left(\frac{\partial F}{\partial p_1}\right)\left(\frac{\partial F}{\partial p_n}\right)\Delta p_n = \sum w\left(\frac{\partial F}{\partial p_1}\right)\Delta F$$

$$\sum w\left(\frac{\partial F}{\partial p_2}\right)\left(\frac{\partial F}{\partial p_1}\right)\Delta p_1 + \sum w\left(\frac{\partial F}{\partial p_2}\right)^2 \Delta p_1 \cdots \sum w\left(\frac{\partial F}{\partial p_2}\right)\left(\frac{\partial F}{\partial p_n}\right)\Delta p_n = \sum w\left(\frac{\partial F}{\partial p_2}\right)\Delta F$$

$$\vdots \qquad\qquad \vdots \qquad\qquad \vdots \qquad\qquad \vdots$$

$$\sum w\left(\frac{\partial F}{\partial p_n}\right)\left(\frac{\partial F}{\partial p_1}\right)\Delta p_1 + \sum w\left(\frac{\partial F}{\partial p_n}\right)\left(\frac{\partial F}{\partial p_2}\right)\Delta p_2 \cdots \sum w\left(\frac{\partial F}{\partial p_n}\right)^2 \Delta p_n = \sum w\left(\frac{\partial F}{\partial p_n}\right)\Delta F \qquad 3\text{-}23$$

summations are over all of the reflections used in the refinement; the weights w reflect the
uncertainty in the value of $|F_o|$; and the derivatives are evaluated from the assumed parameters.
Once the summations have been accumulated for all of the reflections, the equations may be
solved for the unknown Δp_i's, and these used to correct the initial values.   Because of the
approximations made in deriving these equations, the solutions are not exact, but the new values
are more nearly correct than the initial ones and can serve as the basis for a repeated
calculation.

The process of least squares refinement is often discussed in the language of matrix operations. The normal equations may be considered as specific examples of the form given in eq. 3–24, where $a_{ij} = \Sigma w \, (\partial F/\partial p_i) \, (\partial F/\partial p_j)$; $x_i = \Delta p_i$; and $V_i = \Sigma w \, (\partial F/\partial p_i) \, \Delta F$.

$$a_{11}x_1 + a_{12}x_2 + a_{13}x_3 \cdots a_{1n}x_n = V_1$$

$$a_{21}x_1 + a_{22}x_2 + a_{23}x_3 \cdots a_{2n}x_n = V_2$$

$$\vdots \qquad \vdots \qquad \vdots \qquad \vdots$$

$$a_{n1}x_1 + a_{n2}x_2 + a_{n3}x_3 \cdots a_{nn}x_n = V_n \qquad \qquad 3\text{–}24$$

These equations may be regarded as the result of a matrix multiplication of a matrix of coefficients a_{ij} times a vector of unknowns x_i to yield another vector of values V_i (eq. 3–25),

$$
\begin{pmatrix}
a_{11}a_{12} \cdots a_{1n} \\
a_{21}a_{22} \cdots a_{2n} \\
\vdots \quad \vdots \quad \vdots \\
a_{n1}a_{n2} \cdots a_{nn}
\end{pmatrix}
\begin{pmatrix}
x_1 \\
x_2 \\
\\
x_n
\end{pmatrix}
=
\begin{pmatrix}
V_1 \\
V_2 \\
\\
V_n
\end{pmatrix}
\qquad \qquad 3\text{–}25
$$

which may be condensed to eq. 3–26. The process of solving the simultaneous normal

$$\mathbf{AX} = \mathbf{V} \qquad \qquad 3\text{–}26$$

equations when viewed in this way consists of finding an inverse matrix $\mathbf{A}^{-1}$, defined by eq. 3–27, which leads to eq. 3–28 and eq. 3–29. Thus the inverse matrix, when multiplied

$$\mathbf{A}^{-1}\,\mathbf{A} = \mathbf{1} \qquad \qquad 3\text{–}27$$

$$\mathbf{A}^{-1}\,\mathbf{A}\,\mathbf{X} = \mathbf{A}^{-1}\,\mathbf{V} \qquad \qquad 3\text{–}28$$

$$\mathbf{X} = \mathbf{A}^{-1}\,\mathbf{V} \qquad \qquad 3\text{–}29$$

by the value vector, gives exactly the vector of unknown parameter shifts.

The process of computing $\mathbf{A}^{-1}$ from $\mathbf{A}$ is well understood but arithmetically very tedious for arrays of any significant size. Thus the widespread application of this method had to await the appearance of the large-scale computer. Now it is possible to handle routinely (albeit expensively) refinements of 100–400 parameters at one time.

Because most of the time in least squares refinement is spent in carrying out the repeated multiplications and additions needed to compute the array $\mathbf{A}$, various simplified versions of the process have been devised. These are more approximate and do not refine as rapidly, but run faster and require less computer memory. The most extreme approximation notes that the diagonal terms of **A**, i.e., the a_{ii}'s, are sums of squares and so are always positive and large. The off-diagonal terms are the sums of products that may be either positive or negative, and so may be assumed to be smaller. If they are sufficiently small, they may be set to zero and the array reduced to a single diagonal line of nonzero elements. This process reduces greatly the time required both for the construction of **A** and for the *inversion* to $\mathbf{A}^{-1}$, since for a diagonal matrix the nonzero elements b_{ii} of $\mathbf{A}^{-1}$ are simply $1/a_{ii}$. The assumptions prove to be too extreme in practice, however, and despite its apparent attractiveness such *diagonal least squares* methods are now rarely used.

The diagonal approximation fails largely because the magnitude of the off-diagonal term a_{ij} depends on the extent to which the parameters p_i and p_j are correlated, i.e., on the extent to which a change in one affects the value of the other. Thus setting these terms to zero assumes that all of the parameters are completely independent, sometimes a very poor approximation.

Compromise simplifications retain some of the off-diagonal terms, selecting those that are most likely to be significant. These *block diagonal* methods generate arrays that contain independent subarrays arranged along the main diagonal, each subarray containing commonly the positional and/or thermal parameters for a single atom. Still closer approximations to the full matrix may include terms connecting the parameters for those atoms bonded together or all of those in some one region of the molecule.

In all these cases the speed and economy of generating and inverting less than the full array are purchased at the cost of less accurate refinement on each cycle, the need for extra cycles to achieve the same degree of refinement, and greater uncertainty as to the true precision of the final parameters. Consequently, if a high degree of refinement is desired, it is probably best to use full matrix methods or the closest approximation permitted by the available computer (core storage is frequently the limiting factor). For less refinement block diagonal methods may be cheaper, but are more likely to have difficulties for very poor starting parameters.

3-4c Analysis of Results

One of the principal questions asked of a structural analysis after refinement is how accurately the parameters have been determined. Usually the answer is given in terms of an estimate of the accuracy of the final bond lengths and angles. It is, in fact, extremely difficult to determine the *accuracy*, i.e., the error compared with the true value, of a scientific measurement. The best that can usually be done is an estimation of the *precision*—the probable range of values that would be found upon repeated determinations of a quantity—combined with some qualitative arguments as to why the observed value probably is (or is not) correct. The distinction between the two terms arises because repeated measurement gives information about the scatter or *random error* in the value, but says nothing about possible *systematic error* that may cause a reproducible deviation from the truth.

The scatter of a series of measurements of a quantity is expressed in terms of the *standard deviation* σ or its square, the *variance*. The true standard deviation, like the true value, is an unknowable quantity, but it is generally approximated adequately by the *estimated standard deviation* (esd) s given by eq. 3-30. Here the summation is over all of the $i = 1 \to n$ measure-

$$s = \sqrt{\frac{\displaystyle\sum_{i=1}^{n} (x_i - \bar{x})^2}{n - 1}} \qquad\qquad 3\text{-}30$$

ments of x, and $\bar{x}$ is the average value of x obtained from these measurements. For the usual distribution of values and for a moderate number of samples ($n > 15$), one expects to find about 68% of the measurements falling in the range $\bar{x} \pm \sigma$ (or s) and 95% in the range $\bar{x} \pm 2\sigma$.

Note that σ describes the range over which individual measurements scatter and can be used to calculate the probability that another repetition of the measurement will yield a result in a specific range. Thus σ is the *standard deviation of a single measurement*. The mean value, generally the best available estimate of the truth, is known more precisely than any single measurement, however. There is still some uncertainty since carrying out a second independent set of measurements would probably lead to a different mean value. The difference between the two means, however, would be expected to be smaller than the average difference between two single measurements, and this result is expressed in terms of the standard deviation of the mean σ_m (eq. 3-31). In other words, repeating a measurement n times to get an average

$$\sigma_m = \sqrt{\frac{\sum\limits_{i=1}^{n} (x_i - \bar{x})^2}{n(n-1)}} \qquad \text{3-31}$$

improves the precision by $1/\sqrt{n}$.

It is possible to estimate the random errors in the raw crystallographic data by repeated measurement of a given intensity, but this process is very time consuming and is rarely carried out for more than a few reflections per structure. The results from these measurements are generally used to define an equation relating precision to intensity (and occasionally other parameters), and this relationship is used to examine the uncertainty in the other intensity measurements. When available, these values are used to calculate the weights for least squares refinement, since the greatest reliance is reasonably placed on those reflections having the smallest uncertainty. Although approximations are often used, eq. 3-32 gives the proper weight.

$$w = \frac{1}{\sigma_{|F_o|}^{2}} \qquad \text{3-32}$$

Derived quantities such as positional parameters and hence bond lengths are rarely subjected to repeated determinations, and estimates of their error must be obtained indirectly. In general, this is done by relating the uncertainty of the refined parameters to the uncertainty in the data and to the closeness of agreement between $|F_c|$ and $|F_o|$. Methods have been devised for estimating this uncertainty in structures refined by Fourier methods, but since the advent of least squares the usual analysis is based on the diagonal elements of the inverted matrix $\mathbf{A}^{-1}$. It may be shown that for any refined parameter p_i the estimated standard deviation is given by eq. 3-33. Here b_{ii} is the ith diagonal element of $\mathbf{A}^{-1}$, the summation

$$\sigma = \sqrt{\frac{b_{ii}\left(\sum\limits_{1}^{m} w\,|\Delta F|^2\right)}{m - n}} \qquad \text{3-33}$$

is over the weight and ΔF of each of the m reflections, and n is the number of parameters refined. In a structure in which the weights have been correctly assigned and for which the F_c model is an accurate representation of the data, the quantity $\Sigma w |\Delta F|^2/(m - n)$ approaches 1, giving eq. 3-34.

$$\sigma_i \cong \sqrt{b_{ii}} \qquad\qquad 3\text{-}34$$

In general the positional parameters are of interest less for themselves than as the basis for calculations about the geometry of the molecule. Thus, given the atomic coordinates, it is customary to calculate bond lengths, bond angles, and often best planes through various atoms. Programs are also available for preparing computer-directed drawings (including pairs for stereoviewers) of molecules from a list of coordinates.

The precisions of estimations of bond lengths and angles are less than that of the coordinates because two or more coordinates are involved in each calculation and the probability of error compounds. In general, modern analyses provide esd's of 0.003–0.01 Å for bond lengths and 0.1–0.5° for bond angles. These values, obtained from the A^{-1} matrix, are optimistic even for precision because the calculations do not treat properly all sources of random and semisystematic error. A number of comparisons of multiply determined structures have suggested that multiplying the esd's by ~ 1.5 gives values roughly consistent with the observed scatter.

The problem of accuracy in crystal structure determinations is much more difficult. It is well known that there are certain sources of systematic error, e.g., the effects on the apparent bond length of differing thermal motions of the bonded atoms, which cannot be calculated accurately. Estimates of the probable errors arising from these causes nonetheless show that they are sometimes several times the usual esd's. Thus the exact values obtained for susceptible quantities must be treated with reserve. Some crystallographers have suggested multiplying the esd's by as much as 3 in an attempt to avoid these dangers, but such an approach risks the alternative peril of ignoring actually meaningful results.

In addition to the readily tabulatable bond lengths and angles, various other measures are used in attempts to describe molecular shapes quantitatively. The commonest of these are least squares planes, best fit planes calculated through various sets of atoms, and torsional angles, the relative rotations of two atoms about the bond joining them. There are two possible definitions of the sign of such angles but the commoner is that shown in Figure 3-10.

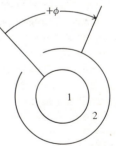

FIGURE 3-10. The common convention for defining torsional angles. The view is from atom 1 toward atom 2.

In fact, none of the various numeric descriptions of a molecule is as useful in providing a general impression of the molecule as is a good stereoview of the sort shown in Figure 3-11. Such a pair of drawings is normally studied with the aid of a viewer that allows each eye to see only one frame. With practice, however, it is often possible to achieve the same effect without mechanical aid. The drawings, which represent two views of the structure seen from slightly different angles are almost always prepared by a computer-directed plotter working from the final molecular parameters. They can be prepared as simple line drawings or embellished as shown with atomic ellipsoids with axes corresponding to the principal directions of thermal vibration.

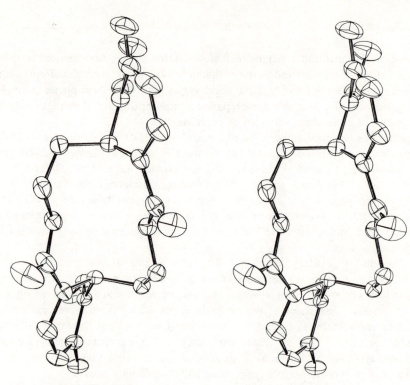

FIGURE 3-11.   *A stereo pair, showing thermal ellipsoids.*

3-4d   Absolute Configuration

One of the great successes of X-ray crystallography has been its ability to reveal the absolute configuration of optically active (chiral) molecules. First exploited as a specialized technique for proving the correctness of the Fischer convention for D and L compounds, the method has now become routine. It is based on the existence of *anomalous dispersion*, an extra change of phase occurring when X-rays are scattered by an element that is also strongly absorbing at the wavelength used. The effect of this anomalous dispersion is to advance the diffracted wave from this atom and thus to increase the phase angle of its contribution to F_{hkl}. The anomalous effect is introduced into structure factor calculations by adding a component $\Delta f''$ to the scattering factor, with a phase angle $90°$ greater than that of f. The result of this asymmetry of the scattering factor is to make $|F_{hkl}| \neq |F_{\bar{h}\bar{k}\bar{l}}|$.

If the molecule is changed from one chirality to another, e.g., from D to L by changing x, y, z to $-x, -y, -z$ in all the coordinates, the effect will be to interchange the intensities of F_{hkl} and $F_{\bar{h}\bar{k}\bar{l}}$.[4] Thus for one chirality $|F_{hkl}|$ will be calculated greater than $|F_{\bar{h}\bar{k}\bar{l}}|$, while for the other it will be smaller. In principle all that is needed is to measure the relative intensities of a number of such *Friedel pairs* and compare them against the calculations for both models. Usually all will favor a particular chirality. The differences in intensity are small, but many reflections can generally be found with easily detectable intensity differences by diffractometer measurement and some even by film.

The magnitudes of the differences can be increased by properly matching the radiation

---

[4]This result obtains since $e^{2\pi i(hx + ky + lz)} \equiv e^{2\pi i(\bar{h}\bar{x} + \bar{k}\bar{y} + \bar{l}\bar{z})}$ and all the terms of the structure factor calculation will remain unchanged.

and the heavy element in the crystal. They are largest when the target source for the X-ray generation is a few atomic numbers higher than the absorbing atom, e.g., Mo (Z = 42) for a Rb (Z = 37) salt. Such concern is not really needed, however, and recent experiments have indicated that even oxygen can provide sufficient anomalous dispersion with Cu radiation to permit the determinations of absolute configurations in favorable cases.

3–5 Summary

The preceding discussion has been concerned with the general techniques by which a crystal structure is determined. As a closing note it is worth considering what can go wrong and what can be learned.

The problems fall into two classes: those that prevent or seriously inhibit the solution of a structure and whose existence (if not whose nature) is obvious, and those that cause systematic errors in the results but may not be perceived. The former are of interest mainly to the experimenter; the latter can creep into the literature and are of wide concern. In the past, difficulties in solution were generally caused by the problem of obtaining suitable trial phases, and many failures occurred. Now the phase problem is well controlled and unsolved structures are fewer. Those difficulties that do arise are principally caused by nonidealities in the crystal, the commonest being disorder, twinning, impurities, decomposition on irradiation, high thermal motion, and uncorrected absorption.

If present in attenuated form, the same problems cause errors that are not accounted for by the standard deviations derived from least squares. These errors may appear in obviously suspicious forms, for example, in significantly different lengths for bonds that should reasonably be equivalent, or the effects may be more subtle. Generally the presence of such errors is signaled by a failure of the R factor (eq. 3–19) to drop to a value consistent with the presumed accuracy of the data. Thus good diffractometer data can generally be refined, using anisotropic temperature factors and including hydrogen atoms, to R's of 3–4%. Consequently, if a refinement hangs up at 10%, there is cause for suspicion. Some years ago such a value would have represented an excellent structure determination; now it is likely to be either incompletely refined or somehow in error.

A common error involves misidentifying an atom type or failing to recognize its existence at all. Misidentification may arise from wrong molecular formula or by making the wrong choice in a symmetric group (e.g., which atom is the oxygen and which the nitrogen in an amide). Usually the temperature factor in a mislabeled atom will be out of line, but this may not always be conclusive. Atoms may be missed if they occur at partially occupied sites in disordered structures or if they exist in unexpected molecules of solvation. It is a good idea to check for such possibilities by calculating and examining a difference Fourier after the final cycles of refinement.

More subtle problems have been seen in structures in which the model used was wrong in small but systematic ways. As a result least squares methods refined to a local minimum from which the molecule did not move, but which had suspiciously irregular bond lengths and sometimes a high R. A difference Fourier map in this case will probably show unjustifiably high and low regions, but these are unlikely to be interpretable. It can be hard to unravel a problem of this kind, and doing so generally involves reexamining the process by which the model was derived.

A somewhat similar problem can arise in those cases in which there is an uncertainty about the space group. It is sometimes difficult to decide whether a structure is a nearly symmetric molecule in a space group of low symmetry or is exactly symmetric and occupying a special position in a more symmetric group.

Given these reservations, a more positive note can be found by considering the sort of information to be gained by crystal structure analyses. Clearly, the first point, and the major direction of this presentation, has been a view of the molecular architecture. Diffraction methods have increasingly replaced deductive chemistry as the method for determining complex organic structures. Only a trace of material is required (a suitable crystal weighs 10–20 μg), and no other knowledge, not even the accurate molecular formula, is required. The results provide not only the gross molecular connectivity but also the relative stereochemistry. Absolute stereochemistry can be obtained in many cases, and even some idea of preferred conformations, although it is not always true that the conformation adopted in the crystal is that favored in solution.

The position of carbon and heavier atoms can be found with a precision of about 0.003 Å and these give bond lengths to 0.005 Å. It is true, however, that the accuracy of the parameters is less (how much less depends on circumstances and is often a matter of personal judgement), and that truly significant differences are generally on the order of a few hundredths of an Ångstrom.

Hydrogen atoms can often be found without great difficulty in cases where their thermal motion is not too great. Their precision is much less, 0.03–0.07 Å, and they are particularly subject to errors that result from shifts in electron density in the bond. Nevertheless, useful information can be obtained about hydrogen bonds and the effects of H–H interactions on conformations and packing.

Beyond the significance of molecular structure as revealed through crystallography, there is interest in the crystal structure itself, especially in terms of the interactions between molecules. Studies of solid state reactions, spectroscopy, catalysis, and intermolecular forces all have benefited from investigation of single crystals. There is growing interest in the ability of crystals to provide a highly concentrated, extremely stereospecific reaction medium, especially for photoexcited molecules. In order to interpret the results, however, a knowledge of the crystal packing is essential.

BIBLIOGRAPHY

1. Data Reduction: G. H. Stout and L. H. Jensen, *X-Ray Structure Determination*, Macmillan, New York, 1968, pp. 195–211; M. M. Woolfson, *X-Ray Crystallography*, Cambridge University Press, Cambridge, 1970, pp. 167–204; M. D. Buerger, *Crystal Structure Analysis*, Wiley, 1960, pp. 152–241.
2. Structure Factors: Stout and Jensen (reference 1), pp. 212–22, 239–46.
3. Fourier Synthesis: Stout and Jensen (reference 1), pp. 222–26, 246–60; Woolfson (reference 1), pp. 84–124.
4. Heavy Atom Method: Stout and Jensen (reference 1), pp. 270–99.
5. Direct Methods: Stout and Jensen (reference 1), pp. 315–38; Woolfson (reference 1), pp. 294–316; J. Karle and T. L. Karle, *Acta Crystallogr.*, **21**, 849 (1966); G. Germain, P. Main, and M. M. Woolfson, *Acta Crystallogr.*, **A27**, 368 (1971).
6. Refinement: Stout and Jensen (reference 1), pp. 353–428; Woolfson (reference 1), pp. 313–52.

Index

579